Power Quality in Power Distribution Systems

The comprehensive textbook will help readers to develop analytic reasoning of power quality aspects in distribution power systems. It will as an ideal study material for senior undergraduate and graduate students in the field of electrical engineering, electronics and communications engineering.

- Provides explanation of transformations and power theories for single phase and three-phase systems.
- Discusses concepts illustrating power quality aspects in power distribution network.
- Examines detailed derivations and analysis of voltage and current compensation techniques.
- Discusses custom power devices such as DSTATCOM, DVR and UPQC.
- Presents solved examples, theoretical and numerical exercises in each chapter.

This textbook comprehensively covers fundamentals concepts of power quality with the help of solved problems. It provides basic understanding of power quality aspects in power systems, especially in power distribution networks and explains issues related to power quality problems, their quantification, analysis and interpretation. It covers important topics including single phase circuits, three phase circuits, theory of fundamental load compensation, instantaneous reactive power theory, theory of instantaneous symmetrical components, dynamic voltage restorer (DVR) and unified power quality conditioner. Pedagogical features including solved problems and unsolved exercises are interspersed throughout the text for better understanding.The textbook is primarily written for senior undergraduate and graduate students in the field of electrical engineering, electronics and communications engineering for courses on power quality/power system/power electronics. The textbook will be accompanied by teaching resource including solution manual for the instructors.

Power Quality in Power Distribution Systems

Concepts and Applications

By Mahesh Kumar Mishra

CRC Press
Taylor & Francis Group
Boca Raton London New York

CRC Press is an imprint of the
Taylor & Francis Group, an **informa** business

First edition published 2024
by CRC Press
2385 NW Executive Center Drive, Suite 320, Boca Raton FL 33431

and by CRC Press
4 Park Square, Milton Park, Abingdon, Oxon, OX14 4RN

CRC Press is an imprint of Taylor & Francis Group, LLC

© 2024 Mahesh Kumar Mishra

ISBN: 978-0-367-75091-6 (hbk)
ISBN: 978-1-032-61729-9 (pbk)
ISBN: 978-1-032-61730-5 (ebk)

DOI: 10.1201/9781032617305

Typeset in Nimbus font
by KnowledgeWorks Global Ltd.

Dedicated to my parents: Late Shri Gendan Lal Mishra
and Late Smt. Ramkali Mishra

Contents

Contents

About the Author

Prof. Mahesh Kumar Mishra received the B.Tech. degree from the College of Technology, Pantnagar, India, in 1991, the M.E. degree from the Indian Institute of Technology, Roorkee, India, in 1993, and the Ph.D. degree from the Indian Institute of Technology, Kanpur, India, in 2002, all in Electrical Engineering. He has 30 years of teaching and research experience. For about ten years (1993–2003), he was with the Department of Electrical Engineering, Visvesvaraya National Institute of Technology, Nagpur, India. Prof. Mahesh has been with the Indian Institute of Technology Madras since 2003 and is currently a Professor in the Department of Electrical Engineering. His research interests include the areas of power quality, power distribution systems, power electronic applications in power systems, microgrids, and renewable energy systems.

Prof. Mahesh is a Life Member of the Indian Society of Technical Education and received IETE Prof. Bimal Bose Award for his outstanding contributions to Power Electronics Applications in Power Systems in 2015. He is a Fellow of the Indian National Academy of Engineering (FNAE) and the Institute of Engineers (India). He has completed a dozen of sponsored projects and consultancies. Under his research supervision, 18 Ph.D. and 12 M.S. have been awarded. Prof. Mahesh and his research scholars have been conferred with many International, National, and Institute level awards. He has 250 research publications in International and National peer-reviewed journals and conferences and delivered numerous expert talks in the areas of power quality, distributed generation, and microgrid systems.

In his leisure, Prof. Mahesh loves to read and write on diverse subjects relating to people, life, culture, the environment, and the world. He is passionate about making music, tunes, and rhythms, and writing songs.

Foreword

Power Quality and Reliability are the two most crucial aspects for customers connected to power distribution networks. While customers would like to have continuity of power supply, they would also like to distortion-free, near sinusoidal voltages at their supply inlets of specified voltage magnitude and frequency. With the increased penetration of converter-interfaced renewable energy resources like rooftop photovoltaics and batteries, the power distribution systems are facing many power quality challenges, such as voltage rise, harmonic distortion, phase imbalance, etc. Therefore, the study of power quality problems are their mitigation techniques are becoming extremely crucial for network planners and practicing engineers. This book is a welcome addition to the literature as it covers all the basics of power quality problems and their mitigation techniques.

The author of the book, Professor Mahesh Kumar Mishra, has over 25 years of research experience in related areas. He has developed the theories of some of the mitigation techniques, developed laboratory prototypes, and implemented converter-based power conditioning devices. His encyclopedic knowledge in the related areas has resulted in this book. The topics are presented in a systematic manner, and the most important topics are covered in depth. The book is organized into seven chapters. Brief descriptions of the chapters are given below.

★ Chapter 1 introduces the book, where different power quality problems in power distribution systems, power quality mitigation techniques using power electronic-based devices, power quality indices, standards, and power acceptability curves are discussed.

★ Chapter 2 presents analyses of single-phase circuits, both in the presence and absence of harmonic distortions. Some important concepts like displacement power factor and active and reactive power in the presence of harmonic distortions are presented.

★ Analyses of three-phase circuits arc presented in Chapter 3, where both balanced and unbalanced circuits are considered. Some important concepts like instantaneous real and reactive power, and apparent power for distorted and unbalanced circuits are discussed.

★ The fundamentals of load compensation are discussed in Chapter 4, where the aspects of voltage regulation, power factor correction, and load balancing for both delta and star-connected loads are discussed.

★ Converter-based load compensation, along with converter control is discussed in Chapter 5, where the load compensation using both the instantaneous reactive power theory and the theory of instantaneous symmetrical components are discussed.

★ Chapter 6 discusses the dynamic voltage regulator (DVR) and its operating principles for both balanced sinusoidal and unbalanced non-sinusoidal cases. The realization of a DVR through a power converter with associated results is also presented.

★ Chapter 7 presents unified power quality conditioner (UPQC), where both left shunt and right shunt configurations are discussed. This chapter also includes reference current and voltage generation and control of shunt current and series voltage through back-to-back connected voltage source converters.

All the mathematical derivations of the underlying concepts are presented in detail. Furthermore, numerous worked-out examples are presented to aid and enhance learning and problem sets are provided at the end of each chapter that are useful for the students and their instructors. The depth and breadth of coverage of the book make this an exceptionally valuable addition to power quality studies that will help senior undergraduate and postgraduate students, instructors, and practicing engineers.

Arindam Ghosh
Professor, School of Electrical Engineering
Computer and Mathematical Sciences
Curtin University, Australia

Preface

In last few decades, there is growing awareness of power quality in power industries and residential usages. There is more concern about reducing the electricity bill for its given usage and at the same time there is demand for high quality and reliability of power supply. To meet these criteria in power system operation, a power system engineer should be well aware of power quality concepts, terms, definitions, methods to enhance power quality, mitigation techniques, and devices. This book is an attempt to provide a comprehensive understanding of the subject matter in a simple way.

The book aims to provide a basic understanding of power quality aspects in power systems, especially in power distribution networks. It brings clarity on various issues related to power quality problems, their quantification, analysis, and interpretation. In this course, students will develop skills to analyze power quality aspects in power systems and provide appropriate solutions using custom power devices. First, the concepts are presented from the basic principles, and then insightful expressions have been derived, followed by examples. At the end of every chapter, there is a sufficient number of questions to strengthen the knowledge. The book acts as an interface between the conceptual understanding of the subject and advanced research. After going through the book and studying it thoroughly, the student will have the confidence to continue further research in this area. All concepts discussed in the book are explained and analytic expressions are derived from the basic principles, without assuming any expressions or formulae from other references. This makes easy and direct understanding of the concepts and their applications. The insights from the derived expressions are discussed and elaborated, which encourages students to think analytically.

Although there are many books on power quality, they are not developed as text books. Here is an attempt to understand power quality in a simple and lucrative way to serve as text book for undergraduate and graduate students in electric power areas at various technical universities and institutions. The book aims to develop a thought process and strong analytic reasoning of power quality aspects in power systems. The book has the following salient features.

■ Insightful and clear understanding of various terms, definitions, transformations, and power theories for single-phase and three-phase systems.

■ Lucid and clear illustration of power quality aspects in power distribution network with examples.

■ Detailed derivations and analysis of voltage and current compensation techniques.

■ Detailed explanation of custom power devices such as reactive network compensators, DSTATCOM, DVR, UPQC.

■ Theoretical and numerical exercises to practice concepts described in each
chapter.

The book is written in seven chapters. Chapter 1 focuses on various aspects of
power quality, such as definitions of various terms in power quality, nature of power
quality mitigation devices, monitoring of power quality, and standards in power quality. Chapter 2 aims to understand various power terms relating to power quality in a
single-phase system with sinusoidal and non-sinusoidal voltage and current. Three-
phase systems, which are widely used in industries due to their high quality, reliability, and efficiency, are explored in Chapter 3. It focuses on understanding the
behavior of the three-phase system under different conditions of voltages and currents. While analyzing three-phase circuits, $\alpha\beta0$ transformation, and instantaneous
symmetrical components transformation are derived and explained from the basic
principles. Various types of apparent powers and the corresponding power factors
are explained with practical applications, such as the design of distribution lines or
feeders.

After having a good understanding of power quality aspects, the focus is moved
toward mitigation techniques. Chapter 4 discusses about voltage regulation and fundamental load compensation using purely reactive networks. The reader will learn
from the discussed concepts and examples, how a delta-connected load or three-
phase ungrounded system can be fully compensated using purely reactive elements
such as inductance and capacitance. The reader's inquisitiveness is arisen by a simple question, whether these methods will work for three-phase grounded system?
The answer to this leads to exploring active power filters as compensators. Thus,
as a natural progression of flow, Chapters 5, 6, and 7 discuss methods employing
active power filters working as Distribution Static Compensator (DSTATCOM), Dynamic Voltage Restorer (DVR), and Unified Power Quality Conditioner (UPQC),
respectively, followed by the Appendix. The content of each chapter is summarized,
followed by set of problems and relevant references. The problems given at the end
of each chapter are thought stimulating and emphasize a deeper understanding of the
concepts described. The keys to numerical problems are provided at the end of the
book to verify the answers for the convenience of the readers.

The book's content, structure, and flow make it ideally suitable as a textbook for
"Power Quality in Distribution Networks: Concepts and Applications" in many universities and institutions worldwide. Additionally, it can also be used as a reference
book on power quality.

ACKNOWLEDGMENTS

First and foremost, I would like to thank my department and the institute for render-
ing their support in writing the book. I thank all faculty colleagues in the Department
of Electrical Engineering, Indian Institute of Technology Madras, Chennai. I am especially thankful to my colleagues Prof. R. Sarathi, Prof. B. Kalyan Kumar, Prof.
S. Krishna, Prof. K. S. Swarup, Prof. Lakshmi Narasamma, Prof. S. Srinivas, Prof.
Arun Karuppaswami, Prof. Krishna Vasudevan, and Prof. Kamalesh Hatua for their

lively discussion on various technical aspects of the subject area, which led to clarity and fluidity in presenting the concepts. Their frank and freewheeling discussions during meetings and tea time have been thought-provoking, thus serving food for mind, body, and soul. I thank Prof. Srikanthan Sridharan, Engineering Design Department, IIT Madras for proofreading the book and inspiring thoughtful discussions in the subject area as well as in life matters.

I have no words to express my heartfelt gratitude to Prof. Arindam Ghosh and Prof. Avinash Joshi, who motivated me on the path of exploration in the form of research and teaching during my Ph.D. days at IIT Kanpur. Without their light of knowledge, I would not have seen the shiner, brighter, and more affluent side of my professional as well as subjective worlds. I am thankful to Prof. S. C. Srivastava and Prof. Santanu Mishra at IIT Kanpur for their long association, encouragement, and inspiration.

For about ten years (1993–2003), I was faculty at the Department of Electrical Engineering, Visvesvaraya National Institute of Technology (VNIT), Nagpur. I continue to have strong association and interaction with the institute and convey thanks to Prof. H. M. Suryawanshi, Prof. M. S. Ballal, and Prof. D. R. Tutakne for many interactive technical sessions through our research collaborations, as well as philosophical discussions. Prof. Yashwant Katpatal at VNIT Nagpur has recorded his friendship in my heart and has always been supportive as my own family.

I would like to thank all my current research students, namely, Rajarshi Basu, Rohan Madnani, Nakka Pruthvi Chaithanya, Lokesh N, Hariharan R., Durga Malleswara Rao, Nafih Mohammad, Tony Thomas, Nimitha Muraleedharan, Ajit Upadhiya, Abhisek Panda, Leelavathi for their contributions to the book, by preparing circuit diagrams, verification of solutions, proofreading of the chapters and many other tasks involved.

My graduated Ph.D. and M.S. scholars, D. K. Karthikeyan, Dr. G. Vincent, Dr. Koteswara U., Dr. S. Sasitharan, Dr. Srinivas B. Karanki, Dr. Siva K. Ganjikunta, Dr. Chandan Kumar, Dr. Sathish Kollimalla, Dr. Narsa Reddy Tummuru, Dr. Nagesh Geddada, Dr. Manoj Kumar M.V., Dr. Sijo Augustine, Dr. T. Sreekanth, Dr. Jakeer Hussain, Dr. Srikanth Kotra, Dr. R. Satish, Dr. S. Srikanthan, Dr. J. Suma, Dr. Nikhil Korada, Dr. P. Harshvardhan, Dr. Linash P. K., Manik Pradhan, V. Leela Krishna, Y.A.P. Ramshankar, Jaganath K., N. Karthikeyan, Anil Ramakuru, and K. Sridhar, deserve their names to be mentioned for their exemplary research works, which directly or indirectly have been a great help to write the book.

I thank Mr. Gauravjeet Singh Reen, Senior Editor-Engineering, CRC Press, and his team members for all their support to make this book in publication form.

My childhood passed in a small village Chausara, in the northern part of the country, surrounded by a beautiful canal and a river, enriched with greenery and agricultural land. It is really difficult to forget childhood days with a large family, relatives, and friends with all golden memories, good or bad. I thank them all from the bottom of my heart for the rich and diverse experiences of my life, which provided me with a subtle sense of music, an intuitive understanding of things, and feelings to know about people, plants, birds, animals, and inanimate beings, endowing the capacity to see the most primitive to the most evolved with the purity of connection among them. For all this journey of life, I am indebted to my parents, late Shri Gendan Lal Mishra and late Smt. Ramkali Mishra for their love and moral support in all my endeavors. I am grateful to my elder brother late Shri Jugal Kishore Mishra, younger brother Dr. Sarvesh Kumar Mishra, and sister Smt. Kumkum Sharma and their families for their love, support, and inspiration in all situations of my life. I also thank my late father-in-law Shri Chhote Lal Sharma, and mother-in-law, Smt. Bina Sharma for their unconditional love, inspiration, and motivation.

I thank my wife, Kumud Mishra, whose inspiration always shows the way to remain focused. I am thankful to my daughter Niharika Mishra and son Tanish Kumar Mishra, who have been a constant source of enthusiasm and inspiration to complete the venture of writing this book.

Last but not least, I am thankful to all known and unknown people, things, and factors who directly or indirectly have led to this creation.

Mahesh Kumar Mishra
Professor, Department of Electrical Engineering
Indian Institute of Technology Madras
Chennai, India

1 Introduction to Power Quality in Power Distribution System

1.1 INTRODUCTION

The evolution of electric power systems has taken over one and a half centuries, and since then it has become an essential aspect of our lives. It started with the development of the dc power system in 1881. The dc power system was simple to realize and dealt with real quantities such as voltage, current, and resistance. However, the power could not be transmitted efficiently at higher voltages, which is required to minimize power losses over long-distance transmission. Around the same period, transformers and three-phase induction machines were developed, which laid the foundation for ac power system all over the world. Throughout the first half of the 19th century, with the advent of synchronous generators, ac power system was in full development with efficient power transfer over long distances at high voltage using step-up transformers. Later on, many such generating units were pooled together using transmission lines to form the ac grid as we know it today. This allowed bulk power transmission over long distances and resulted in more efficient and flexible operation of power system.

Initially, the main objective of the power system was to deliver power to everyone and connect households. But in the late 1900s, with significant development of the electronics industry, the use of power-sensitive electronics devices started increasing rapidly in all sectors of the economy, such as daily household, industry, commerce, business, trade, finance, and healthcare. However, these sensitive devices and products required a clean and reliable supply of power. Subsequently, industrial growth had also started to increase rapidly, which led to increased use of advanced industrial machines and equipment that also needed a clean and reliable supply of electric power. This technological change led to growing concerns for electric utilities and end-users about the quality of electric power. The modern ac power system is quite flexible in terms of voltage an current levels. But at the same time, it has many challenging issues such as unbalance, harmonics, blackout, brownout, steady state and transient stability, reactive power, harmonics power, etc. Over the years, with extensive research in this domain, power engineers developed the existing power system, which allowed us to monitor and control the power system parameters to maintain reliable and good quality of power at all the power system levels, generation, transmission, and distribution.

DOI: 10.1201/9781032617305-1

1.2   POWER QUALITY

Power quality is a very general term used in power systems, and it has different meaning for different parts of the power system. For the generation side, power quality refers to the protection of generator and generation system. It means to protect the generator from over-rating in terms of real and reactive power. For the transmission side, it means to have high transmission efficiency, less outage, good loadability of transmission lines, requisite flow of power, and balanced sinusoidal quantities over transmission network. For the distribution system, it indicates proper regulation of voltage, compensation of unbalance, harmonics, and undesired variations in voltages and currents. In general, power quality refers to a wide variety of electromagnetic phenomena that characterizes the voltage, current at a given time and location in the power system [1]. There are many variants of power quality definition. As per the Institute of Electrical and Electronics Engineers (IEEE) [2], "Power quality is a concept of powering and grounding sensitive equipment in a matter that is suitable to the operation of that equipment." The definition predominantly focuses on the issues affecting the operation and performance of the equipment. On the other hand, the International Electrotechnical Commission (IEC 61000-4-30) defines power quality as "Characteristics of the electricity at a given point of an electrical system evaluated against a set of reference technical parameters" [3]. This definition gives more emphasis on measuring and quantifying power system performance [4]–[17].

In a broad sense, power quality refers to maintaining the sinusoidal nature of voltage and current of the fundamental frequency at all points of the power system network. The power quality is also an important aspect of economic consideration, as it saves electrical energy in appliances. In literature, the term power quality is correlated to electromagnetic compatibility (EMC) and electromagnetic interference (EMI) [3]. Also, voltage quality implies current quality and vice-versa, because voltage and current quantities interact through the transmission and distribution line impedance. The product of voltage quality and current quality reflects the quality of the power supply. While the quality of voltage is determined from the source side, the current quality comes from the load side. To illustrate this, consider a balanced three-phase system represented by a single-line diagram in Fig. 1.1.

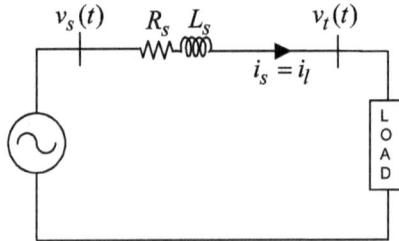

Figure 1.1   Pertaining to voltage, current, and power quality

In Fig. 1.1, v_s is the source voltage which reflects the quality of the source or utility voltage. The current $i_l(t) = i_s(t)$ is the load current (also equals to source current), which reflects the quality of the load current for the given supply voltage, v_s. It depends upon the nature of the load, i.e., balance, unbalance and harmonics, etc. Now the voltage $v_t(t)$ is the voltage at the load terminal, and the quality of $v_t(t)$ depends upon both the source voltage $v_s(t)$ and load current $i_l(t)$. This is clear from the relationship between the $v_s(t)$ and $v_t(t)$, through the voltage drop across the feeder with resistance R_s and inductance L_s, as given in the following (1.1).

$$v_t(t) = v_s(t) - R_s i_l(t) - L_s \frac{di_l(t)}{dt} \qquad (1.1)$$

The instantaneous power $p_l(t)$, which is product of voltage, $v_t(t)$ and current, $i_l(t)$, at the load bus is given by,

$$p_l(t) = v_t(t) i_l(t) \qquad (1.2)$$

As clear from (1.1), $v_t(t)$ is influenced by the quality of $v_s(t)$, $i_l(t)$, and the value of the feeder resistance R_s and inductance L_s. Also, note that the resistive drop, $R_s i_l(t)$, and inductive drop $L_s \frac{di_l(t)}{dt}$ depend upon the load current and its rate of change with time, respectively. The latter has a more serious effect on the quality of $v_t(t)$ due to the derivative term $\frac{di_l(t)}{dt}$, if the value of inductance is high. If $i_l(t)$ contains harmonics, such as the current drawn by a full bridge rectifier or any other converter circuit, it has a worse effect on the quality of $v_t(t)$ producing notches, sharp changes, non-sinusoidal variation, and consequently affecting the quality of power at the load terminal, which is a product of voltage, $v_t(t)$ and current, $i_l(t)$, as given in (1.2). Once the terminal voltage is affected, the load current develops more distortion. This goes on till the three-phase load voltages, and load currents settle to certain distorted waveforms, which may not meet the requirement of the connected load resulting in its malfunctioning and erroneous operation.

It is observed here that the feeder length and its impedance are important parameters in affecting the power quality at any point between the source and the load. For an ideal source, the connecting feeder to the load should have zero resistance and inductance ($R_s = 0$ and $L_s = 0$). Such a system is known as a stiff source. For a non-stiff source, resistance and inductance have some finite non-zero value, leading to a few percentages of voltage drop of the rated voltage. Thus, the extent of "Non-stiffness" of the voltage source is quantified by the magnitude of feeder impedance and its X/R ratio. While non-stiff source has a drawback to affect the power quality at the load bus, at the same time it gives the flexibility to control and regulate the voltage through some compensation schemes at the load bus [18]–[22]. The stiff systems are strong and rigid; therefore, the load has to accept the available voltage quality at its bus. It does not give the flexibility to control the parameters of the affected part of the distribution network. Further, the following points illustrate the nature of the feeder impedance, which helps to correlate the voltage, current, and power quality.

1. The typical value of the per-phase line inductance on a three-phase ac line can be considered to be 1 µH/meter, equivalent to 1 mH/km/phase [5].

2. Also, it is found that the X/R ratio for the distribution lines is often close to unity, with shunt capacitance largely ignored for calculations [23].

3. From the above observations, we can conclude that the typical value of the distribution feeder impedance can be taken around $0.31 + j0.31\ \Omega$/km/phase. It is to be noted that clear standards have not been established for the length of the distribution line feeders. However, experience suggests that the typical feeder mains of 11 kV level (primary distribution) can run between 1 and 25 km in length.

4. In addition, it is suggested that the length of the low tension secondary distribution lines leading to the customer premises (415 V L-L level) be kept to a maximum of 1 km to avoid considerable voltage drops across the lines.

From the above discussion, it is clear that the feeder impedance plays a vital role in co-relating voltage and current and hence power quality problems. For example, if the quality of supply voltage is not good, it will result in poor quality of current on the distribution side, through feeder interaction, as expressed in (1.1). Similarly, poor quality of current on the load side will translate to poor quality of the voltage on the supply side, again through feeder interaction. That is how voltage, current, and power quality relate to each other. Based on this fundamental understanding, the detailed power quality problems and their nature are discussed in the following section.

1.3   POWER QUALITY PROBLEMS

To operate the distribution and transmission network in an acceptable way, the main objective is to maintain power quality in power systems. To achieve this it is important to understand the nature of power quality problems that can occur, how they originate, and their effects on the utility and the consumers. The majority of the power quality problems originate in the distribution grid associated with various kinds of loads used by the end-users. Any deviation in the voltage, current, and supply frequency from its nominal value generates a problem. Thus, it is necessary to define and classify them in a meaningful way. Classification can be done based on the magnitude and duration of the quantities like voltage, current, and frequency. In the following description, power quality issues are classified based on the magnitude and duration of the event. Table 1.1 lists various types of power quality phenomena which are further explained in the following subsections:

- Transients
- Short-duration or rms variations
- Long-duration variations
- Voltage unbalances
- Waveform distortions
- Voltage fluctuations
- Power frequency variations

Table 1.1

Categories and characteristics of power quality phenomena

Categories	Typical spectral content	Typical duration	Typical voltage magnitude
• Transients			
— Impulsive			
• Nanosecond	5 ns rise	< 50 ns	
• Microsecond	1 us rise	50 ns–1 ns	
• Millisecond	0.1 ms rise	> 1 ms	
— Oscillatory			
• Low frequency	< 5 kHz	0.3–50 ms	0–4 p.u.
• Medium frequency	5–500 kHz	20 us	0–8 p.u.
• High frequency	0.5–5 MHz	5us	0–4 p.u.
• Short-duration variations			
— Instantaneous			
• Sag		0.5–30 cycles	0.1–0.9 p.u.
• Swell		0.5–30 cycles	1.1–1.8 p.u.
— Momentary			
• Interruption		0.5 cycles–3 s	< 0.1 p.u.
• Sag		30 cycles–3 s	0.1–0.9 p.u.
• Swell		30 cycles–3 s	1.1–1.4 p.u.
• Voltage imbalance		30 cycles–3 s	2%–15%
— Temporary			
• Interruption		> 3 s–1 min	< 0.1 p.u.
• Sag		> 3 s–1 min	0.1–0.9 p.u.
• Swell		> 3 s–1 min	1.1–1.2 p.u.
• Voltage imbalance		> 3 s–1 min	2–5%
• Long-duration variations			
— Interruption, sustained		> 1 min	0 p.u.
— Undervoltages		> 1 min	0.8–0.9 p.u.
— Overvoltages		> 1 min	1.1–1.2 p.u.
— Current overload		> 1 min	
• Imbalance			
— Voltage		steady state	0.5–5%
— Current		steady state	1–3%
• Waveform distortion			
— DC offset		steady state	0–0.1%
— Harmonics	0–9 kHz	steady state	0–20%
— Interharmonics	0–9 kHz	steady state	0–2%
— Notching		steady state	
— Noise	broadband	steady state	0–1%
• Voltage fluctuations	< 25 Hz	intermittent	0.1–7%
• Power frequency variations		< 10 s	±0.1 Hz

1.3.1 TRANSIENTS

Transients are defined as events which are undesired and momentary in nature in the power system. On quantity-specific definition, transient describes a part of the variable that disappears during the transition from one steady state to another. The utility engineers view transients as surges from lightning strokes which are absorbed by surge arrestors to protect the electrical equipments. The end users view transient as anything unusual that might be observed in the power supply including voltage sags, swells, and interruptions. It is characterized by its duration ranging up to about a few tens of milliseconds, and its spectral content, i.e., its rise time and frequency. Transients are further sub-divided into two categories, impulsive and oscillatory, depending upon the waveform of the quantity involved, either current or voltage.

1.3.1.1 Impulsive Transient

Impulsive transients are defined as a sudden, non-power frequency change in the steady-state characteristics of voltage or current or both, where the change or disturbance is unidirectional in nature, either in positive or negative polarity. These transients exhibit similar characteristics to that of a lightning strike. They are normally characterized by their peak value, rise time, and duration of the disturbance.

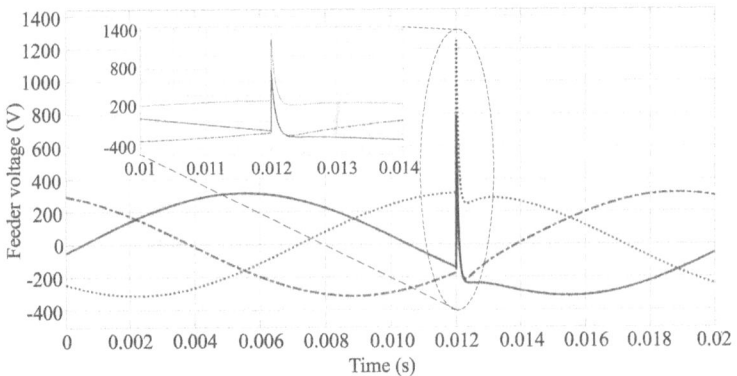

Figure 1.2   Lightning stroke leading to voltage impulsive transients

For example, an impulsive transient described as 1.2/60 is interpreted in the following manner, 1.2 indicates the rise time of the waveform in microseconds, i.e., the duration in which the magnitude rises from 10% to 90% of its peak value, and 60 indicates the duration in microseconds from the start till the point where the waveform magnitude decays to 50% of its peak value. Depending upon the time range in which a transient rises and its duration, there are three categories of impulsive transients, nanosecond, microsecond, and millisecond. Fig. 1.2 depicts the voltage waveform under the occurrence of a lightning strike along with the zoomed portion of the voltage impulse.

Impulsive transients are typically caused due to lightning strikes and normally damp out quickly due to high-impedance circuit elements. It may sometimes lead to lightning flash-over on power line insulators, thus leading to momentary short circuits. Impulsive transients may give rise to oscillatory transients due to the presence of resonance circuits in the power system.

1.3.1.2 Oscillatory Transient

Oscillatory transients can be defined as a sudden, non-power frequency change in the current and voltage steady-state characteristics, with positive and negative values of the amplitude. In these transients, the electrical quantity involved oscillates multiple times, with its peak magnitude decaying over time. These oscillations are mainly characterized by their magnitude, duration, and spectral content (frequency of oscillations). Based on the frequency of spectral content, transients are further classified as high-, medium-, and low-frequency transients. Table 1.1 mentions the frequency range and duration for each category.

Transients with frequency components higher than 500 kHz and a duration measured in few microseconds are considered as high-frequency oscillatory transients. It is caused by some switching events or appears as a local response to an impulsive transient.

Transients with frequencies between 5 and 500 kHz and a duration of a few tens of microseconds are considered medium-frequency transients. Energization of back-to-back capacitor banks results in oscillatory current transients [1]. Transients with frequency components less than 5 kHz and duration from 0.3 to 0.5 ms are classified as low-frequency oscillatory transients. These transients occur very frequently in the distribution system due to capacitor bank energization, which results in an oscillatory voltage transient with a frequency range of 300–1600 Hz. This is illustrated in Fig. 1.3. Ferroresonance phenomenon in transformer energization can

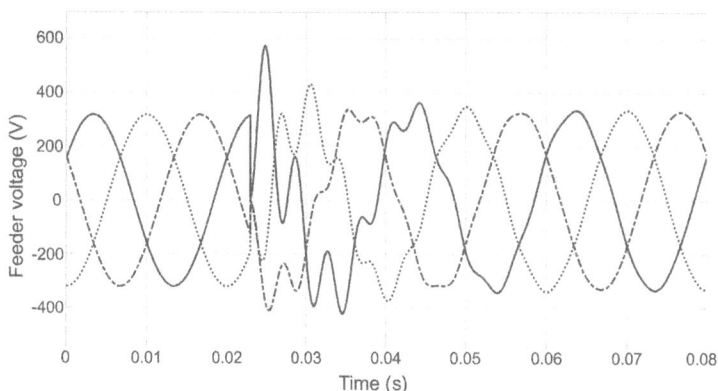

Figure 1.3 Low-frequency oscillatory transient caused by capacitor-bank energization

also lead to low-frequency transients [1]. Ferroresonance refers to a situation where the nonlinear magnetic properties of iron in the transformer iron core interact with capacitance in the electrical network to produce a nonlinear tuned circuit with an unexpected resonant frequency. This results in higher values of voltage and current in the transformer.

1.3.2 SHORT-DURATION VARIATIONS

This category deals with power quality problems with anomalies in the voltage magnitude for a duration ranging from 0.5 cycle to 1 min. Short-duration voltage variations are further categorized into three sets based on the duration of the disturbance, instantaneous, momentary, and temporary. Table 1.1 depicts the time duration for all three subcategories. Now looking into the types, there are three types of short duration voltage variations, namely interruption, sag, and swell.

1.3.2.1   Interruption

An interruption occurs when the rms voltage magnitude decreases to less than 0.1 p.u. of its nominal value with a duration ranging from 0.5 cycle to 1 min. They are caused due to faults occurring in the utility or distribution grids. Interruption can occur as a standalone event or can be preceded by a voltage sag. The duration of an interruption depends on the type and capability of the protective device used in the utility system, i.e., faster re-closing of circuit breakers will lead to an instantaneous interruption, and delayed re-closing can lead to momentary or temporary interruptions. Fig. 1.4 depicts a momentary interruption due to a fault, the upper plot shows the instantaneous voltage variation, and the lower plot shows the rms voltage variation.

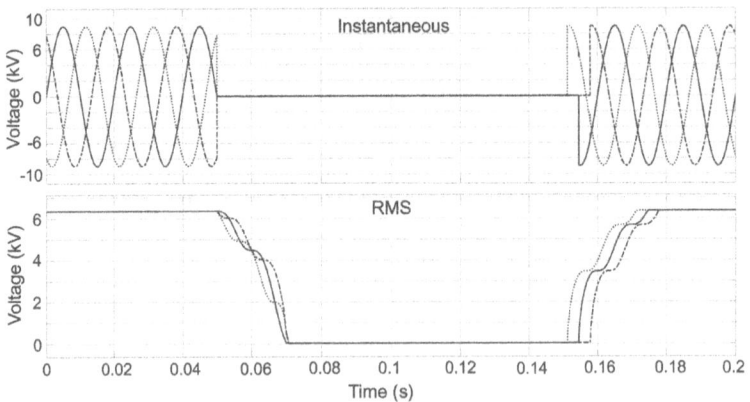

Figure 1.4   Momentary interruption due to fault and subsequent recloser operation

1.3.2.2 Sag

A voltage sag is characterized as a decrease in the rms voltage magnitude to between 0.1 and 0.9 p.u. of its nominal value, for a duration of 0.5 cycle to 1 min. As per recommended practices, 60% voltage sag refers to the voltage drop to 60% of its nominal value. To avoid any confusion, it is better to specify the nominal voltages as well. Voltage sags are usually caused due to faults in the system, i.e., when there is a fault on a phase, it experiences a sag. It can also be due to the starting of heavy industrial loads like motors, which draws huge inrush currents leading to a voltage dip. Many of us might have experienced this phenomenon in the form of flickering of lights when heavy loads like air-conditioners and water pumps are turned on. Fig. 1.5 depicts a voltage sag caused due to a line-line to ground (LLG) fault. Various sub-classifications of sag depending on the duration are given in Table 1.1.

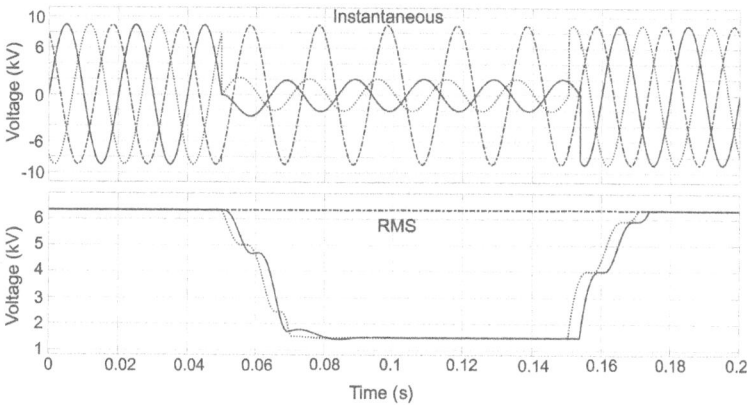

Figure 1.5 Voltage sag caused by an LLG fault

1.3.2.3 Swell

A volatge swell is characterized by an increase in the rms voltage magnitude above 1.1 p.u., usually ranging from 1.1 to 1.8 p.u., for a duration of 0.5 cycle to 1 min. Like sags, swells can also occur due to faults in the system. When faults occur on a single-phase, there may be a rise in voltage in the other unfaulted phases. It can also result from heavy load rejection in the system or switching on capacitor banks. Fig. 1.6 depicts the voltage swell phenomenon and is shown by instantaneous and rms voltage values. The sub-classification of swell depending on the duration is given in Table 1.1. Swells are very less common events and have varied effects on grounded and ungrounded systems.

1.3.3 LONG-DURATION VARIATIONS

This category of power quality problems consists of deviations in voltage at power frequency for more than 1 min. Long-duration voltage variations are of three types:

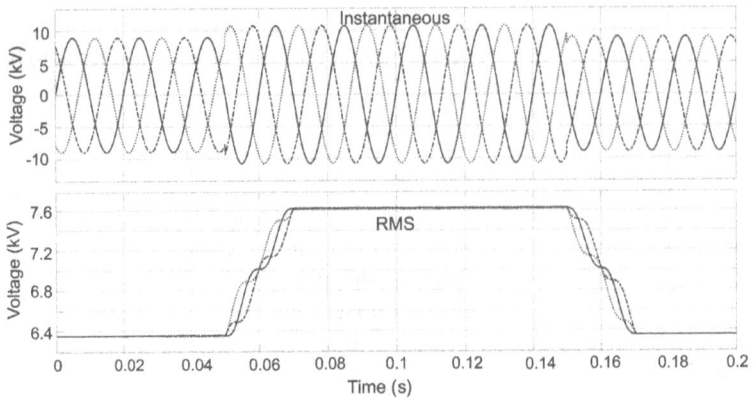

Figure 1.6 Voltage swell

overvoltage, undervoltage, and sustained interruptions. This category is similar to short-duration voltage variations; however, they are caused due to changes in the system load but not as a result of system faults.

1.3.3.1 Overvoltages

Overvoltages are kind of voltage swells with longer duration. They are characterized by an increase in the rms voltage magnitude above 1.1 p.u., usually ranging from 1.1 to 1.2 p.u., for duration more than 1 min. They are caused due to switching-off of large loads or as a result of a change in the reactive power compensation on the system, i.e., injection of more reactive power into the system or due to poor voltage regulation controls.

1.3.3.2 Undervoltages

Undervoltage is similar to voltage sag with a longer duration. They are characterized by a decrease in the rms voltage magnitude to less than 0.9 p.u., usually ranging from 0.8 to 0.9 p.u., for duration more than 1 min. They are caused by the opposite events that cause overvoltages, like switching-on of heavy loads or capacitor banks, or a change in reactive power compensation of the system, i.e., absorption of reactive power from the system.

1.3.3.3 Sustained Interruptions

This type of long-duration variation is similar to interruptions. These are characterized by the decrease in rms voltage magnitude to less than 10% of its nominal value, for a duration longer than 1 min. They are mainly caused by permanent faults in the system, which can be cleared by manual intervention.

1.3.4 IMBALANCE

This category of power quality problem deals with the anomaly in the three-phase sequence components of an ac quantity. As per the IEEE definitions, imbalance in a three-phase system is defined as the ratio of negative to positive sequence component magnitudes, given as a percentage. Depending upon the quantity involved, there can be current or voltage imbalances. Generally, the number of current imbalances are much higher than those of voltage imbalances. Voltage imbalances of 2% or less are mainly caused due to unbalanced single-phase loads on a three-phase circuit. Severe voltage unbalances, those greater than 5%, result from single phasing conditions such as disconnection of one phase of a three-phase motor. Current imbalances are also a result of the presence of single-phase loads.

1.3.5 WAVEFORM DISTORTION

The problems discussed till now deal with deviations in the ac quantity's magnitudes involved, either voltage or current. However, for an ac quantity to be perfectly ideal, its waveform has to be sinusoidal in nature. Therefore, this category of power quality problems deals with the types of anomalies in the waveform of the quantity involved. There are mainly five types of waveform distortion, dc offset, harmonics, interharmonics, notching, and noise.

1.3.5.1 DC Offset

The presence of a dc voltage or current in any ac voltage or current is defined as a dc offset. These problems can result from geomagnetic disturbance on transmission lines or asymmetric operation in power electronic converters, which leads to the undue effect of half-wave rectification. The presence of dc in the ac power system can lead to transformer saturation and associated heating, causing stress in the insulation material.

1.3.5.2 Harmonics

When ac quantities like voltage and current have frequencies that are integer multiples of the system's fundamental frequency, they are termed harmonics. Harmonics, when combined with the fundamental frequency component, leads to waveform distortion. Power electronics converters involving continuous switching are the major source of harmonics. Devices like diode bridges and line-commutated converters can be seen as harmonic current sources. Other devices like pulse width modulated inverters can be seen as harmonic currents reflected as harmonic voltages. As more and more power electronics switching devices are being incorporated into the power system, harmonics are also increasing, which has been a growing concern for both the customers and the utility. Harmonics are characterized by a parameter known as the Total Harmonic Distortion (THD), which is defined as the ratio rms magnitude of the harmonic component to the fundamental component. In many drive applications under lightly loaded conditions, THD gives erroneous results for current harmonics.

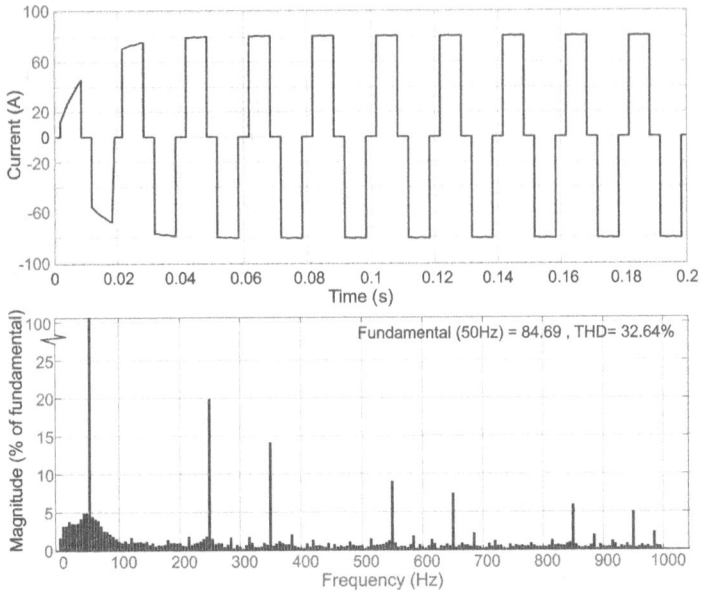

Figure 1.7 Current waveform and harmonic spectrum for a power electronic-based load

IEEE has defined another parameter for current harmonics known as Total Demand Distortion (TDD) to avoid such cases. Fig. 1.7 depicts the waveform and harmonics spectrum for phase-*a* input current of three-phase full bridge rectifier circuit.

1.3.5.3   Interharmonics

In similar notions of harmonics, if the ac quantity has frequency components other than fundamental and is not an integer multiple of the fundamental frequency, they are called as interharmonics. They are mainly caused by devices whose control is not synchronized with the fundamental power frequency like pulse-modulated inverters, cycloconverters, induction furnaces, and arcing devices.

1.3.5.4   Notching

Notching is a kind of voltage disturbance resulting from events occurring in the current waveform, which leads to large periodic spikes in the voltage waveform. It is caused by the normal operation of power electronic converters when switching operation leads to current commutation from one phase to the other, causing a momentary short circuit between the two phases. Notching is periodic and has high-frequency content, which makes a part of transients as well. Voltage notching can cause frequency or timing errors in phase-locked-loop systems due to the detection of additional zero crossings. Fig. 1.8 illustrates an example of voltage notching due to converter operation.

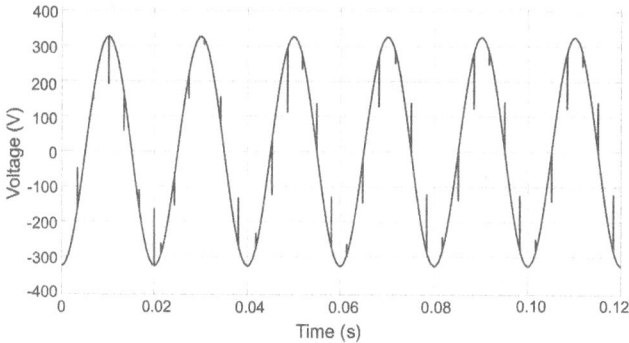

Figure 1.8 Example of voltage notching by converter operation

1.3.5.5 Noise

Distortion or disturbance of any kind that does not fit into transients or any other category of waveform distortion can be categorized as noise. These are unwanted electrical signals with spectral content usually less than 200 kHz and magnitude usually less than 1% of the signal magnitude. These are superimposed on voltage or current signals in phase conductors and are passed on to the neutral conductors. These can be caused by power electronic devices, arcing devices, switching power supplies, control circuits, and improper grounding. The use of filters, isolation transformers, and line conditioners can minimize the effect of noise.

1.3.6 VOLTAGE FLUCTUATIONS

We have observed that for sag or undervoltage, the rms magnitude goes below 0.9 p.u., or in swell or overvoltage, the rms magnitude goes above 1.1 p.u. Now if the voltage magnitude variations are below 1.1 p.u. and above 0.9 p.u., they fall in this category of voltage fluctuations. As per the IEEE definition, voltage fluctuations are defined as variations in the voltage envelope or a series of random changes in the magnitude, which does not exceed the limits of 0.95-1.05 p.u. Voltage fluctuations can also be seen as a form of illumination change in lamps. Voltage fluctuations can be caused by any cyclic variations in the loads' reactive power component or continuous variations in the load current magnitude, such as arc furnaces, welding machines, etc. A decrease in voltage can also be a result of poor load power factor, leading to high reactive current and subsequently more feeder drop.

1.3.7 POWER FREQUENCY VARIATIONS

These are deviations in the fundamental frequency from its predefined nominal value, like 50 or 60 Hz. The nominal value of power frequency is decided solely by the constructional features of the generators producing power. As long as the balance of power generation versus load demand is maintained, the frequency is stable. The only cause of power frequency variation is the shift in the power balance. The magnitude and duration of frequency variations depend upon the dynamic load characteristics and generation system. There is always a minor change in frequency going on, which is acceptable, however severe changes are very rare but highly fatal to the power system as they can lead to complete grid blackouts.

1.4   MITIGATION TECHNIQUES

In the previous section, we discussed various power quality problems. For the efficient and reliable operation of the power system, it is desirable to eliminate or minimize them. Earlier, power quality problems were classified into various categories and subcategories. We now realize that almost all power quality problems arise due to non-sinusoidal variations of voltage, or current over a certain period of time. Thus, it is very likely that every voltage or current quantity should retain the fundamental sinusoidal component after compensation. To achieve this, there are two possible approaches. One approach is to eliminate the problem at the source, i.e., eliminate the cause, leading to such unwanted components. This is achieved through network re-configuration devices. The second approach is to employ different mitigating devices to take care of the unwanted features by injecting desired voltage and currents. These mitigating devices are referred to as Custom Power devices. Based on their functionality, these are broadly divided into two types: network re-configuring type and compensating type. Applications related to various mitigation techniques are listed in Table 1.2.

1.4.1   NETWORK RE-CONFIGURING TYPE

Network re-configuring type devices nowadays used are based on solid-state switches like GTOs or thyristors, unlike the previously used mechanical switches, which were replaced due to their slow operations. They are normally used for fast current limiting and current interruption during faults. There are mainly three devices under this category [9].

1. Solid-State Current Limiter (SSCL): Under a fault condition, this Gate Turn Off Thyristor (GTO) based device inserts a fault current limiting inductance in the circuit, as illustrated in Fig. 1.9. The zinc oxide arrestor (ZnO), current limiting inductor, and GTO units are connected in parallel. The combined circuit is inserted in series with the line. When fault takes place, the current rises to a higher magnitude. This is sensed by the current sensing unit, and accordingly, the turn-off command is sent to the GTOs by applying a negative gate pulse. This will reduce the current through the main line in a matter of few microseconds. During switching-off of the GTOs, the snubber circuit across each GTO switch ramps up the voltage to the level clamped by the ZnO arrestor. Subsequently, it allows the linear rise of current through the inductor till it reaches its voltage limit. Once the fault clears, the turn-on command is given to the GTO units. Thereafter, the current resumes to its rated value. The inductor and ZnO arrestor become non-functional. Several GTOs can be connected back to back in a series to support bidirectional current flow and to match voltage rating.

Table 1.2

Different types of custom power devices and their applications

Custom power devices	Applications
Solid-state current limiter/ Solid-state circuit breaker	Fault current limitation Fault current interruption with further reclosing
Solid-state transfer switch	Protection against voltage sag and swell Transfer of load from one feeder to another
Distribution static compensator (DSTATCOM)	Load current balancing Power factor correction Current harmonic compensation
Dynamic voltage restorer (DVR)	Voltage sag and swell protection Load bus voltage regulation Voltage balancing Voltage Flicker attenuation
Unified power quality conditioner (UPQC)	VAr compensation Voltage regulation Voltage and current balancing Voltage and current harmonic suppression Active and reactive power control

2. Solid-State Circuit Breaker (SSCB): Circuit Breakers are generally mechanical in nature, but SSCB uses a combination of GTOs and Thyristors with the capability to interrupt fault currents very rapidly and perform an auto-reclosing function, as shown in Fig. 1.10. The device allows the fast transfer of power from one supply source to another. The circuit configuration of SSCB is similar to SSCL, except that it uses additional anti-parallel connected thyristor switches in series with the inductor, which eventually allows the cut-off the

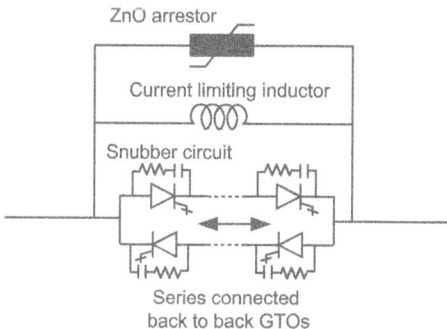

ZnO arrestor

Current limiting inductor

Snubber circuit

Series connected
back to back GTOs

Figure 1.9   Solid-state GTO based current limiter

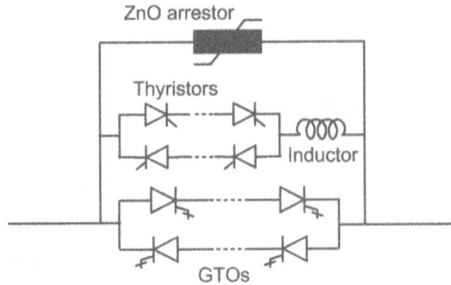

Figure 1.10   Solid-state circuit breaker

inductor current developed through inductor during switching off of GTOs. The number of switches used in the circuit breaker are determined based on its voltage and current ratings.

3. Solid-State Transfer Switch (SSTS): Solid-state transfer switches are thyristor-based devices used for protecting sensitive loads from voltage sag, swell, and interruption. They can rapidly transfer the load from the main source to an auxiliary source within few milliseconds (half a cycle) whenever a voltage sag, swell, or interruption occurs in the primary or main line, acting as an uninterrupted power supply. The schematic diagram of the SSTS is shown in Fig. 1.11. There are two switches: Switch 1 is on the main feeder side, and Switch 2 is on the auxiliary feeder side. Each switch comprises of a pair of series connected thyristors (to match the voltage rating) in an anti-parallel connection. A ZnO arrestor is connected across each switch to provide the voltage limit. Under normal condition the main feeder supplies the load. When fault takes place on the main feeder side, it results in voltage sag, swell, or

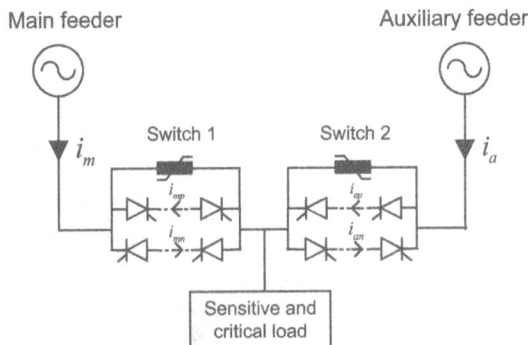

Figure 1.11   Solid-state transfer switch

interruption, and the load is immediately connected to the auxiliary feeder in a make before break action. The currents i_{mp}, i_{mn} and i_{ap}, i_{an} are positive, negative currents on main and auxiliary side of the feeders, respectively.

1.4.2 COMPENSATING TYPE

There are broadly two categories of compensating type devices, namely passive and active compensating devices. Passive compensation is used to mitigate power quality problems by using lossless passive components such as capacitors and inductors. A schematic of a passive filter is shown in Fig. 1.12. In the figure, the current i_f refers to the compensator or filter injected currents at the point of common coupling (PCC) to alleviate unwanted components of load current, i_l, resulting in desired source current, i_s. Inductance-resistance, $L_s - R_s$ represents the feeder impedance, Z_s. Passive compensation methods can achieve the following objectives: reactive power compensation, voltage regulation, and power factor correction. However, these are not suitable for higher order harmonics and dynamic compensation.

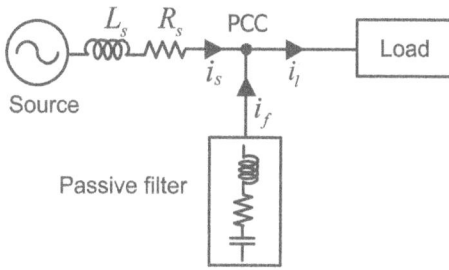

Figure 1.12 Passive filter

Active compensating devices also called as active power filters, which are used for active filtering, load balancing, power factor correction, and voltage regulation with fast dynamic characteristics. There are mainly three types of active compensating devices which we will learn in the subsequent chapters.

1. Distribution Static Compensator (DSTATCOM): This is a shunt type compensating device, also called as shunt active power filter. It injects current at the PCC such that the compensated current nullifies the unbalance and harmonics in the load current. The scheme is mainly used for power factor correction, load balancing, and harmonic eliminations. As a result of this, it improves the quality of the bus voltage. A schematic of DSTATCOM is shown in Fig. 1.13. It consists of a voltage source inverter (VSI) supported by dc link voltage. The VSI is connected to the PCC through interface inductor, L_f.

2. Dynamic Voltage Restorer (DVR): This is a series type compensating device that injects a voltage such that the terminal voltage at the load remains free of voltage fluctuations, sag, swell, interruptions, and harmonic distortions. Thus,

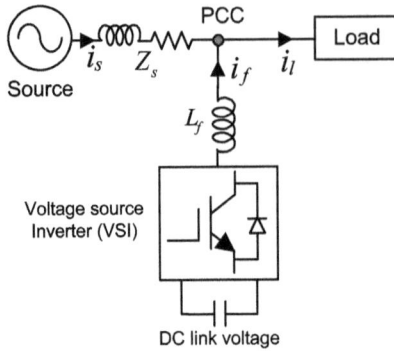

Figure 1.13 Distribution static compensator (DSTATCOM)

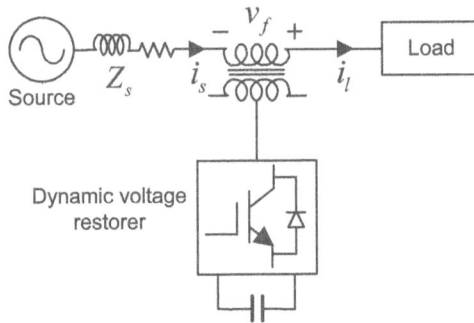

Figure 1.14 Dynamic voltage restorer (DVR)

its main objective is to regulate the voltage across the load terminal bus. The schematic of DVR is shown in Fig. 1.14. It consists of a voltage source inverter supported by the dc link voltage. The output terminals of the inverter are connected to the primary of the transformer and the secondary of the transformer injects voltage in series with the line. The transformer's inductance, along with the filter components shapes the injected voltages, v_f, following the reference DVR voltages.

3. Unified Power Quality Conditioner (UPQC): This is a combination of DSTAT-COM as shunt active power filter and DVR as series active power filter. Thus, it combines the advantage of both the shunt and series compensation. A schematic of UPQC is shown in Fig. 1.15. The device UPQC regulates the voltage at the load bus by injecting series voltage and maintains the supply current by injecting current at the PCC. It has two voltage source inverters and a common dc link to support the operation of the UPQC. Since the dc link is

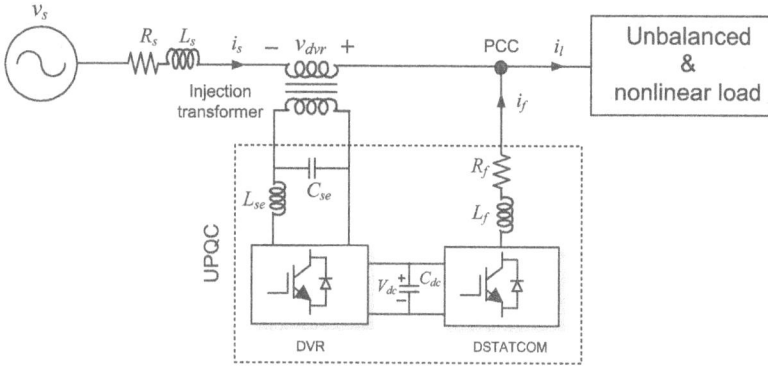

Figure 1.15 Unified power quality conditioner (UPQC)

common, its proper design and rating are very important for satisfactory voltage and current compensation. Its control is more complicated than the other custom power devices as it involves two inverters along with their coordination with the dynamics of the dc link voltage.

1.5 POWER QUALITY MONITORING

Power quality problems are randomly occurring events, and if timely detection and mitigation are not performed can lead to the malfunctioning of critical and sensitive load, which can further cause substantial financial losses. Therefore, monitoring power quality events is an essential part of the mitigation process. There are many aspects of monitoring a power quality event. Identifying events specific to each location of the power system (i.e., a disturbance can have different effects, at a given location as the distance from the point of occurrence changes). Further, power quality monitoring allows to evaluate and predict the performance of loads and choose appropriate mitigation techniques. Table 1.3 depicts some of the parameters used for the detection of power quality events [24]. In general, a power quality monitoring system comprises of four stages,

1. Data Acquisition: In data acquisition, the measured line voltages and currents are sampled and converted to their equivalent digital signals using analog to digital converters (ADCs).
2. Characterization: In characterization, the digital signals of current and voltage are used to determine different parameters using various signal processing algorithms, and are stored.
3. PQ Analysis: At analysis stage, the calculated parameters are analyzed and compared against defined standards using some processing units to detect power quality events and disturbances.
4. Statistical Analysis: In this stage, statistical analysis of the detected power quality events along with previous trends of those events and parameters is performed, and appropriate mitigating actions are selected.

Table 1.3

Stored data used for monitoring various PQ events

Power quality events	Characteristics
Transients	Rate of rise, peak amplitude, duration, spectral content, energy potential, mean absolute variation squared amplitude (MAVSA) index
Short-duration voltage variation (Sag, swell, and interruption)	Duration, rms magnitude (minimum / or maximum), and clearing time
Long-duration voltage variation (Undervoltage, overvoltage)	Duration, rms magnitude (minimum or maximum), and clearing time
Unbalance	Duration, percentage unbalance
Waveform distortions	Harmonic spectrum with magnitude and phase angles, total harmonic distortion (THD), total demand distortion (TDD), frequency spectrum of notching, minimum / maximum amplitude of notch
Voltage fluctuations	Duration, rms magnitude, instantaneous flicker (IFL), perceptibility short term (Pst), perceptibility long term (Plt)
Power frequency variations	Duration, frequency (minimum / maximum)

1.6   POWER QUALITY INDICES

Various power quality indices are used to quantify and characterize the disturbance in electrical systems, industrial plants or residential loads. These indices serve to indicate the complex time and frequency domain waveform into a single number. The important indices used to assess the power quality are listed in Table 1.4 [25]. The indices mentioned in the table are calculated using standardized procedures and are meaningfully interpreted by practicing engineers. The important power quality indices are described in this section for a periodic voltage or current waveform, $x(t)$, given as following.

$$x(t) = X_{dc} + \sqrt{2}X_1 \sin(\omega t - \phi_1) + \sqrt{2}\sum_{h=2}^{\infty} X_h \sin(h\omega t - \phi_h)$$

$$= X_{dc} + x_1 + \sum_{h=2}^{\infty} x_h \qquad (1.3)$$

In the above equation, $x(t)$ represents the instantaneous voltage or current quantity. The terms, X_{dc}, X_1, and X_h are dc component, rms value of fundamental, and

Table 1.4

Commonly used power quality indices

Index	Definition	Application(s)
Total harmonic distortion (THD)	$\sqrt{\sum_{i=2}^{\infty} I_i^2}/I_1$	Measure of waveform distortion
Total demand distortion (TDD)	$\sqrt{\sum_{i=2}^{\infty} I_i^2}/I_L$	Measure of waveform distortion
Power factor (PF)	$P_{tot}/(\lvert V_{rms}\rvert \lvert I_{rms}\rvert)$	Energy metering
Telephone influence factor	$\sqrt{\sum_{i=2}^{\infty} w_i^2 I_i}/I_{rms}$	Audio circuit interference
IT product	$\sqrt{\sum_{i=2}^{\infty} w_i^2 I_i^2}$	Audio circuit interference, Shunt capacitor stress
VT product	$\sqrt{\sum_{i=2}^{\infty} w_i^2 V_i^2}$	Voltage distortion index
K factor	$\left(\sum_{h=1}^{\infty} h^2 I_h^2\right)\left(\sum_{h=1}^{\infty} I_h^2\right)$	Transformer derating
Crest factor	V_{peak}/V_{rms}	Insulation capacity
Unbalance factor	V_-/V_+	Degree of three-phase unbalance
Flicker factor	$\Delta V/\lvert V\rvert$	Incandescent lamp operation, voltage regulation, short circuit capacity

harmonics, respectively. The terms h, ω, t, and ϕ_h represent the harmonic number, angular frequency, time variation, and phase angle of the sinusoidal voltage or current waveform. The dc component, rms, and average values of sinusoidal waveforms are denoted by capital letters such as X_{dc}, X_{rms}, X_{avg} (terms explained below). The instantaneous ac quantities are denoted by small letters such as x_1, x_2, ...x_h as indicated in (1.3).

Harmonics: The sinusoidal voltage or current waveform with frequency being integer multiple of fundamental frequency. Odd integer and even integer multiples result in odd and even harmonics. In (1.3), $x_1, x_2, ...x_h$ components represent the fundamental and harmonic components.

Interharmonics: Sinusoidal voltage or current waveform with frequency which are not integer multiple of fundamental frequency.

Subharmonics: Sinusoidal voltage or current waveform with frequency below the fundamental frequency.

Average value of voltage or current waveform: The average value of a periodic nonsinusoidal voltage or current waveform is defined as following.

$$X_{avg} = \frac{1}{T} \int_0^T x(t)\, dt \qquad (1.4)$$

where $x(t)$ is instantaneous voltage or current quantity and T is the time period of fundamental component of $x(t)$.

rms value of voltage or current waveform: The rms value of a periodic nonsinusoidal voltage or current waveform is defined as following.

$$X_{rms} = \sqrt{\frac{1}{T} \int_0^T x^2(t)\, dt} \qquad (1.5)$$

Crest Factor (CF): It is a measure of ratio of the peak value of voltage or current quantity to its rms value. It is defined as follows.

$$\text{Crest Factor (CF)} = \frac{X_{peak}}{X_{rms}} \qquad (1.6)$$

Form Factor (FF): Form factor indicates the shape of the voltage or current waveform. It is defined as the ratio of the rms to the average value.

$$\text{Form Factor (FF)} = \frac{X_{rms}}{X_{avg}} \qquad (1.7)$$

Ripple Factor (RF): Ripple factor is a measure of the ripple content of the voltage or current waveform. It is defined as the ratio of rms value of ac component to the dc component of voltage or current quantity.

$$\text{Ripple Factor (RF)} = \frac{X_{ac}}{X_{dc}} \qquad (1.8)$$

Since $X_{rms}^2 = X_{ac}^2 + X_{dc}^2$, it implies that $X_{ac} = \sqrt{X_{rms}^2 - X_{dc}^2}$ which leads to $RF = \sqrt{FF^2 - 1}$.

Harmonic Factor (HF): Harmonic factor of h_{th} harmonic is the ratio of the rms values of the h_{th} harmonic to the fundamental. Thus,

$$\text{Harmonic Factor (HF)} = \frac{X_h}{X_1} \qquad (1.9)$$

Total Harmonic Distortion (THD): Total Harmonic Distortion (THD) is the most commonly used index to quantify the harmonic content of the non-sinusoidal waveform. It is defined as the ratio of the rms value of all the harmonic components (X_H to the rms value of the fundamental (X_1). It is given below.

$$\text{Total Harmonic Distortion (THD)} = \frac{X_H}{X_1} = \frac{\sqrt{\sum_{h=2}^{\infty} X_h^2}}{X_1} \tag{1.10}$$

$$= \frac{\sqrt{X^2 - X_1^2}}{X_1} = \sqrt{\left(\frac{X}{X_1}\right)^2 - 1}$$

Total Demand Distortion (TDD): Total Demand Distortion (TDD) is similar to THD except that it is expressed as the ratio of the rms value of the total harmonic components (X_H) to the rms value of the waveform, $x(t)$. It is given as follows.

$$\text{Total Demand Distortion (TDD)} = \frac{X_h}{X} \tag{1.11}$$

Power Factor (pf): Power Factor (pf) is defined as the ratio of real power (P) to the apparent power (S). For a given voltage, $v(t)$ and current, $i(t)$, as given in (1.3), with $x(t)$ representing voltage, $v(t)$ or current, $i(t)$. The power factor is defined as follows.

$$pf = \frac{P}{S} = \frac{\frac{1}{T}\int_0^T v(t)\,i(t)\,dt}{V_{rms}I_{rms}} \tag{1.12}$$

Telephone Influence Factor (TIF): Telephone Influence Factor (TIF) determines the influence of the power system harmonics on telephone communication systems. It is the ratio of weighted rms to the rms value of the quantity and is expressed as below.

$$TIF = \frac{\sqrt{\sum_{i=2}^{\infty} w_i^2 X_i}}{X_{rms}} \tag{1.13}$$

C-Message Weights: C-Message is similar to TIF, except that the weights w_i are replaced with weights c_i. In C-message, C stands for computer. This index is also called as Computer Interference Factor (CIF). It determines the impact of power system harmonics on telephone communication systems. It is the ratio of weighted rms to the rms value of the quantity and is expressed as below.

$$TIF = \frac{\sqrt{\sum_{h=2}^{\infty} w_i^2 X_i}}{X_{rms}} \tag{1.14}$$

IT and VT Product (IT, VT): IT (current product) and VT (voltage product) are weighted rms values of harmonic components of current and voltage, respectively. It is basically the product of rms value of current or voltage with its TIF (hence the terms, $I * T$ or $V * T$). These are kind of alternative indices to the THD. The IT and VT are defined as below.

$$IT = \sqrt{\sum_{i=2}^{\infty} w_i^2 I_i} \qquad (1.15)$$

$$VT = \sqrt{\sum_{i=2}^{\infty} w_i^2 V_i} \qquad (1.16)$$

K-Factor: K-factor is the ratio of sum of weighted squared values of the harmonic load currents according to their effects on transformer heating to the sum of squared rms components. The higher K-factor, implies more heating effect due to harmonic components. Some other loads such as computers, solid-state devices, and motors which generate non linear load currents, cause transformers and system neutrals to overheat, reducing their life span and eventual failure of the transformer. The K-factor is defined as below.

$$K\text{-}factor = \frac{\sum_{h=1}^{h=h_{max}} h^2 I_h^2}{\sum_{h=1}^{h=h_{max}} I_h^2} \qquad (1.17)$$

Unbalance Factor: Unbalance factor is used to measure the unbalance present in the three-phase voltages or currents. This is expressed as the ratio of the negative sequence, (X^-) to positive sequence, (X^+) components as given below.

$$\text{Unbalance Factor} = \frac{X^-}{X^+} \qquad (1.18)$$

Flicker Factor: It is defined as the ratio of change in voltage to its rms value, as expressed below.

$$\text{Flicker Factor} = \frac{\Delta V}{V} \qquad (1.19)$$

1.7 POWER QUALITY STANDARDS

In the preceding sections, we discussed power quality problems and their mitigation devices. However, due to practical constraints, it is not always possible to completely eliminate power quality problems. Thus, it is imperative to alleviate the problems and maintain certain permissible limits of various parameters. If the parameters are beyond the safe operating limits, these start affecting consumers, utilities, and therefore it can be a matter of concern. To address this issue, many regulatory bodies such as the Institute of Electrical and Electronics Engineers (IEEE), International Electrotechnical Commission (IEC), American National Standards Institute (ANSI), British Standards (BS), and Computer Business Equipment Manufacturers Association (CBEMA), etc., have developed certain sets of power quality standards which are used by the electric utilities in generation, transmission and distribution systems to maintain power quality within acceptable levels.

The important limits for voltage and currents at different voltage levels are given in [5]. The allowable voltage limits in the power system are normally within $\pm10\%$ over the rated value. The allowable harmonics in voltage measured as THD should be below 8%, and for the current, THD should be below 5%. When short circuit capacity is high, slightly more distortions can be allowed from these THD values for the voltage and current.

The dc offset of the load currents should be avoided as it leads to the saturation of the transformer and electrical machines leading to their inefficient operation. The dc offset also leads to the misoperation of electronic switching by affecting zero crossing and other control logic. Similarly, there are standards for allowable frequency variations. As per IEEE 1547-2018, allowed frequency variation is normally within ±0.02 p.u. Some of the important power quality standards are mentioned below.

IEEE Power Quality Standards

- IEEE 1159: Monitoring Electric Power Quality

- IEEE 1159.1: Guide For Recorder and Data Acquisition Requirements

- IEEE 1159.2: Power Quality Event Characterization

- IEEE 1159.3: Data File Format for Power Quality Data Interchange

- IEEE 1564: Voltage Sag Indices

- IEEE 1346: Power System Compatibility with Process Equipment

- IEEE 1100: Power and Grounding Electronic Equipment (Emerald Book)

- IEEE 1433: Power Quality Definitions

- IEEE 1453: Voltage flicker

- IEEE 519: Harmonic Control in Electrical Power Systems

- IEEE 519A: Guide for Applying Harmonic Limits on Power Systems

- IEEE 446: Emergency and standby power

- IEEE 1409: Distribution Custom Power

- IEEE 1547: Distributed Resources and Electric Power Systems Interconnection

IEC Power Quality Standards

- IEC Standard 61000-4-7: Deals with the requirements for monitoring and measuring harmonics.

- IEC Standard 61000-4-15: Describes the instrumentation and procedures for monitoring flicker. (The IEEE flicker task force working on Standard IEEE P1453 is set to adopt the IEC Standard 61000-4-15 as its own.)

- IEC Standard 61000-4-30: Plans on providing overall recommendations for monitoring all types of power quality phenomena.

- IEC Standard 61000-2-8: Environment—Voltage Dips and Short Interruptions.

- IEC Standard 60034-1: Rotating electrical machines, Part 1: Rating and performance

1.8   CBEMA AND ITIC CURVES

Computer and Business Equipment Manufacturers' Association, 1970 (CBEMA), Information Technology Industry Council (ITIC) curves are commonly used to indicate power quality in relation to the operation of the electrical equipments [26]. These curves are also known as voltage acceptability or tolerance curves, as shown in Fig. 1.16. The curves basically indicate the tolerance of mainframe computer equipment to the magnitude and duration of the applied voltage. The x-y axes represent the time duration and magnitude of the applied voltage for the equipment, respectively. The key observation from these curves is that under disturbance conditions, there is an energy impact, which here is represented as product of square of rms voltage, V^2 and the time duration, Δt. Thus, every equipment has a certain maximum energy impact that it can tolerate under disturbances. If the impact of the energy level is more,

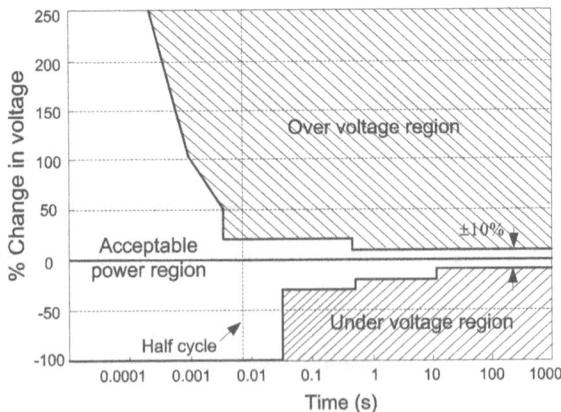

Figure 1.16   CBEMA and ITIC curves

then the operation of the load is disrupted, which is indicated by the outer regions of the curves. The inner regions of the curves indicate the acceptable power range. It can further be observed that 100% change can be normally accepted for about a millisecond, and ±10% is acceptable for the continuous operation of the equipment.

1.9 SUMMARY

In this chapter, the concept of power quality in power systems has been discussed with a focus on distribution power systems. The relation of quality of voltage, current, and power is explained using simple expressions and circuitry. Various power quality problems are discussed and presented with their classification, causes, and effects on the end-users. Further, to eliminate these power quality disturbances, several mitigation techniques like network reconfiguring and compensating typle devices have been discussed, and their operating principles are explained. For monitoring of power quality aspects using power quality analyzers, important power quality indices have been defined and explained. Power quality problems are an inherent part of any electrical network which arises due to the use of certain kinds of electrical equipment and situations within the network. Therefore, it is not possible to completely eliminate all the problems rather maintain the electrical quantities within prescribed limits for which various power quality standards developed by international organizations like IEEE, IEC, ANSI, BS, and CBEMA are described. The standards providing the guidelines to allowable limits of magnitude and harmonic components of voltage and system frequency variations are discussed and explained. Further, CBEMA and Information Technology Industry Council (ITIC) curves illustrate the safe operating range for electrical products and systems.

1.10 PROBLEMS

P 1.1 What is power quality? What does the term power quality refer to in a power distribution system?

P 1.2 Explain how the feeder impedance affects power quality at the load bus.

P 1.3 What are stiff and non-stiff sources? Relate them to the X and R values of the feeder impedance.

P 1.4 Explain how voltage related power quality becomes a current related problem and vice-versa. How the quality of power is related to the quality of voltage and current?

P 1.5 List various power quality problems in the power distribution network. Describe each one of them in detail.

P 1.6 What is the basis of the classification of short term and long term variations of voltage?

P 1.7 Define voltage sag, swell, and interruption along with their classification on time duration.

P 1.8 Describe harmonics and interharmonics in the power system. List down their causes and effects.

P 1.9 What is voltage fluctuation? List a few loads causing voltage fluctuation and explain how these loads generate voltage fluctuation in the power system.

P 1.10 Why is power quality monitoring important for industrial plants/loads? What are the various stages involved in power quality monitoring?

P 1.11 Describe power quality indices and list the important ones to characterize the power system's performance.

P 1.12 Define the following terms: (a) harmonics, (b) interharmonics, (c) subharmonics, (d) harmonic factor, (e) total harmonic distortion (THD), (f) total demand distortion (TDD), (g) power factor, and (h) unbalance factor.

P 1.13 What is the significance of power quality standards? List the important standards concerning power quality.

P 1.14 Discuss the CBEMA and ITIC curves and explain their significance. From these curves, what do you infer about the magnitude and duration of the voltage supplying an electrical device?

REFERENCES

1. IEEE Standard 1159–2019, "IEEE recommended practice for monitoring electric power quality," 2019.

2. IEEE Standard 1100–2005 (Revision of IEEE Std 1100–1999), "IEEE recommended practice for powering and grounding electronic equipment," 2006.

3. IEC, "IEC 61000-3-2 (1995) Electromagnetic Compatibility (EMC) Part 3: Limits- Section 2 Limits for Harmonics Current Emissions (Equipments input currents ¡ 16 A per phase)," 1995.

4. IEEE Standard 1459–2010 (Revision of IEEE Std 1459–2000), "IEEE standard definitions for the measurement of electric power quantities under sinusoidal, nonsinusoidal, balanced, or unbalanced conditions," 2010.

5. IEEE Standard 519–2014 (Revision of IEEE Std 519–1992), "IEEE recommended practices and requirements for harmonic control in electrical power systems," 2014.

6. W. Shepherd and P. Zand, *Energy Flow and Power Factor in Nonsinusoidal Circuits*. Cambridge University Press, 1979.

7. M. H. J. Bollen, *Understanding Power Quality Problems: Voltage Sags and Interruptions*. Wiley-IEEE Press, 1999.

8. Math H. J. Bollen and Irene Y. H. Gu, *Signal Processing of Power Quality Disturbances*. Kluwer Academic Publishers, 2006.

9. Arindam Ghosh and Gerald Ledwich, *Power quality enhancement using custom power systems*. Wiley, 2006.

10. Bhim Singh, Ambrish Chandra, and Kamal Al-Haddad, *Power Quality Problems and Mitigation Techniques*. Wiley, 2015.

11. Roger C. Dugan, Mark F. McGranaghan, Surya Santoso, and H. Wayne Beaty, *Electrical Power System Quality*. Tata McGraw–Hill, 2008.

12. M. A. Masoum and E. F. Fuchs, *Chapter 1 – Introduction to Power Quality*, M. A. Masoum and E. F. Fuchs, Eds. Academic Press, 2015.

13. J. Schalabbach, D. Blume, and T. Stephanblome, *Voltage Quality in Electrical Power Systems*. The Institute of Engineers, London, UK, 2001.

14. S. Chattopadhyay, M. Mitra, and S. Sengupta, *Electric Power Quality*. Springer, 2011.

15. A. Baggini, Ed., *Handbook of Power Quality*. Wiley Sons, England, 2008.

16. B. W. Kennedy, *Power Quality Primer*. Springer, 2000.

17. C. Sankaran, *Power Quality*. CRC Press, 2002.

18. Mahesh K. Mishra, A. Ghosh, and A. Joshi, "Load compensation for systems with non-stiff source using state feedback," *Electric Power Systems Research*, vol. 67, no. 1, pp. 35–44, 2003.

19. Mahesh K. Mishra, A. Ghosh, and A. Joshi, "Operation of a DSTATCOM in voltage control mode," *IEEE Transactions on Power Delivery*, vol. 18, no. 1, pp. 258–264, 2003.

20. L. P. Kunjumuhammed and Mahesh K. Mishra, "A control algorithm for single-phase active power filter under non-stiff voltage source," *IEEE Transactions on Power Electronics*, vol. 21, no. 3, pp. 822–825, 2006.

21. C. Kumar and Mahesh K. Mishra, "A multifunctional DSTATCOM operating under stiff source," *IEEE Transactions on Industrial Electronics*, vol. 61, no. 7, pp. 3131–3136, 2014.

22. C. Kumar, Mahesh K. Mishra, and M. Liserre, "Design of external inductor for improving performance of voltage-controlled DSTATCOM," *IEEE Transactions on Industrial Electronics*, vol. 63, no. 8, pp. 4674–4682, 2016.

23. Brendan Fox, L. Bryans, D. Flynn, N. Jenkins, D. Milborrow, M. O'Malley, R. Watson, and O. Anaya-Lara, *Wind Power Integration: Connection and System Operational Aspects (2nd Edition)*. IET Power and Energy Series, 2014.

24. Dong-Jun Won, Il-Yop Chung, Joong-Moon Kim, Seung-Il Moon, Jang-Cheol Seo, and Jong-Woong Choe, "Development of power quality monitoring system with central processing scheme," vol. 2, pp. 915–919, 2002.

25. G. T. Heydt, *Electric Power Quality*. Stars in a Circle Publications, 1991.

26. Information Technology Industry Council (ITIC), 1250, Eye Street NW, Suite 200 Washington, D.C. (http://www.itic.org).

2 Single-Phase Circuits: Power Definitions and Components

2.1 INTRODUCTION

Single-phase circuits are frequently used in low power applications such as lighting, heating, single-phase motors and few large electric motors in residential and small or medium size industries. The single-phase circuits can be used as part of three-phase circuits or as an independent single-phase system. The single-phase load may form about 10–20% of the total load which is a significant portion to affect the overall performance of the power system. Most single-phase applications assume voltage and current waveform as sinusoidal. When these waveforms are not sinusoidal, the errors may occur in their measurements and in the operation of equipments. Under these conditions, it is necessary to evaluate the measuring techniques used in power system with the increased power electronic-based loads and other non-linear loads. Therefore, the definitions of power and its various components are very important to understand the quantitative and qualitative power quality aspects in power system [1]–[18]. Some quantities, such as reactive power, power factor under non-sinusoidal voltage and current, are not properly elaborated and explained in the literature. The importance of proper definition and quantification of various power terms and components becomes more meaningful in load compensation, voltage regulation, as well as measurement of electrical energy using digital energy meters, when there are harmonics in voltage and current. In this chapter, an attempt is made to explain the concept of instantaneous power, instantaneous active power, instantaneous reactive power, real power, reactive power, and power factor in a systematic and connecting manner. This is not only necessary from the point of view of conceptual clarity, but also very much required for practical applications such as metering, quantification of active power, reactive power, power factor, and other power quality parameters in power system. These aspects become more important when power system is not ideal, i.e., when system deals with unbalance, harmonics, faults and fluctuations in voltage and frequency. We therefore, in this chapter, explore the concepts and fundamentals of single-phase system with some practical applications and illustrations.

2.2 POWER TERMS AND DEFINITIONS IN SINGLE-PHASE SYSTEMS

Let us consider a single-phase system with sinusoidal voltage supplying a linear load as shown in Fig. 2.1(a). A linear load is one which consists of ideal resistive, inductive and capacitive elements. The instantaneous voltage and current are expressed as

DOI: 10.1201/9781032617305-2

below.

$$v(t) = \sqrt{2}V \sin \omega t$$
$$i(t) = \sqrt{2}I \sin(\omega t - \phi) \tag{2.1}$$

The corresponding voltage and current waveforms are shown in Fig. 2.1(b) with peak values $V_m = \sqrt{2}V, I_m = \sqrt{2}I$ and frequency, $\omega = 2\pi f$ for $f = 50$ Hz with V and I are the rms value of voltage and current, respectively. The current lags voltage by ϕ degrees, which depends on the load parameters, i.e., resistance, inductance, and capacitance values. The voltage in the above equation is related to the work done per unit of charge, and the current is related to the rate of change of charge, i.e., $v = dW/dq$, and $i = dq/dt$. Therefore, the product of the voltage and current is the rate of work done, i.e., $vi = dW/dt$ J/s or Watts. The instantaneous power can be expressed as following.

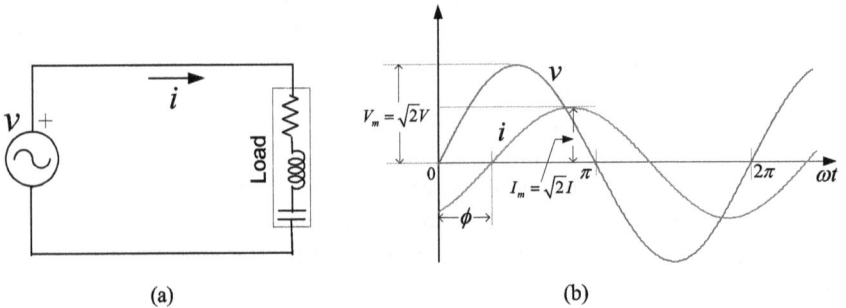

Figure 2.1   (a) A single-phase system and (b) voltage and current waveforms

$$
\begin{aligned}
p(t) = v(t)\, i(t) &= VI[2\sin \omega t \sin(\omega t - \phi)] \\
&= VI[\cos \phi - \cos(2\omega t - \phi)] \\
&= VI\cos \phi (1 - \cos 2\omega t) - VI \sin \phi \sin 2\omega t \\
&= P(1 - \cos 2\omega t) - Q\sin 2\omega t \\
&= p_{active}(t) + p_{reactive}(t)
\end{aligned} \tag{2.2}
$$

The first term in above equation is called as active power, as it is the power that flows in the resistive element. The second term is similarly referred to as reactive power, as this component is present due to reactive elements such as inductance and capacitance.

Here, $P = \frac{1}{T}\int_0^T p_{active}(t)\, dt = \frac{1}{T}\int_0^T p(t)\, dt$ is known by many names such as average active power, real power, useful power or simply power. In above T is the time period of voltage or current waveform. Therefore, real power can be expressed as,

$$P = VI\cos \phi = \frac{1}{T}\int_0^T p_{active}(t)\, dt = \frac{1}{T}\int_0^T p(t)\, dt \tag{2.3}$$

Reactive power, Q is defined as,

$$Q \triangleq \max\{p_{reactive}(t)\} = VI\sin\phi \qquad (2.4)$$

It should be noted that the way Q is defined is different from P. The term Q is defined as maximum value of the second term of (2.2) and not as an average value of the second term, although, it is denoted as average reactive power. This difference should be kept in mind while defining P and Q.

Equation (2.2) shows that instantaneous power can be decomposed into two terms. The first term has an average value of $VI\cos\phi$ and an alternating component of $VI\cos2\omega t$, oscillating at twice the line frequency. This term, i.e., p_{active} is never negative and therefore is called unidirectional or dc power. The second term, i.e., $p_{reactive}$ has an alternating component $VI\sin\phi\sin2\omega t$ oscillating at twice frequency with a peak value of $VI\sin\phi$. The second term has zero average value. Equation (2.2) can further be written in the following form:

$$\begin{aligned} p(t) &= VI\cos\phi - VI\cos(2\omega t - \phi) \\ &= \bar{p}(t) + \tilde{p}(t) \\ &= P_{average} + P_{oscillating} \end{aligned} \qquad (2.5)$$

With the above definitions of P and Q, the instantaneous power $p(t)$ can be re-written as following.

$$p(t) = P(1 - \cos2\omega t) - Q\sin2\omega t \qquad (2.6)$$

Various components of single-phase instantaneous powers are shown in Fig. 2.2. As observed from the figure, the instantaneous power has positive and negative values forming a sinusoidal waveform of twice the fundamental frequency. The positive power is formed when both the voltage and current, are positive or negative, while negative power is formed when one of them is negative. The negative power is shown by a hatched region. Positive power indicates that the power is pumped from the source to the load, while negative power indicates that the power is returned to the source. The sum of average values of the positive and negative powers in a

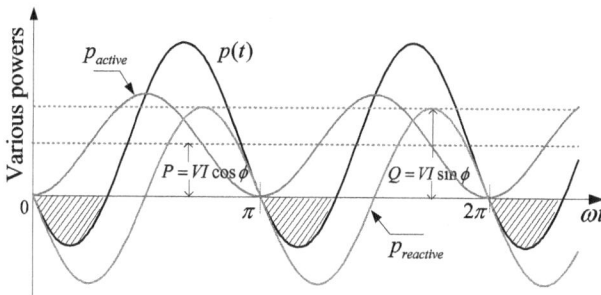

Figure 2.2 Various powers in single-phase system

cycle is the average or real power delivered to the load. As per equation (2.2), the instantaneous power, $p(t)$ has two components, $p_{active}(t)$ and $p_{reactive}(t)$, as shown in the figure. The average value of $p(t)$ and the average value of $p_{active}(t)$ are same and equal to $P = VI\cos\phi$. Also, the maximum value of $p_{reactive}(t)$ is reactive power $Q = VI\sin\phi$. These powers are indicated in the figure by dotted lines. The scalar apparent power, S in single-phase system is defined as the product of rms values of voltage and current as expressed below.

$$S = VI \tag{2.7}$$

The apparent power, S can also be expressed in terms of real and reactive powers as given below.

$$\begin{aligned} S = VI &= \sqrt{V^2 I^2 \cos^2\phi + V^2 I^2 \sin^2\phi} \\ &= \sqrt{P^2 + Q^2} \end{aligned} \tag{2.8}$$

Therefore, we have

$$S^2 = P^2 + Q^2. \tag{2.9}$$

Another important term that indicates the usage of electric power is the power factor (pf). The power factor in single-phase system with sinusoidal nature of voltage and current of the fundamental frequency is defined as the ratio of the real power to the apparent power. This is given below.

$$\text{Power factor} = pf = \frac{P}{S} \tag{2.10}$$

Substituting the value of P from (2.3) and S from (2.7), the above equation can be simplified to,

$$pf = \frac{VI\cos\phi}{VI} = \cos\phi \tag{2.11}$$

The above equation indicates that the power factor can also be defined as cosine angle between the voltage and current. However, this definition holds true only if the voltage and current are of the fundamental frequency. The relationship between P, Q, S and angle ϕ is represented using power triangle shown in Fig. 2.3.

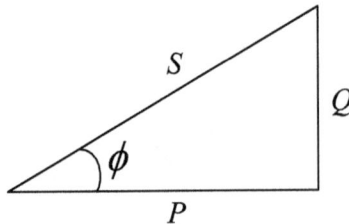

Figure 2.3   Relation between P, Q, S, and ϕ

2.3 PHASOR REPRESENTATION OF ELECTRICAL QUANTITIES

To analyze electrical circuits, we often use phasor notation to represent electrical quantities such as voltage, and current. To compute circuit parameters at different points in the circuit, we make use of voltage and current phasors as well as impedance and apparent power in complex form. In phasor notation, the voltage and currents are represented by a phasor having a magnitude of rms value and phase angle with respect to a reference phasor. A reference phasor is one whose angle is zero. Thus, the voltage and current expressed by (2.1), can be written in phasor form as following.

$$\overline{V} = V \angle 0°$$
$$\overline{I} = I \angle -\phi \tag{2.12}$$

In the above equation, V, I are rms value of voltage and current, respectively. The angle ϕ is an angle between voltage and current. The voltage and current phasors are shown in Fig. 2.4. In case, we have an electrical load represented by linear resistance R, inductance L, and capacitance C parameters, the relationship between voltage and current can be expressed as below:

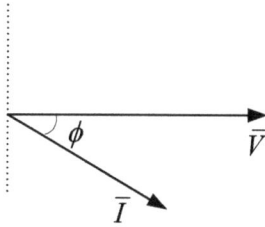

Figure 2.4 Voltage and current phasors

$$\overline{I} = \frac{\overline{V}}{Z} \tag{2.13}$$

where Z is a complex quantity and is known as the impedance of the electrical load. For RLC circuits, it has three components, i.e., resistance R, inductive reactance X_L of inductance L at fundamental frequency and capacitive reactance X_C of capacitance C at fundamental frequency. Further, for series connected RLC load, the impedance can be expressed as,

$$Z = R + jX_L - jX_C. \tag{2.14}$$

Similarly, for RLC parallel connected load, the relationship between admittance, impedance can be expressed as following.

$$Y = \frac{1}{Z} = \frac{1}{R} + \frac{1}{jX_L} + \frac{1}{-jX_C}. \tag{2.15}$$

In the above equations,

$$X_L = 2\pi f L$$
$$X_C = \frac{1}{2\pi f C} \tag{2.16}$$

Where f is the frequency of supply voltage and $\omega = 2\pi f$. With the above description, complex power $\overline{S}$ can be expressed as below.

$$\overline{S} = \overline{V}\overline{I}^* \tag{2.17}$$

The term $\overline{I}^*$ refers to complex conjugate of $\overline{I}$ phasor. This operation ensures that the phase angle difference between the voltage and current is considered to compute complex power $\overline{S}$. Substituting $\overline{V}$ and $\overline{I}$ from (2.12), we get,

$$\begin{aligned} \overline{S} &= V\angle 0° I\angle \phi = VI\angle \phi \\ &= VI\cos\phi + jVI\sin\phi = P + jQ \end{aligned} \tag{2.18}$$

From the above, the magnitude of the complex power, $\overline{S}$ is the apparent power, S as described in (2.8). That is,

$$|\overline{S}| = S = \sqrt{P^2 + Q^2} \tag{2.19}$$

This gives similar relation as given in (2.9).

Example 2.1. Consider a sinusoidal supply voltage $v(t) = 230\sqrt{2}\sin\omega t$ V, supplying a linear load of impedance $Z_L = 12 + j13$ Ω at $\omega = 2\pi f$ rad/s, $f = 50$ Hz. Write the expression for the current, $i(t)$ in the circuit. Based on $v(t)$ and $i(t)$, determine the following.

(a) Instantaneous power $p(t)$, instantaneous active power $p_{active}(t)$ and instantaneous reactive power $p_{reactive}(t)$.

(b) Compute average real power P, reactive power Q, apparent power S, and power factor pf.

(c) Repeat the above when load is $Z_L = 12 - j13$ Ω, $Z_L = 12$ Ω, and $Z_L = j13$ Ω

(d) Comment upon the results.

Solution: A single-phase circuit supplying linear load is shown in Fig. 2.1. In general, the current in the circuit is given as,

$$i(t) = \sqrt{2}I\sin(\omega t - \phi)$$

where $\phi = \tan^{-1}(X_L/R_L)$, and $I = V/|Z_L|$

Case 1: When load is inductive, $Z_L = 12 + j13\ \Omega$

$|Z_L| = \sqrt{R_L^2 + X_L^2} = \sqrt{12^2 + 13^2} = 17.692\ \Omega$, and $I = 230/17.692 = 13$ A

$\phi = \tan^{-1}(X/R) = \tan^{-1}(13/12) = 47.29°$

Therefore, we have

$$\begin{aligned} v(t) &= 230\sqrt{2}\sin\omega t \\ i(t) &= 13\sqrt{2}\sin(\omega t - 47.29°) \end{aligned}$$

The instantaneous power is given as,

$$\begin{aligned} p(t) &= VI\cos\phi(1 - \cos 2\omega t) - VI\sin\phi\sin 2\omega t \\ &= 230 \times 13\cos 47.29°(1 - \cos 2\omega t) - 230 \times 13\sin 47.29°\sin 2\omega t \\ &= 2028.23(1 - \cos 2\omega t) - 2196.9\sin 2\omega t \\ &= p_{active}(t) + p_{reactive}(t) \end{aligned}$$

The above implies that,

$$\begin{aligned} p_{active}(t) &= 2028.23(1 - \cos 2\omega t) \\ p_{reactive}(t) &= -2196.9\sin 2\omega t \end{aligned}$$

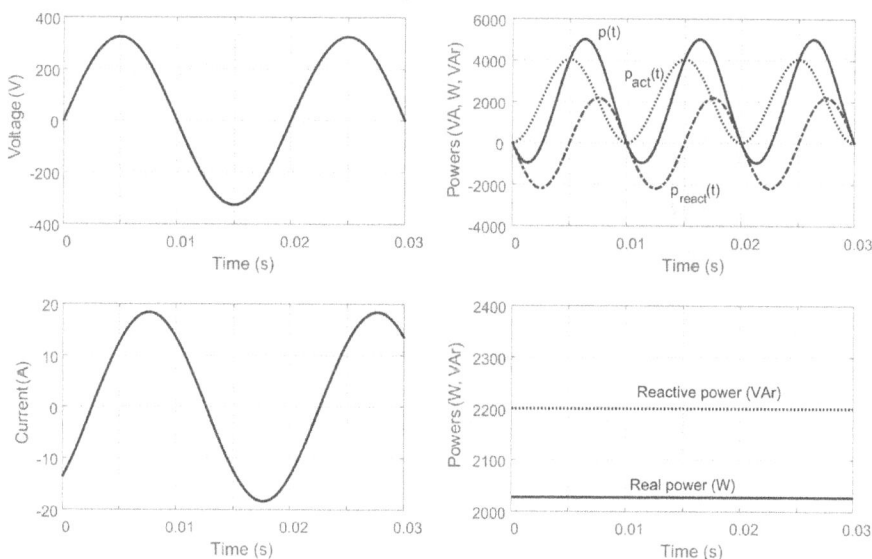

Figure 2.5 Case 1: Voltage, current, and various power components

Average real power (P) is given as,

$$P = \frac{1}{T}\int_0^T p(t)\,dt$$

$$P = VI\cos\phi = 230 \times 13 \times \cos 47.29° = 2028.23\,\text{W}$$

Reactive power (Q) is given as maximum value of $p_{reactive}$, and equals to $VI\sin\phi$ as given below.

$$Q = VI\sin\phi = 230 \times 13 \times \sin 47.2906° = 2196.9\,\text{VAr}$$

$$\text{Apparant power, } S = VI = \sqrt{P^2 + Q^2} = 230 \times 13 = 2990\,\text{VA}$$

$$\text{Power factor, } pf = \frac{P}{S} = \frac{2028.23}{2990} = 0.6783\,(\text{lag})$$

For this case, the voltage, current and various components of the power are shown in Fig. 2.5. As seen from the figure, the current lags the voltage due to inductive load. The p_{active} has an offset of 2028.23 W, which is also indicated as P in the right bottom graph. The $p_{reactive}$ has zero average value and its maximum value is equal to Q, which is 2196.9 VArs.

Case 2: When load is capacitive, $Z_L = 12 - j13$ that implies $|Z_L| = \sqrt{12^2 + 13^2} = 17.692\,\Omega$, and $I = 230/17.692 = 13$ A, $\phi = \tan^{-1}(-13/12) = -47.29°$.

$$v(t) = 230\sqrt{2}\sin\omega t$$
$$i(t) = 13\sqrt{2}\sin(\omega t + 47.29°)$$
$$p(t) = VI\cos\phi(1 - \cos 2\omega t) - VI\sin\phi\sin 2\omega t$$
$$= 230 \times 13\cos(-47.29°)(1 - \cos 2\omega t) - 230 \times 13\sin(-47.29°)\sin 2\omega t$$
$$= 2028.23(1 - \cos 2\omega t) + 2196.9\sin 2\omega t$$

From the above result,

$$p_{active}(t) = 2028.23(1 - \cos 2\omega t)$$
$$p_{reactive}(t) = 2196.9\sin 2\omega t$$

Average active, reactive and apparent powers are calculated as,

$$P = VI\cos\phi = 230 \times 13 \times \cos 47.29° = 2028.23\,\text{W}$$
$$Q = VI\sin\phi = 230 \times 13 \times \sin(-47.29°) = -2196.9\,\text{VAr}$$
$$S = VI = \sqrt{P^2 + Q^2} = 230 \times 13 = 2990\,\text{VA}$$
$$\text{Power factor} = \frac{P}{S} = \frac{2028.23}{2990} = 0.6783\,(\text{lead})$$

For Case 2, the voltage, current and various components of the power are shown in Fig. 2.6. The explanation given earlier also holds true for this case.

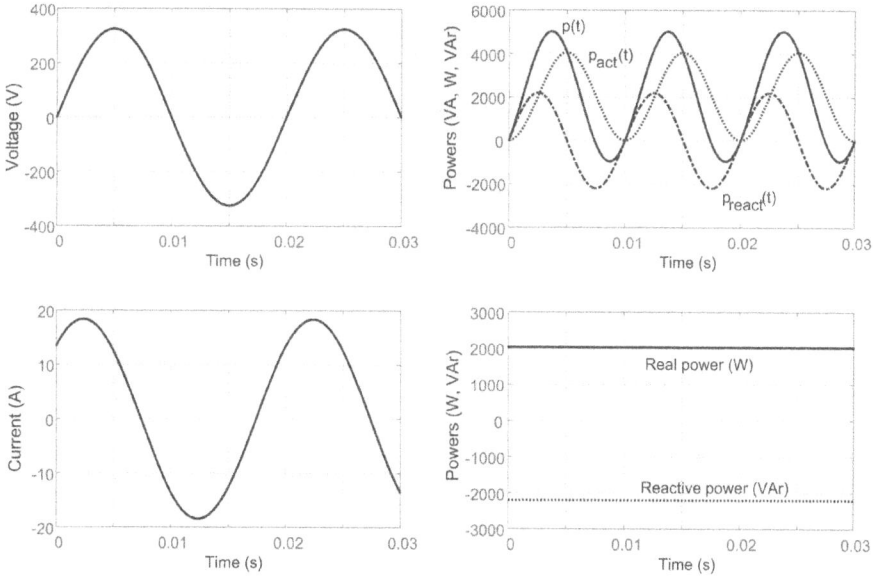

Figure 2.6 Case 2: Voltage, current, and various power components

Case 3: When load is resistive, $Z_L = R_L = 12\,\Omega$, $I = 230/12 = 19.167$ A, and $\phi = 0°$. Therefore, we have

$$v(t) = 230\sqrt{2}\sin\omega t$$
$$i(t) = 19.167\sqrt{2}\sin\omega t$$
$$p(t) = 230 \times 19.167\cos 0°(1 - \cos 2\omega t) - 230 \times 19.167\sin 0° \sin 2\omega t$$
$$= 4408.33(1 - \cos 2\omega t)$$

From the above result,

$$p_{active}(t) = 4408.33(1 - \cos 2\omega t)$$
$$p_{reactive}(t) = 0$$

Average active, reactive and apparent powers are calculated as,

$$P = VI\cos\phi = 230 \times 19.167 \times \cos 0° = 4408.33 \text{ W}$$
$$Q = VI\sin\phi = 230 \times 19.167 \times \sin 0° = 0 \text{ VAr}$$
$$S = VI = \sqrt{P^2 + Q^2} = 230 \times 19.167 = 4408.33 \text{ VA}$$
$$\text{Power factor} = \frac{4408.33}{4408.33} = 1$$

For Case 3, the voltage, current and various components of the power are shown in Fig. 2.7. Since the load is resistive, as seen from the graph $p_{reactive}$

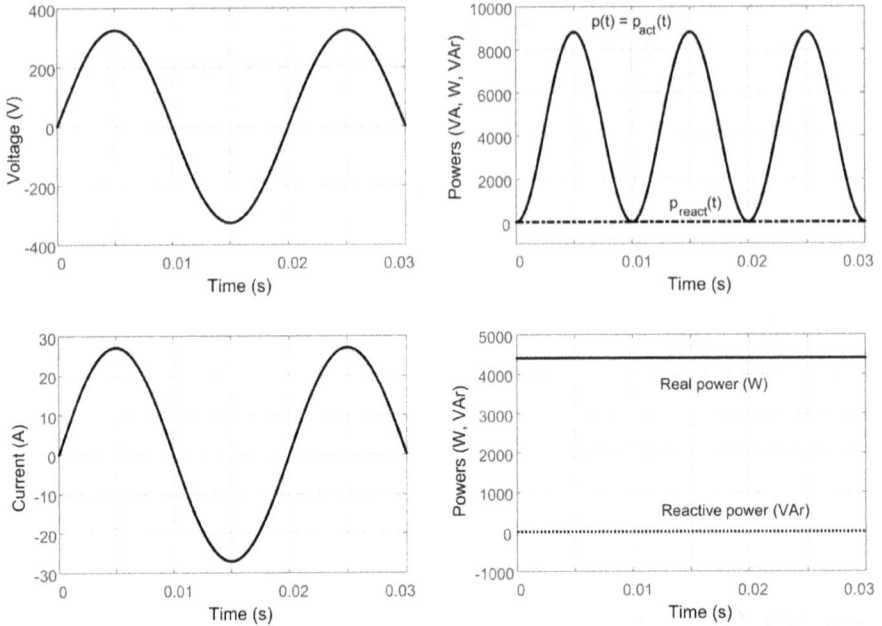

Figure 2.7 Case 3: Voltage, current, and various power components

is zero and $p(t)$ is equal to p_{active}. The average value of $p(t)$ is real power (P), which is equal to 4408.33 W.

Case 4: When the load is purely reactive, $Z_L = j13\ \Omega$, $|Z_L| = 13\ \Omega$, $I = \frac{230}{13} = 17.692$ A, and $\phi = 90°$. Therefore, we have

$$v(t) = 230\sqrt{2}\sin\omega t$$
$$i(t) = 17.692\sqrt{2}\sin(\omega t - 90°)$$
$$p(t) = 230 \times 17.692\cos 90°(1 - \cos 2\omega t) - 230 \times 17.692\sin 90° \sin 2\omega t$$
$$= -4069.16\sin 2\omega t$$

From the above result,

$$P_{active}(t) = 0$$
$$P_{reactive}(t) = -4069.16\sin 2\omega t$$

Average active, reactive and apparent powers are calculated as,

$$P = VI\cos\phi = 230 \times 17.692 \times \cos 90° = 0\ \text{W}$$
$$Q = VI\sin\phi = 230 \times 17.692 \times \sin 90° = 4069.16\ \text{VAr}$$
$$S = VI = \sqrt{P^2 + Q^2} = 230 \times 17.692 = 4069.16\ \text{VA}$$

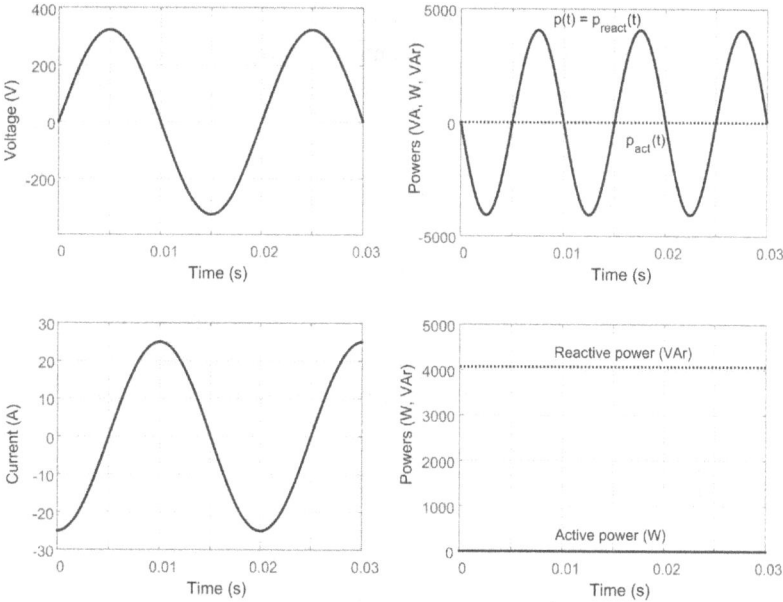

Figure 2.8   Case 4: Voltage, current, and various power components

$$\text{Power factor} \quad = \quad \frac{0}{4069.16} = 0$$

For Case 4, the voltage, current and various components of the power are shown in Fig. 2.8. The load in this case is purely reactive, hence there is no average component of $p(t)$. The maximum value of $p(t)$ is same as $p_{reactive}(t)$ or Q, which is equal to 4069 VArs.

2.4   SINUSOIDAL VOLTAGE SOURCE SUPPLYING NON-LINEAR LOAD CURRENT

Now a nonlinear load is considered, supplied by a sinusoidal voltage source. A non-linear load is the one which consists of switched elements such as diode, transistors, IGBTs (Insulated Gate Bipolar Transistors) MOSFETs (Metal Oxide Semiconductor Field Effect Transistors), etc., in the circuit. In power circuit, non-linear load current exists, when source supplies to power electronics based loads such as rectifier, inverter, cyclo-converters, etc. These loads cause harmonics in the load current. Assuming a general case where all harmonics are present in the load current, the voltage and current are expressed as following.

$$v(t) \quad = \quad \sqrt{2}V \sin \omega t$$

$$i(t) \quad = \quad \sqrt{2} \sum_{n=1}^{\infty} I_n \sin(n\omega t - \phi_n) \qquad (2.20)$$

where I_n and ϕ_n in the above equation represent the rms value and phase angle of nth harmonic component of current. The instantaneous power is therefore given by,

$$
\begin{aligned}
p(t) = v(t)\, i(t) &= \sqrt{2}V \sin \omega t \; \sqrt{2} \sum_{n=1}^{\infty} I_n \sin(n\omega t - \phi_n) \\
&= V \sum_{n=1}^{\infty} [I_n\, 2 \sin \omega t \; \sin(n\omega t - \phi_n)] \\
&= V [I_1\, 2 \sin \omega t \; \sin(\omega t - \phi_1)] \\
&\quad + V \sum_{n=2}^{\infty} [I_n\, 2 \sin \omega t \; \sin(n\omega t - \phi_n)] \qquad (2.21)
\end{aligned}
$$

Note that, $2 \sin A \sin B = \cos(A-B) - \cos(A+B)$, using this, (2.21) can be re-written as the following.

$$
\begin{aligned}
p(t) &= V I_1 [\cos \phi_1 - \cos(2\omega t - \phi_1)] \\
&\quad + V \sum_{n=2}^{\infty} I_n \{\cos[(n-1)\omega t - \phi_n] - \cos[(n+1)\omega t - \phi_n]\} \\
&= V I_1 \cos \phi_1 (1 - \cos 2\omega t) - V I_1 \sin \phi_1 \sin 2\omega t \qquad (2.22) \\
&\quad + \sum_{n=2}^{\infty} V I_n \{\cos[(n-1)\omega t - \phi_n] - \cos[(n+1)\omega t - \phi_n]\}
\end{aligned}
$$

In above equation, average active power P and reactive power Q are given by,

$$
P = P_1 = \text{average value of } p(t) = \frac{1}{T}\int_0^T p(t)\,dt = V I_1 \cos \phi_1 \qquad (2.23)
$$

$$
Q = Q_1 \overset{\Delta}{=} \text{peak value of second term in (2.22)} = V I_1 \sin \phi_1 \qquad (2.24)
$$

The apparent power S is given by

$$
\begin{aligned}
S &= V I \\
S &= V \sqrt{I_1^2 + I_2^2 + I_3^2 + \dots} \qquad (2.25)
\end{aligned}
$$

Equation (2.25) can be re-arranged as given below.

$$
\begin{aligned}
S^2 &= V^2 I_1^2 + V^2 (I_2^2 + I_3^2 + I_4^2 + \dots) \\
&= (V I_1 \cos \phi_1)^2 + (V I_1 \sin \phi_1)^2 + V^2 (I_2^2 + I_3^2 + I_4^2 + \dots) \\
&= P^2 + Q^2 + H^2 \qquad (2.26)
\end{aligned}
$$

In above equation, H is known as harmonic power and represents Volt-Ampere (VAs) rating corresponding to harmonics and is equal to,

$$
H = V \sqrt{I_2^2 + I_3^2 + I_4^2 + \dots} \qquad (2.27)
$$

The following points are observed from the description.

1. P and Q are dependent on the fundamental current components.

2. H is dependent on the current harmonic components.

3. Power components $V I_1 \cos \phi_1 \cos 2\omega t$ and $V I_1 \sin \phi_1 \sin 2\omega t$ are oscillating components with twice the fundamental frequency. The reactive component of power, i.e., $V I_1 \sin \phi_1 \sin 2\omega t$ can be eliminated using appropriately chosen reactive network, such as capacitor or inductor.

4. There are other terms in (2.22), which are functions of multiple integer of fundamental frequency. These terms can be eliminated using tuned LC filters.

Displacement Power Factor (DPF) or Fundamental Power Factor (pf_1) is denoted by $\cos \phi_1$ and is cosine angle between the fundamental voltage and current. This is equal to,

$$\text{DPF} = pf_1 = \cos \phi_1 = \frac{P_1}{S_1}. \tag{2.28}$$

The Power Factor (pf) is defined as ratio of real power to the total apparent power (VI) and is expressed as,

$$
\begin{aligned}
\text{Power Factor} = pf = \cos \phi \quad &= \quad \frac{P}{S} \\
&= \quad \frac{V I_1 \cos \phi_1}{V I} = \left(\frac{I_1}{I} \right) \cos \phi_1 \\
&= \quad \cos \gamma \cos \phi_1 \tag{2.29}
\end{aligned}
$$

Equation (2.29) shows that the power factor becomes less by a factor of $\cos \gamma$, which is the ratio of rms value of the fundamental to the rms value of the total current. This is due to the presence of harmonics in the load current. The nonlinear load current increases the ampere rating of the conductor for the same amount of real power transfer, and therefore such types of loads are not desirable in a power system. The relationship between the real power (P), reactive power (Q), harmonic power (H), DPF angle (ϕ_1), and power factor angle (ϕ) is depicted in Fig. 2.9. This is also known as power tetrahedron.

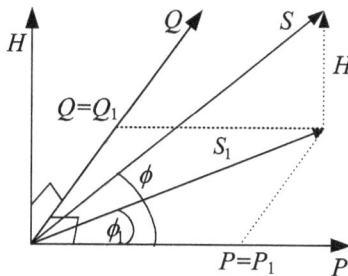

Figure 2.9 Power tetrahedron showing relation among P, Q, H, S, ϕ, and ϕ_1

Example 2.2. Consider an ideal single-phase voltage source supplying a rectifier load as given in Fig. 2.10. Given a supply voltage, $v_s(t) = 230\sqrt{2}\sin\omega t$ and source impedance is negligible. The load inductance is quite large so that the output current is constant. Draw the voltage and current waveforms. Express the source current using Fourier series. Based on that determine the following.

(a) Plot instantaneous power $p(t)$.

(b) Plot components of $p(t)$ i.e., $p_{active}(t)$, $p_{reactive}(t)$.

(c) Compute average real power, reactive power, apparent power, power factor, displacement power factor (or fundamental power factor).

(d) Comment upon the results in terms of VA rating and power output.

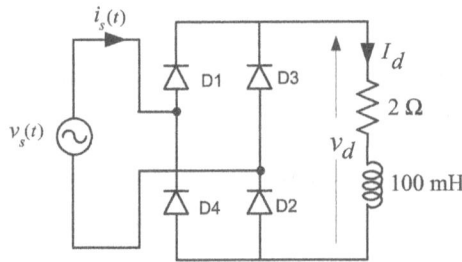

Figure 2.10   A single-phase system with non-linear load

Solution: The supply voltage and current are shown in Fig. 2.11. The current waveform is of the square type and its Fourier series expansion is given below [19]–[22].

$$i_s(t) \;=\; \sum_{h=2n+1}^{\infty} \frac{4I_d}{h\pi}\sin h\omega t \quad \text{where } n = 0,1,2\ldots$$

The instantaneous power is therefore given by,

$$p(t) = v_s(t)\,i_s(t) \;=\; \sqrt{2}V\sin\omega t \sum_{h=2n+1}^{\infty} \frac{4I_d}{h\pi}\sin h\omega t.$$

By expansion of the above equation, the average active power (P) and reactive power (Q) are given as below.

$$
\begin{aligned}
P \;&=\; P_1 = \text{average value of } p_{active}(t) \text{ or } p(t) = V I_1\cos\phi_1 \\
&=\; V I_1 \quad (\text{since, } \phi_1 = 0,\ \cos\phi_1 = 1,\ \sin\phi_1 = 0) \\
Q \;&=\; Q_1 \stackrel{\Delta}{=} \text{peak value of } p_{reactive}(t) = V I_1\sin\phi_1 = 0
\end{aligned}
$$

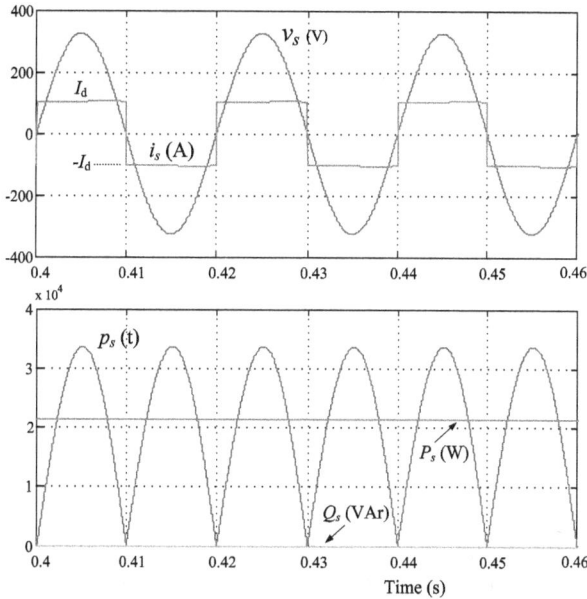

Figure 2.11 Supply voltage, current, and power waveforms

The dc side inductance is relatively large to make output dc current (I_d) as a constant value. This is also clear from the current waveform obtained from the simulation. From the waveform, it is found that $I_d = 103.5$ A. Alternatively, this current can be found by the following expression.

$$I_d = V_d/R = \frac{2\sqrt{2}V}{\pi}\frac{1}{R} = \frac{2\sqrt{2}\,230}{\pi}\frac{1}{2} = 103.5\,\text{A}$$

In the above expression, V_d is the average value of rectifier output voltage, v_d. The rms values of total and fundamental source current are given below.

$$I_{srms} = I_d = 103.5\,\text{A}$$

$$I_{s1} = \frac{2\sqrt{2}}{\pi}I_d = 93.15\,\text{A}$$

(c) The real power (P) is given by

$$P = V I_{s1}\cos\phi_1 = V I_{s1}\cos 0° = V I_{s1}$$

$$= V \times \frac{2\sqrt{2}}{\pi}I_d = 21424.5\,\text{W}$$

The reactive power (Q) is given by

$$Q = Q_1 = V I_{s1}\sin\phi_1 = V I_{s1}\sin 0° = 0$$

The apparent power (S) is given by

$$S = V I_{srms}$$
$$= V I_d = 23805 \text{ VA}$$

The displacement power factor $(\cos \phi_1)$ is,

$$DPF = \cos \phi_1 = 1$$

Therefore power factor is given by,

$$pf = \cos \phi = \frac{P}{S} = \frac{P_1}{S}$$
$$= \frac{V I_{s1}}{V I_{srms}} \cos \phi_1$$
$$= \frac{I_{s1}}{I_{srms}} \times DPF = 0.9 \text{ lag}$$

(d) As seen from the results that the overall power factor reduces due to the presence of harmonics in the load current. This has effect of increasing the VA rating of the system.

Example 2.3. Compute the following for the source current of the system shown in Fig. 2.10 with $R_L = 15 \Omega$. Assume that the load inductor is very high as to cause constant dc current (I_d) at the output terminals and source voltage has negligible impedance.

(a) Power factor, displacement factor.

(b) Total harmonic distortion.

(c) Input real and reactive powers, output power.

Solution:

(a) The average voltage at the dc link (V_d) is given by,

$$V_d = \frac{2V_m}{\pi} = \frac{2 \times 230\sqrt{2}}{\pi} = 207.07 \text{ V}$$

Therefore, the average dc current is computed as follows.

$$I_d = \frac{2V_m}{\pi R_L} = 13.84 \text{ A}$$

The source current can be expressed as,

$$i_s(t) = \frac{4I_d}{\pi} \left[\sin \omega t + \frac{\sin 3\omega t}{3} + \frac{\sin 5\omega t}{5} + ... \right]$$

$$i_s(t) = 17.62 \sin \omega t + 5.87 \frac{\sin 3\omega t}{3} + 3.52 \frac{\sin 5\omega t}{5} + ...$$

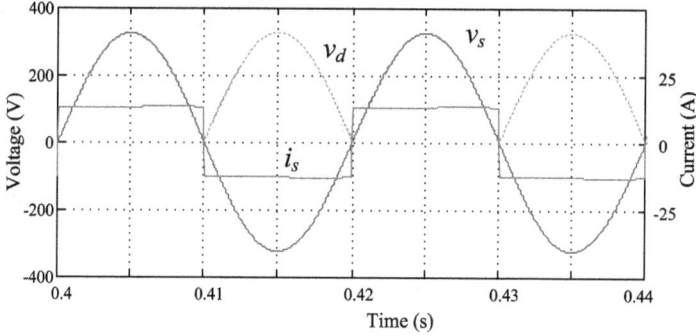

Figure 2.12 Supply voltage and current waveforms

The corresponding voltage and current waveforms are shown in Fig. 2.12 The rms value of nth harmonic component is given as,

$$I_{sn} = \frac{4I_d}{n\pi\sqrt{2}}$$

The rms value of the fundamental component of the source current is given as,

$$I_{s1} = \frac{4 \times 13.84}{\pi\sqrt{2}} = 12.45\,\text{A}$$

Displacement factor can be computed as,

$$\cos\phi_1 = \cos 0° = 1.0$$

(b) The power factor is given as,

$$pf = \frac{V_{s1}I_{s1}\cos\phi_1}{VI} = \frac{230 \times 12.45 \times 1}{230 \times 13.84} = 0.9\,\text{lag}.$$

Total harmonic distortion (THD) of the source current is given as,

$$THD\% = \frac{\sqrt{I_s^2 - I_{s1}^2}}{I_{s1}} \times 100 = \frac{\sqrt{13.84^2 - 12.45^2}}{12.45} \times 100 = 48.55\%$$

(c) The input real power (P_s) is given by

$$P_s = VI_{s1} = 230 \times 12.45 = 2863.5\,\text{W}$$

The input reactive power (Q_s) is given by

$$Q_s = Q_1 = 0\,\text{VAr}$$

The output power (P_o) is given by

$$P_o = V_d I_d = 207.07 \times 13.84 = 2857.56\,\text{W}.$$

2.5   NON-SINUSOIDAL VOLTAGE SOURCE SUPPLYING LINEAR AND NON-LINEAR LOADS

The voltage source too may have harmonics transmitted from generation or produced due to nonlinear loads in presence of feeder impedance. In this case, we shall consider generalized case of non-sinusoidal voltage source supplying nonlinear loads including dc components. For this system, the voltage and current are represented as,

$$v(t) = V_{dc} + \sum_{n=1}^{\infty} \sqrt{2} V_n \sin(n\omega t - \phi_{vn}) \qquad (2.30)$$

and,

$$i(t) = I_{dc} + \sum_{n=1}^{\infty} \sqrt{2} I_n \sin(n\omega t - \phi_{in}) \qquad (2.31)$$

In the above equations, V_n, I_n, ϕ_{vn}, ϕ_{in} are rms magnitude and phase angle of nth harmonic component of voltage and current, respectively. The instantaneous power $p(t)$ is given by,

$$p(t) = \left(V_{dc} + \sum_{n=1}^{\infty} \sqrt{2} V_n \sin(n\omega t - \phi_{vn}) \right) \left(I_{dc} + \sum_{n=1}^{\infty} \sqrt{2} I_n \sin(n\omega t - \phi_{in}) \right) \qquad (2.32)$$

$$p(t) = \underbrace{V_{dc} I_{dc}}_{I} + \underbrace{V_{dc} \sum_{n=1}^{\infty} \sqrt{2} I_n \sin(n\omega t - \phi_{in})}_{II} + \underbrace{I_{dc} \sum_{n=1}^{\infty} \sqrt{2} V_n \sin(n\omega t - \phi_{vn})}_{III}$$

$$+ \underbrace{\sum_{n=1}^{\infty} \sqrt{2} V_n \sin(n\omega t - \phi_{vn}) \sum_{n=1}^{\infty} \sqrt{2} I_n \sin(n\omega t - \phi_{in})}_{IV} \qquad (2.33)$$

$$p(t) = p_{dc-dc} + p_{dc-ac} + p_{ac-dc} + p_{ac-ac} \qquad (2.34)$$

The term I (p_{dc-dc}) contributes to power from dc components of voltage and current. Terms II (p_{dc-ac}) and III (p_{ac-dc}) result from the interaction of dc and ac components of voltage and current, respectively. In case, there are no dc components all these power components are zero. In practical cases, dc components are very less and the first three terms have negligible value compared to IV term. Thus, we shall focus on IV (p_{ac-ac}) term which corresponds to ac components present in power system. The IVth term can be written as,

$$p_{ac-ac} = \left(\sum_{n=1}^{\infty} \sqrt{2} V_n \sin(n\omega t - \phi_{vn}) \right) \left(\sum_{h=1}^{\infty} \sqrt{2} I_h \sin(h\omega t - \phi_{ih}) \right) \qquad (2.35)$$

where $n = h = 1, 2, 3...$, similar frequency terms will interact. When $n \neq h$, dissimilar frequency terms will interact. This is expressed below.

$$
p_{ac-ac}(t) = \underbrace{\sqrt{2}V_1 \sin(\omega t - \phi_{v1}) \sqrt{2}I_1 \sin(\omega t - \phi_{i1})}_{A}
$$

$$
\underbrace{+ \sqrt{2}V_1 \sin(\omega t - \phi_{v1}) \sum_{h=2, h\neq 1}^{\infty} \sqrt{2}I_h \sin(h\omega t - \phi_{ih})}_{B}
$$

$$
\underbrace{+ \sqrt{2}V_2 \sin(2\omega t - \phi_{v2}) \sqrt{2}I_2 \sin(2\omega t - \phi_{i2})}_{A}
$$

$$
\underbrace{+ \sqrt{2}V_2 \sin(2\omega t - \phi_{v2}) \sum_{h=1, h\neq 2}^{\infty} \sqrt{2}I_h \sin(h\omega t - \phi_{ih}) + ... + ...}_{B}
$$

$$
\underbrace{+ \sqrt{2}V_n \sin(n\omega t - \phi_{vn}) \sqrt{2}I_n \sin(n\omega t - \phi_{in})}_{A}
$$

$$
\underbrace{+ \sqrt{2}V_n \sin(n\omega t - \phi_{vn}) \sum_{h=1, h\neq n}^{\infty} \sqrt{2}I_h \sin(h\omega t - \phi_{ih})}_{B} \qquad (2.36)
$$

The terms in A of above equation form similar frequency terms and terms in B form dissimilar frequency terms. We shall denote these terms by $p_{ac-ac-nn}$ and $p_{ac-ac-nh}$, respectively, which are given below.

$$
p_{ac-ac-nn}(t) = \sum_{n=1}^{\infty} 2V_n I_n \sin(n\omega t - \phi_{vn}) \sin(n\omega t - \phi_{in}) \qquad (2.37)
$$

$$
p_{ac-ac-nh}(t) = \sum_{n=1}^{\infty} \sqrt{2}V_n \sin(n\omega t - \phi_{vn}) \sum_{h=1, h\neq n}^{\infty} \sqrt{2}I_n \sin(h\omega t - \phi_{ih}) \qquad (2.38)
$$

Now, let us simplify $p_{ac-ac-nn}$ term in (2.37)

$$
p_{ac-ac-nn}(t) = \sum_{n=1}^{\infty} V_n I_n [\cos(\phi_{in} - \phi_{vn}) - \cos(2n\omega t - \phi_{in} - \phi_{vn})]
$$

$$
= \sum_{n=1}^{\infty} V_n I_n [\cos\phi_n - \cos(2n\omega t - (\phi_{in} - \phi_{vn}) - 2\phi_{vn})]
$$

$$
= \sum_{n=1}^{\infty} V_n I_n [\cos\phi_n - \cos 2(n\omega t - \phi_{vn}) - \phi_n]
$$

$$
= \sum_{n=1}^{\infty} V_n I_n [\cos\phi_n - \cos 2(n\omega t - \phi_{vn}) \cos\phi_n \qquad (2.39)
$$

$$
- \sin 2(n\omega t - \phi_{vn}) \sin\phi_n]
$$

where $\phi_n = (\phi_{in} - \phi_{vn})$ is the phase angle between n^{th} harmonic components of current and voltage. Therefore,

$$p_{ac-ac-nn}(t) = \sum_{n=1}^{\infty} V_n I_n \cos \phi_n \{1 - \cos 2(n\omega t - \phi_{vn})\}$$

$$- \sum_{n=1}^{\infty} V_n I_n \sin \phi_n \sin 2(n\omega t - \phi_{vn}). \qquad (2.40)$$

Thus, the instantaneous power is given by,

$$p(t) = \underbrace{p_{dc-dc}}_{\text{I}} + \underbrace{p_{dc-ac}}_{\text{II}} + \underbrace{p_{ac-dc}}_{\text{III}} + \underbrace{p_{ac-ac-nn}}_{\text{IVA}} + \underbrace{p_{ac-ac-nh}}_{\text{IVB}}$$

$$\underbrace{\phantom{p_{ac-ac-nn} + p_{ac-ac-nh}}}_{\text{IV}}$$

$$p(t) = V_{dc}I_{dc} + V_{dc}\sum_{n=1}^{\infty}\sqrt{2}I_n \sin(n\omega t - \phi_{in}) + I_{dc}\sum_{n=1}^{\infty}\sqrt{2}V_n \sin(n\omega t - \phi_{vn})$$

$$+ \sum_{n=1}^{\infty} V_n I_n \cos \phi_n [1 - \cos 2(n\omega t - \phi_{vn})]$$

$$- \sum_{n=1}^{\infty} V_n I_n \sin \phi_n \sin 2(n\omega t - \phi_{vn})$$

$$+ \left(\sum_{n=1}^{\infty}\sqrt{2}V_n \sin(n\omega t - \phi_{vn})\right)\left(\sum_{h=1, h\neq n}^{\infty}\sqrt{2}I_h \sin(h\omega t - \phi_{ih})\right) \quad (2.41)$$

2.5.1 ACTIVE POWER

Instantaneous active power, $p_{active}(t)$ in (2.40), is expressed as,

$$p_{active}(t) = V_{dc}I_{dc} + \sum_{n=1}^{\infty} V_n I_n \cos \phi_n [1 - \cos 2(n\omega t - \phi_{vn})] \qquad (2.42)$$

It has non-negative value with some average component, giving average active power. Therefore,

$$P = \frac{1}{T}\int_0^T p(t)\,dt = \frac{1}{T}\int_0^T v(t)\,i(t)\,dt$$

$$= V_{dc}I_{dc} + \sum_{n=1}^{\infty} V_n I_n \cos \phi_n \qquad (2.43)$$

The reactive component of the instantaneous power is denoted by $p_{reactive}(t)$ and is given as follows.

$$p_{reactive}(t) = - \sum_{n=1}^{\infty} V_n I_n \sin \phi_n \sin 2(n\omega t - \phi_{vn}) \qquad (2.44)$$

resulting in

$$Q \triangleq \text{sum of maximum value of each term in (2.44)}$$

$$= \sum_{n=1}^{\infty} V_n I_n \sin \phi_n. \qquad (2.45)$$

From (2.43)

$$
\begin{aligned}
P &= P_{dc} + \sum_{n=1}^{\infty} V_n I_n \cos \phi_n \\
&= P_{dc} + V_1 I_1 \cos \phi_1 + V_2 I_2 \cos \phi_2 + V_3 I_3 \cos \phi_3 + \dots \\
&= P_{dc} + P_1 + P_2 + P_3 + \dots \\
&= P_{dc} + P_1 + P_H \qquad (2.46)
\end{aligned}
$$

In the above equation,

P_{dc} = Average active power corresponding to the dc components
P_1 = Average fundamental active power
P_H = Average harmonic active power

Average fundamental active power (P_1) can also be found from fundamentals of voltage and current, i.e.,

$$P_1 = \frac{1}{T} \int_0^T v_1(t) i_1(t) dt \qquad (2.47)$$

and harmonic active power (P_H) can be found as below.

$$P_H = \sum_{n=2}^{\infty} V_n I_n \cos \phi_n = P - P_1. \qquad (2.48)$$

2.5.2 REACTIVE POWER

The reactive power (Q), can be found by summing the maximum value of each term in (2.44). This is given below.

$$
\begin{aligned}
Q &= \sum_{n=1}^{\infty} V_n I_n \sin \phi_n \\
&= V_1 I_1 \sin \phi_1 + V_2 I_2 \sin \phi_2 + V_3 I_3 \sin \phi_3 + \dots \\
&= Q_1 + Q_2 + Q_3 + \dots \\
&= Q_1 + Q_H \qquad (2.49)
\end{aligned}
$$

Usually, this reactive power is referred to as Budeanu's reactive power and we use subscript "B," to indicate this, i.e.,

$$Q_B = Q_{1B} + Q_{HB} \qquad (2.50)$$

The remaining dissimilar frequency terms of (2.41) are accounted using $p_{rest}(t)$. Therefore, we can write,

$$p(t) = p_{dc-dc} + \underbrace{p_{active}(t) + p_{reactive}(t)}_{\text{Similar frequency terms}} + \underbrace{p_{rest}(t)}_{\text{Dissimilar frequency terms}} \qquad (2.51)$$

where,

$$
\begin{aligned}
p_{dc-dc} &= V_{dc}I_{dc} \\
p_{active}(t) &= \sum_{n=1}^{\infty} V_n I_n \cos\phi_n \{1 - \cos 2(n\omega t - \phi_{vn})\} \\
p_{reactive}(t) &= -\sum_{n=1}^{\infty} V_n I_n \sin\phi_n \sin 2(n\omega t - \phi_{vn}) \\
p_{rest}(t) &= V_{dc}\sum_{n=1}^{\infty} \sqrt{2}I_n \sin(n\omega t - \phi_{in}) + I_{dc}\sum_{n=1}^{\infty} \sqrt{2}V_n \sin(n\omega t - \phi_{vn}) \\
&\quad + \sum_{n=1}^{\infty} \sqrt{2}V_n \sin(n\omega t - \phi_{vn}) \sum_{h=1, h\neq n}^{\infty} \sqrt{2}I_h \sin(m\omega t - \phi_{ih})
\end{aligned}
$$

$$(2.52)$$

2.5.3 APPARENT POWER

The scalar apparent power which is defined as product of rms value of voltage and current, is expressed as follows.

$$
\begin{aligned}
S &= VI \\
&= \sqrt{V_{dc}^2 + V_1^2 + V_2^2 + \cdots}\,\sqrt{I_{dc}^2 + I_1^2 + I_2^2 + \cdots} \\
&= \sqrt{V_{dc}^2 + V_1^2 + V_H^2}\,\sqrt{I_{dc}^2 + I_1^2 + I_H^2}
\end{aligned}
\qquad (2.53)
$$

where,

$$
\begin{aligned}
V_H^2 &= V_2^2 + V_3^2 + \cdots = \sum_{n=2}^{\infty} V_n^2 \\
I_H^2 &= I_2^2 + I_3^2 + \cdots = \sum_{n=2}^{\infty} I_n^2
\end{aligned}
\qquad (2.54)
$$

V_H and I_H are denoted as harmonic voltage and harmonic current, respectively. Expanding (2.53) we can write

$$
\begin{aligned}
S^2 &= V^2 I^2 \\
&= (V_{dc}^2 + V_1^2 + V_H^2)(I_{dc}^2 + I_1^2 + I_H^2) \\
&= V_{dc}^2 I_{dc}^2 + V_{dc}^2 I_1^2 + V_{dc}^2 I_H^2 + V_1^2 I_1^2 + V_1^2 I_{dc}^2 + V_1^2 I_H^2 + V_H^2 I_{dc}^2 + V_H^2 I_1^2 + V_H^2 I_H^2 \\
&= V_{dc}^2 I_{dc}^2 + V_1^2 I_1^2 + V_H^2 I_H^2 + V_{dc}^2(I_1^2 + I_H^2) + I_{dc}^2(V_1^2 + V_H^2) + V_1^2 I_H^2 + V_H^2 I_1^2
\end{aligned}
$$

$$
\begin{aligned}
&= S_{dc}^2 + S_1^2 + S_H^2 + S_D^2 \\
&= S_1^2 + \underbrace{S_{dc}^2 + S_H^2 + S_D^2} \\
&= S_1^2 + S_N^2 \tag{2.55}
\end{aligned}
$$

In the above equation, the terms, S_1, S_{dc}, S_H, and S_D are fundamental, dc, harmonic, and distortion (due to dissimilar voltage and current terms) apparent powers, respectively. The term, S_N is known as non-fundamental apparent power and is given as follows.

$$
S_N^2 = V_{dc}^2 I_1^2 + V_{dc}^2 I_H^2 + V_1^2 I_{dc}^2 + V_1^2 I_H^2 + V_H^2 I_{dc}^2 + V_H^2 I_1^2 + V_H^2 I_H^2 + I_{dc}^2 I_H^2 + I_{dc}^2 V_{dc}^2 \tag{2.56}
$$

Practically in power systems dc components are negligible. Therefore neglecting the contribution of V_{dc} and I_{dc} associated terms in (2.56), the following is obtained.

$$
\begin{aligned}
S_N^2 &= I_1^2 V_H^2 + V_1^2 I_H^2 + V_H^2 I_H^2 \\
&= D_V^2 + D_I^2 + S_H^2 \tag{2.57}
\end{aligned}
$$

The terms D_I and D_V in (2.57) are known as apparent powers due to distortion in current and voltage, respectively. These are given below.

$$
\begin{aligned}
D_V &= I_1 V_H \\
D_I &= V_1 I_H \tag{2.58}
\end{aligned}
$$

These are further expressed in terms of THD components of voltage and current, as given below.

$$
\begin{aligned}
THD_V &= \frac{V_H}{V_1} \\
THD_I &= \frac{I_H}{I_1} \tag{2.59}
\end{aligned}
$$

From (2.59), the harmonic components of current and voltage are expressed below.

$$
\begin{aligned}
V_H &= THD_V V_1 \\
I_H &= THD_I I_1 \tag{2.60}
\end{aligned}
$$

Using (2.58) and (2.60),

$$
\begin{aligned}
D_V &= V_1 I_1 THD_V = S_1 THD_V \\
D_I &= V_1 I_1 THD_I = S_1 THD_I \\
S_H &= V_H I_H = S_1 THD_I THD_V \tag{2.61}
\end{aligned}
$$

Therefore using (2.57) and (2.61), S_N could be expressed as following.

$$
S_N^2 = S_1^2 (THD_I^2 + THD_V^2 + THD_I^2 THD_V^2) \tag{2.62}
$$

Normally in power system, $THD_V \ll THD_I$, therefore,

$$S_N \approx S_1 \, THD_I \tag{2.63}$$

The above relationship shows that as the THD content in voltage and current increases, the non fundamental apparent power S_N increases for a given useful transmitted power. This means that there are more losses and hence a less efficient power network.

2.5.4 NON ACTIVE POWER

Non active power is denoted by N and is defined as per the following equation.

$$S^2 = P^2 + N^2 \tag{2.64}$$

This power includes both fundamental as well as non fundamental components and is usually computed by knowing active power (P) and apparent power (S) as given below.

$$N = \sqrt{S^2 - P^2} \tag{2.65}$$

2.5.5 DISTORTION POWER

Due to presence of distortion, the total apparent power S can also be written in terms of active power (P), reactive power (Q), and distortion power (D)

$$S^2 = P^2 + Q^2 + D^2 \tag{2.66}$$

Therefore,

$$D = \sqrt{S^2 - P^2 - Q^2} \tag{2.67}$$

From (2.64) and (2.66), the relationship between N and D is as follows.

$$N^2 = Q^2 + D^2 \text{ or } N = \sqrt{Q^2 + D^2} \tag{2.68}$$

2.5.6 FUNDAMENTAL POWER FACTOR

The fundamental power factor is defined as ratio of fundamental real power (P_1) to the fundamental apparent power (S_1). This is given below.

$$pf_1 = \cos\phi_1 = \frac{P_1}{S_1} \tag{2.69}$$

The fundamental power factor as defined above is also known as the displacement power factor.

2.5.7 POWER FACTOR

Power factor for the single-phase system considered above is the ratio of the total real power (P) to the total apparent power (S) as given by the following equation.

$$
\begin{aligned}
pf &= \frac{P}{S} = \frac{V_1 I_1 \cos\phi_1 + V_2 I_2 \cos\phi_2 + \ldots + V_n I_n \cos\phi_n}{\sqrt{V_1^2 + V_2^2 + \ldots + V_n^2}\,\sqrt{I_1^2 + I_2^2 + \ldots + I_n^2}} \\
&= \frac{P_1 + P_H}{\sqrt{S_1^2 + S_N^2}} \\
&= \frac{(1 + P_H/P_1)}{\sqrt{1 + (S_N/S_1)^2}} \frac{P_1}{S_1}
\end{aligned}
\tag{2.70}
$$

Substituting S_N from (2.62), the power factor can further be simplified to the following equation.

$$
pf = \frac{(1 + P_H/P_1)}{\sqrt{1 + THD_I^2 + THD_V^2 + THD_I^2 THD_V^2}}\, pf_1
\tag{2.71}
$$

Thus, we observe that the power factor of a single-phase system depends upon fundamental (P_1) and harmonic active power (P_H), displacement factor ($DPF = pf_1$) and THDs in voltage and current. Further, we note the following points.

1. P/S is also called as utilization factor indicator as it indicates the usage of real power.

2. The term S_N/S_1 is used to decide the overall degree of harmonic content in the system.

3. The flow of fundamental power can be characterized by measurement of S_1, P_1, pf_1, and Q_1.

For a practical power system $P_1 \gg P_H$ and $THD_V \ll THD_I$, the above expression of power factor is further simplified as given below.

$$
pf = \frac{pf_1}{\sqrt{1 + THD_I^2}}
\tag{2.72}
$$

Example 2.4. Consider a single-phase system with supply voltage as following.

$$
v(t) = \sum_{n=1,3,5} \frac{220\sqrt{2}}{n} \sin(n\,\omega t)
$$

The above supply voltage is connected to the load with $R_L = 3\,\Omega$ and $X_L = 4\,\Omega$ at fundamental frequency 50 Hz, as shown in the given Fig. 2.13. Determine the following.

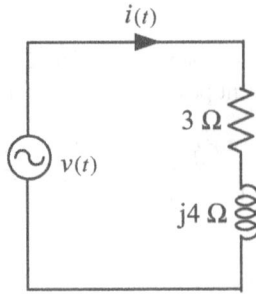

Figure 2.13 A single-phase system with non-sinusoidal voltage and current

(a) Instantaneous current $i(t)$, flowing in the circuit.

(b) Instantaneous power $p(t)$, $p_{active}(t)$, $p_{reactive}(t)$, $p_{rest}(t)$.

(c) Average real power P, reactive power Q, apparent power S, and power factor pf.

(d) Non-active power N and distortion power D.

Solution:

(a) A single-phase circuit supplying linear load is shown in Fig. 2.13. In general, the current in the circuit is computed as follows.

$$v(t) = \sum_{n=1,3,5} \frac{220\sqrt{2}}{n} sin(n\,\omega t).$$

The load impedance at different harmonic components is given as follows.

$$
\begin{aligned}
R_{L1} &= 3\Omega, \ X_{L1} = j4\Omega \\
X_{L3} &= j4 \times 3\Omega \\
X_{L5} &= j4 \times 5\Omega
\end{aligned}
$$

$$
\begin{aligned}
Z_{L1} &= R_{L1} + jX_{L1} = 3 + j4 = 5\angle 53.13°\,\Omega \\
Z_{L3} &= R_{L3} + jX_{L3} = 3 + j12 = 12.37\angle 75.96°\,\Omega \\
Z_{L5} &= R_{L5} + jX_{L5} = 3 + j20 = 20.22\angle 81.47°\,\Omega
\end{aligned}
$$

Therefore, the current is given by,

$$
\begin{aligned}
i(t) &= i_{L1} + i_{L3} + i_{L5} \\
&= \frac{220\sqrt{2}}{5 \times 1} sin(\omega t - 53.13°) + \frac{220\sqrt{2}}{12.37 \times 3} sin(3\omega t - 75.96°)
\end{aligned}
$$

$$+ \frac{220\sqrt{2}}{20.22 \times 5} \sin(5\omega t - 81.47°)$$
$$= 44\sqrt{2} \sin(\omega t - 53.13°) + 5.92\sqrt{2} \sin(3\omega t - 75.96°)$$
$$+ 2.17\sqrt{2} \sin(5\omega t - 81.47°)$$

(b) Instantaneous power $p(t)$,

$$p(t) = v(t)i(t) = \left(\sum_{n=1,3,5..} \frac{220\sqrt{2}}{n} \sin(n\omega t) \right) \times$$
$$\left[44\sqrt{2} \sin(\omega t - 53.13°) + 5.92\sqrt{2} \sin(3\omega t - 75.96°) \right.$$
$$\left. + 2.17\sqrt{2} \sin(5\omega t - 81.47°) \right]$$

The above equation can be simplified as follows.

$$p_{active}(t) = 5808 \left(1 - \cos 2\omega t\right) + 105.45 \left(1 - \cos 6\omega t\right) + 14.2 \left(1 - \cos 10\omega t\right)$$
$$p_{reactive}(t) = -7744 \sin 2\omega t - 421.79 \sin 6\omega t - 94.67 \sin 10\omega t$$

$$p_{rest}(t) = 220\sqrt{2} \sin \omega t \left[5.92\sqrt{2} \sin(3\omega t - 75.96°) + 2.17\sqrt{2} \sin(5\omega t - 81.47°) \right]$$
$$+ 73.33\sqrt{2} \sin 3\omega t \left[44\sqrt{2} \sin(\omega t - 53.13°) + 2.17\sqrt{2} \sin(5\omega t - 81.47°) \right]$$
$$+ 44\sqrt{2} \sin 5\omega t \left[44\sqrt{2} \sin(\omega t - 53.13°) + 5.92\sqrt{2} \sin(3\omega t - 75.96°) \right]$$

(c) P, Q, S, and pf

$$P = P_1 + P_3 + P_5 = 5808 + 105.45 + 14.2 = 5927.65\,\text{W}$$
$$Q = Q_1 + Q_3 + Q_5 = 7744 + 516.45 = 8260.45\,\text{VAr}$$

$$S = V_{rms} I_{rms} = \left[220^2 + \left(\frac{220}{3}\right)^2 + \left(\frac{220}{5}\right)^2 \right]^{\frac{1}{2}} \times \left[44^2 + 5.92^2 + 2.17^2 \right]^{\frac{1}{2}}$$
$$= 236.03 \times 44.45 = 10492\,\text{VA}$$

Therefore, the power factor

$$pf = \frac{P}{S} = \frac{5927.65}{10492} = 0.5650\,(\text{lag})$$

(d) N and D
By definition,

$$N = \sqrt{S^2 - P^2} = 8657.2\,\text{VA}, \qquad D = \sqrt{S^2 - P^2 - Q^2} = 2590.7\,\text{VAr}$$

Example 2.5. Consider the following voltage and current in single-phase system.

$$v_s(t) = 230\sqrt{2}\sin\omega t + 50\sqrt{2}\sin(3\omega t - 30°)$$
$$i(t) = 2 + 10\sqrt{2}\sin(\omega t - 30°) + 5\sqrt{2}\sin(3\omega t - 60°)$$

Determine the following:

(a) Active power, P.

(b) Reactive power, Q.

(c) Apparent power, S.

(d) Power factor, pf.

Solution: Here the source is non-sinusoidal and is feeding a non-linear load. The instantaneous power is given by,

$$p(t) = v(t)i(t)$$
$$= \left(V_{dc} + \sum_{n=1}^{\infty}\sqrt{2}V_n\sin(n\omega t - \phi_{vn})\right)\left(I_{dc} + \sum_{n=1}^{\infty}\sqrt{2}I_n\sin(n\omega t - \phi_{in})\right)$$

(a) The active power, P is given by,

$$P = \frac{1}{T}\int_0^T p(t)\,dt$$
$$= P_{dc} + V_1 I_1\cos\phi_1 + V_2 I_2\cos\phi_2 + \ldots\ldots + V_n I_n\cos\phi_n \quad (2.73)$$
$$= P_{dc} + P_1 + P_H$$

where,

$$\phi_n = \phi_{in} - \phi_{vn}$$
$$P_{dc} = V_{dc} I_{dc}$$
$$P_1 = V_1 I_1\cos\phi_1$$
$$P_H = \sum_{n=2}^{\infty}V_n I_n\cos\phi_n$$

Here, $V_{dc} = 0$, $V_1 = 230$ V, $\phi_{v1} = 0$, $V_3 = 50$ V, $\phi_{v3} = 30°$, $I_{dc} = 2$ A, $I_1 = 10$ A, $\phi_{i1} = 30°$, $I_3 = 5$ A, $\phi_{i3} = 60°$. Therefore, $\phi_1 = \phi_{i1} - \phi_{v1} = 30°$ and $\phi_3 = \phi_{i3} - \phi_{v3} = 30°$. Substituting these values in (2.73), the above equation gives,

$$P = 0\times 2 + 230\times 10\times\cos 30° + 50\times 5\times\cos 30° = 2208.36\,\text{W}.$$

(b) The reactive power (Q) is given by,

$$Q = \sum_{n=1}^{\infty}V_n I_n\sin\phi_n$$
$$= V_1 I_1\sin\phi_1 + V_2 I_2\sin\phi_2 + \ldots..V_n I_n\sin\phi_n$$
$$= 230\times 10\times\sin 30° + 50\times 5\times\sin 30° = 1275\,\text{VAr}$$

(c) The apparent power S is given by,

$$
\begin{aligned}
S &= V_{rms}\, I_{rms} \\
&= \sqrt{V_{dc}^2 + V_1^2 + V_2^2 + \dots V_n^2}\, \sqrt{I_{dc}^2 + I_1^2 + I_2^2 + \dots I_n^2} \\
&= \sqrt{V_{dc}^2 + V_1^2 + V_H^2}\, \sqrt{I_{dc}^2 + I_1^2 + I_H^2}
\end{aligned}
$$

where,

$$
\begin{aligned}
V_H &= \sqrt{V_2^2 + V_3^2 + \dots V_n^2} \\
I_H &= \sqrt{I_2^2 + I_3^2 + \dots I_n^2}
\end{aligned}
$$

Substituting the values of voltage and current components, the apparent power S is computed as follows.

$$
\begin{aligned}
S &= \sqrt{0 + 230^2 + 50^2}\, \sqrt{2^2 + 10^2 + 5^2} \\
&= 235.37 \times 11.357 = 2673.31\ \text{VA}
\end{aligned}
$$

(d) The power factor is given by

$$
pf = \frac{P}{S} = \frac{2208.36}{2673.31} = 0.8261\ (\text{lag})
$$

Example 2.6. Consider following system with distorted supply voltages,

$$
v(t) = V_{dc} + \sum_{n=1,3,5} \frac{\sqrt{2}V_n}{n^2} \sin(n\omega t - \phi_{vn})
$$

with $V_{dc} = 10\,V$, $V_n/n^2 = 230/n^2$ and $\phi_{vn} = 0$ for $n = 1, 3, 5$, and so on. The voltage source supplies a nonlinear current of,

$$
i(t) = I_{dc} + \sum_{n=1,3,5} \frac{\sqrt{2}I_n}{n} \sin(n\omega t - \phi_{in}).
$$

with $I_{dc} = 2$ A, $I_n = 20/n$ A and $\phi_{in} = 30° \times n$ for $n = 1,3,5$. Compute the following.

(a) Plot instantaneous power $p(t)$, $p_{active}(t)$, $p_{reactive}(t)$, P_{dc}, and $p_{rest}(t)$.

(b) Compute P, P_1, P_H.

(c) Compute Q, Q_1, Q_H.

(d) Compute S, S_1, S_H.

(e) Comment upon each result.

Solution: Instantaneous power is given as follows:

$$
\begin{aligned}
p(t) &= v(t)\,i(t) \\[2mm]
&= \left(10 + \sum_{n=1,3,5} \frac{230\sqrt{2}}{n^2}\sin(n\omega t)\right)\left(2 + \sum_{n=1,3,5}\frac{20\sqrt{2}}{n}\sin n(\omega t - 30°)\right) \\[2mm]
&= \underbrace{20}_{\text{I}} + 10\underbrace{\sum_{n=1,3,5}\frac{20\sqrt{2}}{n}\sin n(\omega t - 30°)}_{\text{II}} + 2\underbrace{\sum_{n=1,3,5}\frac{230\sqrt{2}}{n^2}\sin(n\omega t)}_{\text{III}} \\[2mm]
&\quad + \underbrace{\left(\sum_{n=1,3,5}\frac{230\sqrt{2}}{n^2}\sin(n\omega t)\right)\left(\sum_{n=1,3,5}\frac{20\sqrt{2}}{n}\sin n(\omega t - 30°)\right)}_{\text{IV}} \\[2mm]
&= \underbrace{20}_{\text{I}} + \underbrace{\sum_{n=1,3,5}\frac{200\sqrt{2}}{n}\sin n(\omega t - 30°)}_{\text{II}} + \underbrace{\sum_{n=1,3,5}\frac{460\sqrt{2}}{n^2}\sin(n\omega t)}_{\text{III}} \\[2mm]
&\quad + \underbrace{\sum_{n=1,3,5}\frac{4600}{n^3}\left(\cos(30° n)(1 - \cos 2n\omega t) - \sin(2n\omega t)\sin(30° n)\right)}_{\text{IV A}} \\[2mm]
&\quad + \underbrace{\left(\sum_{n=1,3,5}\frac{230\sqrt{2}}{n^2}\sin n\omega t\right)\left(\sum_{h=1,3,5;h\neq n}\frac{20\sqrt{2}}{h}\sin h(\omega t - 30°)\right)}_{\text{IV B}}
\end{aligned}
$$

(a) Computation of $p(t)$, $p_{active}(t)$, $p_{reactive}(t)$, P_{dc}, and $p_{rest}(t)$

$$
P_{dc-dc}(t) = 20\ W
$$

$$
p_{active}(t) = \sum_{n=1,3,5}\frac{4600}{n^3}\cos(30° n)(1 - \cos 2n\omega t)
$$

$$
p_{reactive}(t) = -\sum_{n=1,3,5}\frac{4600}{n^3}\sin(30° n)\sin(2n\omega t)
$$

$$
\begin{aligned}
p_{rest}(t) &= \sum_{n=1,3,5}\frac{200\sqrt{2}}{n}\sin n(\omega t - 30°) + \sum_{n=1,3,5}\frac{460\sqrt{2}}{n^2}\sin(n\omega t) \\[2mm]
&\quad + \left(\sum_{n=1,3,5}\frac{230\sqrt{2}}{n^2}\sin n\omega t\right)\left(\sum_{h=1,3,5;h\neq n}\frac{20\sqrt{2}}{h}\sin h(\omega t - 30°)\right)
\end{aligned}
$$

(b) Computation of P, P_1, P_H

$$
\begin{aligned}
P &= \frac{1}{T}\int_0^T p(t)dt \\
&= 20 + \sum_{n=1,3,5} \frac{4600}{n^3}\cos(30°\,n) \\
&= 20 + 4600\cos 30° + \sum_{n=3,5}\frac{4600}{n^3}\cos(30°\,n) \\
&= 20 + 3983.71 + (-43.48) = 3960.23\,\text{W} \\
&= P_{dc} + P_1 + P_H
\end{aligned}
$$

Thus, active power contributed by dc components of voltage and current, $P_{dc} = 20$ W. Active power contributed by fundamental frequency components of voltage and current, $P_1 = 3983.71$ W. Active power contributed by harmonic frequency components of voltage and current, $P_H = -43.48$ W.

(c) Computation of Q, Q_1, Q_H

$$
\begin{aligned}
Q &= \sum_{n=1,3,5}\frac{4600}{n^3}\sin(30°\,n) \\
&= 4600\sin 30^0 + \sum_{n=3,5}\frac{4600}{n^3}\sin(30°\,n) \\
&= 2300 + 175.75 = 2475.75\,\text{VArs} \\
&= Q_1 + Q_H
\end{aligned}
$$

The above implies that, $Q_1 = 2300$ VArs and $Q_H = \sum_{n=3,5}(4600/n^3)\sin(30°\,n)$ = 175.75 VArs.

(d) Computation of apparent powers and distortion powers.
The apparent power S is expressed as following.

$$
\begin{aligned}
V_{rms} &= \sqrt{V_{dc}^2 + V_1^2 + V_3^2 + V_5^2 + V_7^2 + V_9^2 + ...} \\
&= \sqrt{10^2 + 230^2 + (230/3^2)^2 + (230/5^2)^2 + (230/7^2)^2 + (230/9^2)^2 + ...} \\
&= 231.87\,V \qquad (\text{up to } n = 9)
\end{aligned}
$$

$$
\begin{aligned}
I_{rms} &= \sqrt{I_{dc}^2 + I_1^2 + I_3^2 + I_5^2 + I_7^2 + I_9^2 + ...} \\
&= \sqrt{2^2 + 20^2 + (20/3)^2 + (20/5)^2 + (20/7)^2 + (20/9)^2 + ...} \\
&= 21.85\,\text{A} \qquad (\text{up to } n = 9)
\end{aligned}
$$

The apparent power, $S = V_{rms} I_{rms} = 231.87 \times 21.85 = 5066.36$ VA.
Fundamental apparent power, $S_1 = V_1 I_1 = 4600$ VA.
Apparent power contributed by harmonics $S_H = V_H I_H$.

$$
\begin{aligned}
V_H &= \sqrt{V_3^2 + V_5^2 + V_7^2 + V_9^2 + ...} \\
&= \sqrt{(230/3^2)^2 + (230/5^2)^2 + (230/7^2)^2 + (230/9^2)^2 + ...} \\
&= 27.7\,V \qquad \text{(up to } n = 9\text{)}
\end{aligned}
$$

$$
\begin{aligned}
I_H &= \sqrt{I_3^2 + I_5^2 + I_7^2 + I_9^2 + ...} \\
&= \sqrt{(20/3)^2 + (20/5)^2 + (20/7)^2 + (20/9)^2 + ...} \\
&= 8.57\,A \qquad \text{(up to } n = 9\text{)}.
\end{aligned}
$$

Therefore, the harmonic apparent power, $S_H = V_H I_H = 237.5$ VA.

Non active power, $N = \sqrt{S^2 - P^2} = \sqrt{5067^2 - 3960.2^2} = 3160.8$ VAs (up to $n = 9$).

Distortion Power $D = \sqrt{S^2 - P^2 - Q^2} = \sqrt{5067^2 - 3960.2^2 - 2475.77^2} = 1965.163$ VAs (up to $n = 9$). Displacement power factor ($\cos \phi_1$)

$$
\cos \phi_1 = \frac{P_1}{S_1} = \frac{3983.7}{230 \times 20} = 0.866 \,(\text{lag})
$$

Power factor ($\cos \phi$)

$$
\cos \phi = \frac{P}{S} = \frac{3960.217}{5067} = 0.781 \,(\text{lag})
$$

2.6   SUMMARY

Single-phase load and its operation play an important role in power system, hence its quantitative and qualitative understanding are essential aspects. In this chapter, basic concepts of single-phase system are discussed. Various power terms and components such as instantaneous power, instantaneous active and reactive powers, real and reactive power under sinusoidal and non-sinusoidal voltage and currents are derived from the first principles and are explained. For each conceptual discussion, illustrative examples are presented to develop comprehension of the subject which will enable readers to develop insight and skill to understand and solve the issues related to power quality in single-phase systems.

2.7   PROBLEMS

P 2.1 Why are sinusoidal fundamental components of voltage and current desired in power system?

P 2.2 What are the sources of voltage and current harmonics and how they propagate in the power distribution network?

P 2.3 What are the effects of harmonics in transformers and electrical machines?

P 2.4 Explain that in a single-phase system, where the voltage has fundamental components only and current has fundamental as well harmonics, the real power is contributed by the fundamental components of voltage and current. In other words, harmonics in the current have no effect on the real power. Then, in what way do harmonics affect the performance of the system?

P 2.5 Consider a single-phase system, where voltage and current both have harmonics. Explain that the real power and reactive power are contributed by similar harmonic (frequency) terms. What are the effects of the interaction of dissimilar frequency terms of voltage and current on the system performance?

P 2.6 A single-phase system has the following set of voltage and current,

$$v(t) = \sqrt{2} \sum_{n=1,2,3...}^{N} V_n \sin(n\omega t)$$

$$i(t) = \sqrt{2} \sum_{n=1,2,3...}^{N} I_n \sin(n\omega t - \phi_n).$$

Determine the following.

(a) Total active power (P) and its components, i.e., fundamental active power (P_1) and harmonic active power (P_H).

(b) Apparent power (S) and its components, i.e., fundamental apparent (S_1) and non fundamental apparent power (S_N).

(c) Determine the power factor (pf) in terms of fundamental power factor, pf_1, P_1, P_H, and THDs of voltage and current.

(d) Comment upon the power factor due to the presence of harmonics and voltage and current.

P 2.7 Derive the relation for a power factor of single-phase system containing dc, fundamental, and harmonics both in voltage and currents as given by the following equations.

$$v_s(t) = V_{dc} + V_{m1} \sin(\omega t + \phi_{v1}) + V_{m2} \sin(2\omega t + \phi_{v2})$$
$$+ V_{m3} \sin(3\omega t + \phi_{v3}) + ... + V_{mn} \sin(n\omega t + \phi_{vn}) + ...$$
$$i_s(t) = I_{dc} + I_{m1} \sin(\omega t + \phi_{i1}) + I_{m2} \sin(2\omega t + \phi_{i2})$$
$$+ I_{m3} \sin(3\omega t + \phi_{i3}) + ... + I_{mn} \sin(n\omega t + \phi_{in}) + ...$$

P 2.8 For a single-phase system given in Fig. 2.14, the current, $\bar{I}$ is in phase with the voltage, $\overline{V_s}$ with purely reactive compensator, Z_γ. Compute the value of Z_γ, element of Z_γ, the value of the element, and the current $\bar{I}$.

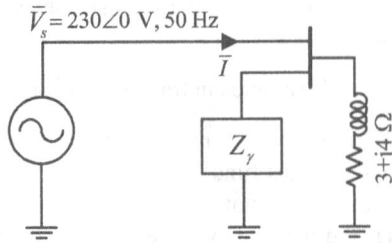

Figure 2.14 Related to problem P 2.8

P 2.9 Develop an expression for instantaneous power for a circuit with the following voltage and current. Then resolve instantaneous active and reactive power expressions and the rest of the power components. From these, find out real power (P), reactive power (Q), apparent power (S), and power factor (pf).

$$v(t) = 230\sqrt{2}\sin\omega t + 50\sqrt{2}\sin(3\omega t + 30°)\,\text{V}$$
$$i(t) = 20\sqrt{2}\sin(\omega t - 30°) + 5\sqrt{2}\sin(3\omega t + 45°)\,\text{A}$$

P 2.10 A single-phase system with supply voltage given in the following is applied to a resistive load as shown in Fig. 2.15.

$$v(t) = \sum_{n=1} \sqrt{2}V_n\sin(n\omega t + \phi_n)$$

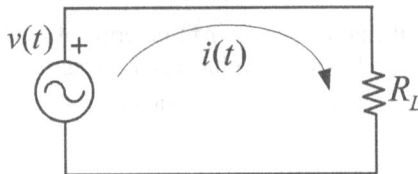

Figure 2.15 Related to problem P 2.10

Determine the following.

(a) The current $i(t)$.

(b) Instantaneous power $p(t)$.

(c) Real power (P), reactive power (Q), apparent power (S), and power factor (pf).

(d) Non-active power (N) and distortion power (D).

P 2.11 A single-phase system with supply voltage given in the following is applied to an inductive load with $R_L = 3\,\Omega$, $X_L = 4\,\Omega$ at the fundamental frequency of 50 Hz, as shown in Fig. 2.16.

$$v(t) = \sum_{n=1,3,5} \frac{220\sqrt{2}}{n} \sin(n\omega t)$$

Determine the following.

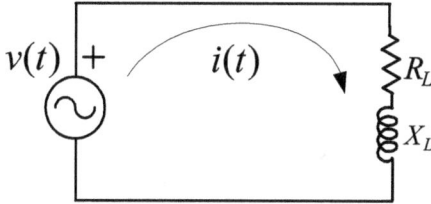

Figure 2.16 Related to problem P 2.11

(a) The current $i(t)$.

(b) Instantaneous power $p(t)$.

(c) Real power (P), reactive power (Q), apparent power (S), and power factor (pf).

(d) Non-active power (N) and distortion power (D).

P 2.12 Consider a residential single-phase circuit with supply voltage with fundamental and harmonics as expressed below. Express instantaneous current $i(t)$, using standard notations and variables, and then compute the power factor for the following cases.

$$v(t) = \sum_{n=1}^{\infty} \sqrt{2} V_n \sin n\omega t$$

(a) When the load is purely resistive, R.

(b) When the load is purely inductive, L.

(c) When the load is purely capacitive, C.

(d) When the load is a series combination of R and L.

(e) When the load is a series combination of R and C.

(f) Comment upon the results for series $R - L$ and $R - C$ cases.

P 2.13 A voltage source has 3rd and 5th harmonics and connected to the series R-L-C load as shown in Fig. 2.17, where

$$v(t) = \sum_{n=1,3,5} \sqrt{2} V_n \sin n\omega t$$

with $R = 20\,\Omega$, $L = 30$ mH, $C = 630\,\mu$ F, $V_1 = 230$ V, $f_1 = 50$ Hz, $V_n = V_1/(2n)$.

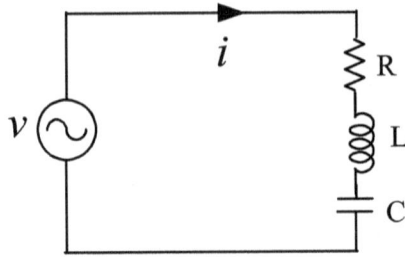

Figure 2.17 Related to problem P 2.13

Compute the following.

(a) Instantaneous expression for current, $i(t)$.

(b) P_1, P_H, P.

(c) Q_1, Q_H, Q.

(d) S, S_1, S_H, N, D.

(e) THD_I, THD_V, power factor (pf).

P 2.14 An ac source having a voltage THD of 2% supplies a nonlinear load which is fully controlled rectifier. The rectifier draws a current with THD of 40% from the ac source, when it is triggered at a firing angle of 30°. The total harmonic power supplied by the source is 3% of the fundamental input power. Calculate the input power factor. What will be the source power factor, if the fully controlled rectifier is replaced by a semi-controlled rectifier operating at the same firing angle? Assume that the rectifier supplies a constant current load.

P 2.15 A single-phase distribution feeder having an impedance of $0.2 + j0.6\,\Omega$ is connected to a source of 230 V, 50 Hz and feeds a load of $3 + j7\,\Omega$. Find the rating of the capacitor to be placed at the load side to improve the source power factor to unity. What will be the source power factor, if the same capacitor is placed at the source side?

P 2.16 Compute the value of displacement factor, power factor, and total harmonic distortion (THD) of input current for the system shown in Fig. 2.18. Assume that the load inductor is very high as to cause constant current with $R = 15\,\Omega$ at the output terminals and source voltage ($V_s = 230\,\text{V}$) has negligible impedance.

P 2.17 An ideal rectifier is connected to the source through an ideal transformer. The load resistance on dc side is 2 Ω. Total THD in the source current should not exceed 35% as per the specification of equipment, and the rms value of harmonic current from the supply is limited to 3 A on the secondary side of the transformer. The current on the dc side is constant due to the large inductance connected in series with the load. Assuming negligible losses, calculate the following.

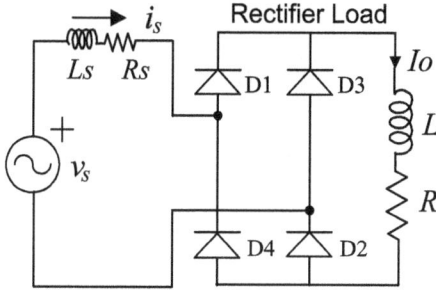

Figure 2.18 Related to problem P 2.16

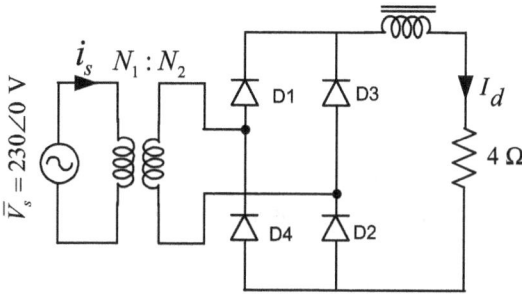

Figure 2.19 Related to problem P 2.17

(a) rms value of maximum allowable fundamental current and total source current.

(b) Considering household supply is 230 V, 50 Hz, find out the voltage at the secondary and turns ratio of the transformer.

(c) Calculate the displacement factor and total power factor on the source side of the transformer.

P 2.18 A dc motor with back EMF 200 V, 5 Ω resistance, and 100 mH leakage inductance is connected as a load to a single-phase full-wave ideal diode rectifier, which is supplied by a sinusoidal voltage source, as shown in Fig. 2.20. If the dc motor draws a constant current of 5 A, what should be the rms value of the supply voltage? Also, compute the real active and average reactive powers, apparent power, and power factor of the source.

P 2.19 Solve the above question with a fully controlled rectifier with firing angle $\alpha = 30°$ (continuous conduction mode).

P 2.20 A single-phase source supplies ac load at 50 Hz and a dc load as shown in Fig. 2.21. The load inductance of the rectifier is quite large to cause to flow of a constant dc current. The source has negligible resistance and inductance.

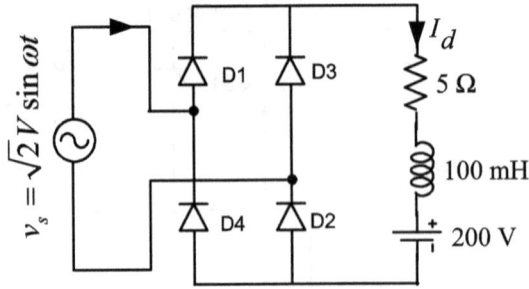

Figure 2.20   Related to problem P 2.18

Considering up to 5th harmonic components in source current, i_{s2}, solve for the following.

(a) Source current, i_s as a function of time.

(b) Real power (P), reactive power (Q), fundamental power factor (pf_1).

(c) The apparent power (S), power factor (pf) at the source.

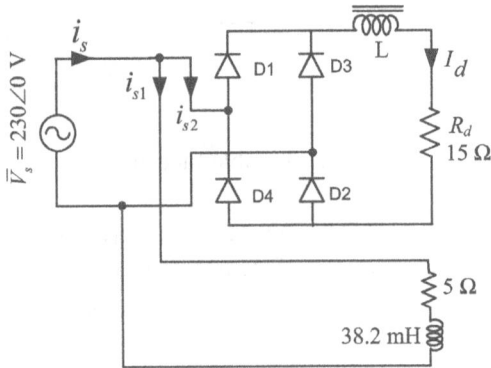

Figure 2.21   Related to problem P 2.20

P 2.21 Repeat the above question for $R_d = 10\,\Omega$, and the ac load consists of a resistance of $4\,\Omega$ in series with capacitance of $200\,\mu F$.

P 2.22 A stiff sinusoidal voltage source supplies power to the incandescent lamp which is rated at 230 V, 1000 W. An ac voltage controller / TRIAC is placed in between the load and source to control the illumination of the lamp, as shown in Fig. 2.22. Find the following for one level of illumination, i.e., $\alpha = 30°$, where $v(t) = 230\sqrt{2}\sin\omega t$ V. Considering up to 5th harmonics, i.e., $n=1,3,5$, answer the following.

(a) Compute $i(t)$.

(b) Compute P, P_1, P_H from the source.

(c) Compute Q, Q_1, Q_H from the source.

(d) For the same amount of fundamental power to be transferred to the load, find the value of resistance to be placed in place of voltage controller. Also, comment upon the performance of the system in the two cases.

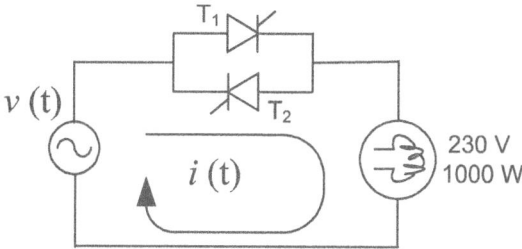

Figure 2.22 Related to problem P 2.22

REFERENCES

1. IEEE Group, "IEEE trial-use standard definitions for the measurement of electric power quantities under sinusoidal, nonsinusoidal, balanced, or unbalanced conditions," 2000.

2. E. Watanabe, R. Stephan, and M. Aredes, "New concepts of instantaneous active and reactive powers in electrical systems with generic loads," *IEEE Transactions on Power Delivery*, vol. 8, no. 2, pp. 697–703, 1993.

3. T. Furuhashi, S. Okuma, and Y. Uchikawa, "A study on the theory of instantaneous reactive power," *IEEE Transactions on Industrial Electronics*, vol. 37, no. 1, pp. 86–90, 1990.

4. A. Emanuel, "Powers in nonsinusoidal situations-a review of definitions and physical meaning," *IEEE Transactions on Power Delivery*, vol. 5, no. 3, pp. 1377–1389, 1990.

5. A. Ferrero and G. Superti-Furga, "A new approach to the definition of power components in three-phase systems under nonsinusoidal conditions," *IEEE Transactions on Instrumentation and Measurement*, vol. 40, no. 3, pp. 568–577, 1991.

6. J. Willems, "A new interpretation of the Akagi-Nabae power components for nonsinusoidal three-phase situations," *IEEE Transactions on Instrumentation and Measurement*, vol. 41, no. 4, pp. 523–527, 1992.

7. J. L. Willems, "Budeanu's reactive power and related concepts revisited," *IEEE Transactions on Instrumentation and Measurement*, vol. 60, no. 4, pp. 1182–1186, 2011.

8. D. Vieira, R. A. Shayani, and M. A. G. de Oliveira, "Reactive power billing under nonsinusoidal conditions for low-voltage systems," *IEEE Transactions on Instrumentation and Measurement*, vol. 66, no. 8, pp. 2004–2011, 2017.

9. M. Erhan Balci and M. Hakan Hocaoglu, "Quantitative comparison of power decompositions," *Electric Power Systems Research*, vol. 78, no. 3, pp. 318–329, 2008.

10. W. Shepherd and P. Zand, *Energy Flow and Power Factor in Nonsinusoidal Circuits*. Cambridge University Press, 1979.

11. S. Stefan, *Power measurement techniques for non-sinusoidal conditions: The significance of harmonics for the measurement of power and other AC quantities*. Doctoral thesis, Department of Electric Power Engineering, Chalmers University of Technology Göteborg Sweden, 1999.

12. R. W. Erickson and D. Maksimović, *Power and Harmonics in Nonsinusoidal Systems*. Boston, MA: Springer US, 2001, pp. 589–607.

13. M. Grady, *Understanding Power System Harmonics*. Department of Electrical and Computer Engineering University of Texas, Austin, 2012.

14. M. H. J. Bollen, *Understanding Power Quality Problems: Voltage Sags and Interruptions*. Wiley-IEEE Press, 1999.

15. Arindam Ghosh and Gerald Ledwich, *Power quality enhancement using custom power systems*. Wiley, 2006.

16. Bhim Singh, Ambrish Chandra, and Kamal Al-Haddad, *Power Quality Problems and Mitigation Techniques*. Wiley, 2015.

17. Roger C. Dugan, Mark F. McGranaghan, Surya Santoso, and H. Wayne Beaty, *Electrical Power System Quality*. Tata McGraw-Hill, 2008.

18. M. A. Masoum and E. F. Fuchs, *Chapter 1 – Introduction to Power Quality*, M. A. Masoum and E. F. Fuchs, Eds. Academic Press, 2015.

19. J. Kassakian, M. Schlecht, and G. Verghese, *Principles of Power Electronics*. Addison-Wesley, 1991.

20. J. Vithayathil, *Power Electronics: Principles and Applications*, ser. McGraw-Hill Series in Electrical and Computer Engineering: Power and Energy. McGraw-Hill, 1995.

21. N. Mohan, T. M. Undeland, and W. P. Robbins, *Power Electronics. Converters, Applications and Design*, 3rd ed. John Wiley and Sons, Inc, 2003.

22. R. Erickson and D. Maksimovic, *Fundamentals of Power Electronics*. Springer US, 2012.

3 Three Phase Circuits: Power Definitions and Various Components

3.1 INTRODUCTION

Three-phase system is an economical and efficient way of transmitting bulk power over long lengths of electrical lines to the distribution system or industrial plant. The striking features of three-phase system are: (i) self starting of ac motors (ii) no need for return path due to $\pm 120°$ phase shift in phases (iii) a constant power (like dc system) (iv) interlinking three-phase fluxes within a closed core (v) supports single-phase operation by adding neutral wire (fourth wire) [1]–[3]. Thus, the usage of three-phase voltage supply is very common for the generation, transmission, and distribution of bulk electrical power and it is most commonly used by electric grids worldwide. Most of the industrial loads are supplied by three-phase power for its advantages over single-phase systems such as cost and efficiency for the same amount of power transmission. In principle, any number of phases can be used in a polyphase electric system, however, three-phase system is simple from the design and construction point of view of a three-phase power system and its components. The three-phase system can be broadly categorized into star and delta connections at various levels of voltage in the power systems such as generation, transmission, and distribution network at 50 or 60 Hz frequency, to maximize the power transfer with minimum losses. The generation system normally uses a medium voltage range, i.e., 11–33 kV, transmission system uses high and extra high voltage ranges, i.e., 33–230 kV and 230–1100 kV, and the distribution power system uses low voltage up to 11 kV. The power system is ideally designed to operate at the rated voltage and frequency with sinusoidal nature of voltages and currents, supplying a three-phase balanced load connected in either star or delta configuration. Any deviations from the allowable limits of these parameters, cause issues that need to be addressed. On the distribution side, the nature of the load such as balanced unbalanced, harmonics, transients, etc., has a significant effect on the performance of the power system in terms of losses, current limits, power factor, and hence it needs to be studied in detail [4]–[11].

In this chapter, the main focus is on understanding the behavior of a three-phase power system in balanced, unbalanced, harmonic components in voltages and currents. Also, important terms related to power and its various components will be explained from basic principles [12]–[16] and their applications are discussed.

DOI: 10.1201/9781032617305-3

3.2   THREE-PHASE BALANCED SYSTEM

A three-phase balanced system is represented in Fig. 3.1.

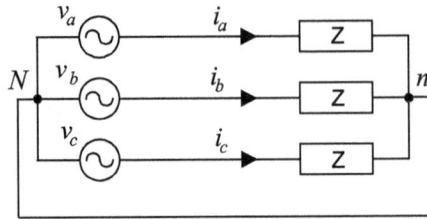

Figure 3.1   A three-phase balanced circuit

A three-phase balanced system is expressed using the following voltages and currents.

$$\begin{aligned}
v_a(t) &= \sqrt{2}V \sin \omega t \\
v_b(t) &= \sqrt{2}V \sin(\omega t - 120°) \\
v_c(t) &= \sqrt{2}V \sin(\omega t + 120°)
\end{aligned} \qquad (3.1)$$

and,

$$\begin{aligned}
i_a(t) &= \sqrt{2}I \sin(\omega t - \phi) \\
i_b(t) &= \sqrt{2}I \sin(\omega t - 120° - \phi) \\
i_c(t) &= \sqrt{2}I \sin(\omega t + 120° - \phi)
\end{aligned} \qquad (3.2)$$

In (3.1) and (3.2) subscripts a, b and c are used to denote three phases that are balanced. Three-phase balanced means that the voltage or current magnitudes (V or I) are the same for all three phases and they have a phase shift of $-120°$ and $120°$ in phase-b and phase-c, respectively. The currents are assumed to have ϕ degree lag with their respective phase voltages for a positive value of ϕ. The three-phase voltage and current waveforms and corresponding phasor diagram are shown in Fig. 3.2(a) and (b), respectively. The balanced three-phase system has certain interesting properties. These will be discussed in the following section.

3.2.1   THREE-PHASE INSTANTANEOUS POWER

Three-phase instantaneous power in three-phase system is given by,

$$\begin{aligned}
p_{3\phi}(t) = p(t) &= v_a(t)i_a(t) + v_b(t)i_b(t) + v_c(t)i_c(t) \\
&= p_a(t) + p_b(t) + p_c(t)
\end{aligned} \qquad (3.3)$$

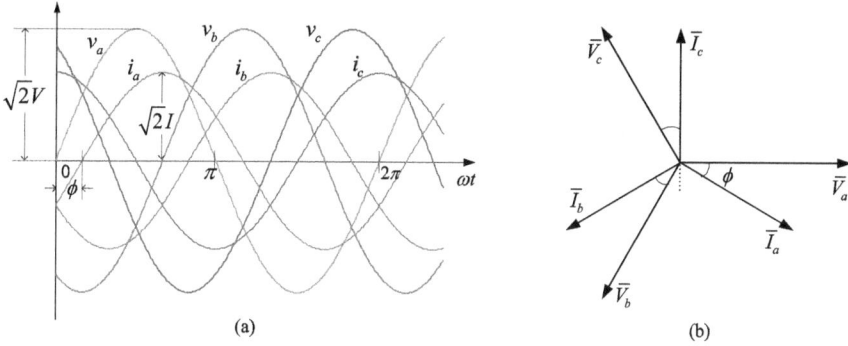

Figure 3.2 (a) A three-phase voltage and current waveforms and (b) phasor diagram

In above equation, $p_a(t)$, $p_b(t)$ and $p_c(t)$ are expressed similar to single-phase system described previously. These are given below.

$$
\begin{aligned}
p_a(t) &= VI\cos\phi\,(1-\cos 2\omega t) - VI\sin\phi\sin 2\omega t \\
p_b(t) &= VI\cos\phi\,[1-\cos 2(\omega t - 120°)] - VI\sin\phi\sin 2(\omega t - 120°) \quad (3.4) \\
p_c(t) &= VI\cos\phi\,[1-\cos 2(\omega t + 120°)] - VI\sin\phi\sin 2(\omega t + 120°)
\end{aligned}
$$

Adding three-phase instantaneous powers given in (3.4), we get total three-phase instantaneous power, as below.

$$
\begin{aligned}
p(t) = 3VI\cos\phi \quad &- \quad VI\cos\phi\,[\cos 2\omega t + \cos 2(\omega t - 120°) + \cos 2(\omega t + 120°)] \\
&- \quad VI\sin\phi\,[\sin 2\omega t + \sin 2(\omega t - 120°) + \sin 2(\omega t + 120°)]
\end{aligned}
$$

$$(3.5)$$

The summation of terms in square brackets equals to zero. Hence, the expression obtained for three-phase instantaneous active power is,

$$
p_{3\phi}(t) = p(t) = P = 3VI\cos\phi \tag{3.6}
$$

This is an interesting result as it infers that for three-phase balanced system, the total instantaneous power is equal to the real power or average active power (P), which is constant. This is the main reason we use three-phase system, as it does not involve the pulsating or oscillating components of power compared to single-phase systems. Thus, it ensures less VA rating for the same amount of power transfer. Here, total three-phase reactive power can be defined as sum of maximum value of $p_{reactive}(t)$ terms in (3.4). Thus,

$$
Q = Q_a + Q_b + Q_c = 3VI\sin\phi. \tag{3.7}
$$

There has been an attempt to define instantaneous reactive power, $q(t)$ similar to $p(t)$, such that Q is the average value of the term $q(t)$ by H. Akagi et al. [17] in the year 1983–84. The definition was facilitated through $\alpha\beta 0$ transformation, which is briefly described in the next subsection.

3.2.2 THREE PHASE INSTANTANEOUS REACTIVE POWER

The abc coordinates and their equivalent $\alpha\beta0$ coordinates are shown in Fig. 3.3. Resolving a, b, c quantities along the $\alpha\beta$ axis, we have,

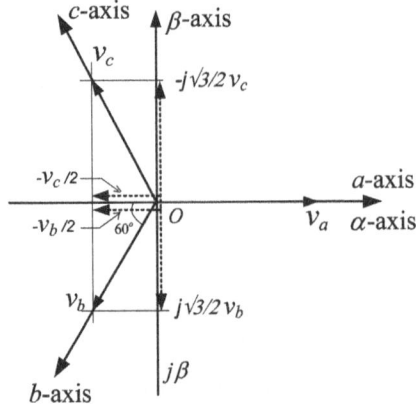

Figure 3.3 A abc to $\alpha\beta0$ transformation

$$v_\alpha = \sqrt{\frac{2}{3}}\left(v_a - \frac{v_b}{2} - \frac{v_c}{2}\right) \tag{3.8}$$

$$v_\beta = \sqrt{\frac{2}{3}}\frac{\sqrt{3}}{2}(v_b - v_c) \tag{3.9}$$

Here, $\sqrt{\frac{2}{3}}$ is a scaling factor, which ensures power invariant transformation. Along with that, we define zero sequence voltage as,

$$v_0 = \sqrt{\frac{2}{3}}\sqrt{\frac{1}{2}}(v_a + v_b + v_c) = \frac{1}{\sqrt{3}}(v_a + v_b + v_c) \tag{3.10}$$

Based on (3.8)–(3.10), we can write the above equations, as follows.

$$\begin{bmatrix} v_0(t) \\ v_\alpha(t) \\ v_\beta(t) \end{bmatrix} = \sqrt{\frac{2}{3}}\begin{bmatrix} \frac{1}{\sqrt{2}} & \frac{1}{\sqrt{2}} & \frac{1}{\sqrt{2}} \\ 1 & -\frac{1}{2} & -\frac{1}{2} \\ 0 & \frac{\sqrt{3}}{2} & -\frac{\sqrt{3}}{2} \end{bmatrix}\begin{bmatrix} v_a(t) \\ v_b(t) \\ v_c(t) \end{bmatrix} \tag{3.11}$$

$$\begin{bmatrix} v_0 \\ v_\alpha \\ v_\beta \end{bmatrix} = [A_{o\alpha\beta}]\begin{bmatrix} v_a \\ v_b \\ v_c \end{bmatrix}$$

The above is known as Clarke-Concordia transformation. Thus, v_a, v_b, and v_c can also be expressed in terms of v_0, v_α, and v_β by pre-multiplying (3.11) by matrix $[A_{0\alpha\beta}]^{-1}$,

$$
\begin{bmatrix} v_a \\ v_b \\ v_c \end{bmatrix} = [A_{0\alpha\beta}]^{-1} \begin{bmatrix} v_0 \\ v_\alpha \\ v_\beta \end{bmatrix}
$$

Furthermore,

$$
[A_{0\alpha\beta}]^{-1} = [A_{abc}] = \left(\sqrt{\frac{2}{3}} \begin{bmatrix} \frac{1}{\sqrt{2}} & \frac{1}{\sqrt{2}} & \frac{1}{\sqrt{2}} \\ 1 & -\frac{1}{2} & -\frac{1}{2} \\ 0 & \frac{\sqrt{3}}{2} & -\frac{\sqrt{3}}{2} \end{bmatrix} \right)^{-1}
$$

$$
[A_{0\alpha\beta}]^{-1} = \left(\sqrt{\frac{2}{3}} \begin{bmatrix} \frac{1}{\sqrt{2}} & 1 & 0 \\ \frac{1}{\sqrt{2}} & -\frac{1}{2} & \frac{\sqrt{3}}{2} \\ \frac{1}{\sqrt{2}} & -\frac{1}{2} & -\frac{\sqrt{3}}{2} \end{bmatrix} \right) = [A_{0\alpha\beta}]^{T} = [A_{abc}] \quad (3.12)
$$

Similarly, we can write down instantaneous symmetrical transformation for currents, which is given below.

$$
\begin{bmatrix} i_0 \\ i_\alpha \\ i_\beta \end{bmatrix} = \sqrt{\frac{2}{3}} \begin{bmatrix} \frac{1}{\sqrt{2}} & \frac{1}{\sqrt{2}} & \frac{1}{\sqrt{2}} \\ 1 & -\frac{1}{2} & -\frac{1}{2} \\ 0 & \frac{\sqrt{3}}{2} & -\frac{\sqrt{3}}{2} \end{bmatrix} \begin{bmatrix} i_a \\ i_b \\ i_c \end{bmatrix} \quad (3.13)
$$

Now based on "$0\alpha\beta$" transformation, the instantaneous power and instantaneous reactive powers are defined as follows [17]. The three-phase instantaneous power, $p(t)$ is expressed as the dot product or inner product of $0\alpha\beta$ components of voltages and currents, such as given below.

$$
\begin{aligned}
p(t) &= v_\alpha i_\alpha + v_\beta i_\beta + v_0 i_0 = \begin{bmatrix} v_0 \\ v_\alpha \\ v_\beta \end{bmatrix}^T \cdot \begin{bmatrix} i_0 \\ i_\alpha \\ i_\beta \end{bmatrix} \\
&= \frac{2}{3} \left[\left(v_a - \frac{v_b}{2} - \frac{v_c}{2} \right) \left(i_a - \frac{i_b}{2} - \frac{i_c}{2} \right) + \frac{\sqrt{3}}{2} (v_b - v_c) \frac{\sqrt{3}}{2} (i_b - i_c) \right. \\
&\quad \left. + \frac{1}{\sqrt{2}} (v_a + v_b + v_c) \frac{1}{\sqrt{2}} (i_a + i_b + i_c) \right] \\
&= v_a i_a + v_b i_b + v_c i_c \quad (3.14)
\end{aligned}
$$

In [17], the instantaneous reactive power $q(t)$ is defined as the cross product of two mutual perpendicular quantities as given below.

$$q(t) = v_\alpha \times i_\beta + v_\beta \times i_\alpha = v_\alpha i_\beta - v_\beta i_\alpha$$

$$= \frac{2}{3}\left[\left(v_a - \frac{v_b}{2} - \frac{v_c}{2}\right)\frac{\sqrt{3}}{2}(i_b - i_c) - \frac{\sqrt{3}}{2}(v_b - v_c)\left(i_a - \frac{i_b}{2} - \frac{i_c}{2}\right)\right]$$

$$= \frac{2}{3}\frac{\sqrt{3}}{2}\left[(-v_b + v_c)i_a + \left(v_a - \frac{v_b}{2} - \frac{v_c}{2} + \frac{v_b}{2} - \frac{v_c}{2}\right)i_b \right. \tag{3.15}$$

$$\left. + \left(-v_a + \frac{v_b}{2} + \frac{v_c}{2} + \frac{v_b}{2} - \frac{v_c}{2}\right)i_c\right]$$

$$= -\frac{1}{\sqrt{3}}\left[(v_b - v_c)i_a + (v_c - v_a)i_b + (v_a - v_b)i_c\right]$$

$$= -\left(v_{bc}i_a + v_{ca}i_b + v_{ab}i_c\right)/\sqrt{3} \tag{3.16}$$

This can also be expressed as the following.

$$q(t) = \frac{1}{\sqrt{3}}\left[(i_b - i_c)v_a + \left(-\frac{i_b}{2} + \frac{i_c}{2} - i_a + \frac{i_b}{2} + \frac{i_c}{2}\right)v_b \right. \tag{3.17}$$

$$\left. + \left(-\frac{i_b}{2} + \frac{i_c}{2} + i_a - \frac{i_b}{2} - \frac{i_c}{2}\right)v_c\right]$$

$$= \frac{1}{\sqrt{3}}\left[(i_b - i_c)v_a + (i_c - i_a)v_b + (i_a - i_b)v_c\right] \tag{3.18}$$

Equations (3.15)–(3.17) represent the instantaneous reactive power using voltage and current quantities in *abc* phases.

3.2.3 POWER INVARIANCE IN *ABC* AND $\alpha\beta 0$ COORDINATES

In order to check the power invariance, we shall compute the energy content of voltage signals in two transformations. The energy associated with the *abc* system is given by $(v_a^2 + v_b^2 + v_c^2)$, and the energy associated with the $\alpha\beta 0$ components is given by $\left(v_0^2 + v_\alpha^2 + v_\beta^2\right)$. The two energies must be equal to ensure power invariance in two transformations. Using, (3.11) and squares of the respective components, we have the following.

$$v_\alpha^2 = \left[\sqrt{\frac{2}{3}}\left(v_a - \frac{v_b}{2} - \frac{v_c}{2}\right)\right]^2$$

$$= \frac{2}{3}\left(v_a^2 + \frac{v_b^2}{4} + \frac{v_c^2}{4} - \frac{2v_a v_b}{2} + \frac{2v_b v_c}{4} - \frac{2v_c v_a}{2}\right)$$

$$= \frac{2}{3}v_a^2 + \frac{v_b^2}{6} + \frac{v_c^2}{6} - \frac{2v_a v_b}{3} + \frac{v_b v_c}{3} - \frac{2v_c v_a}{3} \tag{3.19}$$

Similarly, we can find out square of v_β term as given below.

$$v_\beta^2 = \left[\sqrt{\frac{2}{3}}\frac{\sqrt{3}}{2}(v_b - v_c)\right]^2$$

$$= \frac{1}{2}\left(v_b^2 + v_c^2 - 2v_b v_c\right) = \frac{v_b^2}{2} + \frac{v_c^2}{2} - v_b v_c \quad (3.20)$$

Adding (3.19) and (3.20),

$$v_\alpha^2 + v_\beta^2 = \frac{2}{3}\left(v_a^2 + v_b^2 + v_c^2 - v_a v_b - v_b v_c - v_c v_a\right)$$

$$= \left(v_a^2 + v_b^2 + v_c^2\right) - \left(\frac{v_a^2}{3} + \frac{v_b^2}{3} + \frac{v_c^2}{3} + \frac{2v_a v_b}{3} + \frac{2v_b v_c}{3} + \frac{2v_c v_a}{3}\right)$$

$$= \left(v_a^2 + v_b^2 + v_c^2\right) - \frac{1}{3}\left(v_a + v_b + v_c\right)^2$$

$$= \left(v_a^2 + v_b^2 + v_c^2\right) - \left[\frac{1}{\sqrt{3}}\left(v_a + v_b + v_c\right)\right]^2 \quad (3.21)$$

Since $v_0 = \frac{1}{\sqrt{3}}(v_a + v_b + v_c)$, the above can be written as,

$$v_\alpha^2 + v_\beta^2 + v_0^2 = v_a^2 + v_b^2 + v_c^2. \quad (3.22)$$

From the above, it is implied that the energy associated with the two systems remains the same from instant to instant basis. In general, the instantaneous power $p(t)$ remains same in both transformations. This can also be proved using matrix operations, as given below.

Using (3.14), following can be written.

$$p(t) = v_\alpha i_\alpha + v_\beta i_\beta + v_0 i_0$$

$$= \begin{bmatrix} v_0 \\ v_\alpha \\ v_\beta \end{bmatrix}^T \cdot \begin{bmatrix} i_0 \\ i_\alpha \\ i_\beta \end{bmatrix}$$

$$= \begin{bmatrix} [A_{0\alpha\beta}] \begin{bmatrix} v_a \\ v_b \\ v_c \end{bmatrix} \end{bmatrix}^T \begin{bmatrix} [A_{0\alpha\beta}] \begin{bmatrix} i_a \\ i_b \\ i_c \end{bmatrix} \end{bmatrix}$$

$$= \begin{bmatrix} v_a \\ v_b \\ v_c \end{bmatrix}^T [A_{0\alpha\beta}]^T [A_{0\alpha\beta}] \begin{bmatrix} i_a \\ i_b \\ i_c \end{bmatrix}$$

$$= \begin{bmatrix} v_a \\ v_b \\ v_c \end{bmatrix}^T [A_{0\alpha\beta}]^{-1} [A_{0\alpha\beta}] \begin{bmatrix} i_a \\ i_b \\ i_c \end{bmatrix}$$

$$= \begin{bmatrix} v_a & v_b & v_c \end{bmatrix} \begin{bmatrix} i_a \\ i_b \\ i_c \end{bmatrix}$$

$$= v_a i_a + v_b i_b + v_c i_c \quad (3.23)$$

In the above equation, the following property of matrices of the form (3.12) is used.

$$[A_{0\alpha\beta}]^T [A_{0\alpha\beta}] = [A_{0\alpha\beta}]^{-1} [A_{0\alpha\beta}] = [I] \tag{3.24}$$

where, $[I]$ is an identity matrix.

3.3 INSTANTANEOUS ACTIVE AND REACTIVE POWERS FOR THREE-PHASE CIRCUITS

In the previous section, the instantaneous active and reactive powers were defined using $\alpha\beta0$ transformation. In this section, we shall study these powers for various three-phase circuits such as three-phase balanced, three-phase unbalanced, balanced three-phase with harmonics, and unbalanced three-phase with harmonics. Each case will be considered and analyzed.

3.3.1 THREE-PHASE BALANCE SYSTEM

For three-phase balanced system, three-phase voltages have been expressed by (3.1). For these phase voltages, the line to line voltages are given below.

$$v_{ab} = \sqrt{3}\sqrt{2}V \sin(\omega t + 30°)$$
$$v_{bc} = \sqrt{3}\sqrt{2}V \sin(\omega t - 90°)$$
$$v_{ca} = \sqrt{3}\sqrt{2}V \sin(\omega t + 150°) \tag{3.25}$$

The above relationship between phase and line to line voltages is also illustrated in Fig. 3.4. For the above three-phase system, the instantaneous power $p(t)$ can be

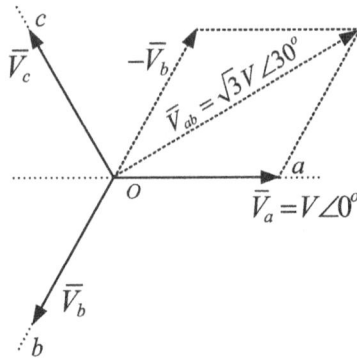

Figure 3.4 Relationship between line-to-line and phase voltages

expressed using (3.23), and it is equal to,

$$\begin{aligned}
p(t) &= v_a i_a + v_b i_b + v_c i_c \\
&= v_\alpha i_\alpha + v_\beta i_\beta + v_0 i_0 \\
&= 3VI\cos\phi.
\end{aligned} \tag{3.26}$$

The instantaneous reactive power $q(t)$ is as follows.

$$
\begin{aligned}
q(t) &= -\left(v_{bc}i_a + v_{ca}i_b + v_{ab}i_c\right)/\sqrt{3} \\
&= -\frac{1}{\sqrt{3}}\left[\sqrt{3}\sqrt{2}V\sin\left(\omega t - 90°\right)\sqrt{2}I\sin\left(\omega t - \phi\right)\right. \\
&\quad + \sqrt{3}\sqrt{2}V\sin\left(\omega t + 150°\right)\sqrt{2}I\sin\left(\omega t - 120° - \phi\right) \\
&\quad + \left.\sqrt{3}\sqrt{2}V\sin\left(\omega t + 30°\right)\sqrt{2}I\sin\left(\omega t + 120° - \phi\right)\right] \\
&= -VI\left[\cos\left(90° - \phi\right) - \cos\left(2\omega t - 90° - \phi\right)\right. \\
&\quad + \cos\left(90° - \phi\right) - \cos\left(2\omega t + 30° - \phi\right) \\
&\quad + \left.\cos\left(90° - \phi\right) - \cos\left(2\omega t + 150° - \phi\right)\right] \\
&= -VI\left[3\sin\phi - \cos\left(2\omega t - \phi + 30°\right) - \cos\left(2\omega t - \phi + 30° + 120°\right)\right. \\
&\quad \left.- \cos\left(2\omega t - \phi + 30° - 120°\right)\right] \\
q(t) &= -3VI\sin\phi \quad\quad\quad\quad\quad (3.27)
\end{aligned}
$$

The above value of instantaneous reactive power is same as defined by Budeanu's [12] and is given in (3.7). Thus, instantaneous reactive power given in (3.15) matches the conventional definition of reactive power defined in (3.7). However the time varying part of second terms of each phase in (3.4) has no relevance with the definition given in (3.15). Another interpretation of line to line voltages in (3.15) is that the voltages v_{ab}, v_{bc}, and v_{ca} have, 90° phase shift with respect to voltages v_c, v_a, and v_b, respectively. These are expressed below.

$$
\begin{aligned}
v_{ab} &= \sqrt{3}v_c \angle - 90° \\
v_{bc} &= \sqrt{3}v_a \angle - 90° \quad\quad\quad (3.28) \\
v_{ca} &= \sqrt{3}v_b \angle - 90°
\end{aligned}
$$

In the above equation, $v_c\angle - 90°$, $v_a\angle - 90°$, $v_b\angle - 90°$ denote v_c, v_a, v_b voltages shifted by $-90°$, respectively. Analyzing each term in (3.15) contributes to,

$$
\begin{aligned}
v_{bc}i_a &= \sqrt{3}v_a\angle -90°\ i_a \\
&= \sqrt{3}\sqrt{2}V\sin\left(\omega t - 90°\right)\sqrt{2}I\sin\left(\omega t - \phi\right) \\
&= \sqrt{3}VI\left[2\sin\left(\omega t - 90°\right)\sin\left(\omega t - \phi\right)\right] \\
&= \sqrt{3}VI\left[\cos\left(90° - \phi\right) - \cos\left(2\omega t - 90° - \phi\right)\right] \\
&= \sqrt{3}VI\left[\sin\phi - \sin\left(2\omega t - \phi\right)\right] \\
&= \sqrt{3}VI\left[\sin\phi - \sin 2\omega t\cos\phi + \cos 2\omega t\sin\phi\right] \\
v_{bc}i_a/\sqrt{3} &= VI\left[\sin\phi\left(1 + \cos 2\omega t\right) - \cos\phi\sin 2\omega t\right]
\end{aligned}
$$

Similarly,

$$
\begin{aligned}
v_{ca}i_b/\sqrt{3} &= VI\left\{\sin\phi\left[1 + \cos 2\left(\omega t - \frac{2\pi}{3}\right)\right]\right\} - VI\cos\phi\sin 2\left(\omega t - \frac{2\pi}{3}\right) \\
v_{ab}i_c/\sqrt{3} &= VI\left\{\sin\phi\left[1 + \cos 2\left(\omega t + \frac{2\pi}{3}\right)\right]\right\} - VI\cos\phi\sin 2\left(\omega t + \frac{2\pi}{3}\right)
\end{aligned}
$$

$$(3.29)$$

Thus, we see that the role of the coefficients of $\sin \phi$ and $\cos \phi$ have reversed. Now, if we take the average value of (3.29), it is not equal to zero, but $VI \sin \phi$ in each phase. Thus, three-phase reactive power will be $3VI \sin \phi$. The maximum value of second term in (3.29) represents active average power i.e., $VI \cos \phi$. However, this is not normally convention about the notation of the powers. Thus, the important contribution of this definition is that the reactive power could be defined as the average value of terms in (3.29).

3.3.2 THREE-PHASE UNBALANCED SYSTEM

Three-phase unbalanced system is not uncommon in power system. Three-phase unbalance may result from single-phasing, faults, different loads in three phases. To study three-phase system with fundamental unbalance, the voltages and currents are expressed as following.

$$
\begin{aligned}
v_a &= \sqrt{2}V_a \sin(\omega t - \phi_{va}) \\
v_b &= \sqrt{2}V_b \sin(\omega t - 120° - \phi_{vb}) \\
v_c &= \sqrt{2}V_c \sin(\omega t + 120° - \phi_{vc})
\end{aligned}
\tag{3.30}
$$

and,

$$
\begin{aligned}
i_a &= \sqrt{2}I_a \sin(\omega t - \phi_{ia}) \\
i_b &= \sqrt{2}I_b \sin(\omega t - 120° - \phi_{ib}) \\
i_c &= \sqrt{2}I_c \sin(\omega t + 120° - \phi_{ic})
\end{aligned}
\tag{3.31}
$$

For the above system, the three-phase instantaneous power is given by,

$$
\begin{aligned}
p_{3\phi}(t) = p(t) &= v_a i_a + v_b i_b + v_c i_c \\
&= \sqrt{2}V_a \sin(\omega t - \phi_{va})\sqrt{2}I_a \sin(\omega t - \phi_{ia}) \\
&\quad + \sqrt{2}V_b \sin(\omega t - 120° - \phi_{vb})\sqrt{2}I_b \sin(\omega t - 120° - \phi_{ib}) \\
&\quad + \sqrt{2}V_c \sin(\omega t + 120° - \phi_{vc})\sqrt{2}I_c \sin(\omega t + 120° - \phi_{ic})
\end{aligned}
\tag{3.32}
$$

Simplifying above expression we get,

$$
\begin{aligned}
p_{3\phi}(t) = &\underbrace{V_a I_a \cos\phi_a [1 - \cos 2(\omega t - \phi_{va})]}_{P_{a,active}} - \underbrace{V_a I_a \sin\phi_a \sin 2(\omega t - \phi_{va})}_{P_{a,reactive}} \\
&+ \underbrace{V_b I_b \cos\phi_b \{1 - \cos 2[(\omega t - 120°) - \phi_{vb}]\}}_{P_{b,active}} \\
&- \underbrace{V_b I_b \sin\phi_b \sin 2[(\omega t - 120°) - \phi_{vb}]}_{P_{b,active}} \\
&+ \underbrace{V_c I_c \cos\phi_c \{1 - \cos 2[(\omega t + 120°) - \phi_{vc}]\}}_{P_{c,active}} \\
&- \underbrace{V_c I_c \sin\phi_c \sin 2[(\omega t + 120°) - \phi_{vc}]}_{P_{c,reative}}
\end{aligned}
\tag{3.33}
$$

where, $\phi_a = (\phi_{ia} - \phi_{va})$
$\phi_b = (\phi_{ib} - \phi_{vb})$
$\phi_c = (\phi_{ic} - \phi_{vc})$

Therefore,

$$p_{3\phi}(t) = p_{a,active} + p_{b,active} + p_{c,active} + p_{a,reactive} + p_{b,reactive} + p_{c,reactive} \quad (3.34)$$

The above power components can also be written in average and oscillating components, as given below.

$$p_{3\phi}(t) = \overline{P}_a + \overline{P}_b + \overline{P}_c + \widetilde{p}_a + \widetilde{p}_b + \widetilde{p}_c \quad (3.35)$$

where,

$$\overline{P}_a = P_a = V_a I_a \cos\phi_a$$
$$\overline{P}_b = P_b = V_b I_b \cos\phi_b \quad (3.36)$$
$$\overline{P}_c = P_c = V_c I_c \cos\phi_c$$

and

$$\widetilde{p}_a = -V_a I_a \cos[2(\omega t - \phi_{va}) - \phi_a]$$
$$\widetilde{p}_b = -V_b I_b \cos[2(\omega t - 120° - \phi_{vb}) - \phi_b] \quad (3.37)$$
$$\widetilde{p}_c = -V_c I_c \cos[2(\omega t + 120° - \phi_{vc}) - \phi_c]$$

Also, it is noted that,

$$\overline{P}_a + \overline{P}_b + \overline{P}_c = \frac{1}{T}\int_0^T (v_a i_a + v_b i_b + v_c i_c)\,dt = P \quad (3.38)$$

and,

$$\widetilde{p}_a + \widetilde{p}_b + \widetilde{p}_c \neq 0 \quad (3.39)$$

This implies that, we have no longer the advantage of getting constant power, $3VI\cos\phi$ from interaction of three-phase voltages and currents. Now, let us analyze three-phase instantaneous reactive power $q(t)$ as per definition given in (3.15).

$$
\begin{aligned}
q(t) &= -\frac{1}{\sqrt{3}}[(v_b - v_c)i_a + (v_c - v_a)i_b + (v_a - v_b)i_c] \\
&= -\frac{2}{\sqrt{3}}\Big\{ [V_b sin(\omega t - 120° - \phi_{vb}) - V_c sin(\omega t + 120° - \phi_{vc})] I_a sin(\omega t - \phi_{ia}) \\
&\quad + [V_c sin(\omega t + 120° - \phi_{vc}) - V_a sin(\omega t - \phi_{va})] I_b sin(\omega t - 120° - \phi_{ib}) \\
&\quad + [V_a sin(\omega t - \phi_{va}) - V_b sin(\omega t - 120° - \phi_{vb})] I_c sin(\omega t + 120° - \phi_{ic}) \Big\}
\end{aligned}
$$
$$(3.40)$$

From the above,

$$\sqrt{3}q(t) = -\Big\{ V_bI_a\left[\cos(\phi_{ia}-\phi_{vb}-120°)-\cos(2\omega t-\phi_{ia}-\phi_{vb}-120°)\right]$$
$$-V_cI_a\left[\cos(\phi_{ia}-\phi_{vc}+120°)-\cos(2\omega t-\phi_{ia}-\phi_{vc}+120°)\right]$$
$$+V_cI_b\left[\cos(\phi_{ib}-\phi_{vc}-120°)-\cos(2\omega t-\phi_{ib}-\phi_{vc})\right]$$
$$-V_aI_b\left[\cos(\phi_{ib}-\phi_{va}+120°)-\cos(2\omega t-\phi_{va}-\phi_{ib}-120°)\right]$$
$$+V_aI_c\left[\cos(\phi_{ic}-\phi_{va}-120°)-\cos(2\omega t-\phi_{va}-\phi_{ic}+120°)\right]$$
$$-V_bI_c\left[\cos(\phi_{ic}-\phi_{vb}+120°)-\cos(2\omega t-\phi_{ic}-\phi_{vb})\right]\Big\} \qquad (3.41)$$

Now looking this expression, we can say that,

$$\frac{1}{T}\int_0^T q(t)dt = -\frac{1}{\sqrt{3}}\Big[V_bI_a\cos(\phi_{ia}-\phi_{vb}-120°)$$
$$-V_cI_a\cos(\phi_{ia}-\phi_{vc}+120°)$$
$$+V_cI_b\cos(\phi_{ib}-\phi_{vc}-120°)$$
$$-V_aI_b\cos(\phi_{ib}-\phi_{va}+120°)$$
$$+V_aI_c\cos(\phi_{ic}-\phi_{va}-120°)$$
$$-V_bI_c\cos(\phi_{ic}-\phi_{vb}+120°)\Big]$$
$$= \bar{q}_a(t)+\bar{q}_b(t)+\bar{q}_c(t)$$
$$\neq V_aI_a\sin\phi_a+V_bI_b\sin\phi_b+V_cI_c\sin\phi_c \qquad (3.42)$$

Hence, the definition of instantaneous reactive power does not match the definition of Budeanue's reactive power [12] for three-phase unbalanced circuit. If only voltages or currents are distorted, the above holds true. This is illustrated below.
Let us consider that only currents are unbalanced, as expressed in the following:

$$v_a(t) = \sqrt{2}V\sin(\omega t)$$
$$v_b(t) = \sqrt{2}V\sin(\omega t-120°) \qquad (3.43)$$
$$v_c(t) = \sqrt{2}V\sin(\omega t+120°)$$

and

$$i_a(t) = \sqrt{2}I_a\sin(\omega t-\phi_a)$$
$$i_b(t) = \sqrt{2}I_b\sin(\omega t-120°-\phi_b) \qquad (3.44)$$
$$i_c(t) = \sqrt{2}I_c\sin(\omega t+120°-\phi_c)$$

And the instantaneous reactive power is given by,

$$
\begin{aligned}
q(t) &= -\frac{1}{\sqrt{3}}(v_{bc}i_a + v_{ca}i_b + v_{ab}i_c) \\
&= -[\sqrt{2}V\sin(\omega t - \pi/2)\sqrt{2}I_a\sin(\omega t - \phi_{ia}) \\
&\quad + \sqrt{2}V\sin(\omega t - 2\pi/3 - \pi/2)\sqrt{2}I_b\sin(\omega t - 2\pi/3 - \phi_{ib}) \\
&\quad + \sqrt{2}V\sin(\omega t + 2\pi/3 - \pi/2)\sqrt{2}I_c\sin(\omega t + 2\pi/3 - \phi_{ic})] \\
&= -\{VI_a[\cos(\pi/2 - \phi_{ia}) - \cos(2\omega t - \pi/2 - \phi_{ia})] \\
&\quad + VI_b[\cos(\pi/2 - \phi_{ib}) - \cos(2\omega t - 4\pi/3 - \pi/2 - \phi_{ib})] \\
&\quad + VI_c[\cos(\pi/2 - \phi_{ic}) - \cos(2\omega t + 4\pi/3 - \pi/2 - \phi_{ic})]\} \\
&= -[(VI_a\sin\phi_{ia} + VI_b\sin\phi_{ib} + VI_c\sin\phi_{ic}) \\
&\quad - VI_a\sin(2\omega t - \phi_{ia}) - VI_b\sin(2\omega t - 4\pi/3 - \phi_{ib}) \\
&\quad - VI_c\sin(2\omega t + 4\pi/3 - \phi_{ic})]
\end{aligned}
$$

Thus, the average reactive power, Q is given by,

$$
Q = \frac{1}{T}\int_0^T q(t)dt = -(VI_a\sin\phi_{ia} + VI_b\sin\phi_{ib} + VI_c\sin\phi_{ic}) \tag{3.45}
$$

In the above equation, the average value of three-phase instantaneous reactive power is same as Budeanu's reactive power.

The oscillating term of $q(t)$ which is equal to $\tilde{q}(t)$, is given below.

$$
\begin{aligned}
\tilde{q}(t) &= VI_a\sin(2\omega t - \phi_{ia}) + VI_b\sin(2\omega t - 240° - \phi_{ib}) \\
&\quad + VI_c\sin(2\omega t + 240° - \phi_{ic}) \tag{3.46}
\end{aligned}
$$

which is not similar to what is being defined as reactive component of power in (3.4).

3.4 SYMMETRICAL COMPONENTS

In the previous section, the fundamental unbalance in three-phase voltages and currents have been considered. Ideal power systems are not designed for unbalance quantities as it makes power system components over rated and inefficient. Thus, to understand unbalance three-phase systems, a concept of symmetrical components introduced by C. L. Fortescue, will be discussed. In 1918, C. L Fortescue, wrote a paper [18] presenting that an unbalanced system of n-related phasors can be resolved into n system of balanced phasors, called the symmetrical components of the original phasors. The n phasors of each set of components are equal in length and the angles. Although, the method is applicable to any unbalanced polyphase system, we shall discuss three-phase systems, as it is commonly used.

For the discussion of symmetrical components, a complex operator denoted as a is defined as,

$$\begin{aligned}
a &= 1\angle 120^\circ = e^{j2\pi/3} = \cos 2\pi/3 + j\sin 2\pi/3 \\
&= -1/2 + j\sqrt{3}/2 \\
a^2 &= 1\angle 240^\circ = 1\angle -120^\circ = e^{j4\pi/3} = e^{-j2\pi/3} = \cos 4\pi/3 + j\sin 4\pi/3 \\
&= -1/2 - j\sqrt{3}/2 \\
a^3 &= 1\angle 360^\circ = e^{j2\pi} = 1
\end{aligned}$$

Also note an interesting property relating 1, a, and a^2,

$$1 + a + a^2 = 0 \tag{3.47}$$

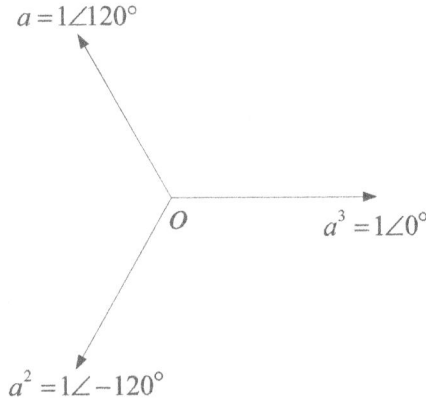

Figure 3.5 Phasor representation of 1, a, and a^2

These quantities, i.e., a, a^2, and a^3 also represent three phasors which are shifted by 120° from each other. This is shown in Fig. 3.5. Knowing the above and using Fortescue theorem, three unbalanced phasors of a three-phase unbalanced system can be resolved into three balanced system phasors, as described below.

1. Positive sequence components, that are composed of three phasors, equal in magnitude with a phase shift of -120° and 120° and a phase sequence similar to the original phasors.

2. Negative sequence components, consist of three phasors equal in magnitude with a phase shift of 120° and -120° and a phase sequence opposite to that of the original phasors.

3. Zero sequence components, consisting of three phasors equal in magnitude with zero phase shift from each other.

These sequence components are denoted as following.

Positive sequence components: $\overline{V}_{a+}, \overline{V}_{b+}, \overline{V}_{c+}$

Negative sequence components: $\overline{V}_{a-}, \overline{V}_{b-}, \overline{V}_{c-}$

Zero sequence components: $\overline{V}_{a0}, \overline{V}_{b0}, \overline{V}_{c0}$.

Thus, we can write,

$$
\begin{aligned}
\overline{V}_a &= \overline{V}_{a+} + \overline{V}_{a-} + \overline{V}_{a0} \\
\overline{V}_b &= \overline{V}_{b+} + \overline{V}_{b-} + \overline{V}_{b0} \\
\overline{V}_c &= \overline{V}_{c+} + \overline{V}_{c-} + \overline{V}_{c0}
\end{aligned}
\tag{3.48}
$$

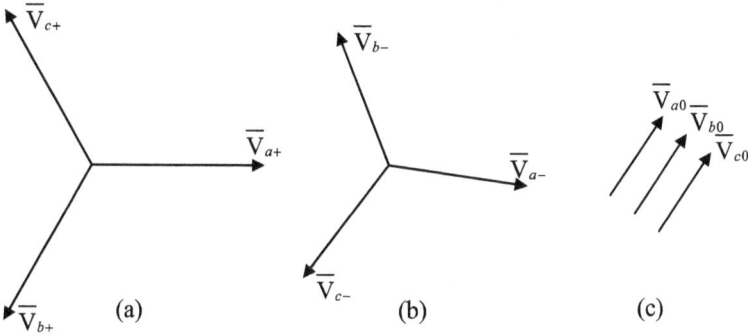

Figure 3.6 Sequence components (a) Positive sequence, (b) Negative sequence, (c) Zero sequence

Graphically, these are represented in Fig. 3.6. Thus, if we add the sequence components of each phase using phasors, we shall get $\overline{V}_a$, $\overline{V}_b$, and $\overline{V}_c$ as per (3.48). This is illustrated in Fig. 3.7. Now knowing all these preliminaries, we can proceed as follows. Considering $\overline{V}_{a+}$ as reference phasor, voltages, $\overline{V}_{b+}$ and $\overline{V}_{c+}$ can be written as,

$$
\begin{aligned}
\overline{V}_{b+} &= a^2 \overline{V}_{a+} = \overline{V}_{a+} \angle -120° \\
\overline{V}_{c+} &= a \overline{V}_{a+} = \overline{V}_{a+} \angle 120°
\end{aligned}
\tag{3.49}
$$

Similarly, $\overline{V}_{b-}$ and $\overline{V}_{c-}$ can be expressed in terms of $\overline{V}_{a-}$ as following.

$$
\begin{aligned}
\overline{V}_{b-} &= a \overline{V}_{a-} = \overline{V}_{a-} \angle 120° \\
\overline{V}_{c-} &= a^2 \overline{V}_{a-} = \overline{V}_{a-} \angle -120°
\end{aligned}
\tag{3.50}
$$

The zero sequence components have same magnitude and phase angle and therefore these are expressed as,

$$
\overline{V}_{b0} = \overline{V}_{c0} = \overline{V}_{a0}
\tag{3.51}
$$

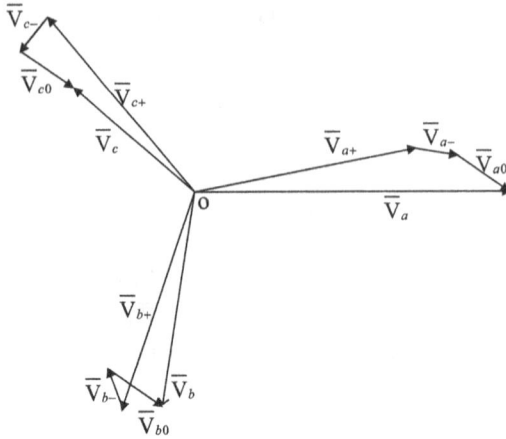

Figure 3.7 Unbalanced phasors as vector sum of positive, negative, and zero sequence phasors

Using (3.49), (3.50), and (3.51) we have,

$$\overline{V}_a = \overline{V}_{a0} + \overline{V}_{a+} + \overline{V}_{a-} \tag{3.52}$$

$$\begin{aligned}
\overline{V}_b &= \overline{V}_{b0} + \overline{V}_{b+} + \overline{V}_{b-} \\
&= \overline{V}_{a0} + a^2 \overline{V}_{a+} + a\overline{V}_{a-}
\end{aligned} \tag{3.53}$$

$$\begin{aligned}
\overline{V}_c &= \overline{V}_{c0} + \overline{V}_{c+} + \overline{V}_{c-} \\
&= \overline{V}_{a0} + a\overline{V}_{a+} + a^2 \overline{V}_{a-}
\end{aligned} \tag{3.54}$$

Equations (3.52)–(3.54) can be written in matrix form as given below.

$$\begin{bmatrix} \overline{V}_a \\ \overline{V}_b \\ \overline{V}_c \end{bmatrix} = \begin{bmatrix} 1 & 1 & 1 \\ 1 & a^2 & a \\ 1 & a & a^2 \end{bmatrix} \begin{bmatrix} \overline{V}_{a0} \\ \overline{V}_{a+} \\ \overline{V}_{a-} \end{bmatrix} \tag{3.55}$$

Premultipling by inverse of matrix $[A_{abc}] = \begin{bmatrix} 1 & 1 & 1 \\ 1 & a^2 & a \\ 1 & a & a^2 \end{bmatrix}$, the symmetrical components are expressed as given below.

$$\begin{aligned}
\begin{bmatrix} \overline{V}_{a0} \\ \overline{V}_{a+} \\ \overline{V}_{a-} \end{bmatrix} &= \frac{1}{3} \begin{bmatrix} 1 & 1 & 1 \\ 1 & a & a^2 \\ 1 & a^2 & a \end{bmatrix} \begin{bmatrix} \overline{V}_a \\ \overline{V}_b \\ \overline{V}_c \end{bmatrix} \tag{3.56} \\
&= [A_{0+-}] \begin{bmatrix} \overline{V}_a \\ \overline{V}_b \\ \overline{V}_c \end{bmatrix}
\end{aligned}$$

The symmetrical transformation matrices $[A_{0+-}]$ and $[A_{abc}]$ are related by the following expression.

$$[A_{0+-}] = [A_{abc}]^{-1} = \frac{1}{3}[A_{abc}]^* = \frac{1}{3}[A_{abc}]^T \tag{3.57}$$

Note that in the above equation, the matrix $[A_{abc}]$ has following property.

$$[A_{abc}]^T = [A_{abc}]^*$$

From (3.56), the symmetrical components can therefore be expressed as the following.

$$\overline{V}_{a0} = \frac{1}{3}(\overline{V}_a + \overline{V}_b + \overline{V}_c)$$
$$\overline{V}_{a+} = \frac{1}{3}(\overline{V}_a + a\overline{V}_b + a^2\overline{V}_c) \tag{3.58}$$
$$\overline{V}_{a-} = \frac{1}{3}(\overline{V}_a + a^2\overline{V}_b + a\overline{V}_c)$$

The other components, i.e., $\overline{V}_{b0}$, $\overline{V}_{c0}$, $\overline{V}_{b+}$, $\overline{V}_{c+}$, $\overline{V}_{b-}$, $\overline{V}_{c-}$ can be found from voltages, $\overline{V}_{a0}$, $\overline{V}_{a+}$, $\overline{V}_{a-}$. It should be noted that the quantity $\overline{V}_{a0}$ does not exist if sum of unbalanced phasors is zero. Since sum of line to line voltage phasors, i.e., $\overline{V}_{ab} + \overline{V}_{bc} + \overline{V}_{ca} = (\overline{V}_a - \overline{V}_b) + (\overline{V}_b - \overline{V}_c) + (\overline{V}_c - \overline{V}_a)$ is always zero, hence zero sequence voltage components are never present in the line voltage, regardless of amount of unbalance. The sum of the three-phase voltages, i.e., $\overline{V}_a + \overline{V}_b + \overline{V}_c$ is not necessarily zero and hence zero sequence voltage exists. Similarly, sequence components can be written for currents. Denoting three-phase currents by $\overline{I}_a$, $\overline{I}_b$, and $\overline{I}_c$, respectively, the sequence components in matrix form are given below.

$$\begin{bmatrix} \overline{I}_{a0} \\ \overline{I}_{a+} \\ \overline{I}_{a-} \end{bmatrix} = \frac{1}{3} \begin{bmatrix} 1 & 1 & 1 \\ 1 & a & a^2 \\ 1 & a^2 & a \end{bmatrix} \begin{bmatrix} \overline{I}_a \\ \overline{I}_b \\ \overline{I}_c \end{bmatrix} \tag{3.59}$$

Thus,

$$\overline{I}_{a0} = \frac{1}{3}(\overline{I}_a + \overline{I}_b + \overline{I}_c)$$

$$\overline{I}_{a+} = \frac{1}{3}(\overline{I}_a + a\overline{I}_b + a^2\overline{I}_c)$$

$$\overline{I}_{a-} = \frac{1}{3}(\overline{I}_a + a^2\overline{I}_b + a\overline{I}_c)$$

In three-phase four-wire system, the sum of line currents is equal to the neutral current $(\overline{I}_n)$. Thus,

$$\overline{I}_n = \overline{I}_a + \overline{I}_b + \overline{I}_c = 3\overline{I}_{a0} \tag{3.60}$$

This neutral current flows in the fourth wire or neutral wire. For a three-phase three-wire system, in which the neutral wire is absent, the zero sequence current is always zero irrespective of unbalance in phase currents. This is illustrated below.

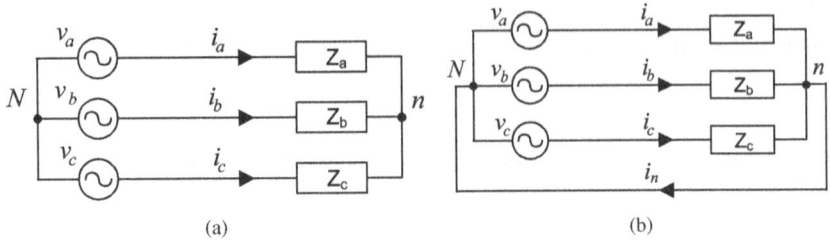

(a)   (b)

Figure 3.8 Various three-phase systems (a) three-phase three-wire system and (b) three-phase four-wire system

In Fig. 3.8(a), there is no neutral current due to the absence of a neutral wire. However, in this configuration, the neutral voltage, V_{Nn}, may or may not be equal to zero depending upon whether the system is balanced or unbalanced. In Fig. 3.8(b), i_n, may or may not be zero depending upon the load currents. However, the neutral voltage (V_{Nn}) between the system and load neutral is always equal to zero.

Example 3.1. Consider a balanced three-phase system with the following phase voltages with respect to system neutral.

$$\overline{V}_a = 100\angle 0° \text{ V}$$
$$\overline{V}_b = 100\angle -120° \text{ V}$$
$$\overline{V}_c = 100\angle 120° \text{ V}$$

Solution: Using (3.58), it can be easily seen that the zero and negative sequence components are zero, indicating that there is no unbalance in voltages. However, the converse may not be true. For the following phase voltages, compute the sequence components and show that the energy associated with the voltage components in both systems remains constant.

$$\overline{V}_a = 100\angle 0° \text{ V}$$
$$\overline{V}_b = 150\angle -100° \text{ V}$$
$$\overline{V}_c = 75\angle 100° \text{ V}$$

Solution: Using (3.58), sequence components are computed as,

$$\overline{V}_{a0} = \frac{1}{3}(\overline{V}_a + \overline{V}_b + \overline{V}_c) = 31.91\angle -50.48° \text{ V}$$

$$\overline{V}_{a+} = \frac{1}{3}(\overline{V}_a + a\overline{V}_b + a^2\overline{V}_c) = 104.16\angle 4.7° \text{ V}$$

$$\overline{V}_{a-} = \frac{1}{3}(\overline{V}_a + a^2\overline{V}_b + a\overline{V}_c) = 28.96\angle 146.33° \text{ V}$$

If you find energy content of two frames that is "abc" and "$0+-$" system, it is found to be constant.

$$E_{abc} = k[V_a^2 + V_b^2 + V_c^2] = 381.25\,k$$
$$E_{0+-} = 3k[V_{a0}^2 + V_{a+}^2 + V_{a-}^2] = 381.25\,k$$

Thus, $E_{abc} = E_{0+-}$ (with k as some constant of proportionality)

3.4.1 POWER INVARIANCE IN SYMMETRICAL COMPONENTS TRANSFORMATION, VECTOR AND ARITHMETIC APPARANT POWERS

The invariance of power can be further shown by the following proof. The vector apparent power can be expressed as the inner product of the voltage vector and current vector conjugate as given in the following:

$$\bar{S}_v = P + jQ = \bar{V}_a \bar{I}_a^* + \bar{V}_b \bar{I}_b^* + \bar{V}_c \bar{I}_c^* = \begin{bmatrix} \bar{V}_a \\ \bar{V}_b \\ \bar{V}_c \end{bmatrix}^T \begin{bmatrix} \bar{I}_a \\ \bar{I}_b \\ \bar{I}_c \end{bmatrix}^*$$

$$= \left([A_{abc}] \begin{bmatrix} \bar{V}_{a0} \\ \bar{V}_{a+} \\ \bar{V}_{a-} \end{bmatrix}\right)^T \left([A_{abc}] \begin{bmatrix} \bar{I}_{a0} \\ \bar{I}_{a+} \\ \bar{I}_{a-} \end{bmatrix}\right)^*$$

$$= \begin{bmatrix} \bar{V}_{a0} \\ \bar{V}_{a+} \\ \bar{V}_{a-} \end{bmatrix}^T [A_{abc}]^T [A_{abc}]^* \begin{bmatrix} \bar{I}_{a0} \\ \bar{I}_{a+} \\ \bar{I}_{a-} \end{bmatrix}^* \tag{3.61}$$

The term $\bar{S}_v$ is referred to as vector or geometric complex apparent power. Further,

$$[A_{abc}]^T [A_{abc}]^* = 3[I] \tag{3.62}$$

The matrix, $[I]$, is identity matrix. Using (3.62), (3.61) can be written as the following.

$$\bar{S}_V = P + jQ = \begin{bmatrix} \bar{V}_{a0} \\ \bar{V}_{a+} \\ \bar{V}_{a-} \end{bmatrix}^T 3[I] \begin{bmatrix} \bar{I}_{a0} \\ \bar{I}_{a+} \\ \bar{I}_{a-} \end{bmatrix}^*$$

$$= 3 \begin{bmatrix} \bar{V}_{a0} \\ \bar{V}_{a+} \\ \bar{V}_{a-} \end{bmatrix}^T \begin{bmatrix} \bar{I}_{a0} \\ \bar{I}_{a+} \\ \bar{I}_{a-} \end{bmatrix}^*$$

$$\bar{S}_v = P + jQ = \bar{V}_a \bar{I}_a^* + \bar{V}_b \bar{I}_b^* + \bar{V}_c \bar{I}_c^*$$
$$= 3[\bar{V}_{a0}\bar{I}_{a0}^* + \bar{V}_{a+}\bar{I}_{a+}^* + \bar{V}_{a-}\bar{I}_{a-}^*] \tag{3.63}$$
$$= \bar{S}_0 + \bar{S}_+ + \bar{S}_-$$

Equation (3.63) indicates that power invariance holds true for both "abc" and "$0+-$" frames of reference. Further, (3.63) implies that,

$$\bar{S}_v = P + jQ = 3[(V_{a0}I_{a0}\cos\phi_{a0} + V_{a+}I_{a+}\cos\phi_{a+} + V_{a-}I_{a-}\cos\phi_{a-})$$
$$+ j(V_{a0}I_{a0}\sin\phi_{a0} + V_{a+}I_{a+}\sin\phi_{a+} + V_{a-}I_{a-}\sin\phi_{a-})] \tag{3.64}$$

The power terms in (3.64) accordingly form positive sequence, negative sequence and zero sequence powers denoted as follows. The positive sequence real power is given as,

$$P_+ = V_{a+}I_{a+}\cos\phi_{a+} + V_{b+}I_{b+}\cos\phi_{b+} + V_{c+}I_{c+}\cos\phi_{c+}$$
$$= 3V_{a+}I_{a+}\cos\phi_{a+} \tag{3.65}$$

Negative sequence real power is expressed as,

$$P_- = 3V_{a-}I_{a-}\cos\phi_{a-} \tag{3.66}$$

The zero sequence real power is,

$$P_0 = 3V_{a0}I_{a0}\cos\phi_{a0}.$$

Similarly, sequence reactive power are denoted by the following expressions.

$$Q_+ = 3V_{a+}I_{a+}\sin\phi_{a+}$$
$$Q_- = 3V_{a-}I_{a-}\sin\phi_{a-}$$
$$Q_0 = 3V_{a0}I_{a0}\sin\phi_{a0}$$

Thus, following holds true for active and reactive powers.

$$P = P_a + P_b + P_c = P_0 + P_+ + P_-$$
$$Q = Q_a + Q_b + Q_c = Q_0 + Q_+ + Q_-$$

Here, positive sequence, negative sequence, and zero sequence apparent powers are denoted as the following.

$$S_+ = |\bar{S}_+| = \sqrt{P_+^2 + Q_+^2} = 3V_{a+}I_{a+}$$
$$S_- = |\bar{S}_-| = \sqrt{P_-^2 + Q_-^2} = 3V_{a-}I_{a-}$$
$$S_0 = |\bar{S}_0| = \sqrt{P_0^2 + Q_0^2} = 3V_{a0}I_{a0} \tag{3.67}$$

The scalar value of vector apparent power $(\bar{S}_v)$ is given as following.

$$S_v = |\bar{S}_v| = |\bar{S}_a + \bar{S}_b + \bar{S}_c| = |\bar{S}_0 + \bar{S}_+ + \bar{S}_-|$$
$$= |(P_a + P_b + P_c) + j(Q_a + Q_b + Q_c)| \tag{3.68}$$
$$= |(P_0 + P_+ + P_-) + j(Q_0 + Q_+ + Q_-)| \tag{3.69}$$
$$= \sqrt{P^2 + Q^2}$$

Similarly, arithmetic apparent power (S_A) is defined as the algebraic sum of each phase or sequence apparent power, i.e.,

$$
\begin{aligned}
S_A &= |\overline{S}_a| + |\overline{S}_b| + |\overline{S}_c| = |\overline{S}_0| + |\overline{S}_+| + |\overline{S}_-| \\
&= |P_a + jQ_a| + |P_b + jQ_b| + |P_c + jQ_c| \\
&= \sqrt{P_a^2 + Q_a^2} + \sqrt{P_b^2 + Q_b^2} + \sqrt{P_c^2 + Q_c^2}
\end{aligned}
\tag{3.70}
$$

The vector (S_v) and arithmetic (S_A) apparent powers in terms of active reactive and

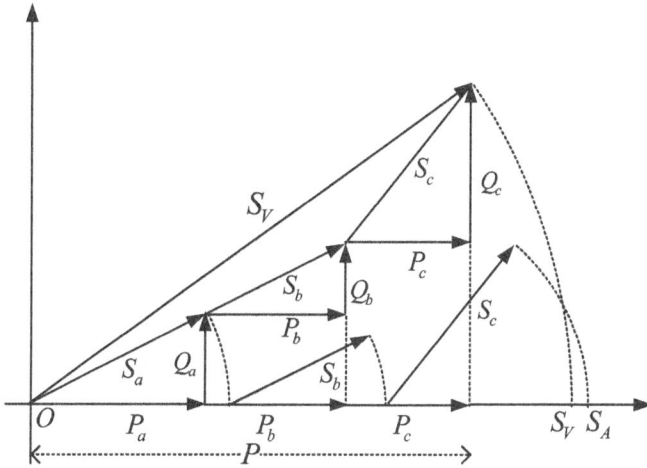

Figure 3.9 Interpretation of arithmetic and vector apparent powers in three-phase system

apparent of each phase are shown in Fig. 3.9. It is evident from the geometry that $S_v \leq S_A$. In terms of sequence component apparent power,

$$
\begin{aligned}
S_A &= |\overline{S}_0| + |\overline{S}_+| + |\overline{S}_-| \\
&= |P_0 + jQ_0| + |P_+ + jQ_+| + |P_- + jQ_-| \\
&= \sqrt{P_0^2 + Q_0^2} + \sqrt{P_+^2 + Q_+^2} + \sqrt{P_-^2 + Q_-^2}
\end{aligned}
\tag{3.71}
$$

Based on these two definitions of the apparent powers, the power factors are defined as the following.

$$
\text{Vector apparent power factor} = pf_v = \frac{P}{S_v}
\tag{3.72}
$$

$$
\text{Arithmetic apparent power factor} = pf_A = \frac{P}{S_A}
\tag{3.73}
$$

Example 3.2. Consider a three-phase four-wire system supplying resistive load between phase-a and phase-b, as shown in Fig. 3.10. Determine the load power and feeder losses. It is given that the voltages $\overline{V}_a$, $\overline{V}_b$ and $\overline{V}_c$ are balanced at the fundamental frequency.

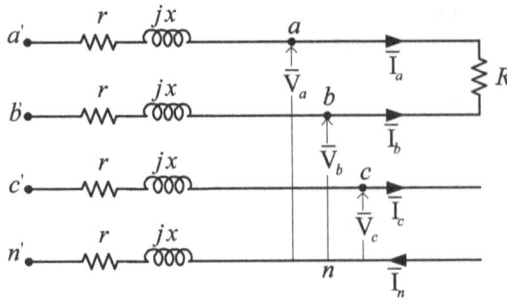

Figure 3.10   A three-phase unbalanced load

Solution:

$$\text{Power dissipated by the load} = \frac{(\sqrt{3}V)^2}{R} = \frac{3V^2}{R}$$

$$\text{The current flowing in the line} = \frac{\sqrt{3}V}{R} = \left|\frac{\overline{V}_a - \overline{V}_b}{R}\right|$$

$$\text{and } \overline{I}_b = -\overline{I}_a$$

$$\text{Therefore losses in the feeder} = \left(\frac{\sqrt{3}V}{R}\right)^2 \times r + \left(\frac{\sqrt{3}V}{R}\right)^2 \times r$$

$$= 2\left(\frac{r}{R}\right)\left(\frac{3V^2}{R}\right) = 2\left(\frac{r}{R}\right)P$$

Now, consider another example of a three-phase balanced system supplying a three-phase load, consisting of three resistors (R) in star as shown in Fig. 3.11. Let us find out the above parameters.

$$\text{Power supplied to load} = 3\left(\frac{V}{R}\right)^2 \times R = \frac{3V^2}{R}$$

$$\text{Losses in the feeder} = 3\left(\frac{V}{R}\right)^2 \times r = \left(\frac{r}{R}\right)\left(\frac{3V^2}{R}\right)$$

Thus, it is interesting to see that the power loss in the unbalanced system is twice that of the balanced circuit for the same output power. This leads to a conclusion that the power factor in phases would become less than unity in unbalanced circuit, while for balanced circuit, the power factor is unity. Power analysis of the unbalanced circuit is shown in Fig. 3.10 is given below.

$$\text{The current in phase-}a, \overline{I}_a = \frac{\overline{V}_a - \overline{V}_b}{R} = \frac{\overline{V}_{ab}}{R} = \frac{\sqrt{3}V_a}{R}\angle 30°$$

$$\text{The current in phase-}b, \overline{I}_b = -\overline{I}_a = \frac{\sqrt{3}V}{R}\angle(30 - 180)°$$

$$= \frac{\sqrt{3}V}{R}\angle - 150°$$

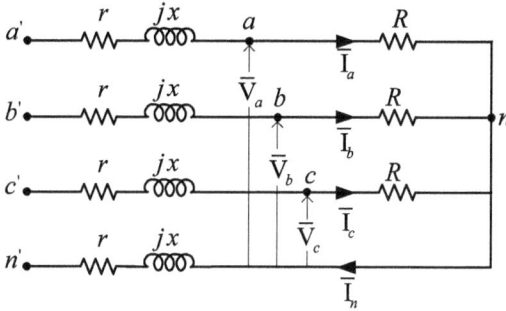

Figure 3.11 A three-phase balanced load

The current in phase-c and neutral are zero, $\bar{I}_c = \bar{I}_n = 0$

The phase voltages are: $\bar{V}_a = V\angle 0°, \bar{V}_b = V\angle -120°, \bar{V}_c = V\angle 120°$ and the rms value of phase-a and phase-b currents are $I = \frac{\sqrt{3}V}{R}$

The phase active and reactive and apparent powers are as follows.

$$\bar{S}_a = \bar{V}_a \bar{I}_a^* = V_a I_a \cos 30° - jV_a I_a \sin 30° = \frac{\sqrt{3}}{2}VI - j\frac{1}{2}VI$$

$$P_a = \frac{\sqrt{3}}{2}VI \text{ and } Q_a = -\frac{1}{2}VI$$

$$\bar{S}_b = \bar{V}_b \bar{I}_b^* = V_b I_b \cos 30° + jV_b I_b \sin 30° = \frac{\sqrt{3}}{2}VI + j\frac{1}{2}VI$$

$$P_b = \frac{\sqrt{3}}{2}VI \text{ and } Q_b = \frac{1}{2}VI$$

$$P_c = Q_c = S_c = 0$$

Total active power $P = P_a + P_b + P_c = 2 \times \frac{\sqrt{3}}{2}VI = \sqrt{3}VI = \frac{3V^2}{R}$

Total reactive power $Q = Q_a + Q_b + Q_c = 0$

The vector apparent power, $S_v = \sqrt{P^2 + Q^2} = 3V^2/R = P$
The arithmetic apparent power, $S_A = S_a + S_b + S_c = 2VI = (2/\sqrt{3})P$

From the values of S_v and S_A, it implies that,

$$pf_v = \frac{P}{S_v} = \frac{P}{P} = 1$$

$$pf_A = \frac{P}{S_A} = \frac{P}{(2/\sqrt{3})P} = \frac{\sqrt{3}}{2} = 0.866$$

What is the nature of the above arithmetic power factor, lagging or leading? This can be understood by looking phase wise power factor. In the considered circuit, power factor of phase-a is 0.866 leading, while for phase-b, it is 0.866 lagging.

For balanced load $S_A = S_v$, therefore, $pf_A = pf_V = 1.0$. Thus, for three-phase electrical system, the following equation holds true.

$$pf_A \leq pf_V \tag{3.74}$$

3.4.2 EFFECTIVE APPARENT POWER

For unbalanced three-phase circuits, there is one more definition of apparent power, which is known as effective apparent power [12], [19]. The concept assumes a virtual balanced circuit that has the same power output and losses as that of the actual unbalanced circuit. This equivalence leads to the definition of effective line current, I_e and effective line to neutral voltage, V_e. The equivalent three-phase unbalanced and balanced circuits with same power output and losses are shown in Fig. 3.12.

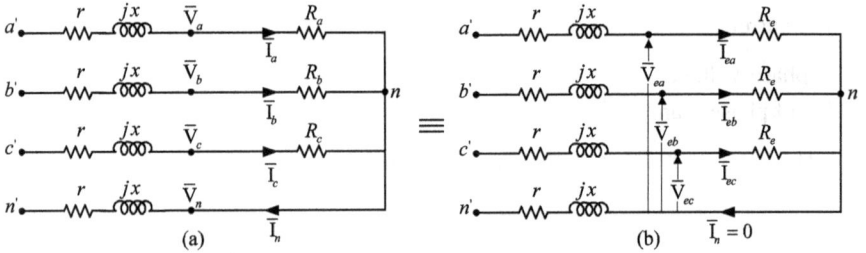

Figure 3.12 (a) Three-phase with unbalanced voltage and currents. (b) Effective equivalent three-phase system

To maintain the same losses for these two systems, as shown in Fig. 3.12,

$$rI_a^2 + rI_b^2 + rI_c^2 + rI_n^2 = 3rI_e^2 \tag{3.75}$$

In the above, $I_e = I_{ea} = I_{eb} = I_{ec}$ as shown in Fig. 3.12(b). The above equation implies that the effective rms current in each phase is given as follows.

$$I_e = \sqrt{\frac{I_a^2 + I_b^2 + I_c^2 + I_n^2}{3}} \tag{3.76}$$

For the original circuit shown in Fig. 3.10, the effective current I_e is computed using the above equation and is given below.

$$I_e = \sqrt{\frac{I_a^2 + I_b^2}{3}} \quad \text{since, } I_c = 0 \quad \text{and } I_n = 0$$

$$= \sqrt{\frac{2I_a^2}{3}} = \sqrt{\frac{2(\sqrt{3}V/R)^2}{3}} = \frac{\sqrt{2}V}{R}$$

To account same power output in circuits shown above, the following identity is used with $R_a = R_b = R_c = R_e = R$ in Fig. 3.12.

$$\frac{V_a^2}{R} + \frac{V_b^2}{R} + \frac{V_c^2}{R} + \frac{V_{ab}^2 + V_{bc}^2 + V_{ca}^2}{3R} = \frac{3V_e^2}{R} + \frac{9V_e^2}{3R} \tag{3.77}$$

From (3.77), the effective rms value of voltage is expressed as,

$$V_e = \sqrt{\frac{1}{18}\{3(V_a^2 + V_b^2 + V_c^2) + V_{ab}^2 + V_{bc}^2 + V_{ca}^2\}} \tag{3.78}$$

Assuming, $3(V_a^2 + V_b^2 + V_c^2) \approx V_{ab}^2 + V_{bc}^2 + V_{ca}^2$, (3.78) can be written as,

$$V_e = \sqrt{\frac{V_a^2 + V_b^2 + V_c^2}{3}} = V \tag{3.79}$$

Therefore, the effective apparent power (S_e), using the values of V_e and I_e, is given by,

$$S_e = 3V_e I_e = \frac{3\sqrt{2}V^2}{R}$$

Thus, the effective power factor based on the definition of effective apparent power (S_e), for the circuit shown in Fig. 3.10 is given by,

$$pf_e = \frac{P}{S_e} = \frac{3V^2/R}{3\sqrt{2}V^2/R} = \frac{1}{\sqrt{2}} = 0.707$$

Thus, we observe that,

$$S_V \leq \qquad S_A \leq \qquad S_e$$
$$pf_e\,(0.707) \leq \quad pf_A\,(0.866) \leq \quad pf_V\,(1.0)$$

When the system is balanced,

$$\begin{aligned}
V_a &= V_b = V_c = V_{en} = V_e \\
I_a &= I_b = I_c = I_e \\
I_n &= 0 \\
\text{and, } S_V &= S_A = S_e
\end{aligned}$$

3.4.3 POSITIVE SEQUENCE AND UNBALANCE POWERS

The unbalance power S_u can be expressed in terms of fundamental positive sequence powers P_+, Q_+ and S_+ as given below.

$$S_u = \sqrt{S_e^2 - S_+^2} \tag{3.80}$$

where $S_+ = 3V_+I_+$ and $S_+^2 = P_+^2 + Q_+^2$

Example 3.3. Consider the following three-phase system. It is given that voltages $\overline{V}_a, \overline{V}_b$ and $\overline{V}_c$ are balanced sinusoids with rms value of 220 V and frequency 50 Hz. The feeder impedance is $r_f + jx_f = 0.02 + j0.1\,\Omega$. The unbalanced load parameters are: $R_L = 12\,\Omega$ and $X_L = 13\,\Omega$. Compute the following.

(a) The currents in each phase, i.e., $\bar{I}_a$, $\bar{I}_b$, and $\bar{I}_c$ and neutral current, $\bar{I}_n$.

(b) Losses in the system.

(c) The active and reactive powers in each phase and total three-phase active and reactive powers.

(d) Arithmetic, vector, and effective apparent powers and power factors based on them.

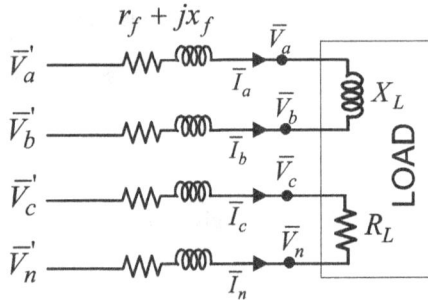

Figure 3.13   An unbalanced three-phase circuit

Solution:

(a) Computation of currents
For the following voltages,

$$\begin{aligned}
v_a(t) &= 220\sqrt{2}\sin(\omega t) \\
v_b(t) &= 220\sqrt{2}\sin(\omega t - 120°) \\
v_c(t) &= 220\sqrt{2}\sin(\omega t + 120°) \\
v_{ab}(t) &= 220\sqrt{6}\sin(\omega t + 30°)
\end{aligned}$$

the line currents are given below.

$$\begin{aligned}
\bar{I}_a &= \frac{220\sqrt{3}\angle 30}{13\angle 90°} = 29.31\angle{-60°}\ \text{A} \\
\bar{I}_b &= -\bar{I}_a = -29.311\angle{-60°} = 29.31\angle 120°\ \text{A} \\
\bar{I}_c &= \frac{220\angle 120°}{12} = 18.33\angle 120°\ \text{A} \\
\bar{I}_n &= -\bar{I}_c = 18.33\angle{-60°}\ \text{A}
\end{aligned}$$

Thus, the instantaneous expressions of phase currents can be given as following.

$$\begin{aligned}
i_a(t) &= 41.45\sin(\omega t - 60°)\,\text{A}\\
i_b(t) &= -i_a(t) = -41.45\sin(\omega t - 60°) = 41.45\sin(\omega t + 120°)\,\text{A}\\
i_c(t) &= 25.93\sin(\omega t + 120°)\,\text{A}\\
i_n(t) &= 25.93\sin(\omega t - 60°)\,\text{A}
\end{aligned}$$

Thus, we observe that due to severe unbalance, the currents, i_a and i_b have phase shift and the neutral current is negative of the phase-c current.

(b) Computation of losses

The losses occur due to resistance of the feeder impedance. These are computed as below.

$$\begin{aligned}
\text{Losses} &= r_f(I_a^2 + I_b^2 + I_c^2 + I_n^2)\\
&= 0.02\,(29.31^2 + 29.31^2 + 18.33^2 + 18.33^2) = 47.80\,\text{W}
\end{aligned}$$

(c) Computation of powers components

Phase-a active and reactive power:

$$\overline{S}_a = \overline{V}_a \overline{I}_a^* = 220\angle 0° \times 29.31\angle 60° = 6448.12\angle 60° = 3224.21 + j5584.49$$
$$P_a = 3224.1\,\text{W},\ Q_a = 5584.30\,\text{VAr}$$

Similarly,

$$\begin{aligned}
\overline{S}_b &= \overline{V}_b \overline{I}_b^* = 220\angle -120° \times 29.31\angle -120°\\
&= 6448.12\angle -240° = -3224.21 + j5584.49\\
P_b &= -3224.1\,\text{W},\ Q_b = 5584.30\,\text{VAr}
\end{aligned}$$

For phase-c,

$$\begin{aligned}
\overline{S}_c &= \overline{V}_c \overline{I}_c^* = 220\angle 120° \times 18.33\angle -120° = 4032.6\angle 0° = 4032.6 + j0\\
P_c &= 4032.6\,\text{W},\ Q_c = 0\,\text{VAr}
\end{aligned}$$

Total three-phase active and reactive powers are given by,

$$\begin{aligned}
P_{3\phi} &= P_a + P_b + P_c = 3224.1 - 3224.1 + 4032.6 = 4032.6\,\text{W}\\
Q_{3\phi} &= Q_a + Q_b + Q_c = 5584.3 + 5584.3 + 0 = 11168.6\,\text{VAr}
\end{aligned}$$

(d) Apparent powers and power factors

The arithmetic, vector, and effective apparent powers are computed as below.

$$\begin{aligned}
S_A &= |\overline{S}_a| + |\overline{S}_b| + |\overline{S}_c| = V_a I_a + V_b I_b + V_c I_c\\
&= 6448.12 + 6448.12 + 4032.6 = 16928.84\,\text{VA}
\end{aligned}$$

$$S_v = |\overline{S_a} + \overline{S_b} + \overline{S_c}|$$
$$= |4032.6 + j11168.6| = |11874.32\angle70.14| = 11874.32\,\text{VA}$$

$$S_e = 3V_e I_e = 3 \times 220 \times \sqrt{\frac{I_a^2 + I_b^2 + I_c^2 + I_n^2}{3}}$$

$$= 3 \times 220 \times \sqrt{\frac{29.31^2 + 29.31^2 + 18.33^2 + 18.33^2}{3}} = 3 \times 220 \times 28.22$$

$$= 18629.19\,\text{VA}$$

Based on the above apparent powers, the arithmetic, vector and effective apparent power factors are computed as below.

$$pf_A = \frac{P_{3\phi}}{S_A} = \frac{4032.6}{16928.84} = 0.2382$$

$$pf_v = \frac{P_{3\phi}}{S_v} = \frac{4032.6}{11874.32} = 0.3396$$

$$pf_e = \frac{P_{3\phi}}{S_e} = \frac{4032.6}{18629.19} = 0.2165$$

In the above computation, the effective voltage and current are found as given in the following:

$$V_e = \sqrt{\frac{V_a^2 + V_b^2 + V_c^2}{3}} = 220\,\text{V}$$

$$I_e = \sqrt{\frac{I_a^2 + I_b^2 + I_c^2 + I_n^2}{3}} = 28.226\,\text{A}$$

Example 3.4. A three-phase, three-wire system is shown in Fig. 3.14. The three-phase voltages are balanced sinusoids with rms value of 230 V. The three-phase loads connected in star are given as follows.

$$Z_a = 5 + j12\,\Omega, \; Z_b = 6 + j8\,\Omega, \text{ and } Z_c = 12 - j5\,\Omega$$

Compute the following.

(a) Line currents, i.e., $\bar{I}_a, \bar{I}_b,$ and $\bar{I}_c$ and their instantaneous expressions.

(b) Load active and reactive powers and power factor of each phase.

(c) Compute various apparent powers and power factors based on them.

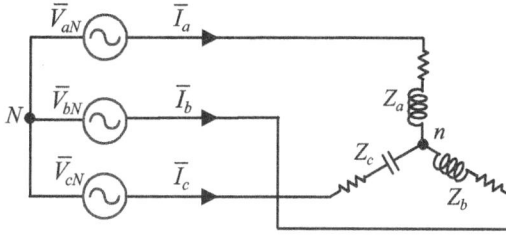

Figure 3.14 A star connected three-phase unbalanced load

Solution:

(a) Computation of currents

Given that $Z_a = 5 + j\,12\,\Omega$, $Z_b = 6 + j\,8\,\Omega$, $Z_c = 12 - j\,5\,\Omega$.

$$
\begin{aligned}
\overline{V}_{aN} &= 230\angle 0^\circ\,\text{V} \\
\overline{V}_{bn} &= 230\angle -120^\circ\,\text{V} \\
\overline{V}_{cN} &= 230\angle 120^\circ\,\text{V}
\end{aligned}
$$

Applying Kirchoff's current law at node n, we get $\overline{I}_a + \overline{I}_b + \overline{I}_c = 0$. Therefore, from the above equation,

$$
\frac{\overline{V}_{aN} - \overline{V}_{nN}}{Z_a} + \frac{\overline{V}_{bN} - \overline{V}_{nN}}{Z_b} + \frac{\overline{V}_{cN} - \overline{V}_{nN}}{Z_c} = 0 \tag{3.81}
$$

Which implies that,

$$
\begin{aligned}
\overline{V}_{nN} &= \frac{1}{\frac{1}{Z_a} + \frac{1}{Z_b} + \frac{1}{Z_c}} \left(\frac{\overline{V}_{aN}}{Z_a} + \frac{\overline{V}_{bN}}{Z_b} + \frac{\overline{V}_{cN}}{Z_c} \right) \\
&= \frac{1}{\frac{1}{5+j12} + \frac{1}{6+j8} + \frac{1}{12-j5}} \left(\frac{230\angle 0^\circ}{5+j12} + \frac{230\angle -120^\circ}{6+8j} + \frac{230\angle 120^\circ}{12-j5} \right) \\
&= \frac{1}{0.2013\angle -37.09^\circ}\, 31.23\angle -164.50^\circ \\
&= -94.22 - j123.18 = 155.09\angle -127.41^\circ\,\text{V}
\end{aligned}
$$

Now the line currents are computed as below.

$$
\overline{I}_a = \frac{\overline{V}_{aN} - \overline{V}_{nN}}{Z_a} = \frac{230\angle 0^\circ - 155.09\angle -127.41^\circ}{5+j12} = 26.67\angle -46.56^\circ\,\text{A}
$$

$$
\overline{I}_b = \frac{\overline{V}_{bN} - \overline{V}_{nN}}{Z_b} = \frac{230\angle -120^\circ - 155.09\angle -127.41^\circ}{6+j8} = 7.88\angle -158.43^\circ\,\text{A}
$$

$$
\overline{I}_c = \frac{\overline{V}_{cN} - \overline{V}_{nN}}{Z_c} = \frac{230\angle 120^\circ - 155.09\angle -127.41^\circ}{12-j5} = 24.85\angle 116.3^\circ\,\text{A}
$$

Thus, the instantaneous expressions of line currents are given as follows.

$$\begin{aligned}
i_a(t) &= 37.72\sin(\omega t - 46.56°) \\
i_b(t) &= 11.14\sin(\omega t - 158.43°) \\
i_c(t) &= 35.14\sin(\omega t + 116.3°)
\end{aligned}$$

(b) Computation of load active and reactive powers

$$\begin{aligned}
S_a &= \overline{V}_{aN}\overline{I}_a^* = 230\angle0° \times 26.67\angle46.56° = 4218.03 + j4456.8 \\
S_b &= \overline{V}_{bN}\overline{I}_b^* = 230\angle-120° \times 7.88\angle158.43° = 1419.82 + j1126.06 \\
S_c &= \overline{V}_{cN}\overline{I}_c^* = 230\angle120° \times 24.85\angle-116.3° = 5703.43 + j368.11
\end{aligned}$$

implies that,

$$\begin{aligned}
P_a &= 4218.03\,\text{W}, \quad Q_a = 4456.8\,\text{VAr} \\
P_b &= 1419.82\,\text{W}, \quad Q_b = 1126.06\,\text{VAr} \\
P_c &= 5703.43\,\text{W}, \quad Q_c = 368.11\,\text{VAr}
\end{aligned}$$

Total three-phase active and reactive powers are given by,

$$\begin{aligned}
P_{3\phi} &= P_a + P_b + P_c = 4218.03 + 1419.82 + 5703.43 = 11341.29\,\text{W} \\
Q_{3\phi} &= Q_a + Q_b + Q_c = 4456.8 + 1126.06 + 368.11 = 5950.99\,\text{VAr}
\end{aligned}$$

The power factors for phases a, b, and c are given as follows.

$$\begin{aligned}
pf_a &= \frac{P_a}{|S_a|} = \frac{4218.03}{\sqrt{4218.03^2 + 4456.8^2}} = \frac{4218.03}{6136.3} = 0.6873\,(\text{lag}) \\
pf_b &= \frac{P_b}{|S_b|} = \frac{1419.82}{1419.82^2 + 1126.06^2} = \frac{1419.82}{1812.16} = 0.7835\,(\text{lag}) \\
pf_c &= \frac{P_c}{|S_c|} = \frac{5703.43}{5703.43^2 + 368.11^2} = \frac{5703.43}{5715.30} = 0.9979\,(\text{lag})
\end{aligned}$$

(c) Computation of apparent powers and power factors

The arithmetic, vector, and effective apparent powers are computed as below.

$$\begin{aligned}
S_A &= |\overline{S}_a| + |\overline{S}_b| + |\overline{S}_c| \\
&= 6136.3 + 1812.16 + 5715.30 = 13663.82\,\text{VA} \\
S_v &= |\overline{S}_a + \overline{S}_b + \overline{S}_c| \\
&= |11341.29 + j5909.92| = 12807.78\,\text{VA} \\
S_e &= 3V_e I_e = 3 \times 230 \times \sqrt{\frac{I_{la}^2 + I_{lb}^2 + I_{lc}^2 + I_{ln}^2}{3}} \\
&= 3 \times 220 \times \sqrt{\frac{26.67^2 + 7.88^2 + 24.85^2 + 0^2}{3}} = 3 \times 230 \times 21.53 \\
&= 14859.7\,\text{VA}
\end{aligned}$$

The arithmetic, vector and effective apparent power factors are computed as below.

$$pf_A = \frac{P_{3\phi}}{S_A} = \frac{11341.29}{13663.82} = 0.8300$$

$$pf_v = \frac{P_{3\phi}}{S_v} = \frac{11341.29}{12807.78} = 0.8855$$

$$pf_e = \frac{P_{3\phi}}{S_e} = \frac{11341.29}{14859.7} = 0.7632$$

Example 3.5. Consider a three-phase balanced system shown in Fig. 3.15. The supply voltages are: $\overline{V}_a' = 230\angle 0°$ V, $\overline{V}_b' = 230\angle -120°$ V, $\overline{V}_c' = 230\angle 120°$ V with balanced load impedances $Z_a = Z_b = Z_c = 14 + j16\,\Omega$.The feeder impedance for all phases is given as $Z_s = 1 + j2\,\Omega$.

(a) Compute the load currents, voltages at the load points, load real powers, reactive powers, and the real and reactive power losses in the system.

(b) Now, an unbalanced load is considered by connecting an impedance between phases a and b for the same supply system and the same output real, reactive powers. Compute the load impedance in this scenario, the various apparent powers, their corresponding power factors, and the losses in the system.

(c) With the same system losses and real power output as in the unbalanced case, find an equivalent three-phase balanced circuit. Compute the apparent power and power factor. Comment on the results obtained.

(d) Will the results in (a)–(c) change if the system were a three-phase four-wire system?

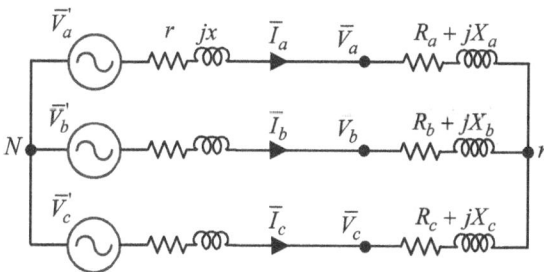

Figure 3.15   A three-phase star connected balanced load

Solution:

(a) The system under consideration is a three-phase three-wire balanced system. Hence the voltage between the two neutral is zero, and the analysis can be done on a per-phase basis. Using the data given, the load currents are computed as follows.

$$\bar{I}_a = \frac{\bar{V}'_a}{Z_s + Z_a} = \frac{230\angle 0°}{1 + j2 + 14 + j16} = 9.2\angle -53.13° \text{ A}$$

$$\bar{I}_b = \frac{\bar{V}'_b}{Z_s + Z_b} = \frac{230\angle -120°}{1 + j2 + 14 + j16} = 9.2\angle -173.13° \text{ A}$$

$$\bar{I}_c = \frac{\bar{V}'_c}{Z_s + Z_c} = \frac{230\angle 120°}{1 + j2 + 14 + j16} = 9.2\angle 66.87° \text{ A}$$

Thus, the load point voltages are computed as,

$$\bar{V}_a = \bar{V}'_a - \bar{I}_a Z_s = 230\angle 0° - 9.2\angle -53.13°(1 + j2) = 209.79\angle -1° \text{ V}$$

$$\bar{V}_b = \bar{V}'_b - \bar{I}_b Z_s = 230\angle -120° - 9.2\angle -173.13°(1 + j2)$$
$$= 209.79\angle -121° \text{ V}$$

$$\bar{V}_c = \bar{V}'_c - \bar{I}_c Z_s = 230\angle 120° - 9.2\angle 66.87°(1 + j2) = 209.79\angle 119° \text{ V}$$

The load apparent powers are calculated as,

$$\begin{aligned}
\bar{S}_a &= \bar{V}_a \bar{I}_a^* = 1185 + j1523.5 \text{ VA} \\
\bar{S}_b &= \bar{V}_b \bar{I}_b^* = 1185 + j1523.5 \text{ VA} \\
\bar{S}_c &= \bar{V}_c \bar{I}_c^* = 1185 + j1523.5 \text{ VA}
\end{aligned}$$

Therefore, the complex load apparent power is given as follows.

$$\bar{S} = \bar{S}_a + \bar{S}_b + \bar{S}_c = 3554.9 + j4570.6 \text{ VA}$$

Hence, $P_o = 3554.9$ W, and $Q_o = 4570.6$ VAr. The power loss in the system is computed as,

$$S_{loss} = I_a^2 Z_s + I_b^2 Z_s + I_c^2 Z_s = 3I_a^2(r + jx) = 253.92 + j507.84 \text{ VA}$$

Thus, $P_{loss} = 253.92$ W, and $Q_{loss} = 507.84$ VAr. The total input power from the source is computed as,

$$\begin{aligned}
\bar{S}_{sa} &= \bar{V}'_a \bar{I}_a^* = 1269.6 + j1692.8 \text{ VA} \\
\bar{S}_{sb} &= \bar{V}'_b \bar{I}_b^* = 1269.6 + j1692.8 \text{ VA} \\
\bar{S}_{sc} &= \bar{V}'_c \bar{I}_c^* = 1269.6 + j1692.8 \text{ VA}
\end{aligned}$$

Therefore, the complex source apparent power is given as follows.

$$\bar{S}_s = \bar{S}_{sa} + \bar{S}_{sb} + \bar{S}_{sc} = 3808.8 + j5078.4 \text{ VA}$$

From the above analysis, we can observe that P_{loss} is 6.66% of the total input real power, and Q_{loss} is 10% of the total input reactive power.

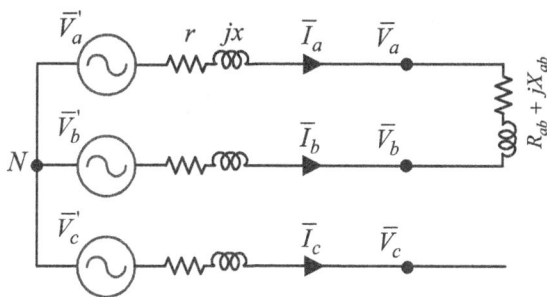

Figure 3.16 A three-phase unbalanced load

(b) Now, an unbalanced circuit of the same power output is considered, as shown in Fig. 3.16. The impedance Z_{ab} is computed as follows.

$$\bar{I}_a = \frac{\bar{V}'_a - \bar{V}'_b}{(R_{ab} + 2r) + j(X_{ab} + 2x)}$$

$$\bar{I}_a = \frac{230\sqrt{3}}{\sqrt{(R_{ab} + 2)^2 + (X_{ab} + 4)^2}}$$

To maintain the same power output, the following two conditions are implemented.

$$P_o = 3554.9 = I_a^2 R_{ab} = \frac{3 \times 230^2}{(R_{ab} + 2)^2 + (X_{ab} + 4)^2} R_{ab}$$

$$Q_o = 4570.6 = I_a^2 X_{ab} = \frac{3 \times 230^2}{(R_{ab} + 2)^2 + (X_{ab} + 4)^2} X_{ab}$$

Dividing P_o/Q_o, we get,

$$\frac{R_{ab}}{X_{ab}} = \frac{3554.9}{4570.6} = 0.7778 \implies R_{ab} = 0.7778 X_{ab}$$

Substituting the above relation in the expression of P_o,

$$3554.9 = \frac{3 \times 230^2 \times 0.7778 X_{ab}}{(0.7778 X_{ab} + 2)^2 + (X_{ab} + 4)^2}$$

$$\implies 1.6049 X_{ab}^2 - 23.6117 X_{ab} + 20 = 0$$

Solving for X_{ab}, we get two solutions, $X_{ab} = 13.81, 0.90$. A higher value of X_{ab} is chosen to have lesser currents. Therefore,

$$Z_{ab} = R_{ab} + jX_{ab} = 10.74 + j13.81 \, \Omega$$

Thus, the load currents are,

$$
\begin{aligned}
\bar{I}_a &= \frac{230\sqrt{3}}{\sqrt{10.74+2)^2+(13.81+4)^2}} = 18.19\angle-24.42°\,\text{A} \\
\bar{I}_b &= -\bar{I}_a = 18.1919\angle155.58°\,\text{A} \\
\bar{I}_c &= 0\,\text{A}
\end{aligned}
$$

The load point voltages are,

$$
\begin{aligned}
\bar{V}_a &= \bar{V}'_a - \bar{I}_a Z_s = 200.04\angle-7.35°\,\text{V} \\
\bar{V}_b &= \bar{V}'_b - \bar{I}_b Z_s = 192.57\angle-115.66°\,\text{V} \\
\bar{V}_c &= 230\angle120°\,\text{V}
\end{aligned}
$$

Load apparent powers are computed as,

$$
\begin{aligned}
\bar{S}_a &= \bar{V}_a \bar{I}_a^* = 3478.9 + j1068\,\text{VA} \\
\bar{S}_b &= \bar{V}_b \bar{I}_b^* = 75.868 + j3502.4\,\text{VA} \\
\bar{S}_c &= \bar{V}_c \bar{I}_c^* = 0\,\text{VA}
\end{aligned}
$$

The arithmetic, vector, and effective apparent powers are computed as,

$$
\begin{aligned}
S_v &= |S_a + S_b + S_c| = 5790.1\,\text{VA} \\
S_A &= |S_a| + |S_b| + |S_c| = 7142.4\,\text{VA} \\
S_e &= 3V_e I_e = 3\sqrt{\frac{V_a^2+V_b^2+V_c^2}{3}}\sqrt{\frac{I_a^2+I_b^2+I_c^2}{3}} = 9276.1\,\text{VA}
\end{aligned}
$$

The corresponding power factors are given as follows.

$$
\begin{aligned}
pf_v &= 0.6139 \\
pf_A &= 0.4977 \\
pf_e &= 0.3832
\end{aligned}
$$

The losses in the system are computed as,

$$
S_{loss} = I_a^2 Z_s + I_b^2 Z_s + I_c^2 Z_s = 661.9 + j1323.8\,\text{VA}
$$

The total input power is calculated as,

$$
\bar{S}_s = \bar{V}'_a \bar{I}_a^* + \bar{V}'_b \bar{I}_b^* + \bar{V}'_c \bar{I}_c^* = 4216.7 + j5894.2\,\text{VA}
$$

Thus, for the unbalanced circuit, it can be observed that P_{loss} is 17.38% of the total input real power, and Q_{loss} is 26.06% of the total input reactive power. This shows that system losses have drastically increased due to the unbalanced load.

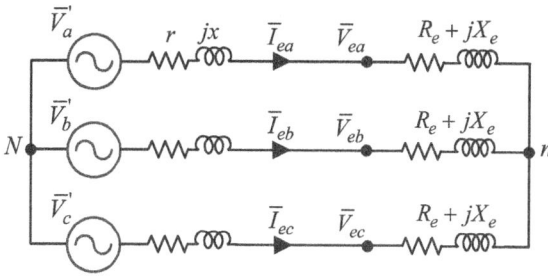

Figure 3.17 Equivalent three-phase star connected balanced load

(c) Now, keeping the same system losses and the same real power output, an equivalent balanced circuit is designed as shown in Fig. 3.17. Keeping the same real and reactive feeder losses, we obtain the load current magnitude for the equivalent balanced case as,

$$P_{loss} = 661.9 = 3I_{ea}^2 r \implies I_{ea} = \sqrt{\frac{661.9}{3 \times 1}} = 14.85\,\text{A}$$

Keeping the load real power output the same, we get

$$P_o = 3554.9 = 3I_{ea}^2 R_e \implies R_e = \frac{3554.9}{3 \times 14.85^2} = 5.37\,\Omega$$

The total real power input from the source will also remain the same as in the case of the unbalanced circuit.

$$P_s = 4216.7 = 3V_a' I_{ea} \cos\phi \implies \cos\phi = \frac{4216.7}{3 \times 230 \times 14.85} = 0.4114$$
$$\implies \phi = 65.71°$$

The power factor is considered to be lagging in this problem which is generally the scenario in the power distribution system, as well as in this case. The effective load reactance is calculated from the power triangle theory as follows.

$$\tan\phi = \frac{X_e + x}{R_e + r} \implies X_e = (R_e + r)\tan\phi - x = 12.11\,\Omega$$

Hence, the effective balanced load impedance is

$$Z_e = R_e + jX_e = 5.37 + j12.11\,\Omega$$

Therefore, the load currents are given as,

$$\bar{I}_{ea} = \frac{\bar{V}'_a}{Z_e + Z_s} = 14.85\angle - 65.71°\,\text{A}$$

$$\bar{I}_{eb} = \frac{\bar{V}'_b}{Z_e + Z_s} = 14.85\angle 174.29°\,\text{A}$$

$$\bar{I}_{ec} = \frac{\bar{V}'_c}{Z_e + Z_s} = 14.85\angle 54.29°\,\text{A}$$

The load point voltages are as follow.

$$\bar{V}_a = \bar{V}'_a - \bar{I}_{ea}Z_s = 196.82\angle 0.38°\,\text{V}$$
$$\bar{V}_b = \bar{V}'_b - \bar{I}_{eb}Z_s = 196.82\angle - 119.62°\,\text{V}$$
$$\bar{V}_c = \bar{V}'_b - \bar{I}_{ec}Z_s = 196.82\angle 120.38°\,\text{V}$$

The load powers are computed as,

$$P_o = 3I_{ea}^2 R_e = 3 \times 14.85^2 \times 5.37 = 3554.9\,\text{W}$$
$$Q_o = 3I_{ea}^2 X_e = 3 \times 14.85^2 \times 12.11 = 8017.98\,\text{VAr}$$

The arithmetic, vector, and effective apparent powers are the same. Thus, the apparent power for the system is calculated as,

$$S = 3V_a I_{ea} = 8770.7\,\text{VA}$$

The load power factor is given as,

$$pf = \frac{3554.9}{2770.7} = 0.4053$$

Thus, from the entire analysis of this example, the following points are observed

1. The power factor of the original balanced circuit is 0.6140. The system losses, in this case, are $P_{loss} = 253.92$ W, and $Q_{loss} = 507.84$ VAr.

2. The power factor of the unbalanced circuit with the same load power output is 0.3822. The reduction in power factor indicates an increase in system losses ($P_{loss} = 661.9$ W, and $Q_{loss} = 1323.8$ VAr) due to the unbalanced load.

3. The power factor of the effective balanced circuit is 0.4053, which is better than the unbalanced circuit owing to its balanced nature, but poorer than the original circuit because of the consideration of the same system losses ($P_{loss} = 661.9$ W, and $Q_{loss} = 1323.8$ VAr) as in the unbalanced case.

4. In the case of the effective balanced circuit, the load output reactive power increases to 8017.98 VAr (from 4570.6 VAr, in case of unbalance). This extra reactive power highlights the fact that to make an equivalent balanced circuit from an unbalanced circuit, extra reactive power needs to be drawn from the source.

(d) The results obtained in (a)–(c) will not change for a three-phase four-wire system. The circuits in (a) and (c) are balanced, hence there will be no current in the neutral wire, leaving the results unchanged. In (b), the load is between phases a and b, and phases c and n are open, hence the presence of neutral does not make any difference in the results.

3.5   THREE-PHASE BALANCED NONSINUSOIDAL SYSTEM

A three-phase balanced nonsinusoidal system is represented by the following set of voltage and current equations.

$$v_a(t) = \sqrt{2}\sum_{n=1}^{\infty} V_n \sin(n\omega t - \alpha_n)$$

$$v_b(t) = \sqrt{2}\sum_{n=1}^{\infty} V_n \sin\{n(\omega t - 120°) - \alpha_n\} \qquad (3.82)$$

$$v_c(t) = \sqrt{2}\sum_{n=1}^{\infty} V_n \sin\{n(\omega t + 120°) - \alpha_n\}$$

Similarly, the line currents can be expressed as,

$$i_a(t) = \sqrt{2}\sum_{n=1}^{\infty} I_n \sin(n\omega t - \beta_n)$$

$$i_b(t) = \sqrt{2}\sum_{n=1}^{\infty} I_n \sin\{n(\omega t - 120°) - \beta_n\} \qquad (3.83)$$

$$i_c(t) = \sqrt{2}\sum_{n=1}^{\infty} I_n \sin\{n(\omega t + 120°) - \beta_n\}$$

In this case,

$$S_a = S_b = S_c$$
$$P_a = P_b = P_c \qquad (3.84)$$
$$Q_a = Q_b = Q_c$$
$$D_a = D_b = D_c$$

In above, the terms D_a, D_b, and D_c are known as distortion powers in phase-a, b, and c, respectively. The definition of distortion power, D, was given in Section 2.5.5 of Chapter 2. The equation (3.85) suggests that such a system has potential to increase the significant rating of the neutral conductor and the corresponding power losses in the system, as illustrated in the following section.

3.5.1 NEUTRAL CURRENT

The neutral current for three-phase balanced system with harmonics can be given by the following equation:

$$
\begin{aligned}
i_n = \quad & i_a+ & i_b+ & & i_c \\
= \quad & \sqrt{2}\,[I_1 \sin(\omega t - \beta_1)+ & I_2 \sin(2\omega t - \beta_2)+ & & I_3 \sin(3\omega t - \beta_3) \\
+ \quad & I_1 \sin(\omega t - 120^\circ - \beta_1)+ & I_2 \sin[2(\omega t - 120^\circ) - \beta_2]+ & & I_3 \sin[3(\omega t - 120^\circ) - \beta_3] \\
+ \quad & I_1 \sin(\omega t + 120^\circ - \beta_1)+ & I_2 \sin[2(\omega t + 120^\circ) - \beta_2]+ & & I_3 \sin[3(\omega t + 120^\circ) - \beta_3] \\
+ \quad & I_4 \sin(4\omega t - \beta_4)+ & I_5 \sin(5\omega t - \beta_5)+ & & I_6 \sin(6\omega t - \beta_6) \\
+ \quad & I_4 \sin[4(\omega t - 120^\circ) - \beta_4]+ & I_5 \sin[5(\omega t - 120^\circ) - \beta_5]+ & & I_6 \sin[6(\omega t - 120^\circ) - \beta_6] \\
+ \quad & I_4 \sin[4(\omega t + 120^\circ) - \beta_4]+ & I_5 \sin[5(\omega t + 120^\circ) - \beta_5]+ & & I_6 \sin[6(\omega t + 120^\circ) - \beta_6] \\
+ \quad & I_7 \sin(7\omega t - \beta_7)+ & I_8 \sin(8\omega t - \beta_8)+ & & I_9 \sin(9\omega t - \beta_9) \\
+ \quad & I_7 \sin[7(\omega t - 120^\circ) - \beta_7]+ & I_8 \sin[8(\omega t - 120^\circ) - \beta_8]+ & & I_9 \sin[9(\omega t - 120^\circ) - \beta_9] \\
+ \quad & I_7 \sin[7(\omega t + 120^\circ) - \beta_7]+ & I_8 \sin[8(\omega t + 120^\circ) - \beta_8]+ & & I_9 \sin[9(\omega t + 120^\circ) - \beta_9]]
\end{aligned}
$$

$$(3.85)$$

From the above equation, we observe that the triplen harmonics are added up in the neutral current. All other harmonics except triplen harmonics do not contribute to the neutral current due to their balanced nature. Therefore the neutral current is given by,

$$
i_n = i_a + i_b + i_c = \sum_{h=3,6,9\ldots}^{\infty} 3\sqrt{2}I_h \sin(h\omega t - \beta_h). \tag{3.86}
$$

The rms value of the current in neutral wire is therefore given by,

$$
I_n = 3 \left[\sum_{h=3,6,9\ldots}^{\infty} I_h^2 \right]^{1/2} \tag{3.87}
$$

Due to dominant triplen harmonics in electrical loads such as Uninterrupted Power Supplies (UPS), rectifiers, inverters, and other power electronic-based loads, the current rating of the neutral wire may be comparable to the phase wires.

It is worth mentioning that all harmonics in three-phase balanced systems can be categorized in three groups, i.e., $(3n + 1)$, $(3n + 2)$ and $3n + 3$ (for $n = 0, 1, 2, 3, \ldots$) called positive, negative, and zero sequence harmonics, respectively. This means that balanced fundamental, $4^{th}, 7^{th}, 10^{th}, \ldots$ form positive sequence only. Balanced $2^{nd}, 5^{th}, 8^{th}, 11^{th}, \ldots$ form negative sequence only and the balanced triplen harmonics, i.e., $3^{rd}, 6^{th}, 9^{th}, \ldots$ form zero sequence only. But in case of unbalanced three-phase systems with harmonics, $(3n + 1)$ components start forming negative and zero sequence components. Similarly, $(3n + 2)$ components start forming positive and zero sequence components, and $3n + 3$ components start forming positive and negative sequence components.

3.5.2 LINE TO LINE VOLTAGE

For the three-phase balanced system with harmonics, the line-to-line voltages are denoted as v_{ab}, v_{bc}, and v_{ca}. Let us consider line-to-line voltage between phases a

and b. It is given as follows.

$$
\begin{aligned}
v_{ab}(t) &= v_a(t) - v_b(t) \\
&= \sum_{n=1}^{\infty} \sqrt{2}V_n \sin(n\omega t - \alpha_n) - \sum_{n=1}^{\infty} \sqrt{2}V_n \sin[n(\omega t - 120°) - \alpha_n] \\
&= \sum_{n=1}^{\infty} \sqrt{2}V_n \sin(n\omega t - \alpha_n) - \sum_{n=1}^{\infty} \sqrt{2}V_n \sin[(n\omega t - \alpha_n) - n \times 120°] \\
&= \sum_{n=1}^{\infty} \sqrt{2}V_n [\sin(n\omega t - \alpha_n) - \sin(n\omega t - \alpha_n)\cos(n \times 120°) \\
&\qquad\qquad\qquad + \cos(n\omega t - \alpha_n)\sin(n \times 120°)] \\
&= \sum_{n \neq 3,6,9...}^{\infty} \sqrt{2}V_n [\sin(n\omega t - \alpha_n) - \sin(n\omega t - \alpha_n)(-1/2) \\
&\qquad\qquad\qquad + \cos(n\omega t - \alpha_n)(\pm\sqrt{3}/2)] \\
&= \sqrt{2} \sum_{n \neq 3,6,9...}^{\infty} V_n \left[(3/2)\sin(n\omega t - \alpha_n) + (\pm\sqrt{3}/2)\cos(n\omega t - \alpha_n)\right] \\
&= \sqrt{3}\sqrt{2} \sum_{n \neq 3,6,9...}^{\infty} V_n \left[(\sqrt{3}/2)\sin(n\omega t - \alpha_n) + (\pm 1/2)\cos(n\omega t - \alpha_n)\right]
\end{aligned}
$$

$$(3.88)$$

In (3.88), subscript n in V_n denotes the harmonic number. Considering $\sqrt{3}/2 = r_n \cos\phi_n$ and $\pm 1/2 = r_n \sin\phi_n$, we get $r_n = 1$ and $\phi_n = \pm 30°$. Thus, $v_{ab}(t)$ in (3.88), can be written as follows.

$$
v_{ab}(t) = \sqrt{3}\sqrt{2} \sum_{n \neq 3,6,9...}^{\infty} V_n [\sin(n\omega t - \alpha_n \pm 30°)]. \qquad (3.89)
$$

In (3.88) and (3.89), $v_{ab} = 0$ for $n = 3, 6, 9, \ldots$ and the $\pm$ sign of 30° changes alternatively with $n = 1, 2, 4, 5, 7$, and so on. Thus, it is observed that triplen harmonics are missing in the line to line voltages, in spite of their presence in phase voltages. In general, the following identity holds true for this system.

$$
V_{LL} \leq \sqrt{3}V_{Ln} \qquad (3.90)
$$

Above equation further implies that,

$$
\sqrt{3}V_{LL}I_L \leq 3V_{Ln}I_L \qquad (3.91)
$$

In above, I_L refers to the rms value of the line current. This implies that line to line kVA is less than single-phase kVA rating due to missing triplen harmonics in line to line voltages. For the above case, $I_a = I_b = I_c = I_L$ and $I_n = 3\sqrt{\sum_{h=3,6,9...}^{\infty} I_h^2}$. Therefore, effective rms current, I_e is given by the following.

$$
\begin{aligned}
I_e &= \sqrt{\frac{3I^2 + 3\sum_{n=3,6,9...}^{\infty} I_n^2}{3}} \\
&= \sqrt{I^2 + \sum_{n=3,6,9...}^{\infty} I_n^2} \geq I
\end{aligned}
\tag{3.92}
$$

The above relation shows that it is better to consider effective rating for design consideration of the conductor used in power system. This accounts the unbalance in supply voltages and the load currents, giving more realistic value of the power factor.

3.5.3 APPARENT POWER WITH BUDEANU RESOLUTION: BALANCED DISTORTION CASE

The apparent power is given as,

$$
\begin{aligned}
S &= 3V_{ln}I = \sqrt{P^2 + Q_B^2 + D_B^2} \\
&= \sqrt{P^2 + Q^2 + D^2}
\end{aligned}
\tag{3.93}
$$

where,

$$
\begin{aligned}
P &= P_1 + P_H = P_1 + P_2 + P_3 + ... \\
&= 3V_1 I_1 \cos\phi_1 + 3\sum_{n=2}^{\infty} V_n I_n \cos\phi_n
\end{aligned}
$$

where, $\phi_n = \beta_n - \alpha_n$. Similarly,

$$
\begin{aligned}
Q &= Q_B = Q_{B1} + Q_{BH} \\
&= Q_1 + Q_H
\end{aligned}
\tag{3.94}
$$

where Q in (3.93) is called Budeanu's reactive power (VAr) or simply reactive power which is detailed below.

$$
\begin{aligned}
Q &= Q_1 + Q_H = Q_1 + Q_2 + Q_3 + ... \\
&= 3V_1 I_1 \sin\phi_1 + 3\sum_{n=2}^{\infty} V_n I_n \sin\phi_n
\end{aligned}
\tag{3.95}
$$

3.5.4 EFFECTIVE APPARENT POWER FOR BALANCED NON-SINUSOIDAL SYSTEM

The effective apparent power S_e for the above system is given by,

$$
S_e = 3V_e I_e
\tag{3.96}
$$

For a three-phase, three-wire balanced system, the effective apparent power is found after calculating effective voltage and current as given below.

$$V_e = \sqrt{\frac{V_{ab}^2 + V_{bc}^2 + V_{ca}^2}{9}} = \frac{V_{eL}}{\sqrt{3}} \tag{3.97}$$

Where V_{eL} is effective line to line voltage. The effective current for the system is given as following.

$$I_e = \sqrt{\frac{I_a^2 + I_b^2 + I_c^2}{3}} = I \tag{3.98}$$

Therefore,

$$S_e = \sqrt{3} V_{eL} I_e \tag{3.99}$$

For a four-wire system, V_e is same is given (3.97) and I_e is given by (3.92). Therefore, the effective apparent power is given below.

$$\sqrt{3} V_L I \le 3 V_e I_e \tag{3.100}$$

The above implies that,

$$S_e \ge S_A \tag{3.101}$$

Therefore, it can be further concluded that,

$$pf_e \,(= P/S_e) \le pf_A \,(= P/S_A) \tag{3.102}$$

Example 3.6. Consider a three-phase four-wire system as shown in Fig. 3.18 with the following balanced supply voltages with harmonics.

$$
\begin{aligned}
v_a &= 230\sqrt{2}\sin\omega t + 80\sqrt{2}\sin 3\omega t + 50\sqrt{2}\sin 5\omega t \text{ V} \\
v_b &= 230\sqrt{2}\sin(\omega t - 120°) + 80\sqrt{2}\sin[3(\omega t - 120°)] \\
&\quad + 50\sqrt{2}\sin[5(\omega t - 120°)] \text{ V} \\
v_c &= 230\sqrt{2}\sin(\omega t + 120°) + 80\sqrt{2}\sin[3(\omega t + 120°)] \\
&\quad + 50\sqrt{2}\sin[5(\omega t + 120°)] \text{ V}
\end{aligned}
$$

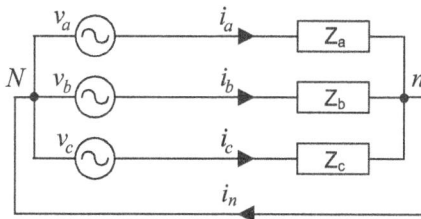

Figure 3.18 A three-phase four-wire system

The load impedance at the fundamental frequency are $Z_a = Z_b = Z_c = 3 + j4\,\Omega$. For this system,

(a) Determine the instantaneous expressions for line currents, i.e., i_a, i_b, i_c.

(b) Determine fundamental active (P_1) and reactive (Q_1) powers and harmonic active (P_H), and reactive (Q_H) powers.

(c) Compute vector, arithmetic, effective apparent powers, and their corresponding power factors.

(d) Find the rating of the neutral conductor.

Solution:

(a) Computation of currents
   It is given $Z_{a1} = Z_{b1} = Z_{c1} = 3 + j4\,\Omega$. We can compute, $Z_{a3} = Z_{b3} = Z_{c3} = 3 + j12\,\Omega$, $Z_{a5} = Z_{b5} = Z_{c5} = 3 + j20\,\Omega$. The line currents are computed below as,

$$\overline{I}_{ah} = \frac{\overline{V}_{ah}}{Z_{ah}}$$

$$\overline{I}_{bh} = \frac{\overline{V}_{bh}}{Z_{bh}}$$

$$\overline{I}_{ch} = \frac{\overline{V}_{ch}}{Z_{ch}}$$

Hence for $h = 1$,

$$\overline{I}_{a1} = \frac{\overline{V}_{a1}}{Z_{a1}} = \frac{230\angle 0°}{3 + j4} = 46\angle - 53.13°\,\text{A}$$

$$\overline{I}_{b1} = \frac{\overline{V}_{b1}}{Z_{b1}} = \frac{230\angle - 120°}{3 + j4} = 46\angle - 173.13°\,\text{A}$$

$$\overline{I}_{c1} = \frac{\overline{V}_{c1}}{Z_{c1}} = \frac{230\angle 120°}{3 + j4} = 46\angle 66.87°\,\text{A}$$

$$\overline{I}_{n1} = \overline{I}_{a1} + \overline{I}_{b1} + \overline{I}_{c1} = 0\,\text{A}$$

For $h = 3$,

$$\overline{I}_{a3} = \frac{\overline{V}_{a3}}{Z_{a3}} = \frac{80\angle 0°}{3 + j12} = 6.47\angle - 75.96°\,\text{A}$$

$$\overline{I}_{b3} = \frac{\overline{V}_{b3}}{Z_{b3}} = \frac{80\angle - (3 \times 120)°}{3 + j12} = 6.47\angle - 75.96°\,\text{A}$$

$$\overline{I}_{c3} = \frac{\overline{V}_{c3}}{Z_{c3}} = \frac{80\angle (3 \times 120)°}{3 + j12} = 6.47\angle - 75.96°\,\text{A}$$

$$\overline{I}_{n3} = \overline{I}_{a3} + \overline{I}_{b3} + \overline{I}_{c3} = 19.4\angle - 75.96°\,\text{A}$$

For $h = 5$,

$$\bar{I}_{a5} = \frac{\bar{V}_{a5}}{Z_{a5}} = \frac{50\angle 0°}{3 + j20} = 2.47\angle -81.47° \text{ A}$$

$$\bar{I}_{b5} = \frac{\bar{V}_{b5}}{Z_{b5}} = \frac{50\angle -(5 \times 120)°}{3 + j20} = 2.47\angle 38.53° \text{ A}$$

$$\bar{I}_{c5} = \frac{\bar{V}_{c5}}{Z_{c5}} = \frac{50\angle(5 \times 120)°}{3 + j20} = 2.47\angle 158.53° \text{ A}$$

$$\bar{I}_{n5} = \bar{I}_{a5} + \bar{I}_{b5} + \bar{I}_{c5} = 0 \text{ A}$$

Thus, the instantaneous expressions of line currents are given as follows.

$$
\begin{aligned}
i_a(t) &= 46\sqrt{2}\sin(\omega t - 53.13°) + 6.47\sqrt{2}\sin(3\omega t - 75.96°) \\
&\quad + 2.47\sqrt{2}\sin(5\omega t - 81.46°) \\
i_b(t) &= 46\sqrt{2}\sin(\omega t - 173.13°) + 6.47\sqrt{2}\sin(3\omega t - 75.96°) \\
&\quad + 2.47\sqrt{2}\sin(5\omega t + 38.53°) \\
i_c(t) &= 46\sqrt{2}\sin(\omega t + 66.86°) + 6.47\sqrt{2}\sin(3\omega t - 75.96°) \\
&\quad + 2.47\sqrt{2}\sin(5\omega t + 158.53°)
\end{aligned}
$$

(b) Computation of load active and reactive powers

For $h = 1$,

$$
\begin{aligned}
\bar{S}_{a1} &= \bar{S}_{b1} = \bar{S}_{c1} = \bar{V}_{a1}\bar{I}_{a1}^* = 230\angle 0° \times 46\angle 53.13° \\
&= 6348 + j8464 \text{ VA}
\end{aligned}
$$

Thus,

$$
\begin{aligned}
P_{a1} &= P_{b1} = P_{c1} = 6348 \text{ W} \\
Q_{a1} &= Q_{b1} = Q_{c1} = 8464 \text{ VAr}
\end{aligned}
$$

Total three-phase fundamental active and reactive powers are given by,

$$
\begin{aligned}
P_1 &= P_{a1} + P_{b1} + P_{c1} = 3 \times 6348 - 19044 \text{ W} \\
Q_1 &= Q_{a1} + Q_{b1} + Q_{c1} = 3 \times 8464 = 25392 \text{ VAr}
\end{aligned}
$$

For $h = 3, 5$,

$$
\begin{aligned}
S_{a35} &= S_{b35} = S_{c35} = \bar{V}_{a3}\bar{I}_{a3}^* + \bar{V}_{a5}\bar{I}_{a5}^* \\
&= 80\angle 0° \times 6.47\angle 75.96° + 50\angle 0° \times 2.47\angle 81.47° \\
&= 143.83 + j624.2 \text{ VA}
\end{aligned}
$$

This gives,

$$
\begin{aligned}
P_{a35} &= P_{b35} = P_{c35} = 143.83 \text{ W} \\
Q_{a35} &= Q_{b35} = Q_{c35} = 624.2 \text{ VAr}
\end{aligned}
$$

Total three-phase harmonic active and reactive powers are given by,

$$
\begin{aligned}
P_h &= P_{a35} + P_{b35} + P_{c35} = 3 \times 143.83 = 431.49\,\text{W} \\
Q_h &= Q_{a35} + Q_{b35} + Q_{c35} = 3 \times 624.2 = 1872.6\,\text{VAr}
\end{aligned}
$$

The powers in phases are given as follows.

$$
\begin{aligned}
P_a = P_b = P_c &= P_{a1} + P_{a35} = 6491.8\,\text{W} \\
Q_a = Q_b = Q_c &= Q_{a1} + Q_{a35} = 9088.2\,\text{VAr}
\end{aligned}
$$

RMS voltage of each phase

$$
\begin{aligned}
V_a = V_b = V_c &= \sqrt{V_{a1}^2 + V_{a3}^2 + V_{a5}^2} \\
&= \sqrt{230^2 + 80^2 + 50^2} = 248.59\,\text{V}
\end{aligned}
$$

RMS current of each phase

$$
\begin{aligned}
I_{la} = I_{lb} = I_{lc} &= \sqrt{I_{la1}^2 + I_{la3}^2 + I_{la5}^2} \\
&= \sqrt{46^2 + 6.47^2 + 2.47^2} = 46.52\,\text{A}
\end{aligned}
$$

The power factors for phases a, b, and c are as follows.

$$
\begin{aligned}
pf_a = pf_b = pf_c &= \frac{P_a}{|S_a|} \\
&= \frac{6491.8}{248.59 \times 46.52} = 0.5614\,(\text{lag})
\end{aligned}
$$

(c) Computation of apparent powers and power factors
Distortion power is calculated as below.

$$
\begin{aligned}
D_a = D_b = D_c &= \sqrt{|\overline{S_a}|^2 - P_a^2 - Q_a^2} \\
&= \sqrt{11564^2 - 6491.8^2 - 9088.2^2} = 2998.7
\end{aligned}
$$

The rms value of the neutral current is given as,

$$
I_n = \sqrt{I_{n1}^2 + I_{n3}^2 + I_{n5}^2} = \sqrt{0^2 + 19.4^2 + 0^2} = 19.4\,\text{A}
$$

The vector, arithmetic, and effective apparent powers are computed below.

$$
\begin{aligned}
S_v &= \sqrt{(P_a + P_b + P_c)^2 + (Q_a + Q_b + Q_c)^2 + (D_a + D_b + D_c)^2} = 34693\,\text{VA} \\
S_A &= |\overline{S_a}| + |\overline{S_b}| + |\overline{S_c}| = V_a I_{la} + V_b I_{lb} + V_c I_{lc} = 3 \times 11564 = 34693\,\text{VA}
\end{aligned}
$$

$$S_e = 3V_eI_e = 3 \times \sqrt{\frac{V_a^2 + V_b^2 + V_c^2}{3}} \times \sqrt{\frac{I_{la}^2 + I_{lb}^2 + I_{lc}^2 + I_{ln}^2}{3}}$$

$$= 3\sqrt{\frac{248.59^2 + 248.59^2 + 248.59^2}{3}}$$

$$\times \sqrt{\frac{46.52^2 + 46.52^2 + 46.52^2 + 19.4^2}{3}}$$

$$= 3 \times 248.59 \times 46.52 = 35684 \, \text{VA}$$

The vector, arithmetic, and effective apparent power factors are computed as below.

$$pf_v = \frac{P_{3\phi}}{S_v} = \frac{19475}{34693} = 0.5614$$

$$pf_A = \frac{P_{3\phi}}{S_A} = \frac{19475}{34693} = 0.5614$$

$$pf_e = \frac{P_{3\phi}}{S_e} = \frac{19475}{35684} = 0.5458$$

(d) The neutral current rating is 19.4 A.

3.6 THREE-PHASE UNBALANCED AND NON-SINUSOIDAL SYSTEM

In this system, we shall consider the most general case, i.e., three-phase system with voltage and current quantities which are unbalanced and non-sinusoidal. These voltages and currents are expressed as follows.

$$
\begin{aligned}
v_a(t) &= \sum_{n=1}^{\infty} \sqrt{2}V_{an} \sin(n\omega t - \alpha_{an}) \\
v_b(t) &= \sum_{n=1}^{\infty} \sqrt{2}V_{bn} \sin[n(\omega t - 120°) - \alpha_{bn}] \qquad (3.103) \\
v_c(t) &= \sum_{n=1}^{\infty} \sqrt{2}V_{cn} \sin[n(\omega t + 120°) - \alpha_{cn}]
\end{aligned}
$$

Similarly, currents can be expressed as,

$$
\begin{aligned}
i_a(t) &= \sum_{n=1}^{\infty} \sqrt{2}I_{an} \sin(n\omega t - \beta_{an}) \\
i_b(t) &= \sum_{n=1}^{\infty} \sqrt{2}I_{bn} \sin[n(\omega t - 120°) - \beta_{bn}] \qquad (3.104) \\
i_c(t) &= \sum_{n=1}^{\infty} \sqrt{2}I_{cn} \sin[n(\omega t + 120°) - \beta_{cn}]
\end{aligned}
$$

For the above voltages and currents in three-phase system, the instantaneous power is given as follows.

$$
\begin{aligned}
p(t) &= v_a(t)i_a(t) + v_b(t)i_b(t) + v_c(t)i_c(t) \\
&= p_a(t) + p_b(t) + p_c(t) \\
&= 2 \left\{ \sum_{n=1}^{\infty} V_{an} \sin(n\omega t - \alpha_{an}) \right\} \left\{ \sum_{n=1}^{\infty} I_{an} \sin(n\omega t - \beta_{an}) \right\} \\
&+ 2 \left\{ \sum_{n=1}^{\infty} V_{bn} \sin\left[n(\omega t - 120°) - \alpha_{bn}\right] \right\} \left\{ \sum_{n=1}^{\infty} I_{bn} \sin\left[n(\omega t - 120°) - \beta_{bn}\right] \right\} \\
&+ 2 \left\{ \sum_{n=1}^{\infty} V_{cn} \sin\left[n(\omega t + 120°) - \alpha_{cn}\right] \right\} \left\{ \sum_{n=1}^{\infty} I_{cn} \sin\left[n(\omega t + 120°) - \beta_{cn}\right] \right\}
\end{aligned}
\tag{3.105}
$$

In (3.105), each phase power can be found using expressions derived in Section 1.4 of Unit 1. The direct result is written as following.

$$
\begin{aligned}
p_a(t) &= \sum_{n=1}^{\infty} V_{an}I_{an} \cos\phi_{an} \left[1 - \cos 2(n\omega t - \alpha_{an})\right] \\
&- \sum_{n=1}^{\infty} V_{an}I_{an} \sin\phi_{an} \sin 2(n\omega t - \alpha_{an}) \\
&+ \left\{ \sum_{n=1}^{\infty} \sqrt{2}V_{an} \sin(n\omega t - \alpha_{an}) \right\} \left\{ \sum_{m=1, m\neq n}^{\infty} \sqrt{2}I_{am} \sin(m\omega t - \beta_{am}) \right\} \\
&= \sum_{n=1}^{\infty} P_{an} \left[1 - \cos 2(n\omega t - \alpha_{an})\right] - \sum_{n=1}^{\infty} Q_{an} \sin 2(n\omega t - \alpha_{an}) \\
&+ \left\{ \sum_{n=1}^{\infty} \sqrt{2}V_{an} \sin(n\omega t - \alpha_{an}) \right\} \left\{ \sum_{m=1, m\neq n}^{\infty} \sqrt{2}I_{am} \sin(m\omega t - \beta_{am}) \right\}
\end{aligned}
\tag{3.106}
$$

In the above equation, $\phi_{an} = (\beta_{an} - \alpha_{an})$. Similarly, for phases b and c, the instantaneous power is expressed as below.

$$
\begin{aligned}
p_b(t) &= \sum_{n=1}^{\infty} P_{bn} \left\{1 - \cos 2[n(\omega t - 120°) - \alpha_{bn}]\right\} \\
&- \sum_{n=1}^{\infty} Q_{bn} \cos 2[n(\omega t - 120°) - \alpha_{bn}] \\
&+ \left\{ \sum_{n=1}^{\infty} \sqrt{2}V_{bn} \sin\left[n(\omega t - 120°) - \alpha_{bn}\right] \right\} \\
&\times \left\{ \sum_{m=1, m\neq n}^{\infty} \sqrt{2}I_{bm} \sin\left[m(\omega t - 120°) - \beta_{bm}\right] \right\}
\end{aligned}
\tag{3.107}
$$

and,

$$
\begin{aligned}
p_c(t) &= \sum_{n=1}^{\infty} P_{cn} \{1 - \cos 2[n(\omega t + 120°) - \alpha_{cn}]\} \\
&\quad - \sum_{n=1}^{\infty} Q_{cn} \cos 2[n(\omega t + 120°) - \alpha_{cn}] \\
&\quad + \left\{ \sum_{n=1}^{\infty} \sqrt{2} V_{cn} \sin [n(\omega t + 120°) - \alpha_{cn}] \right\} \\
&\quad \times \left\{ \sum_{m=1, m\neq n}^{\infty} \sqrt{2} I_{cm} \sin [m(\omega t + 120°) - \beta_{cm}] \right\} \quad (3.108)
\end{aligned}
$$

From equations (3.106), (3.107), and (3.108), the real powers in three phases are given as follows.

$$
\begin{aligned}
P_a &= \sum_{n=1}^{\infty} V_{an} I_{an} \cos \phi_{an} \\
P_b &= \sum_{n=1}^{\infty} V_{bn} I_{bn} \cos \phi_{bn} \quad (3.109) \\
P_c &= \sum_{n=1}^{\infty} V_{cn} I_{cn} \cos \phi_{cn}
\end{aligned}
$$

Similarly, the reactive powers in three phases are given as follows.

$$
\begin{aligned}
Q_a &= \sum_{n=1}^{\infty} V_{an} I_{an} \sin \phi_{an} \\
Q_b &= \sum_{n=1}^{\infty} V_{bn} I_{bn} \sin \phi_{bn} \quad (3.110) \\
Q_c &= \sum_{n=1}^{\infty} V_{cn} I_{cn} \sin \phi_{cn}
\end{aligned}
$$

Therefore, the total active and reactive powers are computed by summing the phase powers using equations (3.109) and (3.110), which are given below.

$$
\begin{aligned}
P &= P_a + P_b + P_c = \sum_{n=1}^{\infty} (V_{an} I_{an} \cos \phi_{an} + V_{bn} I_{bn} \cos \phi_{bn} + V_{cn} I_{cn} \cos \phi_{cn}) \\
&= V_{a1} I_{a1} \cos \phi_{a1} + V_{b1} I_{b1} \cos \phi_{b1} + V_{c1} I_{c1} \cos \phi_{c1} \\
&\quad + \sum_{n=2}^{\infty} (V_{an} I_{an} \cos \phi_{an} + V_{bn} I_{bn} \cos \phi_{bn} + V_{cn} I_{cn} \cos \phi_{cn}) \\
&= P_{a1} + P_{b1} + P_{c1} + \sum_{n=2}^{\infty} (P_{an} + P_{bn} + P_{cn}) \\
&= P_1 + P_H \quad (3.111)
\end{aligned}
$$

and,

$$
\begin{aligned}
Q &= Q_a + Q_b + Q_c = \sum_{n=1}^{\infty} \left(V_{an} I_{an} \sin \phi_{an} + V_{bn} I_{bn} \sin \phi_{bn} + V_{cn} I_{cn} \sin \phi_{cn} \right) \\
&= V_{a1} I_{a1} \sin \phi_{a1} + V_{b1} I_{b1} \sin \phi_{b1} + V_{c1} I_{c1} \sin \phi_{c1} \\
&\quad + \sum_{n=2}^{\infty} \left(V_{an} I_{an} \sin \phi_{an} + V_{bn} I_{bn} \sin \phi_{bn} + V_{cn} I_{cn} \sin \phi_{cn} \right) \\
&= Q_{a1} + Q_{b1} + Q_{c1} + \sum_{n=2}^{\infty} \left(Q_{an} + Q_{bn} + Q_{cn} \right) \\
&= Q_1 + Q_H
\end{aligned}
\tag{3.112}
$$

3.6.1 ARITHMETIC AND VECTOR APPARENT POWER WITH BUDEANU'S RESOLUTION

Using Budeanu's resolution, the arithmetic apparent power for phase-a, b, and c are expressed as following.

$$
\begin{aligned}
S_a &= \sqrt{P_a^2 + Q_a^2 + D_a^2} = V_a I_a \\
S_b &= \sqrt{P_b^2 + Q_b^2 + D_b^2} = V_b I_b \\
S_c &= \sqrt{P_c^2 + Q_c^2 + D_c^2} = V_c I_c
\end{aligned}
\tag{3.113}
$$

The three-phase arithmetic apparent power is arithmetic sum of S_a, S_b, and S_c in the above equation. This is given below.

$$
S_A = S_a + S_b + S_c \tag{3.114}
$$

The three-phase vector apparent power is given as follows.

$$
S_v = \sqrt{P^2 + Q^2 + D^2} \tag{3.115}
$$

where P and Q are given in (3.111) and (3.112), respectively. The total distortion power D is given as,

$$
D = D_a + D_b + D_c \tag{3.116}
$$

In above equation,

$$
\begin{aligned}
D_a &= \sqrt{S_a^2 - P_a^2 - Q_a^2} \\
D_b &= \sqrt{S_b^2 - P_b^2 - Q_b^2} \\
D_c &= \sqrt{S_c^2 - P_c^2 - Q_c^2}
\end{aligned}
\tag{3.117}
$$

Based on the above definitions of the apparent powers, the arithmetic and vector power factors are given below.

$$pf_A = \frac{P}{S_A}$$

$$pf_v = \frac{P}{S_v} \tag{3.118}$$

From equations (3.114), (3.115), and (3.118), it can be inferred that

$$S_A \geq S_v$$

$$pf_A \leq pf_v \tag{3.119}$$

3.6.2 EFFECTIVE APPARENT POWER

Effective apparent power ($S_e = 3V_e I_e$) for the three-phase unbalanced systems with harmonics can be found by computing V_e and I_e as following. The effective rms current (I_e) can be resolved into two parts i.e., effective fundamental and effective harmonic components as given below.

$$I_e = \sqrt{I_{e1}^2 + I_{eH}^2} \tag{3.120}$$

Similarly,

$$V_e = \sqrt{V_{e1}^2 + V_{eH}^2} \tag{3.121}$$

For three-phase four-wire system,

$$I_e = \sqrt{\frac{I_a^2 + I_b^2 + I_c^2 + I_n^2}{3}} \tag{3.122}$$

$$= \sqrt{\frac{I_{a1}^2 + I_{a2}^2 + \dots + I_{b1}^2 + I_{b2}^2 + \dots + I_{c1}^2 + I_{c2}^2 + \dots + I_{n1}^2 + I_{n2}^2 + \dots}{3}}$$

$$= \sqrt{\frac{I_{a1}^2 + I_{b1}^2 + I_{c1}^2 + I_{n1}^2 + \dots + I_{a2}^2 + I_{b2}^2 + I_{c2}^2 + I_{n2}^2 + \dots}{3}}$$

$$= \left[\frac{I_{a1}^2 + I_{b1}^2 + I_{c1}^2 + I_{n1}^2}{3} \right.$$

$$\left. + \frac{I_{a2}^2 + I_{a3}^2 + \dots + I_{b2}^2 + I_{b3}^2 + \dots + I_{c2}^2 + I_{c3}^2 + \dots + I_{n2}^2 + I_{n3}^2 \dots}{3} \right]^{(0.5)}$$

$$I_e = \sqrt{I_{e1}^2 + I_{eH}^2}$$

In the above equation,

$$I_{e1} = \sqrt{\frac{I_{a1}^2 + I_{b1}^2 + I_{c1}^2 + I_{n1}^2}{3}}$$

$$I_{eH} = \sqrt{\frac{I_{aH}^2 + I_{bH}^2 + I_{cH}^2 + I_{nH}^2}{3}} = \sqrt{\frac{1}{3}\sum_{h=2}^{\infty}(I_{ah}^2 + I_{bh}^2 + I_{ch}^2 + I_{nh}^2)} \quad (3.123)$$

Similarly, the effective rms voltage V_e is given as,

$$V_e = \sqrt{\frac{1}{18}[3(V_a^2 + V_b^2 + V_c^2) + (V_{ab}^2 + V_{bc}^2 + V_{ca}^2)]}$$

$$= \sqrt{V_{e1}^2 + V_{eH}^2} \quad (3.124)$$

where,

$$V_{e1} = \sqrt{\frac{1}{18}[3(V_{a1}^2 + V_{b1}^2 + V_{c1}^2) + (V_{ab1}^2 + V_{bc1}^2 + V_{ca1}^2)]} \approx \sqrt{\frac{V_{a1}^2 + V_{b1}^2 + V_{c1}^2}{3}}$$

$$V_{eH} = \sqrt{\frac{1}{18}[3(V_{aH}^2 + V_{bH}^2 + V_{cH}^2) + (V_{abH}^2 + V_{bcH}^2 + V_{caH}^2)]}$$

$$\approx \sqrt{\frac{V_{aH}^2 + V_{bH}^2 + V_{cH}^2}{3}}$$

$$= \sqrt{\frac{1}{18}\sum_{h=2}^{\infty}[3(V_{ah}^2 + V_{bh}^2 + V_{ch}^2) + (V_{abh}^2 + V_{bch}^2 + V_{cah}^2)]}$$

For three-phase three-wire system, $I_n = I_{n1} = I_{nH} = 0$.

$$I_{e1} = \sqrt{\frac{I_{a1}^2 + I_{b1}^2 + I_{c1}^2}{3}}$$

$$I_{eH} = \sqrt{\frac{I_{aH}^2 + I_{bH}^2 + I_{cH}^2}{3}} \quad (3.125)$$

Similarly,

$$V_{e1} = \sqrt{\frac{V_{ab1}^2 + V_{bc1}^2 + V_{ca1}^2}{9}}$$

$$V_{eH} = \sqrt{\frac{V_{abH}^2 + V_{bcH}^2 + V_{caH}^2}{9}} \quad (3.126)$$

The expression for effective apparent power S_e is given as follows.

$$
\begin{aligned}
S_e &= 3V_e I_e \\
&= 3\sqrt{V_{e1}^2 + V_{eH}^2}\,\sqrt{I_{e1}^2 + I_{eH}^2} \\
&= \sqrt{9V_{e1}^2 I_{e1}^2 + 9V_{e1}^2 I_{eH}^2 + 9V_{eH}^2 I_{e1}^2 + 9V_{eH}^2 I_{eH}^2} \\
&= \sqrt{S_{e1}^2 + S_{eN}^2}
\end{aligned}
\tag{3.127}
$$

In the above equation,

$$
S_{e1} = 3V_{e1} I_{e1}
\tag{3.128}
$$

$$
\begin{aligned}
S_{eN} &= \sqrt{S_e^2 - S_{e1}^2} = \sqrt{D_{eV}^2 + D_{eI}^2 + S_{eH}^2} \\
&= 3\sqrt{I_{e1}^2 V_{eH}^2 + V_{e1}^2 I_{eH}^2 + V_{eH}^2 I_{eH}^2}
\end{aligned}
\tag{3.129}
$$

In (3.128), distortion powers D_{eI}, D_{eV} and harmonic apparent power S_{eH} are given as following.

$$
\begin{aligned}
D_{eI} &= 3V_{e1} I_{eH} \\
D_{eV} &= 3V_{eH} I_{e1} \\
S_{eH} &= 3V_{eH} I_{eH}
\end{aligned}
\tag{3.130}
$$

By defining the above effective voltage and current quantities, the effective total harmonic distortion (THD_e) is expressed in the following.

$$
THD_{eV} = \frac{V_{eH}}{V_{e1}}
$$

$$
THD_{eI} = \frac{I_{eH}}{I_{e1}}
\tag{3.131}
$$

Substituting V_{eH} and I_{eH} in (3.128),

$$
S_{eN} = S_{e1}\sqrt{THD_{eI}^2 + THD_{eV}^2 + THD_{eI}^2 THD_{eV}^2}
\tag{3.132}
$$

In the above equation,

$$
\begin{aligned}
D_{eI} &= S_{e1} THD_{eI} \\
D_{eV} &= S_{e1} THD_{eV} \\
S_{eH} &= S_{e1} THD_{eI} THD_{eV}
\end{aligned}
\tag{3.133}
$$

Using (3.127) and (3.132), the effective apparent power is given by,

$$
S_e = \sqrt{S_{e1}^2 + S_{eN}^2} = S_{e1}\sqrt{1 + THD_{eV}^2 + THD_{eI}^2 + THD_{eV}^2 THD_{eI}^2}
\tag{3.134}
$$

Based on above equation, the effective power factor is given as,

$$pf_e = \frac{P}{S_e} = \frac{P_1 + P_H}{S_{e1}\sqrt{1 + THD_{ev}^2 + THD_{el}^2 + THD_{ev}^2 THD_{el}^2}}$$

$$= \frac{(1 + P_H/P_1)}{\sqrt{1 + THD_{ev}^2 + THD_{el}^2 + THD_{ev}^2 THD_{el}^2}} \frac{P_1}{S_{e1}}$$

$$= \frac{(1 + P_H/P_1)}{\sqrt{1 + THD_{ev}^2 + THD_{el}^2 + THD_{ev}^2 THD_{el}^2}} pf_{e1} \qquad (3.135)$$

Practically, the THDs in voltages are far less than those of currents THDs, therefore $THD_{ev} \ll THD_{el}$. Using this practical constraint and assuming $P_H \ll P_1$, the above equation can be simplified to,

$$pf_e \approx \frac{pf_{e1}}{\sqrt{1 + THD_{el}^2}} \qquad (3.136)$$

In the above context, there is one more useful term to denote unbalance in the system. The term is defined as fundamental unbalanced power, as given below.

$$S_{U1} = \sqrt{S_{e1}^2 - (S_1^+)^2} \qquad (3.137)$$

where, S_1^+ is the fundamental positive sequence apparent power, which is given below.

$$S_1^+ = \sqrt{(P_1^+)^2 + (Q_1^+)^2} \qquad (3.138)$$

In above, $P_1^+ = 3V_1^+ I_1^+ \cos\phi_1^+$ and $Q_1^+ = 3V_1^+ I_1^+ \sin\phi_1^+$. Fundamental positive sequence power factor can thus be expressed as a ratio of P_1^+ and S_1^+ as given below.

$$P_{f1}^+ = \frac{P_1^+}{S_1^+} \qquad (3.139)$$

Example 3.7. Consider a three-phase four-wire system, as shown in Fig. 3.18, with the following unbalanced supply voltages with harmonics.

$$
\begin{aligned}
v_a &= 200\sqrt{2}\sin\omega t + 50\sqrt{2}\sin 3\omega t + 10\sqrt{2}\sin 5\omega t \text{ V} \\
v_b &= 150\sqrt{2}\sin(\omega t - 120°) + 25\sqrt{2}\sin[3(\omega t - 120°)] \\
&\quad + 8\sqrt{2}\sin(5(\omega t - 120°)) \text{ V} \\
v_c &= 230\sqrt{2}\sin(\omega t + 120°) + 100\sqrt{2}\sin[3(\omega t + 120°)] \\
&\quad + 12\sqrt{2}\sin[5(\omega t + 120°)] \text{ V}
\end{aligned}
$$

The load impedance at the fundamental frequency are $Z_a = 3 + j4\ \Omega$, $Z_b = 8 + j5\ \Omega$, and $Z_c = 12 + j4\ \Omega$. For this system,

(a) Determine the instantaneous expressions for line currents, i.e., i_a, i_b, i_c.

(b) Determine fundamental active (P_1) and reactive (Q_1) powers and harmonic active (P_H) and reactive (Q_H) powers.

(c) Compute vector, arithmetic, effective apparent powers, and their corresponding power factors.

(d) Find the rating of the neutral conductor.

Solution:

(a) Computation of currents

$$\overline{I}_{ah} = \frac{\overline{V}_{ah}}{Z_{ah}}$$

$$\overline{I}_{bh} = \frac{\overline{V}_{bh}}{Z_{bh}}$$

$$\overline{I}_{ch} = \frac{\overline{V}_{ch}}{Z_{ch}}$$

For $h = 1$, $Z_{a1} = 3 + j4\,\Omega$, $Z_{b1} = 8 + j5\,\Omega$, $Z_{c1} = 12 + j4\,\Omega$

$$\overline{I}_{a1} = \frac{\overline{V}_{a1}}{Z_{a1}} = \frac{200\angle 0°}{3 + j4} = 40\angle - 53.13° \text{ A}$$

$$\overline{I}_{b1} = \frac{\overline{V}_{b1}}{Z_{b1}} = \frac{150\angle - 120°}{8 + j5} = 15.9\angle - 152° \text{ A}$$

$$\overline{I}_{c1} = \frac{\overline{V}_{c1}}{Z_{c1}} = \frac{230\angle 120°}{12 + j4} = 18.18\angle 101.56° \text{ A}$$

$$\overline{I}_{n1} = \overline{I}_{a1} + \overline{I}_{b1} + \overline{I}_{c1} = 22.55\angle - 73.73° \text{ A}$$

For $h = 3$, $Z_{a1} = 3 + j12\,\Omega$, $Z_{b1} = 8 + j15\,\Omega$, $Z_{c1} = 12 + j12\,\Omega$

$$\overline{I}_{a3} = \frac{\overline{V}_{a3}}{Z_{a3}} = \frac{50\angle 0°}{3 + j12} = 4.04\angle - 75.96° \text{ A}$$

$$\overline{I}_{b3} = \frac{\overline{V}_{b3}}{Z_{b3}} = \frac{25\angle - (3 \times 120)°}{8 + j15} = 1.47\angle - 61.92° \text{ A}$$

$$\overline{I}_{c3} = \frac{\overline{V}_{c3}}{Z_{c3}} = \frac{100\angle (3 \times 120)°}{12 + j12} = 5.89\angle - 45° \text{ A}$$

$$\overline{I}_{n3} = \overline{I}_{a3} + \overline{I}_{b3} + \overline{I}_{c3} = 11.05\angle - 58.11° \text{ A}$$

For $h = 5$, $Z_{a1} = 3 + j20\,\Omega$, $Z_{b1} = 8 + j25\,\Omega$, $Z_{c1} = 12 + j20\,\Omega$

$$\bar{I}_{a5} = \frac{\bar{V}_{a5}}{Z_{a5}} = \frac{10\angle 0°}{3 + j20} = 0.495\angle -81.46°\,\text{A}$$

$$\bar{I}_{b5} = \frac{\bar{V}_{b5}}{Z_{b5}} = \frac{8\angle -(5 \times 120)°}{8 + j25} = 0.305\angle 47.74°\,\text{A}$$

$$\bar{I}_{c5} = \frac{\bar{V}_{c5}}{Z_{c5}} = \frac{12\angle (5 \times 120)°}{12 + j20} = 0.515\angle -179.04°\,\text{A}$$

$$\bar{I}_{n5} = \bar{I}_{a5} + \bar{I}_{b5} + \bar{I}_{c5} = 0.3602\angle -130.95°\,\text{A}$$

Thus, the instantaneous expressions of line currents are given as follows.

$$
\begin{aligned}
i_a(t) &= 40\sqrt{2}\sin(\omega t - 53.13°) + 4.04\sqrt{2}\sin(3\omega t - 75.96°) \\
&\quad + 0.495\sqrt{2}\sin(5\omega t - 81.46°) \\
i_b(t) &= 15.9\sqrt{2}\sin(\omega t - 152°) + 1.47\sqrt{2}\sin(3\omega t - 61.92°) \\
&\quad + 0.305\sqrt{2}\sin(5\omega t + 47.74°) \\
i_c(t) &= 18.18\sqrt{2}\sin(\omega t + 101.56°) + 5.89\sqrt{2}\sin(3\omega t - 45°) \\
&\quad + 0.515\sqrt{2}\sin(5\omega t - 179.04°)
\end{aligned}
$$

(b) Computation of load active and reactive powers

For $h = 1$,

$$
\begin{aligned}
S_{a1} &= \bar{V}_{a1}\bar{I}_{a1}^* = 200\angle 0° \times 40\angle 53.13° = 4800 + j6400\,\text{VA} \\
S_{b1} &= \bar{V}_{b1}\bar{I}_{b1}^* = 150\angle -120° \times 15.9\angle 152° = 2022.5 + j1264\,\text{VA} \\
S_{c1} &= \bar{V}_{c1}\bar{I}_{c1}^* = 230\angle 120° \times 18.18\angle -101.56° = 3967.5 + j1322.5\,\text{VA}
\end{aligned}
$$

Thus,

$$
\begin{aligned}
P_{a1} &= 4800\,\text{W}, \quad Q_{a1} = 6400\,\text{VAr} \\
P_{b1} &= 2022.5\,\text{W}, \quad Q_{b1} = 1264\,\text{VAr} \\
P_{c1} &= 3967.5\,\text{W}, \quad Q_{c1} = 1322.5\,\text{VAr}
\end{aligned}
$$

Total three-phase fundamental active and reactive powers are given by,

$$
\begin{aligned}
P_1 &= P_{a1} + P_{b1} + P_{c1} = 4800 + 2022.5 + 3967.5 = 10790\,\text{W} \\
Q_1 &= Q_{a1} + Q_{b1} + Q_{c1} = 6400 + 1264 + 1322.5 = 8986.5\,\text{VAr}
\end{aligned}
$$

For $h = 3, 5$,

$$
\begin{aligned}
S_{a35} &= \bar{V}_{a3}\bar{I}_{a3}^* + \bar{V}_{a5}\bar{I}_{a5}^* \\
&= 50\angle 0° \times 4.04\angle 75.96° + 10\angle 0° \times 0.495\angle 81.46° \\
&= 49.75 + j200.97
\end{aligned}
$$

$$S_{b35} = \overline{V}_{b3}\overline{I}_{b3}^* + \overline{V}_{b5}\overline{I}_{b5}^*$$
$$= 25\angle-360° \times 1.47\angle61.92° + 8\angle-600° \times 0.305\angle-47.74°$$
$$= 18.044 + j34.761$$

$$S_{c35} = \overline{V}_{c3}\overline{I}_{c3}^* + \overline{V}_{c5}\overline{I}_{c5}^*$$
$$= 100\angle360° \times 5.89\angle45° + 12\angle600° \times 0.515\angle179.04°$$
$$= 419.84 + j421.96$$

Therefore,

$$P_{a35} = 49.75\,\text{W}, \quad Q_{a35} = 200.97\,\text{VAr}$$
$$P_{b35} = 18.044\,\text{W}, \quad Q_{b35} = 34.761\,\text{VAr}$$
$$P_{c35} = 419.84\,\text{W}, \quad Q_{c35} = 421.96\,\text{VAr}$$

Total three-phase harmonic active and reactive powers are given by,

$$P_h = P_{a35} + P_{b35} + P_{c35} = 49.75 + 18.044 + 419.84 = 487.64\,\text{W}$$
$$Q_h = Q_{a35} + Q_{b35} + Q_{c35} = 200.97 + 34.761 + 421.96 = 657.69\,\text{VAr}$$

The powers in phases are given as follow.

$$P_a = P_{a1} + P_{a35} = 4849.8\,\text{W}, Q_a = Q_{a1} + Q_{a35} = 6601\,\text{VAr}$$
$$P_b = P_{b1} + P_{b35} = 2040.5\,\text{W}, Q_b = Q_{b1} + Q_{b35} = 1298.8\,\text{VAr}$$
$$P_c = P_{c1} + P_{c35} = 4387.3\,\text{W}, Q_c = Q_{c1} + Q_{c35} = 1744.5\,\text{VAr}$$

RMS voltage of each phase is as follows.

$$V_a = \sqrt{V_{a1}^2 + V_{a3}^2 + V_{a5}^2} = \sqrt{200^2 + 50^2 + 10^2} = 206.39\,\text{V}$$
$$V_b = \sqrt{V_{b1}^2 + V_{b3}^2 + V_{b5}^2} = \sqrt{150^2 + 25^2 + 8^2} = 152.27\,\text{V}$$
$$V_c = \sqrt{V_{c1}^2 + V_{c3}^2 + V_{c5}^2} = \sqrt{230^2 + 100^2 + 12^2} = 251.08\,\text{V}$$

RMS current of each phase is computed as given below.

$$I_{la} = \sqrt{I_{la1}^2 + I_{la3}^2 + I_{la5}^2} = \sqrt{40^2 + 4.04^2 + 0.495^2} = 40.21\,\text{A}$$
$$I_{lb} = \sqrt{I_{lb1}^2 + I_{lb3}^2 + I_{lb5}^2} = \sqrt{15.9^2 + 1.47^2 + 0.305^2} = 15.97\,\text{A}$$
$$I_{lc} = \sqrt{I_{lc1}^2 + I_{lc3}^2 + I_{lc5}^2} = \sqrt{18.18^2 + 5.89^2 + 0.515^2} = 19.12\,\text{A}$$

The power factors for phases a, b and c are given as follows.

$$pf_a = \frac{P_a}{|\overline{S_a}|} = \frac{4849.8}{206.39 \times 40.21} = \frac{4849.8}{8298.9} = 0.5844\,(\text{lag})$$

$$pf_b = \frac{P_b}{|\overline{S_b}|} = \frac{2040.5}{152.27 \times 15.97} = \frac{2040.5}{2432} = 0.839\,(\text{lag})$$

$$pf_c = \frac{P_c}{|\overline{S_c}|} = \frac{4387.34}{251.08 \times 19.12} = \frac{4387.34}{4801} = 0.914\,(\text{lag})$$

(c) Computation of apparent powers and power factors
Distortion power is calculated as

$$
\begin{aligned}
D_a &= \sqrt{|\overline{S_a}|^2 - P_a^2 - Q_a^2} \\
&= \sqrt{8298.9^2 - 4849.8^2 - 6601^2} \\
&= 1331.8 \\
D_b &= \sqrt{|\overline{S_b}|^2 - P_b^2 - Q_b^2} \\
&= \sqrt{2432^2 - 2040.5^2 - 1298.8^2} \\
&= 253.18 \\
D_c &= \sqrt{|\overline{S_c}|^2 - P_c^2 - Q_c^2} \\
&= \sqrt{4801^2 - 4387.34^2 - 1744.5^2} \\
&= 870.46
\end{aligned}
$$

The rms value of the neutral current is given as,

$$
\begin{aligned}
I_n &= \sqrt{I_{n1}^2 + I_{n3}^2 + I_{n5}^2} \\
&= \sqrt{22.55^2 + 11.05^2 + 0.3602^2} = 25.12\,\text{A}
\end{aligned}
$$

The arithmetic, vector, and effective apparent powers are computed below.

$$S_v = \sqrt{(P_a + P_b + P_c)^2 + (Q_a + Q_b + Q_c)^2 + (D_a + D_b + D_c)^2} = 15041\,\text{VA}$$

$$
\begin{aligned}
S_A &= |\overline{S_a}| + |\overline{S_b}| + |\overline{S_c}| \\
&= 8298.9 + 2432 + 4801 = 15532\,\text{VA}
\end{aligned}
$$

$$S_e = 3V_e I_e = 3 \times \sqrt{\frac{V_a^2 + V_b^2 + V_c^2}{3}} \times \sqrt{\frac{I_{la}^2 + I_{lb}^2 + I_{lc}^2 + I_{ln}^2}{3}}$$

$$
\begin{aligned}
&= 3 \times \sqrt{\frac{206.39^2 + 152.27^2 + 251.08^2}{3}} \\
&\quad \times \sqrt{\frac{40.21^2 + 15.97^2 + 19.12^2 + 25.12^2}{3}} \\
&= 3 \times 207.23 \times 30.92 = 19223\,\text{VA}
\end{aligned}
$$

The vector, arithmetic, and effective apparent power factors are computed as below.

$$pf_v = \frac{P_{3\phi}}{S_v} = \frac{11278}{15041} = 0.7498$$

$$pf_A = \frac{P_{3\phi}}{S_A} = \frac{11278}{15532} = 0.7261$$

$$pf_e = \frac{P_{3\phi}}{S_e} = \frac{11278}{19223} = 0.5867$$

(d) The neutral current rating is 25.12 A

3.7 SUMMARY

In this chapter, the behavior of the three-phase system under different conditions of voltages and currents is analyzed, and related important power terms are explained in detail. The components of three-phase power under balanced, unbalanced, and harmonic conditions are described and illustrated through examples. A component of power, i.e., instantaneous reactive power is introduced using $\alpha\beta0$ transformation, and its relation with Budeanu's reactive power is discussed. An unbalanced three-phase system is described in terms of three balance sets of voltage components, i.e., positive, negative, and zero sequence components, and accordingly, sequence powers are derived using symmetrical components. Based on the definition of apparent power, the expressions for various power factors such as vector, arithmetic, and effective power factors are developed. The equivalence of a three-phase balanced and three-phase unbalanced load based on the same output power and losses in the feeder is explained. From this equivalence, it is inferred that the power factor of a balanced equivalent circuit is greater or equal to the effective power factor, and hence effective power factor is considered for designing the feeder in electrical systems.

3.8 PROBLEMS

P 3.1 List the reasons why three-phase system is preferred over three, single-phase systems?

P 3.2 Why $\alpha\beta0$ transformation is used to define instantaneous reactive power using in Akagi's reactive power definition? Derive the expression for instantaneous reactive power in $\alpha\beta0$ as well as abc systems.

P 3.3 Prove the power and energy invariance in $\alpha\beta0$ and abc systems. Is it necessary while making use of transformations?

P 3.4 What do you understand by unbalance voltages and currents in three-phase systems? Do unbalances exist at the fundamental level as well harmonics level? Give some practical examples of unbalanced systems.

P 3.5 When system voltages have fundamental balanced components and currents have unbalanced fundamental components, derive expression of instantaneous three-phase real and reactive power. How do these expressions differ from the balanced systems?

P 3.6 Using symmetrical transformation, derive expressions for positive, negative, and zero sequence real and reactive powers. Prove that the power remains invariant in the transformation.

P 3.7 Define the following terms.

    (a) Arithmetic apparent power

    (b) Vector apparent power

    (c) Effective apparent power

Based on their definitions, define power factors and comment upon their values using some practical examples. Differentiate between them using two practical considerations. Which power factor definition is more appropriate from a practical point of view to design an electrical line of feeder?

P 3.8 When three-phase voltages and currents both contain unbalanced voltages and currents at fundamental and harmonic levels derive expressions for real power (P), average reactive power (Q), arithmetic (S_A), vector (S_v) and effective apparent (S_e) powers.

P 3.9 Prove that for balanced three-phase systems, the neutral current carries the triplen harmonics and line to line voltage is less than $\sqrt{3}$ times the phase voltage.

P 3.10 Mention the three-phase loads which generate odd harmonics. Why even harmonics are not significantly present in power systems?

P 3.11 Consider a three-phase, four-wire balanced supply with rms voltage of 230 V, 50 Hz, and load impedance of 12+j13 Ω per phase as shown in Fig. 3.19. Determine the following.

    (a) The three-phase instantaneous active power, $p(t)$ and instantaneous reactive power, $q(t)$ as defined by the following equations. $p(t) = v_a i_a + v_b i_b + v_c i_c$
$$q(t) = \frac{1}{\sqrt{3}} \left((i_b - i_c) v_a + (i_c - i_a) v_b + (i_a - i_b) v_c \right)$$

    (b) Verify that the values of $p(t)$ and $q(t)$ are the same as those of average real and reactive powers i.e., $3VI\cos\phi$ and $3VI\sin\phi$.

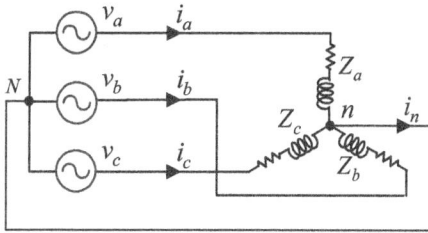

Figure 3.19 Related to problem P 3.11

P 3.12 In the above question, consider now unbalanced load impedances, i.e., $Z_a = 3 + j4$, $Z_b = 12 + j13$, $Z_c = 6 - j8\,\Omega$. Compute the following.

(a) The line currents $(\bar{I}_a, \bar{I}_b, \bar{I}_c)$ and neutral current $\bar{I}_n$.

(b) The real and reactive powers of each phase and the power factors. Also, find out the total active and reactive power.

(c) The arithmetic, vector, and effective apparent powers. Calculate the power factors based on these definitions of the apparent powers. Comment upon the result.

P 3.13 Consider a three-phase three-wire system with load impedances, i.e., $Z_a = 3 + j4$, $Z_b = 12 + j13$, $Z_c = 6 - j8\,\Omega$ supplied by a three-phase balanced voltage source with line to line voltage as 440 V rms, 50 Hz. Compute the following.

(a) The line currents $(\bar{I}_a, \bar{I}_b, \bar{I}_c)$ and their time domain expressions (i_a, i_b, i_c).

(b) The real and reactive powers of each phase and the real and three-phase real and reactive power.

(c) The arithmetic, vector, and effective apparent powers. Calculate the power factors based on these definitions of the apparent powers.

P 3.14 In the above question, consider the source voltages to be unbalanced, i.e., $\bar{V}_a = 230\angle 0°$ V, $\bar{V}_b = 200\angle -150°$ V, $\bar{V}_c = 180\angle 90°$ V, supplying same load impedances. Answer (a) to (c) as in the above question.

P 3.15 For problem P 3.14, calculate the following.

(a) Positive, negative, and zero sequence components of voltages and currents.

(b) Positive, negative, and zero sequence active and reactive powers and power factor for each sequence component.

(c) Verify the values of total active and reactive powers to those obtained in the above problem.

P 3.16 The following three-phase three-wire system as given in Fig. 3.20 shows a balanced star-connected source with rms value 230 V. The delta-connected load containing impedances $Z_{ab} = 12 + j15\,\Omega$, $Z_{bc} = 5 + j4\,\Omega$ and $Z_{ca} = 4 + j10\,\Omega$.

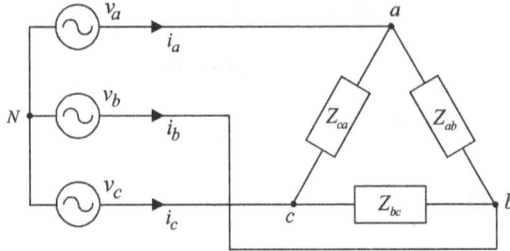

Figure 3.20 Related to problem P 3.16

    (a) Find the line currents $\bar{I}_a, \bar{I}_b, \bar{I}_c$. Write their instantaneous expressions.

    (b) Draw the corresponding phasor diagram of the line and phase current and voltages.

    (c) Calculate active (P) and reactive (Q) powers and power factor angle of the loads connected between points ab, bc, and ca.

    (d) Calculate active (P) and reactive (Q) powers in each phase drawn from the source. Also, calculate the power factor angle for each phase.

    (e) Comment upon the results obtained in (c) and (d). Are these values same? If not, then explain why is it so?

    (f) Calculate the various apparent powers (arithmetic, vector and effective) and the corresponding power factors.

    (g) What can you conclude about the relationship between these apparent powers and their corresponding power factors?

P 3.17 In the above question, if the delta-connected inductive loads are replaced by resistive loads with values $R_{ab} = 7\,\Omega$, $R_{bc} = 4\,\Omega$ and $R_{ca} = 11\,\Omega$. Compute the following.

    (a) The line currents $\bar{I}_a, \bar{I}_b, \bar{I}_c$. Explain why these are not in phase with the respective phase voltages, although the loads are purely resistive.

    (b) Calculate active (P) and reactive (Q) powers in each phase drawn from the source. Also, calculate the power factor for each phase.

    (c) Compute arithmetic, vector, and effective apparent powers and the corresponding power factors.

    (d) Comment upon the total reactive power from the source. Explain why it is equal to zero.

P 3.18 A three-phase, four-wire system supplies an unbalanced load as shown in Fig. 3.21. The voltages $\overline{V}_a, \overline{V}_b$, and $\overline{V}_c$ are balanced (with V as rms value) at fundamental frequency. The impedance $r + jx$, is the feeder impedance. Assume suitable standard, notations along with variables, V, R, and r.

(a) Compute the currents $\overline{I}_a, \overline{I}_b$, and $\overline{I}_c$.

(b) Calculate real, reactive powers, and power factor in each phase.

(c) Calculate total real and reactive powers. Compute vector and arithmetic apparent powers and the corresponding power factors.

(d) Find the equivalent three-phase four-wire system with the same power output. Compare the losses in the two systems.

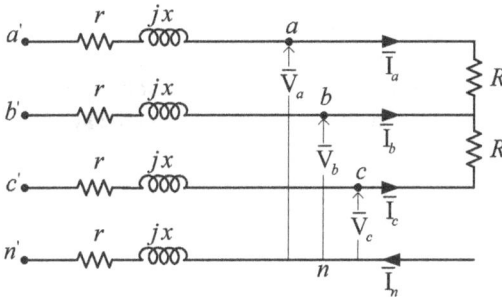

Figure 3.21   Related to problem P 3.18

P 3.19 Consider a three-phase, three-wire system supplied by three-phase balanced voltages with a phase voltage of 230 V (rms), 50 Hz, as shown in Fig. 3.16. The feeder impedance of each line is $Z_s = r_s + jx_s = 1 + j2\Omega$. The load connected between the phases a and b is $Z_{ab} = R_{ab} + jX_{ab}$. The real and reactive powers drawn by the load are 4000 W and 5000 VAr, respectively.

(a) Determine the load impedance Z_{ab} and line currents, $\overline{I}_a, \overline{I}_b, \overline{I}_c$.

(b) Determine the voltages at the load points, $\overline{V}_a, \overline{V}_b, \overline{V}_c$ and compute the real and reactive power losses due to the feeder in the system

(c) Based on the calculation of load voltages and line currents, determine vector, arithmetic, and effective apparent powers and the corresponding power factors.

(d) Find out the three-phase balanced equivalent circuit (i.e., $Z_a = Z_b = Z_c = Z$), with the same real and reactive power output. For this balanced circuit, determine the new real and reactive power losses and compare their values with those in (b). Comment upon the result.

(e) For case (d), determine the vector, arithmetic, and effective apparent powers and the corresponding power factors at the load terminals. Comment upon the result.

(f) Now develop three-phase balanced equivalent circuit with the same feeder losses and real power output as in (a). Compute the vector, arithmetic, effective apparent powers, and the corresponding power factors at the load terminals. Comment upon the result.

P 3.20 Consider a three-phase balanced system shown in Fig. 3.22. The supply voltages are: $\overline{V}_a' = 230\angle 0°$ V, $\overline{V}_b' = 230\angle -120°$ V, $\overline{V}_c' = 230\angle 120°$ V with balance load resistance, $R_a = R_b = R_c = 2\,\Omega$. The output power, P_o is 200 kW, and the losses in the feeder are 5% of the output power. The reactance of the feeder is not considered in the study.

  (a) Determine voltages at the load points, phase currents, and feeder resistance.

  (b) Now, let us say that the load is unbalanced by connecting a resistive load between phases a and b with the same power output. With this configuration, compute active, reactive, various apparent powers and power factors based on them. Compute losses in the system and comment upon the result.

  (c) With the same losses and power output, find an equivalent three-phase balanced circuit and repeat (b). Comment on the result.

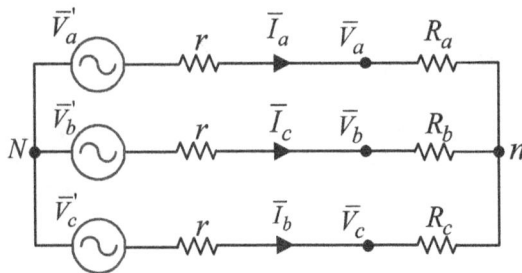

Figure 3.22 Related to problem P 3.20

P 3.21 Consider a three-phase system shown in Fig. 3.23. It is given that voltages $\overline{V}_a$, $\overline{V}_b$ and $\overline{V}_c$ are balanced sinusoids with rms value of 220 V (L-n). The feeder impedance is $r_f + jx_f = 0.02 + j0.1\,\Omega$. The unbalanced load parameters are: $R_L = 12\,\Omega$ and $X_L = 13\,\Omega$.

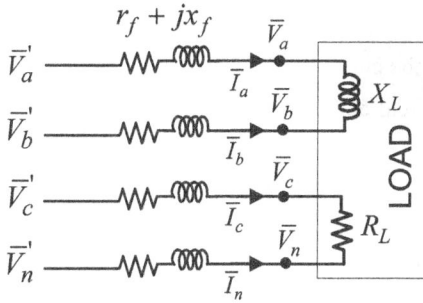

Figure 3.23   Related to problem P 3.21

Compute the following.

(a) The current in each phase $(\bar{I}_a, \bar{I}_b, \bar{I}_c)$ and neutral $(\bar{I}_n)$.

(b) Losses in the system.

(c) The active (P), reactive (Q) power and power factor in each phase as well as total active $(P_{3\phi})$ and reactive $(Q_{3\phi})$ power.

(d) Arithmetic, vector and effective apparent powers (S_A, S_v, S_e) and power factors based on the definitions of these apparent powers.

P 3.22 A three-phase, four-wire system supplies an unbalanced load as shown in Fig. 3.24. The voltages v_a, v_b, and v_c are balanced with V = 230 V rms at the fundamental frequency of 50 Hz. The feeder impedance of each line is, $r_f + jx_f = 0.02 + j0.1\,\Omega$. The load impedances are: $Z_{Lab} = 6 + j8\,\Omega$, $Z_{Lbc} = 3 - j4\,\Omega$ and $Z_{Lca} = 10\,\Omega$. Also, consider the neutral line at zero potential.

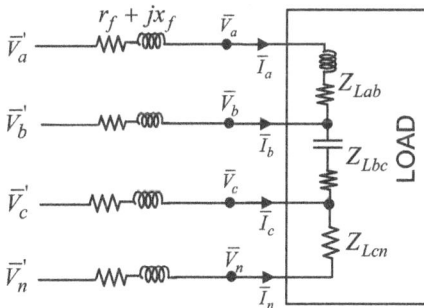

Figure 3.24   Related to problem P 3.22

Compute the following.

(a) Compute the currents $\bar{I}_a$, $\bar{I}_b$, $\bar{I}_c$, and $\bar{I}_n$.

(b) Calculate real and reactive powers in each phase and three-phase at the load bus.

(c) Compute vector and arithmetic and effective apparent powers and corresponding power factors at the load bus.

P 3.23 A three-phase four-wire system has the following balanced supply voltages with harmonics.

$$
\begin{aligned}
v_a &= 230\sqrt{2}\sin\omega t + 120\sqrt{2}\sin(3\omega t - 30°) + 40\sqrt{2}\sin 5\omega t \\
v_b &= 230\sqrt{2}\sin(\omega t - 120°) + 120\sqrt{2}\sin[3(\omega t - 120°) - 30°] \\
&\quad + 40\sqrt{2}\sin[5(\omega t - 120°)] \\
v_c &= 230\sqrt{2}\sin(\omega t + 120°) + 120\sqrt{2}\sin[3(\omega t + 120°) - 30°] \\
&\quad + 40\sqrt{2}\sin 5(\omega t + 120°)
\end{aligned}
$$

The load impedance at the fundamental frequency are $Z_a = Z_b = Z_c = 8 + j5\,\Omega$. For this system,

(a) Determine the instantaneous expressions for line currents, i.e., i_a, i_b, i_c.

(b) Determine fundamental active (P_1) and reactive (Q_1) powers and harmonic active (P_H) and reactive (Q_H) powers.

(c) Find the power factor for each phase and the overall power factor based on vector, arithmetic, and effective apparent powers and the corresponding power factors.

(d) Find the rating of the neutral conductor.

P 3.24 Consider a three-phase four-wire system with the following unbalanced supply voltages with harmonics.

$$
\begin{aligned}
v_a &= 230\sqrt{2}\sin\omega t + 40\sqrt{2}\sin 3\omega t + 10\sqrt{2}\sin 5\omega t \\
v_b &= 200\sqrt{2}\sin(\omega t - 120°) + 20\sqrt{2}\sin 3(\omega t - 120°) \\
&\quad + 8\sqrt{2}\sin 5(\omega t - 120°) \\
v_c &= 250\sqrt{2}\sin(\omega t + 120°) + 50\sqrt{2}\sin 3(\omega t + 120°) \\
&\quad + 12\sqrt{2}\sin 5(\omega t + 120°)
\end{aligned}
$$

The load impedances at the fundamental frequency are $Z_a = 12 + j13\,\Omega$, $Z_b = 4 + j3\,\Omega$, and $Z_c = 6 - j8\,\Omega$. For this system,

(a) Determine the instantaneous expressions for line currents, i.e., i_a, i_b, i_c.

(b) Determine fundamental active (P_1) and reactive (Q_1) powers and harmonic active (P_H) and reactive (Q_H) powers.

(c) Find the power factor for each phase and the overall power factor based on vector, arithmetic, and effective apparent powers and the corresponding power factors.

(d) Find the rating of the neutral conductor.

REFERENCES

1. J. R. Stewart, "History of electrical power Cohoes and Niagara: Mills, canals, and hydropower," *IEEE Power Engineering Review*, vol. 11, no. 6, pp. 31–31, 1991.

2. T. Blalock, "The first polyphase system: a look back at two-phase power for ac distribution," *IEEE Power and Energy Magazine*, vol. 2, no. 2, pp. 63–66, 2004.

3. G. Neidhofer, "Early three-phase power (history)," *IEEE Power and Energy Magazine*, vol. 5, no. 5, pp. 88–100, 2007.

4. J. D. Glover and M. S. Sarma, *Power System Analysis and Design*, 3rd ed. USA: Brooks/Cole Publishing Co., 2001.

5. J. Grainger and W. Stevenson, *Power System Analysis*, ser. Electrical Engineering Series. McGraw-Hill, 1994.

6. W. Stevenson, *Elements of Power System Analysis*, ser. Electrical Power and Energy Series. McGraw-Hill, 1982.

7. C. Wadhwa, *Electrical Power Systems*. New Age International (P) Limited, 2006.

8. A. Ghosh and G. Ledwich, *Power Quality Enhancement Using Custom Power Devices*, ser. Power Electronics and Power Systems. Springer US, 2012.

9. Roger C. Dugan, Mark F. McGranaghan, Surya Santoso, and H. Wayne Beaty, *Electrical Power System Quality*. Tata McGraw-Hill, 2008.

10. M. A. Masoum and E. F. Fuchs, *Chapter 1 – Introduction to Power Quality*, M. A. Masoum and E. F. Fuchs, Eds. Academic Press, 2015.

11. Bhim Singh, Ambrish Chandra, and Kamal Al-Haddad, *Power Quality Problems and Mitigation Techniques*. Wiley, 2015.

12. IEEE Group, "IEEE trial-use standard definitions for the measurement of electric power quantities under sinusoidal, nonsinusoidal, balanced, or unbalanced conditions," 2000.

13. E. Watanabe, R. Stephan, and M. Aredes, "New concepts of instantaneous active and reactive powers in electrical systems with generic loads," *IEEE Transactions on Power Delivery*, vol. 8, no. 2, pp. 697–703, 1993.

14. T. Furuhashi, S. Okuma, and Y. Uchikawa, "A study on the theory of instantaneous reactive power," *IEEE Transactions on Industrial Electronics*, vol. 37, no. 1, pp. 86–90, 1990.

15. A. Ferrero and G. Superti-Furga, "A new approach to the definition of power components in three-phase systems under nonsinusoidal conditions," *IEEE Transactions on Instrumentation and Measurement*, vol. 40, no. 3, pp. 568–577, 1991.

16. J. Willems, "A new interpretation of the akagi-nabae power components for nonsinusoidal three-phase situations," *IEEE Transactions on Instrumentation and Measurement*, vol. 41, no. 4, pp. 523–527, 1992.

17. H. Akagi, Y. Kanazawa, and A. Nabae, "Instantaneous reactive power compensators comprising switching devices without energy storage components," *IEEE Transactions on Industry Applications*, vol. IA-20, no. 3, pp. 625–630, 1984.

18. C. L. Fortesque, "Method of symmetrical co-ordinates applied to the solution of polyphase networks," *AIEE*, 1918.

19. L. Czarnecki, "On some misinterpretations of the instantaneous reactive power p-q theory," *IEEE Transactions on Power Electronics*, vol. 19, no. 3, pp. 828–836, 2004.

4 Theory of Fundamental Load Compensation

4.1 INTRODUCTION

In general, the loads which give rise to poor power factor, unbalance, harmonics, and dc components in currents, require compensation. These loads are arc and induction furnaces, sugar plants, steel rolling mills with high usage of adjustable speed drives, power electronics based devices, large motors with frequent start and stop, etc. All these kinds of problematic loads can be broadly classified into three basic categories.

1. Unbalanced load with fundamental components.

2. Unbalanced load with harmonics (nonlinear loads).

3. Unbalanced load with harmonics and dc components.

The unbalanced loads with fundamental frequency are present due to single-phasing, faults, etc. The unbalanced loads with harmonics, for example, arc furnaces, generate a significant amount of harmonics at the load bus. Other loads which degrade power quality are adjustable speed drives, power electronics based converters such as thyristor controlled drives, rectifiers, cyclo converters, etc. These unbalanced nonlinear loads generate harmonics in addition to fundamental components in voltages and currents. The dc component is normally generated by the usage of half-wave rectifiers. In general, compensation of load depends upon the nature of load, and its requirement of harmonic and reactive powers [1]–[11].

In this chapter, we shall discuss the fundamental load compensation techniques for unbalanced linear loads such as resistance, inductance, capacitance, and their combinations. The objective here will be to maintain source currents balanced sinusoidal and in-phase with their respective phase voltages to achieve a unity power factor. The mitigation techniques for three-phase load with harmonics will be discussed in the next chapter.

4.2 THEORY OF FUNDAMENTAL LOAD COMPENSATION

In this section, a relationship between supply system, the load, and the compensator at fundamental frequency is discussed. We shall start with the principle of power factor correction [12]–[15]. Here, the supply system is modeled as a Thevenin's equivalent circuit with an open circuit voltage and a series impedance across the load. The compensator is modeled as variable impedance or as a variable source (or sink) of reactive current. The modeling and analysis presented here is on the basis of steady state behavior of power circuit using phasors of voltages and currents.

DOI: 10.1201/9781032617305-4

4.2.1 POWER FACTOR CORRECTION

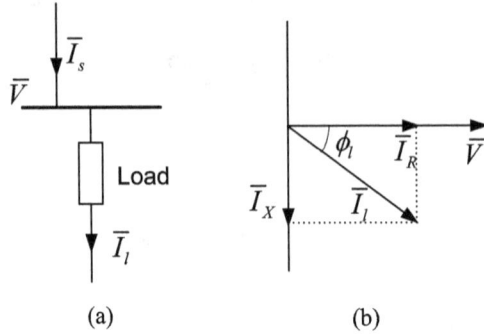

Figure 4.1   (a) Single-line diagram of electrical system and (b) phasor diagram

Consider a single-phase system shown in Fig. 4.1(a). The load admittance is represented by $Y_l = G_l + jB_l$ supplied from a load bus at voltage $\overline{V} = V\angle 0$. The load current, $\overline{I}_l$ is given as,

$$
\begin{aligned}
\overline{I}_l &= \overline{V}(G_l + jB_l) = VG_l + jVB_l \\
&= I_R + jI_X = I_l \cos\phi_l + jI_l \sin\phi_l
\end{aligned} \tag{4.1}
$$

In above equation, ϕ_l denotes the phase angle between the load bus voltage and the load current. The load current has two components, i.e., the resistive or in phase component and reactive component or phase quadrature component and are represented by I_R and I_X, respectively. The current, I_X will lag $90°$ for inductive load and it will lead $90°$ for capacitive load with respect to the reference voltage phasor. This is shown in 4.1(b). The load apparent power can be expressed in terms of bus voltage V and load current I_l as given below.

$$
\begin{aligned}
\overline{S}_l &= \overline{V}\overline{I}_l^* = V(I_R + jI_X)^* = V(I_R - jI_X) \\
&= V(I_l \cos\phi_l - jI_l \sin\phi_l) = VI_l \cos\phi_l - jVI_l \sin\phi_l \\
&= S_l \cos\phi_l - jS_l \sin\phi_l
\end{aligned} \tag{4.2}
$$

From (4.1), $\overline{I}_l = \overline{V}(G_l + jB_l) = VG_l + jVB_l$, (4.2) can also be written as following.

$$
\begin{aligned}
\overline{S}_l &= \overline{V}(\overline{I}_l)^* = V(VG_l + jVB_l)^* = V^2G_l - jV^2B_l \\
&= P_l + jQ_l
\end{aligned} \tag{4.3}
$$

From (4.3), load active power (P_l) and reactive power (Q_l) are given as,

$$
\begin{aligned}
P_l &= V^2G_l = S_l \cos\phi_l \\
Q_l &= -V^2B_l = -S_l \sin\phi_l
\end{aligned} \tag{4.4}
$$

Now, a compensator is connected across the load such that the compensator current, I_γ is equal to $-I_X$, thus,

$$\begin{aligned}
\bar{I}_\gamma &= \bar{V} Y_\gamma = V(G_\gamma + jB_\gamma) = -jI_X \\
&= -jVB_l \tag{4.5}
\end{aligned}$$

The above condition implies that $G_\gamma = 0$ and $B_\gamma = -B_l$. The source current, I_s can be given as,

$$\bar{I}_s = \bar{I}_l + \bar{I}_\gamma = I_R \tag{4.6}$$

Thus, due to the compensator, the source supplies only in-phase component of the load current, which makes the source power factor unity. This reduces the rating of the power conductor and losses due to the feeder impedance. The rating of the compensator is given by the following expression.

$$\begin{aligned}
\bar{S}_\gamma &= P_\gamma + jQ_\gamma = \bar{V}(\bar{I}_\gamma)^* = \bar{V}(-jVB_l)^* \\
&= jV^2 B_l = -jQ_l \tag{4.7}
\end{aligned}$$

The above equation implies that $P_\gamma = 0$ and $Q_\gamma = -Q_l$, indicating that the compensator generates reactive power, which is equal and opposite to the load reactive power, without affecting the real power of the load. This is illustrated in Fig. 4.2. Using (4.2) and (4.7), the compensator rating can further be expressed as,

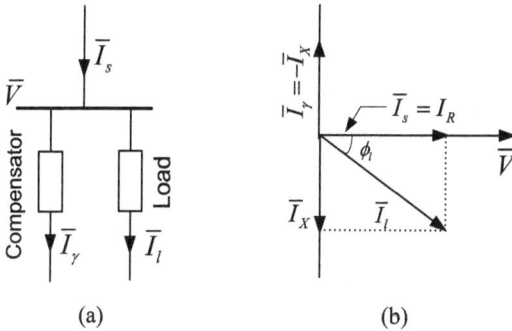

(a)   (b)

Figure 4.2 (a) Single-line diagram of compensated system and (b) phasor diagram

$$Q_\gamma = -Q_l = S_l \sin \phi_l = S_l \sqrt{1 - \cos^2 \phi_l} \text{ VAr} \tag{4.8}$$

The above equation shows that as the power factor varies from unity to lagging or leading, the requirement of compensator reactive power, Q_γ increases. If $|Q_\gamma| < |Q_l|$, then the load is partially compensated. The compensator of fixed admittance will not be able to compensate variations in the reactive power requirement of the load. In practical, however a compensator such as a bank of capacitors can be divided into parallel sections, each section switched separately, so that discrete changes in the reactive power compensation can be made according to the load requirement. Some advanced compensators can be used to provide smooth and dynamic control of reactive power [13]–[15].

Here, the supply voltage is assumed to be constant. In general, if supply voltage varies, the Q_γ will vary with voltage as well as load conditions. In the following discussion, voltage variations are examined, and some additional features of the ideal compensator are studied.

4.2.2 VOLTAGE REGULATION

Voltage regulation can be defined as the proportional change in voltage magnitude at the load bus due to change in load current (say from no load to full load). The voltage drop is caused due to feeder impedance carrying the load current, as illustrated in Fig. 4.3(a). If the supply voltage is represented by Thevenin's equivalent, then the voltage regulation (VR) is given by,

$$VR = \frac{|\overline{E}| - |\overline{V}|}{|\overline{E}|} \tag{4.9}$$

In the above equation, $\overline{V}$ is a reference phasor and $\overline{E}$ is source voltage, which appears as load voltage when the load current is zero. In absence of compensator, the source and load currents are same and the voltage drop due to the feeder is given by,

$$\Delta \overline{V} = \overline{E} - \overline{V} = Z_s \overline{I}_s = Z_s \overline{I}_l \tag{4.10}$$

The feeder impedance, $Z_s = R_s + jX_s$. The relationship between the load powers and its voltage and current is expressed below.

$$\overline{S}_l \quad = \quad \overline{V}(\overline{I}_l)^* = P_l + jQ_l \tag{4.11}$$

Since $\overline{V} = V\angle 0° = V$, the load current is expressed as following.

$$\overline{I}_l \quad = \quad \frac{P_l - jQ_l}{V} \tag{4.12}$$

Substituting, $\overline{I}_l$ from above equation into (4.10), we get

$$\begin{aligned} \Delta \overline{V} = \overline{E} - \overline{V} \quad &= \quad (R_s + jX_s)\left(\frac{P_l - jQ_l}{V}\right) \\ &= \quad \frac{R_s P_l + X_s Q_l}{V} + j\frac{X_s P_l - R_s Q_l}{V} \\ &= \quad \Delta V_R + j\Delta V_X \end{aligned} \tag{4.13}$$

Thus, the voltage drop across the feeder has two components, one in phase (ΔV_R) and another is in phase quadrature (ΔV_X), as illustrated in Fig. 4.3(b). From the above, it is evident that load bus voltage ($\overline{V}$) is dependent on the value of the feeder impedance, magnitude, and phase angle of the load current. In other words, voltage change ($\Delta \overline{V}$)

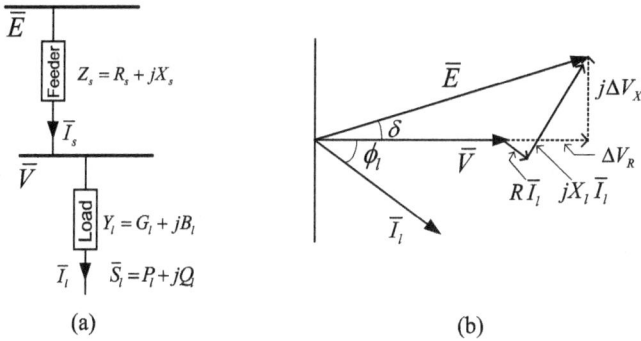

Figure 4.3 (a) Single-phase system with feeder impedance and (b) phasor diagram

depends upon the real and reactive power flow of the load and the value of the feeder impedance.

Now let us add compensator in parallel with the load, as shown in Fig. 4.4(a). The question is, whether it is possible to make $|\overline{E}| = |\overline{V}|$ in order to achieve zero voltage regulation irrespective of change in the load. The answer is yes if the compensator consists of purely reactive components and has enough capacity to supply the required amount of the reactive power. This situation is shown using the phasor diagram in Fig. 4.4(b). The net reactive power at the load bus is now $Q_s = Q_\gamma + Q_l$. The compensator reactive power (Q_γ) has to be adjusted in such a way as to rotate the phasor $\Delta\overline{V}$ until $|\overline{E}| = |\overline{V}|$.

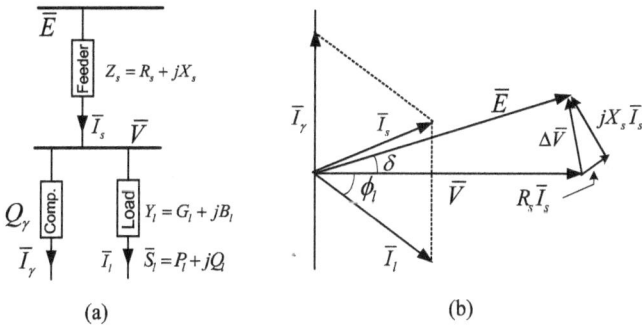

Figure 4.4 (a) Single-line diagram of compensated system with feeder impedance and (b) phasor diagram

The voltage drop, $\Delta\overline{V}$ across the feeder with compensator, is given as,

$$\Delta\overline{V} = \overline{E} - \overline{V} = \frac{R_s P_l + X_s Q_s}{V} + j\frac{X_s P_l - R_s Q_s}{V} \tag{4.14}$$

From (4.14) and Fig. 4.3(b),

$$E\angle\delta = \left(V + \frac{R_s P_l + X_s Q_s}{V}\right) + j\left(\frac{X_s P_l - R_s Q_s}{V}\right) \tag{4.15}$$

The above equation implies that,

$$E^2 = \left(V + \frac{R_s P_l + X_s Q_s}{V}\right)^2 + \left(\frac{X_s P_l - R_s Q_s}{V}\right)^2$$

The above equation can be simplified to,

$$E^2 V^2 = (V^2 + R_s P_l)^2 + X_s^2 Q_s^2 + 2(V^2 + R_s P_l) X_s Q_s$$
$$+ X_s^2 P_l^2 + R_s^2 Q_s^2 - 2X_s P_l R_s Q_s \tag{4.16}$$

The above equation, rearranged in the powers of Q_s, is written as follows.

$$(R_s^2 + X_s^2) Q_s^2 + 2V^2 X_s Q_s + (V^2 + R_s P_l)^2 + (X_s P_l)^2 - E^2 V^2 = 0 \tag{4.17}$$

Thus, the above equation is quadratic in Q_s and can be represented using coefficients of Q_s as given below.

$$a Q_s^2 + b Q_s + c = 0 \tag{4.18}$$

where $a = R_s^2 + X_s^2$, $b = 2V^2 X_s$ and $c = (V^2 + R_s P_l)^2 + X_s^2 P_l^2 - E^2 V^2$.

Thus, the solution to the above equation is as follows.

$$Q_s = \frac{-b \pm \sqrt{(b^2 - 4ac)}}{2a} \tag{4.19}$$

In the actual compensator, this value would be determined automatically by the control loop. The equation also indicates that we can find the value of Q_s by subjecting a condition such as $E = V$ irrespective of the requirement of the load powers (P_l, Q_l). This leads to the following conclusion that a purely reactive compensator can eliminate supply voltage variation caused by changes in both the real and reactive power of the load, provided that there is sufficient range and rate of Q_γ with respect to voltage. This compensator, therefore acts as an ideal voltage regulator. It is mentioned here that we are regulating the magnitude of voltage and not its phase angle. In fact, its phase angle is continuously varying depending on the load current.

    We now look at this aspect from a different perspective. We have seen that compensator can be made to supply all or a part of the load reactive power, which makes it act like a power factor correction device. In this mode, the compensator is designed to compensate the required power factor by providing the net reactive power at the load as $Q_s = Q_l + Q_\gamma$. By expanding (4.17), we obtain the following.

$$(R_s^2 + X_s^2) Q_l^2 + 2V^2 X_s Q_l + V^4 + R_s^2 P_l^2 + 2V^2 R_s P_l + X_s^2 P_l^2 - E^2 V^2 = 0$$

In the above equation, arranging the terms in the power of V^2 terms, we get the following.

$$V^4 + \{2(R_s P_l + X_s Q_l) - E^2\}V^2 + (R_s^2 + X_s^2)(P_l^2 + Q_l^2) = 0 \tag{4.20}$$

The above equation gives the load bus voltage for the given power factor correction. Once the bus voltage is known, the voltage regulation at the load bus can be computed using (4.13). For a unity power factor, $Q_s = Q_l + Q_\gamma = 0$, implying that $Q_\gamma = -Q_l$. Thus, substituting $Q_s = 0$ in (4.20), the following equation is obtained.

$$V^4 + (2R_s P_l - E^2)V^2 + (R_s^2 + X_s^2)P_l^2 = 0 \tag{4.21}$$

The corresponding voltage regulation for a unity power factor at the load bus can be computed by substituting $Q_s = 0$ in (4.14), as given below.

$$\Delta \overline{V} = \frac{(R_s + jX_s)}{V}P_l = (R_s + jX_s)\overline{I}_s = Z_s \overline{I}_s \tag{4.22}$$

In the above equation, note that $\overline{I}_s = \frac{P_l - jQ_s}{V} = \frac{P_l}{V}$, since $Q_s = 0$. Further, it is observed that $\Delta \overline{V}$ is independent of Q_l for this case. Thus, we conclude that a purely reactive compensator cannot maintain both constant voltage and unity power factor simultaneously. Of course, the exception to this rule is a trivial case when $P_l = 0$. It is to be noted that the upstream source powers and net powers at the load bus are related as below.

$$
\begin{aligned}
\overline{S}_s' &= P_s' + jQ_s' = \overline{E}\,\overline{I}_s^* = (\overline{V} + \Delta \overline{V})\overline{I}_s^* = \overline{V}\,\overline{I}_s^* + (\overline{I}_s Z_s)\overline{I}_s^* \\
&= P_s + jQ_s + (R_s + jX_s)I_s^2 = (P_s + I_s^2 R_s) + j(Q_s + I_s^2 X_s)
\end{aligned} \tag{4.23}
$$

The above equation implies that,

$$
\begin{aligned}
P_s' &= P_s + I_s^2 R_s \\
Q_s' &= Q_s + I_s^2 X_s
\end{aligned} \tag{4.24}
$$

In (4.24), P_s and Q_s are the net real and reactive powers at the load bus. From these expressions, it is inferred that for unity power factor operation at the load bus, i.e., $Q_s = 0$ $(Q_\gamma = -Q_l)$, the power factor at the upstream source will not be unity due to the reactive power loss $(I_s^2 X_s)$ in the feeder reactance. Similarly, the real power at the upstream source is the sum of real power at the load bus $(P_s = P_l)$ and power loss $(I_s^2 R_s)$ in the feeder resistance.

4.2.3 AN APPROXIMATION EXPRESSION FOR THE VOLTAGE REGULATION

Consider a supply system with Short Circuit Capacity (S_{sc}) at the load bus, as illustrated in Fig. 4.5. This short circuit capacity can be expressed in terms of short circuit active and reactive powers as given below.

$$\overline{S}_{sc} = P_{sc} + jQ_{sc} = \overline{E}\,\overline{I}_{sc}^* = \overline{E}\left(\frac{\overline{E}}{Z_{sc}}\right)^* = \frac{E^2}{Z_{sc}^*} \tag{4.25}$$

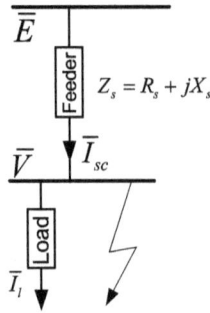

Figure 4.5 System under fault condition

where $Z_{sc} = R_s + jX_s$ and $\bar{I}_{sc}$ is the short circuit current. Assuming $\frac{X}{R}$ ratio as $\tan\phi_{sc}$, i.e., $\frac{X_{sc}}{R_{sc}} = \frac{X_s}{R_s} = \tan\phi_{sc}$, the following expressions can be written.

$$|Z_{sc}| = \frac{E^2}{S_{sc}}$$

$$R_s = \frac{E^2}{S_{sc}}\cos\phi_{sc}$$

$$X_s = \frac{E^2}{S_{sc}}\sin\phi_{sc}$$

Substituting the above values of R_s and X_s in (4.13), the following can be obtained.

$$\frac{\Delta\bar{V}}{V} = \left(\frac{P_l\cos\phi_{sc} + Q_l\sin\phi_{sc}}{V^2} + j\frac{P_l\sin\phi_{sc} - Q_l\cos\phi_{sc}}{V^2}\right)\frac{E^2}{S_{sc}}$$

$$\frac{\Delta\bar{V}}{V} = \frac{\Delta V_R}{V} + j\frac{\Delta V_X}{V} \tag{4.26}$$

Using an approximation that $E \approx V$, the above equation is written as follows.

$$\frac{\Delta\bar{V}}{V} \approx \left(\frac{P_l\cos\phi_{sc} + Q_l\sin\phi_{sc}}{S_{sc}} + j\frac{P_l\sin\phi_{sc} - Q_l\cos\phi_{sc}}{S_{sc}}\right) \tag{4.27}$$

The above implies that,

$$\frac{\Delta V_R}{V} \approx \frac{P_l\cos\phi_{sc} + Q_l\sin\phi_{sc}}{S_{sc}}$$

$$\frac{\Delta V_X}{V} \approx \frac{P_l\sin\phi_{sc} - Q_l\cos\phi_{sc}}{S_{sc}}$$

Often $(\Delta V_X/V)$ is ignored on the ground that the phase quadrature component has negligible contribution to the magnitude of the overall phasor. It mainly contributes to the phase angle. Therefore, (4.27) is simplified to the following.

$$\frac{\Delta\bar{V}}{V} = \frac{\Delta V_R}{V} = \frac{P_l\cos\phi_{sc} + Q_l\sin\phi_{sc}}{S_{sc}} \tag{4.28}$$

The approximate expression in (4.28) is quite useful in terms of short circuit capacity (S_{sc}), X_s/R_s ratio, and active and reactive powers of the load. Further, feeder reactance (X_s) is far greater than feeder resistance (R_s), i.e., $X_s >> R_s$. This implies that $\phi_{sc} \rightarrow 90°$, $\sin\phi_{sc} \rightarrow 1$ and $\cos\phi_{sc} \rightarrow 0$. Using this approximation, the voltage regulation is given as follows.

$$\frac{E-V}{V} \approx \frac{Q_l}{S_{sc}}. \tag{4.29}$$

The above leads to the following expression,

$$V \simeq \frac{E}{(1+\frac{Q_l}{S_{sc}})} \simeq E(1-\frac{Q_l}{S_{sc}}) \tag{4.30}$$

with the assumption that, $Q_l/S_{sc} << 1$. Although the above relationship is obtained with approximations, however, it is very useful in visualizing the action of compensator on the voltage. The above equation is graphically represented in Fig. 4.6. The nature of voltage variation is drooping with increase in inductive reactive power of the load. This is shown by the negative slope $-E/S_{sc}$ as indicated in the figure.

The above characteristics also explain that when load is capacitive, Q_l is negative. This makes $V > E$. This is similar to the Ferranti effect due to lightly loaded electric lines.

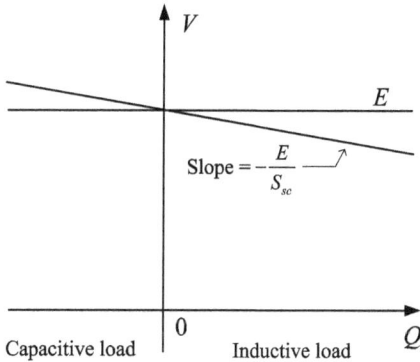

Figure 4.6   Voltage variation with reactive power of the load

Example 4.1. Consider a three-phase supply with 6.9 kV (L-L), 50 Hz, feeding a three-phase induction motor with real and reactive power demand of 25 MW and 40 MVAr, respectively. The supply bus has short-circuit level of 300 MVA and feeder with X_s/R_s ratio of 5. Compute the following,

(a) Find the load (induction motor) bus voltage $(\overline{V})$ and the voltage drop $(\Delta\overline{V})$ in the supply feeder. Thus, determine system voltage $(\overline{E})$, and voltage regulation.

(b) It is required to maintain the load bus voltage to be the same as the supply bus voltage, i.e., $V_{LL} = 6.9$ kV. Calculate reactive power supplies by the compensator.

(c) What should be the load bus voltage and compensator current if it is required to maintain the unity power factor at the load bus?

(d) For parts (b) and (c), find the real and reactive powers at the upstream source.

Solution: Per phase quantities, feeder resistance, and reactance are computed as in the following.

$E_s = 6.9/\sqrt{3} = 3.9837$ kV, $S_{sc} = 300/3 = 100$ MVA, $P_l = 25/3 = 8.33$ MW/phase, $Q_l = 40/3 = 13.33$ MVAr/phase, $Z_s = E_s^2/S_{sc} = (3.9837\,\text{kV})^2/100 = 0.1587\,\Omega$/phase. It is given that, $X_s/R_s = \tan \phi_{sc} = 5$, therefore $\phi_{sc} = \tan^{-1}5 = 78.69°$. From this,

$$R_s \quad = \quad Z_s \cos \phi_{sc} = 0.1587 \cos(78.69°) = 0.0311\,\Omega/\text{phase}$$
$$X_s \quad = \quad Z_s \sin \phi_{sc} = 0.1587 \sin(78.69°) = 0.1556\,\Omega/\text{phase}$$

Therefore, the feeder impedance can be expressed as following.

$$Z_s \quad = \quad 0.0311 + j0.1556 = 0.1587\angle 78.69\,\Omega/\text{phase}$$

(a) Without compensation ($Q_s = Q_l$, $Q_\gamma = 0$)
   To know $\Delta \overline{V}$, first the voltage at the load bus has to be computed using (4.20), as given below.

$$V^4 + \left\{2(R_sP_l + X_sQ_l) - E^2\right\}V^2 + (R_s^2 + X_s^2)(P_l^2 + Q_l^2) = 0$$

Now substituting values of R_s, X_s, P_l, Q_l and E in above equation, we get,

$$V^4 \quad + \quad \{2[0.0311 \times 8.33 + 0.1556 \times 13.33] - 3.9837^2\}V^2$$
$$+ \quad (0.0311^2 + 0.1556^2)(8.33^2 + 13.33^2) = 0$$

After simplifying the above, we have the following equation.

$$V^4 - 11.2015V^2 + 6.2265 = 0$$

Therefore,

$$V^2 \quad = \quad \frac{11.2015 \pm \sqrt{11.2015^2 - 4 \times 6.2265}}{2}$$
$$= \quad 10.6149, 0.5866$$
$$\text{and} \quad V \quad = \quad \pm 3.2580 \text{ kV}, \pm 0.7659 \text{ kV}$$

Since rms value cannot be negative and the maximum rms value must be a feasible solution, therefore $V = 3.2580$ kV. The voltage regulation is thus given as follows.

$$\text{VR}(\%) = \frac{|\overline{E}| - |\overline{V}|}{|\overline{E}|} \times 100 = \frac{3.9837 - 3.2580}{3.9837} = 18.21\%.$$

Since V is known, the load current can be computed as following.

$$
\begin{aligned}
\bar{I}_l &= \frac{P_l - jQ_l}{V} = \frac{8.33 - j13.33}{3.2580} \\
&= 2.5578 - j4.0924 \text{ kA} \\
&= 4.8260\angle - 57.9946° \text{ kA}
\end{aligned}
$$

The term $\Delta\bar{V}$ is now calculated using $\Delta\bar{V} = \bar{I}_l Z_s$.

$$
\Delta\bar{V} = \bar{I}_l Z_s = 4.8260\angle - 57.9946° \times 0.1587\angle 78.69° = 0.7654\angle 20.69° \text{ kV}
$$

The term $\Delta\bar{V}$ can also be computed using (4.13), as it is given below.

$$
\begin{aligned}
\Delta\bar{V} &= \frac{R_s P_l + X_s Q_l}{V} + j\frac{X_s P_l - R_s Q_l}{V} \\
&= \frac{0.0311 \times 8.33 + 0.1556 \times 13.33}{3.2580} + j\frac{0.1556 \times 8.33 - 0.0311 \times 13.33}{3.2580} \\
&= 0.7165 + j0.2707 \text{ kV} = 0.7659\angle 20.69° \text{ kV}
\end{aligned}
$$

Now the system voltage, $\bar{E}$, is calculated as follows.

$$
\bar{E} = \bar{V} + \Delta\bar{V} = 3.2580\angle 0° + 0.7659\angle 20.6955° = 3.9837\angle 3.8958° \text{ kV}
$$

The above are illustrated in phasor diagram similar to what is shown in Fig. 4.3(b).

(b) Compensator as a voltage regulator

Now it is required to maintain $V = E = 3.98$ kV at the load bus. For this, let there be reactive power, Q_γ supplied by the compensator at the load bus. Therefore, the net reactive power at the load bus is equal to Q_s, which is given below.

$$
Q_s = Q_l + Q_\gamma
$$

Thus from (4.17), we get,

$$
(R_s^2 + X_s^2)Q_s^2 + 2V^2 X_s Q_s + (V^2 + R_s P_l)^2 + X_s^2 P_l^2 - E^2 V^2 = 0
$$

From the above we have,

$$
0.0252\, Q_s^2 + 4.9393\, Q_s + 9.9812 = 0.
$$

Solving the above equation we get,

$$
\begin{aligned}
Q_s &= \frac{-4.9393 \pm \sqrt{4.9393^2 - 4 \times 0.0252 \times 9.9812}}{2 \times 0.0252} \\
&= -194.0741 \quad \text{or} - 2.0420 \text{ MVAr}.
\end{aligned}
$$

The feasible solution is $Q_s = -2.0420$ MVAr because it requires less rating of the compensator. Therefore the reactive power of the compensator (Q_γ) is,

$$Q_\gamma = Q_s - Q_l = -2.0420 - 13.3333 = -15.37 \text{ MVAr}.$$

With $Q_s = -2.0420$ MVAr, the net supply current (source current) at the load bus is computed is as following.

$$
\begin{aligned}
\bar{I}_s &= \frac{P_l - jQ_s}{V} = \frac{8.33 - j(-2.0420)}{3.9837} \\
&= 2.0402 + j0.6900 \text{ kA} = 2.1537\angle 18.6861° \text{ kA}.
\end{aligned}
$$

Therefore, the $\Delta \bar{V}$ is computed as given below.

$$\Delta \bar{V} = \bar{I}_s Z_s = 2.1537\angle 18.6861° \times 0.1587\angle 78.69° = 0.3418\angle 92.4587° \text{ kV}$$

This can be further verified by replacing Q_s for Q_l in (4.13) as given below.

$$
\begin{aligned}
\Delta \bar{V} &= \frac{R_s P_l + X_s Q_s}{V} + j\frac{X_s P_l - R_s Q_s}{V} \\
&= \frac{0.0311 \times 8.33 + 0.1556 \times (-2.0420)}{3.9837} \\
&\quad + j\frac{0.1556 \times 8.33 - 0.0311 \times (-2.0420)}{3.9837} \\
&= -0.0147 + j0.3415 \text{ kV} = 0.3418\angle 92.4587° \text{ kV}
\end{aligned}
$$

Now, we can find supply voltage $\bar{E}$ as given below.

$$
\begin{aligned}
\bar{E} &= \bar{V} + \Delta \bar{V} \\
&= 3.9837 - 0.0147 + j0.3415 \\
&= 3.9691 + j0.3415 = 3.9837\angle 4.9174° \text{ kV}
\end{aligned}
$$

This indicates that power factor is not unity for perfect voltage regulation, i.e., $E = V$. For this case the compensator current is given below.

$$
\begin{aligned}
\bar{I}_\gamma &= \frac{\bar{S}_\gamma^*}{V} = \frac{(P_\gamma + jQ_\gamma)^*}{V} = \frac{-jQ_\gamma}{V} \\
&= \frac{-j(-15.3754)}{3.9837} = j3.8596 \text{ kA}
\end{aligned}
$$

The load current is computed as below.

$$
\begin{aligned}
\bar{I}_l &= \frac{P_l - jQ_l}{V} = \frac{8.33 - j13.33}{3.9837} \\
&= 2.0918 - j3.3470 = I_{lR} + jI_{lX} = 3.9469\angle -57.9946° \text{ kA}
\end{aligned}
$$

The phasor diagram is similar to the one shown in Fig. 4.4(b). The voltage at the load bus is maintained to 1.0 p.u. It is observed that the reactive power of the compensator Q_γ is not equal to load reactive power (Q_l), i.e., $Q_s = Q_l + Q_\gamma = -2.042$ MVAr/phase = -6.12 MVAr/three-phase. As a result of this compensation, the voltage regulation is zero. However, power factor is not unity. The phase angle between $\overline{V}$ and $\overline{I}_s$ is 18.68°, which gives a power factor of $\cos 18.68° = 0.9473$ (lead). The angle between $\overline{E}$ and $\overline{I}_s$ is $(18.68° - 4.91° = 13.77°)$. Thus, source power factor (ϕ_s) is $\cos(13.77°) = 0.9712$ (lead).

(c) Compensation for unity power factor

To achieve a unity power factor at the load bus, the condition $Q_\gamma = -Q_l$ must be satisfied, which further implies that the net reactive power at the load bus is zero. Therefore, substituting $Q_s = 0$ in (4.20), we get the following.

$$V^4 + (2R_sP_l - E^2)V^2 + (R_s^2 + X_s^2)P_l^2 = 0$$

Substituting values of parameters in the above equation we get,

$$V^4 + (2 \times 0.0311 \times 8.3333 - 3.9837^2)V^2 + (0.0311^2 + 0.1556^2)8.3333^2 = 0$$

From the above,

$$V^4 - 15.3513V^2 + 1.7490 = 0$$

The solution of the above equation is,

$$V^2 = \frac{96.08 \pm 93.97}{2} = 15.2365, 0.1148$$
$$V = \pm 3.9034 \,\text{kV}, \pm 0.3388 \,\text{kV}$$

Since rms value cannot be negative and maximum rms value must be a feasible solution, therefore V = 3.9034 kV. Thus, it is seen that obtaining unity power factor at the load bus does not ensure desired voltage regulation. Now, the other quantities are computed as given below.

$$\overline{I}_l = \frac{P_l - jQ_l}{V} = \frac{8.33 - j13.33}{3.9034} = 2.1349 - j3.4158 = 4.0281\angle -57.9946° \,\text{kA}$$

$$\overline{I}_s = \frac{P_l}{V} = \frac{8.33}{3.9034} = 2.1349\angle 0° \,\text{kA}$$

Thus, the $\Delta \overline{V}$ is computed as given below.

$$\Delta \overline{V} = \overline{I}_s Z_s = 2.1349\angle 0° \times 0.1587\angle 78.69° = 0.3388\angle 78.6901° \,\text{kV}$$

Since $Q_\gamma = -Q_l$, this implies that $I_\gamma = -jQ_\gamma/V = jQ_l/V = j3.4158$ kA. The voltage drop across the feeder can also be calculated as follows.

$$\Delta \overline{V} = \frac{R_sP_l + X_sQ_s}{V} + j\frac{X_sP_l - R_sQ_s}{V}$$
$$= \frac{(0.0311 \times 8.33 + j0.1556 \times 8.33)}{3.9034}$$
$$= 0.0664 + j0.3322 = 0.3388\angle 78.6901° \,\text{kV}$$

The supply voltage $\bar{E}$ as given below.

$$\bar{E} = 3.9034\angle 0° + 0.3388\angle 78.6901° = 3.9837\angle 4.7838° \text{ kV}$$

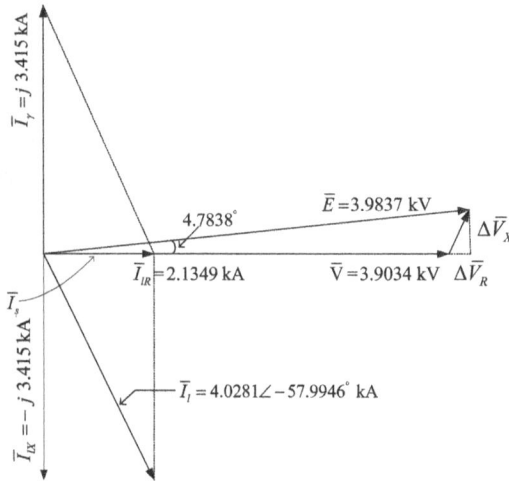

Figure 4.7   Phasor diagram for a system with a compensator in power factor correction mode

For this case as explained earlier, the power factor at the load bus is unity. The power factor at the source bus is the cosine angle between the supply voltage, $\bar{E}$ and the source current, $\bar{I}_s$, which is $\cos(-4.7838°) = 0.9966$ lag. The phasor diagram for the above case is shown in Fig. 4.7. The percentage voltage change $= (3.9837 - 3.9034)/3.9837 \times 100 = 2.0157\%$. Thus, we see that power factor correction improves voltage regulation compared with the uncompensated case. In many cases, the degree of improvement is adequate and the compensator can be designed to provide reactive power requirement of load rather than as an ideal voltage regulator.

(d) For part (b),

$$\bar{S}'_s = P'_s + jQ'_s = \bar{E}\bar{I}^*_s = 3.9837\angle 4.9174° \times 2.1537\angle -18.6861°$$
$$= 8.4777 - j1.3202$$

This gives, $P'_s = 8.4777$ MW/phase and $Q'_s = 1.3202$ MVAr/phase (capacitive). For part (c),

$$\bar{S}'_s = P'_s + jQ'_s = \bar{E}\bar{I}^*_s = 3.9837\angle 4.7838° \times 2.1349\angle 0° = 8.4752 + j0.7093$$

This gives, $P'_s = 8.4752$ MW/phase and $Q'_s = 0.7093$ MVAr/phase (inductive). These powers can also be verified using (4.24).

4.3 SOME PRACTICAL ASPECTS OF COMPENSATOR USED AS VOLTAGE REGULATOR

In this section, some practical aspects of the compensator in voltage regulation mode will be discussed. The important parameters of the compensator which play significant role in obtaining desired voltage regulation are: Knee point (V_k), maximum or rated reactive power ($Q_{\gamma max}$) and the compensator gain (K_γ).

The compensator gain K_γ is defined as the rate of change of compensator reactive power Q_γ with the change in the voltage (V), as given below.

$$K_\gamma = \frac{dQ_r}{dV} \tag{4.31}$$

For a linear relationship between Q_γ and V with incremental change, the above equation is written as the following.

$$\Delta Q_\gamma = \Delta V\, K_\gamma \tag{4.32}$$

Assuming compensator characteristics to be linear with $Q_\gamma \leq Q_{\gamma max}$ limit, the voltage can be represented as,

$$V = V_k + \frac{Q_\gamma}{K_\gamma} \tag{4.33}$$

This is re-written as,

$$Q_\gamma = K_\gamma(V - V_k) \tag{4.34}$$

Flat V-Q characteristics imply that $K_\gamma \to \infty$. This means that the compensator can absorb or generate the exactly right amount of reactive power to maintain the supply voltage constant as the load varies without any constraint. We shall now see the regulating properties of the compensator, when the compensator has finite gain K_γ operating on supply system with a finite short circuit level, S_{sc}. The net reactive power at the load bus is sum of the load and the compensator reactive power as given below.

$$Q_l + Q_\gamma = Q_s \tag{4.35}$$

Using earlier voltage and reactive power relationship from equation (4.30), it can be written as the following.

$$V \simeq E\left(1 - \frac{Q_s}{S_{sc}}\right) \tag{4.36}$$

The compensator voltage represented by (4.33) and system voltage represented by (4.36) are shown in Fig. 4.8(a) and (b), respectively. Differentiating V with respect to Q_s, we get, the intrinsic sensitivity of the supply voltage with variation in Q_s as given below.

$$\frac{dV}{dQ_s} = -\frac{E}{S_{sc}} \tag{4.37}$$

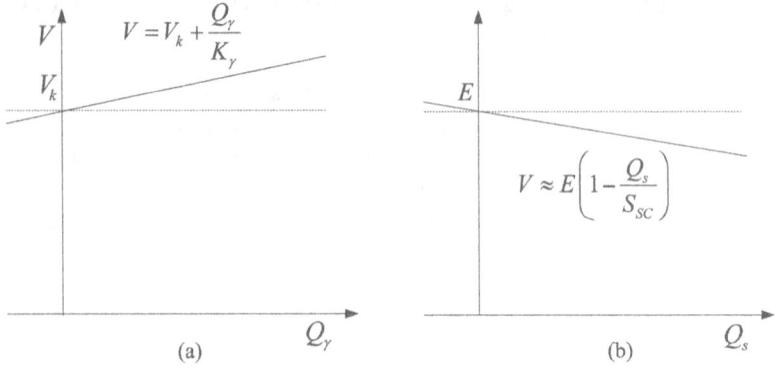

Figure 4.8 (a) Voltage characteristics of compensator and (b) system voltage characteristics

It is seen from the above equation that the high value of short circuit level S_{sc} reduces the voltage sensitivity, making voltage variation flat irrespective of Q_l. With compensator replacing $Q_s = Q_\gamma + Q_l$ in (4.36), we have the following.

$$V \simeq E\left(1 - \frac{Q_l + Q_\gamma}{S_{sc}}\right) \tag{4.38}$$

Substituting Q_γ from (4.34), we get the following equation.

$$V \simeq E\left[\frac{1 + K_\gamma V_k/S_{sc}}{1 + E K_\gamma/S_{sc}} - \frac{Q_l/S_{sc}}{1 + E K_\gamma/S_{sc}}\right] \tag{4.39}$$

Although approximate, the above equation gives the effects of all the major parameters such as load reactive power, the compensator characteristics V_γ, K_γ, and the system characteristics E and S_{sc}. As we discussed, V-Q characteristics is flat for high or infinite value K_γ. However, the higher value of the gain K_γ means a large rating and a quick rate of change of the reactive power with variation in the system voltage. This makes the design of the compensator challenging and the cost of the compensator is high.

The compensator has two effects, as seen from (4.39), i.e., it alters the no load supply voltage (E) and it modifies the sensitivity of supply point voltage to the variation in the load reactive power.

Differentiating (4.39) with respect to Q_l, we get,

$$\frac{dV}{dQ_l} = -\frac{E/S_{sc}}{1 + K_r E/S_{sc}} \tag{4.40}$$

which is voltage sensitivity of supply voltage to the load reactive power. It can be seen that the voltage sensitivity is reduced as compared to the voltage sensitivity without compensator as indicated in (4.37). It is useful to express the slope $(-E/S_{sc})$ by a term in a form similar to $K_\gamma = dQ_\gamma/dV$, as given below.

$$K_s = \frac{S_{sc}}{E}$$

$$\text{Thus,} \quad \frac{1}{K_s} = \frac{E}{S_{sc}} \quad\quad\quad (4.41)$$

Substituting V from (4.38) into (4.34), the following is obtained.

$$Q_\gamma = K_\gamma \left[E \left(1 - \frac{Q_l + Q_\gamma}{S_{sc}} \right) - V_k \right] \quad\quad\quad (4.42)$$

Collecting the coefficients of Q_γ from both sides of the above equation, we get

$$Q_\gamma = \frac{K_\gamma}{1 + K_\gamma(E/S_{sc})} \left[E \left(1 - \frac{Q_l}{S_{sc}} \right) - V_k \right] \quad\quad\quad (4.43)$$

Setting knee voltage V_k of the compensator equal to system voltage E, i.e., $V_k = E$, the above equation is simplified to,

$$Q_\gamma = -\frac{K_\gamma(E/S_{sc})}{1 + K_\gamma(E/S_{sc})} Q_l$$

$$= -\left(\frac{K_\gamma/K_s}{1 + K_\gamma/K_s} \right) Q_l. \quad\quad\quad (4.44)$$

From the above equation, it is observed that when the compensator gain $K_\gamma \to \infty$ then $Q_\gamma \to -Q_l$. This indicates perfect compensation of the load reactive power in order to regulate the load bus voltage. For the finite value of K_γ, the compensation leads to partial compensation of the load.

Example 4.2. Consider a three-phase system with line-line voltage 11 kV and short circuit capacity of 480 MVA. With a compensator gain of 100 p.u., determine voltage sensitivity with and without the compensator. For each case, if a load reactive power changes by 10 MVAr, find out the change in load bus voltage assuming a linear relationship between V-Q characteristics. Also, find the relationship between compensator and load reactive powers.

Solution: The voltage sensitivity can be computed using the following equation.

$$\frac{dV}{dQ_l} = -\frac{E/S_{sc}}{1 + K_\gamma E/S_{sc}}$$

Without compensator, $K_\gamma = 0, E = (11/\sqrt{3}) = 6.35$ kV and $S_{sc} = 480/3 = 160$ MVA. Substituting these values in the above equation, the voltage sensitivity is given below.

$$\frac{dV}{dQ_l} = -\frac{6.35/160}{1 + 0 \times 6.35/160} = -0.039$$

The change in voltage due to variation of reactive power by 10 MVAr, $\Delta V = -0.039 \times 10 = -0.39$ kV.

With compensator, $K_\gamma = 100$

$$\frac{dV}{dQ_l} = -\frac{6.35/160}{1 + 100 \times 6.35/160} = -0.0078$$

The change in voltage due to variation of reactive power by 10 MVAr, $\Delta V = -0.0078 \times 10 = -0.078$ kV. Thus, it is seen that even with the finite compensator gain, there is a significant reduction in the voltage sensitivity, which means that the load bus voltage is fairly constant for a considerable change in the load reactive power. The compensator reactive power Q_γ and load reactive power Q_l are related by (4.44) and is given below.

$$\begin{aligned} Q_\gamma &= -\frac{K_\gamma (E/S_{sc})}{1 + K_\gamma (E/S_{sc})} Q_l = -\frac{100 \times (6.35/160)}{1 + 100 \times (6.35/160)} Q_l \\ &= -0.79 Q_l \end{aligned}$$

It can be observed that when compensator gain (Q_γ) is quite large, then compensator reactive power, Q_γ is equal and opposite to that of load reactive power, i.e., $Q_\gamma = -Q_l$. It is further observed that due to finite compensator gain, i.e., $K_\gamma = 100$, reactive power is partially compensated. The compensator reactive power varies from 0 to 7.9 MVAr for 0 to 10 MVAr change in the load reactive power.

4.4   PHASE BALANCING AND POWER FACTOR CORRECTION OF UNBALANCED LOADS

So far, we have discussed voltage regulation and power factor correction for single-phase systems. In this section, we will focus on the balancing of three-phase unbalanced loads at the fundamental frequency. In considering unbalanced loads, both load and compensator are modeled in terms of their impedances or admittances.

4.4.1   THREE-PHASE UNBALANCED LOADS

Consider a three-phase three-wire system supplying an unbalanced load as shown in Fig. 4.9. Applying KVL, in the two loops shown in the above figure, we get

$$\begin{aligned} -\overline{V}_{aN} + Z_a \overline{I}_1 + Z_b (\overline{I}_1 - \overline{I}_2) + \overline{V}_{bN} &= 0 \\ -\overline{V}_{bN} + Z_c \overline{I}_2 + Z_b (\overline{I}_2 - \overline{I}_1) + \overline{V}_{cN} &= 0 \end{aligned} \tag{4.45}$$

Rearranging the above, we get the following.

$$\begin{aligned} \overline{V}_{aN} - \overline{V}_{bN} &= (Z_a + Z_b) \overline{I}_1 - Z_b \overline{I}_2 \\ \overline{V}_{bN} - \overline{V}_{cN} &= (Z_b + Z_c) \overline{I}_2 - Z_b \overline{I}_1 \end{aligned} \tag{4.46}$$

The above can be represented in matrix form as given below.

$$\begin{bmatrix} \overline{V}_{aN} - \overline{V}_{bN} \\ \overline{V}_{bN} - \overline{V}_{cN} \end{bmatrix} = \begin{bmatrix} (Z_a + Z_b) & -Z_b \\ -Z_b & (Z_b + Z_c) \end{bmatrix} \begin{bmatrix} \overline{I}_1 \\ \overline{I}_2 \end{bmatrix} \tag{4.47}$$

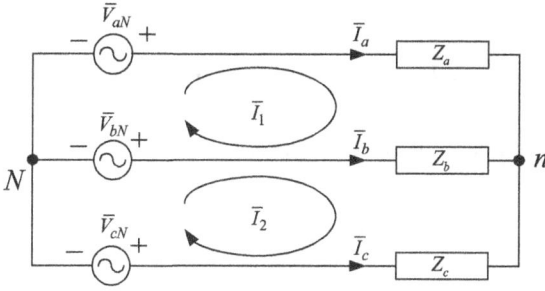

Figure 4.9 Three-phase unbalanced load

Therefore the currents are given as below.

$$
\begin{bmatrix} \bar{I}_1 \\ \bar{I}_2 \end{bmatrix} = \begin{bmatrix} (Z_a + Z_b) & -Z_b \\ -Z_b & (Z_b + Z_c) \end{bmatrix}^{-1} \begin{bmatrix} \bar{V}_{aN} - \bar{V}_{bN} \\ \bar{V}_{bN} - \bar{V}_{cN} \end{bmatrix}
$$

$$
= \frac{1}{\Delta_Z} \begin{bmatrix} (Z_b + Z_c) & Z_b \\ Z_b & (Z_a + Z_b) \end{bmatrix} \begin{bmatrix} \bar{V}_{aN} - \bar{V}_{bN} \\ \bar{V}_{bN} - \bar{V}_{cN} \end{bmatrix}. \tag{4.48}
$$

where, $\Delta_Z = (Z_b + Z_c)(Z_a + Z_b) - Z_b^2 = Z_a Z_b + Z_b Z_c + Z_c Z_a$. The current $\bar{I}_1$ is given below.

$$
\bar{I}_1 = \frac{1}{\Delta_Z} \left[(Z_b + Z_c)(\bar{V}_{aN} - \bar{V}_{bN}) + Z_b(\bar{V}_{bN} - \bar{V}_{cN}) \right]
$$

$$
= \frac{1}{\Delta_Z} \left[(Z_b + Z_c)\bar{V}_{aN} - Z_c \bar{V}_{bN} - Z_b \bar{V}_{cN} \right] \tag{4.49}
$$

Similarly,

$$
\bar{I}_2 = \frac{1}{\Delta_Z} \left[Z_b(\bar{V}_{aN} - \bar{V}_{bN}) + (Z_a + Z_b)(\bar{V}_{bN} - \bar{V}_{cN}) \right]
$$

$$
= \frac{1}{\Delta_Z} \left[Z_b \bar{V}_{aN} + Z_a \bar{V}_{bN} - (Z_a + Z_b)\bar{V}_{cN} \right] \tag{4.50}
$$

Now,

$$
\bar{I}_a = \bar{I}_1 = \frac{1}{\Delta_Z} \left[(Z_b + Z_c)\bar{V}_{aN} - Z_c \bar{V}_{bN} - Z_b \bar{V}_{cN} \right]
$$

$$
\bar{I}_b = \bar{I}_2 - \bar{I}_1
$$

$$
= \frac{1}{\Delta_Z} \left[Z_b \bar{V}_{aN} + Z_a \bar{V}_{bN} - (Z_a + Z_b)\bar{V}_{cN} - (Z_b + Z_c)\bar{V}_{aN} + Z_c \bar{V}_{bN} + Z_b \bar{V}_{cN} \right]
$$

$$
= \frac{(Z_c + Z_a)\bar{V}_{bN} - Z_a \bar{V}_{cN} - Z_c V_{aN}}{\Delta_Z} \tag{4.51}
$$

and,

$$\bar{I}_c = -\bar{I}_2 = -\bar{I}_b - \bar{I}_a = \frac{(Z_a + Z_b)\overline{V}_{cN} - Z_b\overline{V}_{aN} - Z_a\overline{V}_{bN}}{\Delta_Z} \tag{4.52}$$

Alternatively, phase currents can be expressed as following.

$$
\begin{aligned}
\bar{I}_a &= \frac{\overline{V}_{aN} - \overline{V}_{nN}}{Z_a} \\[2mm]
\bar{I}_b &= \frac{\overline{V}_{bN} - \overline{V}_{nN}}{Z_b} \\[2mm]
\bar{I}_c &= \frac{\overline{V}_{cN} - \overline{V}_{nN}}{Z_c}
\end{aligned}
\tag{4.53}
$$

Applying Kirchoff's current law at node n, we get $\bar{I}_a + \bar{I}_b + \bar{I}_c = 0$. Therefore, from the above equation,

$$\frac{\overline{V}_{aN} - \overline{V}_{nN}}{Z_a} + \frac{\overline{V}_{bN} - \overline{V}_{nN}}{Z_b} + \frac{\overline{V}_{cN} - \overline{V}_{nN}}{Z_c} = 0. \tag{4.54}$$

This implies that,

$$\frac{\overline{V}_{aN}}{Z_a} + \frac{\overline{V}_{bN}}{Z_b} + \frac{\overline{V}_{cN}}{Z_c} = \left(\frac{1}{Z_a} + \frac{1}{Z_b} + \frac{1}{Z_c} \right) \overline{V}_{nN} = \frac{Z_aZ_b + Z_bZ_c + Z_cZ_a}{Z_aZ_bZ_c} \overline{V}_{nN} \tag{4.55}$$

From the above equation the voltage between the load and system neutral can be found. It is given below.

$$
\begin{aligned}
\overline{V}_{nN} &= \frac{Z_aZ_bZ_c}{\Delta_Z} \left(\frac{\overline{V}_{aN}}{Z_a} + \frac{\overline{V}_{bN}}{Z_b} + \frac{\overline{V}_{cN}}{Z_c} \right) \\[3mm]
&= \frac{1}{\frac{1}{Z_a} + \frac{1}{Z_b} + \frac{1}{Z_c}} \left(\frac{\overline{V}_{aN}}{Z_a} + \frac{\overline{V}_{bN}}{Z_b} + \frac{\overline{V}_{cN}}{Z_c} \right)
\end{aligned}
\tag{4.56}
$$

Some interesting points are observed from the above formulation.

1. If both source voltage and load impedances are balanced, i.e., $Z_a = Z_b = Z_c = Z$, then $\overline{V}_{nN} = \frac{1}{3}\left(\overline{V}_{aN} + \overline{V}_{bN} + \overline{V}_{cN}\right) = 0$. Thus, there will not be any voltage between the two neutrals.

2. If supply voltages are balanced and load impedances are unbalanced, then $\overline{V}_{nN} \neq 0$ and is given by the above equation.

3. If supply voltages are not balanced but load impedances are identical, then the zero sequence voltage, $\overline{V}_0 = \overline{V}_{nN} = \frac{1}{3}\left(\overline{V}_{aN} + \overline{V}_{bN} + \overline{V}_{cN}\right)$.

It is interesting to note that if the two neutrals are connected together, i.e., $\overline{V}_{nN} = 0$, then each phase becomes independent through neutral. Such a configuration is called three-phase four-wire system. In general, three-phase four-wire system has the following properties.

$$\overline{V}_{nN} = 0$$
$$\overline{I}_{nN} = \overline{I}_a + \overline{I}_b + \overline{I}_c \tag{4.57}$$

The neutral current, $\overline{I}_{nN}$ in above may or may not be equal to zero depending upon the load currents. The current $\overline{I}_{nN}$ is equivalent to zero sequence current ($\overline{I}_0$) and it will flow in the neutral wire.

For three-phase three-wire system, the zero sequence current is always zero and therefore following properties are satisfied.

$$\left.\begin{array}{l} \overline{V}_{nN} = \dfrac{1}{\frac{1}{Z_a} + \frac{1}{Z_b} + \frac{1}{Z_c}} \left(\dfrac{\overline{V}_{aN}}{Z_a} + \dfrac{\overline{V}_{bN}}{Z_b} + \dfrac{\overline{V}_{cN}}{Z_c} \right) \\[4mm] \overline{I}_{nN} = \overline{I}_a + \overline{I}_b + \overline{I}_c = 0 \end{array}\right\} \tag{4.58}$$

The neutral voltage, $\overline{V}_{nN}$ may or may not be equal to zero depending upon the load impedances and the source voltages. Thus, it is interesting to observe that three-phase three-wire and three-phase four-wire systems have dual properties in regard to the neutral voltage and the neutral current.

4.4.2 REPRESENTATION OF THREE-PHASE DELTA CONNECTED UNBAL-ANCED LOAD

A three-phase delta connected unbalanced and its equivalent star connected load are shown in Fig. 4.10(a) and (b), respectively. The three-phase load is represented by line-line admittances as given below.

$$Y_l^{ab} = G_l^{ab} + jB_l^{ab}$$
$$Y_l^{bc} = G_l^{bc} + jB_l^{bc} \tag{4.59}$$
$$Y_l^{ca} = G_l^{ca} + jB_l^{ca}$$

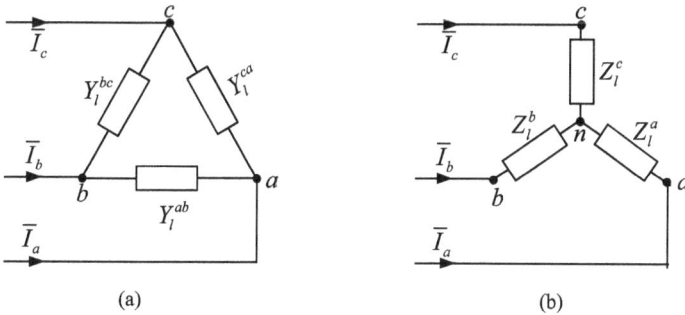

(a)   (b)

Figure 4.10   (a) An unbalanced delta connected load and (b) equivalent star connected load

The delta connected load can be equivalently converted to star connected load using the following expressions.

$$Z_l^a = \frac{Z_l^{ab} Z_l^{ca}}{Z_l^{ab} + Z_l^{bc} + Z_l^{ca}}$$

$$Z_l^b = \frac{Z_l^{bc} Z_l^{ab}}{Z_l^{ab} + Z_l^{bc} + Z_l^{ca}} \qquad (4.60)$$

$$Z_l^c = \frac{Z_l^{ca} Z_l^{bc}}{Z_l^{ab} + Z_l^{bc} + Z_l^{ca}}$$

where $Z_l^{ab} = 1/Y_l^{ab}$, $Z_l^{bc} = 1/Y_l^{bc}$ and $Z_l^{ca} = 1/Y_l^{ca}$. The above equation can also be written in admittance form as given below.

$$Y_l^a = \frac{Y_l^{ab} Y_l^{bc} + Y_l^{bc} Y_l^{ca} + Y_l^{ca} Y_l^{ab}}{Y_l^{bc}}$$

$$Y_l^b = \frac{Y_l^{ab} Y_l^{bc} + Y_l^{bc} Y_l^{ca} + Y_l^{ca} Y_l^{ab}}{Y_l^{ca}} \qquad (4.61)$$

$$Y_l^c = \frac{Y_l^{ab} Y_l^{bc} + Y_l^{bc} Y_l^{ca} + Y_l^{ca} Y_l^{ab}}{Y_l^{ab}}$$

Example 4.3. Consider three-phase system supply a delta connected unbalanced load with $Z_l^a = R_a = 10\,\Omega$, $Z_l^b = R_b = 15\,\Omega$ and $Z_l^c = R_c = 30\,\Omega$ as shown in Fig. 4.10. Determine the voltage between neutrals and find the line currents. Assume a balance supply voltage with rms value of 230 V. Find out the vector and arithmetic power factor. Comment upon the results.

Solution: The voltage between neutrals $\overline{V}_{nN}$ is given as following.

$$
\begin{aligned}
V_{nN} &= \frac{R_a R_b R_c}{R_a R_b + R_b R_c + R_c R_a} \left(\frac{\overline{V}_{aN}}{R_a} + \frac{\overline{V}_{bN}}{R_b} + \frac{\overline{V}_{cN}}{R_c} \right) \\
&= \frac{10 \times 15 \times 30}{10 \times 15 + 15 \times 30 + 30 \times 10} \left(\frac{V\angle 0^\circ}{10} + \frac{V\angle -120^\circ}{15} + \frac{V\angle 120^\circ}{30} \right) \\
&= \frac{4500}{900} \left(\frac{3V\angle 0^\circ + 2V\angle -120^\circ + V\angle 120^\circ}{30} \right) \\
&= \frac{4500}{900} \frac{1}{30} V \left[3 + 2 \left(-\frac{1}{2} - j\frac{\sqrt{3}}{2} \right) + \left(-\frac{1}{2} + j\frac{\sqrt{3}}{2} \right) \right] \\
&= \frac{V}{6} \left(3 - 1 - \frac{1}{2} - j2 \times \frac{\sqrt{3}}{2} + j\frac{\sqrt{3}}{2} \right) \\
&= \frac{V}{6} \left(\frac{3}{2} - j\frac{\sqrt{3}}{2} \right) = V \left(\frac{1}{4} - j\frac{1}{4\sqrt{3}} \right) = \frac{V}{2\sqrt{3}} \angle -30^\circ = 66.39 \angle -30^\circ \text{ V}
\end{aligned}
$$

Knowing this voltage, we can find phase currents as following.

$$\bar{I}_a = \frac{\bar{V}_{aN} - \bar{V}_{nN}}{R_a} = \frac{V\angle 0° - \frac{V}{2\sqrt{3}}\angle -30°}{10}$$

$$= \frac{V(1 - 0.25 + j0.1443)}{10}$$

$$= 230 \times (0.075 + j0.01443)$$

$$= 17.56\angle 10.89° \, A$$

Similarly,

$$\bar{I}_b = \frac{\bar{V}_{bN} - \bar{V}_{nN}}{R_b} = \frac{V\angle -120° - \frac{V}{2\sqrt{3}}\angle -30°}{15}$$

$$= 230 \times (-0.05 - j0.04811)$$

$$= 15.94\angle -136.1° \, A$$

and,

$$\bar{I}_c = \frac{\bar{V}_{cN} - \bar{V}_{nN}}{Z_c} = \frac{V\angle 120° - \frac{V}{2\sqrt{3}}\angle -30°}{30}$$

$$= 230 \times (-0.025 + j0.03367)$$

$$= 9.64\angle 126.58° \, A$$

It can be seen that $\bar{I}_a + \bar{I}_b + \bar{I}_c = 0$. The phase powers are computed as below.

$$\bar{S}_a = \bar{V}_a(\bar{I}_a)^* = P_a + jQ_a = 230 \times 17.56\angle -10.81° = 3967.12 - j757.48 \, VA$$

$$\bar{S}_b = \bar{V}_b(\bar{I}_b)^* = P_b + jQ_b = 230 \times 15.94\angle(-120° + 136.1°)$$

$$= 3522.4 + j1016.69 \, VA$$

$$\bar{S}_c = \bar{V}_c(\bar{I}_c)^* = P_c + jQ_c = 230 \times 9.64\angle(120° - 126.58°)$$

$$= 2202.59 - j254.06 \, VA$$

From the above the total apparent power $\bar{S}_V = S_a + S_b + S_c = 9692.11 + j0 \, VA$.
Therefore, $S_V = |\bar{S}_a + \bar{S}_b + \bar{S}_c| = 9692.11 \, VA$.
The total arithmetic apparent power $S_A = |\bar{S}_a| + |\bar{S}_b| + |\bar{S}_c| = 9922.2 \, VA$.
The efffective apparent power is given as following.

$$S_e = 3V_e I_e = 3V_e\sqrt{\frac{I_a^2 + I_b^2 + I_c^2}{3}}$$

$$= 3 \times 230 \times 14.78 = 10198.36 \, VA$$

Therefore, the vector, arithmetic, and effective apparent power factors are given as in the following.

$$pf_V = \frac{P}{S_V} = \frac{9622.11}{9622.11} = 1.00$$

$$pf_A = \frac{P}{S_A} = \frac{9692.11}{9922.2} = 0.9768$$

$$pf_e = \frac{P}{S_e} = \frac{9622.11}{10198.36} = 0.9435$$

It is interesting to note that although the load in each phase is resistive, each phase has some reactive power. This is due to unbalance of the load currents. This apparently increases the rating of power conductors for a given amount of power transfer. It is also to be noted that the net reactive power $Q = Q_a + Q_b + Q_c = 0$, leading to the unity vector apparent power factor. However, the arithmetic and effective power factors are less than unity showing the effect of the unbalance loads on the power factor. The effective power factor has the least value, and hence is used to design the feeder for unbalanced loads.

4.4.3 AN ALTERNATE APPROACH TO DETERMINE PHASE CURRENTS AND POWERS

In this section, an alternate approach will be discussed to solve phase currents and powers directly without computing the neutral voltage for the system shown in Fig.4.10(a). First, we express three-phase supply voltage with respect to its neutral, in the following form.

$$\begin{aligned}
\overline{V}_a &= V \angle 0° \\
\overline{V}_b &= V \angle -120° = \alpha^2 V \\
\overline{V}_c &= V \angle 120° = \alpha V
\end{aligned} \tag{4.62}$$

where, in the above equation, α is known as complex operator and value of α and α^2 are given below.

$$\begin{aligned}
\alpha &= e^{j2\pi/3} = 1\angle 120° = -1/2 + j\sqrt{3}/2 \\
\alpha^2 &= e^{j4\pi/3} = 1\angle 240° = 1\angle -120° = -1/2 - j\sqrt{3}/2
\end{aligned} \tag{4.63}$$

Also note the following property,

$$1 + \alpha + \alpha^2 = 0. \tag{4.64}$$

Using the above property, the line to line voltages can be expressed as following.

$$\begin{aligned}
\overline{V}_{ab} &= \overline{V}_a - \overline{V}_b = (1 - \alpha^2)V \\
\overline{V}_{bc} &= \overline{V}_b - \overline{V}_c = (\alpha^2 - \alpha)V \\
\overline{V}_{ca} &= \overline{V}_c - \overline{V}_a = (\alpha - 1)V
\end{aligned} \tag{4.65}$$

Therefore, currents in line ab, bc and ca are given as,

$$
\begin{aligned}
\bar{I}_{abl} &= Y_l^{ab}\bar{V}_{ab} = Y_l^{ab}(1-\alpha^2)V \\
\bar{I}_{bcl} &= Y_l^{bc}\bar{V}_{bc} = Y_l^{bc}(\alpha^2-\alpha)V \\
\bar{I}_{cal} &= Y_l^{ca}\bar{V}_{ca} = Y_l^{ca}(\alpha-1)V
\end{aligned}
\tag{4.66}
$$

Hence line currents are given as,

$$
\begin{aligned}
\bar{I}_{al} &= \bar{I}_{abl}-\bar{I}_{cal} = [(1-\alpha^2)Y_l^{ab}-(\alpha-1)Y_l^{ca}]V \\
\bar{I}_{bl} &= \bar{I}_{bcl}-\bar{I}_{abl} = [(\alpha^2-\alpha)Y_l^{bc}-(1-\alpha^2)Y_l^{ab}]V \\
\bar{I}_{cl} &= \bar{I}_{cal}-\bar{I}_{bcl} = [(\alpha-1)Y_l^{ca}-(\alpha^2-\alpha)Y_l^{bc}]V
\end{aligned}
\tag{4.67}
$$

Example 4.4. Compute line currents directly using (4.67) for the problem in Example 4.3.

Solution: To compute line currents directly from the above expressions, we need to compute Y_l^{ab}, Y_l^{bc} and Y_l^{ca}. These are given below.

$$
\begin{aligned}
Y_l^{ab} &= \frac{1}{Z_l^{ab}} = \frac{Z_l^c}{Z_l^a Z_l^b + Z_l^b Z_l^c + Z_l^c Z_l^a} \\
Y_l^{bc} &= \frac{1}{Z_l^{bc}} = \frac{Z_l^a}{Z_l^a Z_l^b + Z_l^b Z_l^c + Z_l^c Z_l^a} \\
Y_l^{ca} &= \frac{1}{Z_l^{ca}} = \frac{Z_l^b}{Z_l^a Z_l^b + Z_l^b Z_l^c + Z_l^c Z_l^a}
\end{aligned}
\tag{4.68}
$$

Substituting, $Z_l^a = R_a = 10\,\Omega$, $Z_l^b = R_b = 15\,\Omega$ and $Z_l^c = R_c = 30\,\Omega$ into above equation, we get the following.

$$
\begin{aligned}
Y_l^{ab} &= G_l^{ab} = \frac{1}{30}\,\Omega \\
Y_l^{bc} &= G_l^{bc} = \frac{1}{90}\,\Omega \\
Y_l^{ca} &= G_l^{ca} = \frac{1}{60}\,\Omega
\end{aligned}
$$

Substituting above values of the admittances in (4.67) , line currents are computed as below.

$$
\begin{aligned}
\bar{I}_a &= \left\{ \frac{1}{30}\left[1-(-\frac{1}{2}-j\frac{\sqrt{3}}{2})\right] - \frac{1}{60}\left[(-\frac{1}{2}+j\frac{\sqrt{3}}{2})-1\right] \right\}V \\
&= V\,(0.075 + j0.0144) = 0.07637\,V\angle 10.89^\circ \\
&= 17.56\angle 10.89^\circ\,\text{A, for } V = 230\,\text{V}
\end{aligned}
$$

Similarly for phase-b current,

$$
\begin{aligned}
\bar{I}_b &= \left\{ \frac{1}{90}\left[(-\frac{1}{2}-j\frac{\sqrt{3}}{2})-(-\frac{1}{2}+j\frac{\sqrt{3}}{2})\right] - \frac{1}{30}\left[1-(-\frac{1}{2}-j\frac{\sqrt{3}}{2})\right]\right\}V \\
&= V(-0.05 - j0.0481) = 0.06933\,V\angle - 136.1° \\
&= 15.94\angle - 136.1°\,\text{A, for } V = 230\,\text{V}
\end{aligned}
$$

Similarly for phase-c current,

$$
\begin{aligned}
\bar{I}_c &= \left\{ \frac{1}{60}\left[(-\frac{1}{2}+j\frac{\sqrt{3}}{2})-1)\right] - \frac{1}{90}\left[(-\frac{1}{2}-j\frac{\sqrt{3}}{2})-(-\frac{1}{2}+j\frac{\sqrt{3}}{2})\right]\right\}V \\
&= V(-0.025 + j0.0336) = 0.04194\,V\angle126.58° \\
&= 9.64\angle126.58°\,\text{A, for } V = 230\,\text{V}
\end{aligned}
$$

Thus, it is found that the above values are the same to those calculated in the previous example. The other quantities such as powers and power factors can also be verified.

4.4.4 AN EXAMPLE OF BALANCING AN UNBALANCED DELTA CONNECTED LOAD

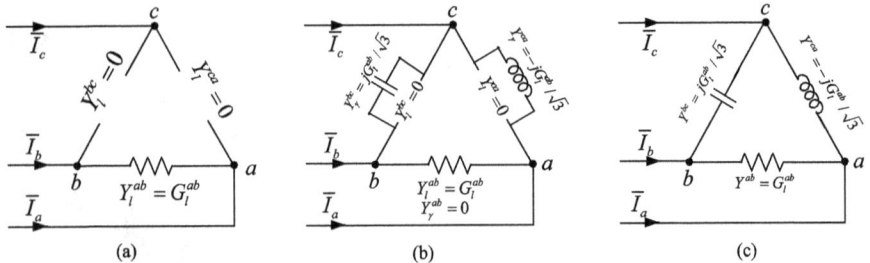

Figure 4.11 (a) An unbalanced three-phase load, (b) with compensator, and (c) compensated system

An unbalanced delta connected load is shown in Fig. 4.11(a). As can be seen from the figure that between phase-a and b, there is admittance $Y_l^{ab} = G_l^{ab}$ and the other two branches are open. This is an example of extreme unbalanced load. Obviously, for this load, line currents will be highly unbalanced. The aim is to make these line currents to be balanced and in phase with their phase voltages. Let us assume that we add admittances Y_γ^{ab}, Y_γ^{bc}, and Y_γ^{ca} between phases ab, bc, and ca, respectively as shown in Fig. 4.11(b) and (c). Let the values of compensator susceptances be given by following expressions.

$$
\begin{aligned}
Y_\gamma^{ab} &= 0 \\
Y_\gamma^{bc} &= jG_l^{ab}/\sqrt{3} \\
Y_\gamma^{ca} &= -jG_l^{ab}/\sqrt{3}.
\end{aligned}
$$

Thus, total admittances between lines are given by,

$$
\begin{aligned}
Y^{ab} &= Y_l^{ab} + Y_\gamma^{ab} = G_l^{ab} + 0 = G_l^{ab} \\
Y^{bc} &= Y_l^{bc} + Y_\gamma^{bc} = 0 + jG_l^{ab}/\sqrt{3} = jG_l^{ab}/\sqrt{3} \\
Y^{ca} &= Y_l^{ca} + Y_\gamma^{ca} = 0 - jG_l^{ab}/\sqrt{3} = -jG_l^{ab}/\sqrt{3}.
\end{aligned}
$$

Therefore the impedances between load lines are given by,

$$
Z^{ab} = \frac{1}{Y^{ab}} = \frac{1}{G_l^{ab}}
$$

$$
Z^{bc} = \frac{1}{Y^{bc}} = \frac{-j\sqrt{3}}{G_l^{ab}}
$$

$$
Z^{ca} = \frac{1}{Y^{ca}} = \frac{j\sqrt{3}}{G_l^{ab}}
$$

Note that $Z^{ab} + Z^{bc} + Z^{ca} = 1/G_l^{ab} - j\sqrt{3}/G_l^{ab} + j\sqrt{3}/G_l^{ab} = 1/G_l^{ab}$.

The impedances, Z_a, Z_b and Z_c of equivalent star connected load are given as follows.

$$
Z_a = \frac{Z^{ab} \times Z^{ca}}{Z^{ab} + Z^{bc} + Z^{ca}} = \frac{\frac{1}{G_l^{ab}} \times \frac{j\sqrt{3}}{G_l^{ab}}}{\frac{1}{G_l^{ab}}} = \frac{j\sqrt{3}}{G_l^{ab}}
$$

$$
Z_b = \frac{Z^{bc} \times Z^{ab}}{Z^{ab} + Z^{bc} + Z^{ca}} = \frac{\frac{1}{G_l^{ab}} \times \frac{-j\sqrt{3}}{G_l^{ab}}}{\frac{1}{G_l^{ab}}} = \frac{-j\sqrt{3}}{G_l^{ab}}
$$

$$
Z_c = \frac{Z^{ca} \times Z^{bc}}{Z^{ab} + Z^{bc} + Z^{ca}} = \frac{\frac{-j\sqrt{3}}{G_l^{ab}} \times \frac{j\sqrt{3}}{G_l^{ab}}}{\frac{1}{G_l^{ab}}} = \frac{3}{G_l^{ab}}
$$

The above impedances as seen from the load side are shown in Fig. 4.12(a). Using (4.56), the voltage between load and system neutral of delta equivalent star load as shown in Fig. 4.12, is computed as below.

$$
\begin{aligned}
V_{nN} &= \frac{1}{\frac{1}{Z_a} + \frac{1}{Z_b} + \frac{1}{Z_c}} \left(\frac{\overline{V}_{aN}}{Z_a} + \frac{\overline{V}_{bN}}{Z_b} + \frac{\overline{V}_{cN}}{Z_c} \right) \\
&= \frac{1}{\frac{1}{(j\sqrt{3}/G_l^{ab})} + \frac{1}{(-j\sqrt{3}/G_l^{ab})} + \frac{1}{3/G_l^{ab}}} \left(\frac{V\angle 0°}{j\sqrt{3}/G_l^{ab}} + \frac{V\angle -120°}{-j\sqrt{3}/G_l^{ab}} + \frac{V\angle 120°}{3/G_l^{ab}} \right) \\
&= \frac{3V}{G_l^{ab}} \left(\frac{G_l^{ab}}{3} - j\frac{G_l^{ab}}{\sqrt{3}} \right) = V(1 - j\sqrt{3}) = 2V \left(\frac{1}{2} - j\frac{\sqrt{3}}{2} \right) \\
&= 2V\angle -60°
\end{aligned}
$$

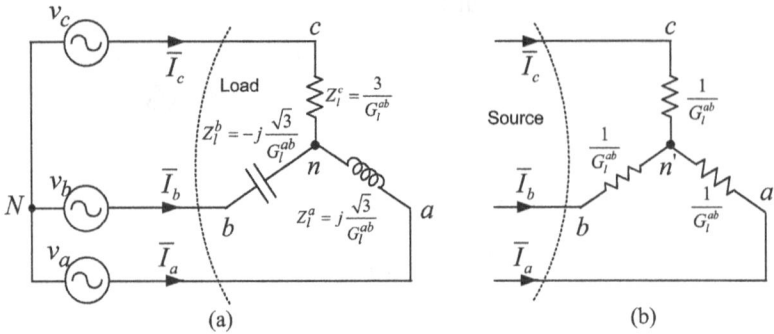

Figure 4.12   Compensated system (a) load side and (b) source side

Using the above value of neutral voltage, the line currents are computed as follows.

$$\bar{I}_a = \frac{\bar{V}_{aN} - \bar{V}_{nN}}{Z_a} = \frac{V\angle 0° - 2V\angle -60°}{j\frac{\sqrt{3}}{G_l^{ab}}} = G_l^{ab}V = G_l^{ab}\bar{V}_a$$

$$\bar{I}_b = \frac{\bar{V}_{bN} - \bar{V}_{nN}}{Z_b} = \frac{V\angle -120° - 2V\angle -60°}{-j\frac{\sqrt{3}}{G_l^{ab}}} = G_l^{ab}V\angle 240° = G_l^{ab}\bar{V}_b$$

$$\bar{I}_c = \frac{\bar{V}_{cN} - \bar{V}_{nN}}{Z_c} = \frac{V\angle 120° - 2V\angle -60°}{\frac{3}{G_l^{ab}}} = G_l^{ab}V\angle 120° = G_l^{ab}\bar{V}_c$$

From the above discussion, it is seen that the three-phase currents are balanced and in phase with their respective voltages. From the load side, the compensated circuit consists of inductance, capacitance and resistance in phase-a, b, and c, respectively. However, from the source side, it is seen as a balanced resistive load. This happens due to the presence of neutral voltage, $\bar{V}_{nN}$. This is illustrated in Fig. 4.12(b).

   It is to be observed that the two load neutrals, i.e., n and n' in Fig. 4.12 (a) and (b), respectively are not at same potential, as we know, $\bar{V}_{nN} = 2V\angle -60°$ and $\bar{V}_{n'N} = 0$. However, the reader may be curious to know why $Y_\gamma^{ab} = 0$, $Y_\gamma^{bc} = jG_l^{ab}/\sqrt{3}$ and $Y_\gamma^{ca} = -jG_l^{ab}/\sqrt{3}$ have been chosen as compensator admittances. The answer to this question can be found in the following section.

4.5   A GENERALIZED APPROACH FOR LOAD COMPENSATION USING SYMMETRICAL COMPONENTS

In the previous section, we have expressed line currents $\bar{I}_a$, $\bar{I}_b$ and $\bar{I}_c$, in terms of load admittances and the voltage V for a delta connected unbalanced load as shown in Fig 4.13(a). For the sake of completeness, these are reproduced below.

$$\bar{I}_{al} = \bar{I}_{abl} - \bar{I}_{cal} = [(1 - \alpha^2)Y_l^{ab} - (\alpha - 1)Y_l^{ca}]V$$
$$\bar{I}_{bl} = \bar{I}_{bcl} - \bar{I}_{abl} = [(\alpha^2 - \alpha)Y_l^{bc} - (1 - \alpha^2)Y_l^{ab}]V \qquad (4.69)$$
$$\bar{I}_{cl} = \bar{I}_{cal} - \bar{I}_{bcl} = [(\alpha - 1)Y_l^{ca} - (\alpha^2 - \alpha)Y_l^{bc}]V$$

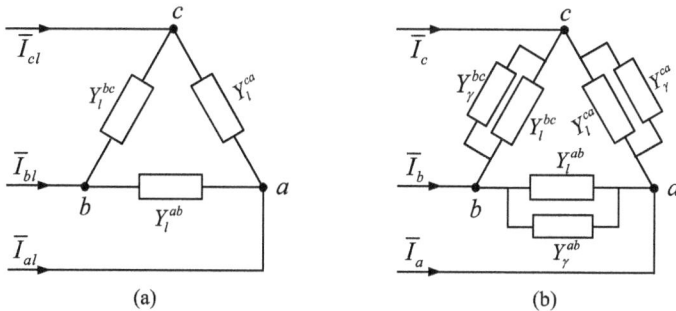

Figure 4.13 (a) An unbalanced delta connected load and (b) compensated system

Since load currents are unbalanced, they will have positive and negative sequence currents. The zero sequence current will be zero, as it is three-phase three-wire system. These symmetrical components of the load currents are expressed as following.

$$\begin{bmatrix} \bar{I}_{0l} \\ \bar{I}_{1l} \\ \bar{I}_{2l} \end{bmatrix} = \frac{1}{\sqrt{3}} \begin{bmatrix} 1 & 1 & 1 \\ 1 & \alpha & \alpha^2 \\ 1 & \alpha^2 & \alpha \end{bmatrix} \begin{bmatrix} \bar{I}_{al} \\ \bar{I}_{bl} \\ \bar{I}_{cl} \end{bmatrix} \qquad (4.70)$$

In (4.70), a factor of $1/\sqrt{3}$ is considered to have an unitary symmetrical transformation. From the above equation, zero sequence current is given below.

$$\bar{I}_{0l} = (\bar{I}_{al} + \bar{I}_{bl} + \bar{I}_{cl})/\sqrt{3} \qquad (4.71)$$

The positive sequence current component is found as follows.

$$\begin{aligned} \bar{I}_{1l} &= \frac{1}{\sqrt{3}}(\bar{I}_{al} + \alpha\bar{I}_{bl} + \alpha^2\bar{I}_{cl}) \\ &= \frac{1}{\sqrt{3}}\left\{ (1 - \alpha^2)Y_l^{ab} - (\alpha - 1)Y_l^{ca} + \alpha\left[(\alpha^2 - \alpha)Y_l^{bc} - (1 - \alpha^2)Y_l^{ab}\right] \right. \\ &\quad \left. + \alpha^2\left[(\alpha - 1)Y_l^{ca} - (\alpha^2 - \alpha)Y_l^{bc}\right] \right\}V \\ &= \frac{1}{\sqrt{3}}[(1 - \alpha^2 - \alpha + \alpha^3)Y_l^{ab} \qquad (4.72) \\ &\quad + (\alpha^3 - \alpha^2 - \alpha^4 + \alpha^3)Y_l^{bc} + (-\alpha + 1 + \alpha^3 - \alpha^2)Y_l^{ca}]V \\ &= \left(Y_l^{ab} + Y_l^{bc} + Y_l^{ca}\right)\sqrt{3}V \end{aligned}$$

Similarly, negative sequence component of the current is,

$$
\begin{aligned}
\bar{I}_{2l} &= \frac{1}{\sqrt{3}}\left(\bar{I}_{al} + \alpha^2 \bar{I}_{bl} + \alpha \bar{I}_{cl}\right) \\
&= \frac{1}{\sqrt{3}}\left\{(1-\alpha^2)Y_l^{ab} - (\alpha-1)Y_l^{ca} + \alpha^2\left[(\alpha^2-\alpha)Y_l^{bc} - (1-\alpha^2)Y_l^{ab}\right]\right. \\
&\quad \left. + \alpha\left[(\alpha-1)Y_l^{ca} - (\alpha^2-\alpha)Y_l^{bc}\right]\right\}V \\
&= \frac{1}{\sqrt{3}}\left[(1-\alpha^2-\alpha^2+\alpha^4)Y_l^{ab} + (\alpha^4-\alpha^3-\alpha^3+\alpha^2)Y_l^{bc}\right. \\
&\quad \left. + (-\alpha+1+\alpha^2-\alpha)Y_l^{ca}\right] \\
&= \frac{1}{\sqrt{3}}\left(-3\alpha^2 Y_l^{ab} - 3Y_l^{bc} - 3\alpha Y_l^{ca}\right)V \\
&= -\left(\alpha^2 Y_l^{ab} + Y_l^{bc} + \alpha Y_l^{ca}\right)\sqrt{3}V
\end{aligned}
\tag{4.73}
$$

From the above, it can be written that,

$$
\begin{aligned}
\bar{I}_{0l} &= 0 \\
\bar{I}_{1l} &= \left(Y_l^{ab} + Y_l^{bc} + Y_l^{ca}\right)\sqrt{3}V \\
\bar{I}_{2l} &= -\left(\alpha^2 Y_l^{ab} + Y_l^{bc} + \alpha Y_l^{ca}\right)\sqrt{3}V
\end{aligned}
\tag{4.74}
$$

When compensator is used, three delta branches Y_γ^{ab}, Y_γ^{bc} and Y_γ^{ca} are added, as shown in Fig. 4.13(b). Using above analysis, the sequence components of the compensator currents can be given as below.

$$
\begin{aligned}
\bar{I}_{0\gamma} &= 0 \\
\bar{I}_{1\gamma} &= \left(Y_\gamma^{ab} + Y_\gamma^{bc} + Y_\gamma^{ca}\right)\sqrt{3}V \\
\bar{I}_{2\gamma} &= -\left(\alpha^2 Y_\gamma^{ab} + Y_\gamma^{bc} + \alpha Y_\gamma^{ca}\right)\sqrt{3}V
\end{aligned}
\tag{4.75}
$$

Since, compensator currents are purely reactive, i.e., $G_\gamma^{ab} = G_\gamma^{bc} = G_\gamma^{ca} = 0$,

$$
\begin{aligned}
Y_\gamma^{ab} &= G_\gamma^{ab} + jB_\gamma^{ab} = jB_\gamma^{ab} \\
Y_\gamma^{bc} &= G_\gamma^{bc} + jB_\gamma^{bc} = jB_\gamma^{bc} \\
Y_\gamma^{ca} &= G_\gamma^{ca} + jB_\gamma^{ca} = jB_\gamma^{ca}.
\end{aligned}
\tag{4.76}
$$

Using the above relations, the compensated sequence currents can be written as,

$$
\begin{aligned}
\bar{I}_{0\gamma} &= 0 \\
\bar{I}_{1\gamma} &= j\left(B_\gamma^{ab} + B_\gamma^{bc} + B_\gamma^{ca}\right)\sqrt{3}V \\
\bar{I}_{2\gamma} &= -j(\alpha^2 B_\gamma^{ab} + B_\gamma^{bc} + \alpha B_\gamma^{ca})\sqrt{3}V
\end{aligned}
\tag{4.77}
$$

On knowing the nature of compensator and load currents, we can set compensation objectives as following.

1. All negative sequence component of the load current must be supplied from the compensator negative current, i.e.,

$$\bar{I}_{2\gamma} = -\bar{I}_{2l} \qquad (4.78)$$

The above further implies that,

$$\mathrm{Re}\left(\bar{I}_{2\gamma}\right) + j\mathrm{Im}\left(\bar{I}_{2\gamma}\right) = -\mathrm{Re}\left(\bar{I}_{2l}\right) - j\mathrm{Im}\left(\bar{I}_{2l}\right). \qquad (4.79)$$

The above equation gives,

$$\mathrm{Re}\left(\bar{I}_{2\gamma}\right) = -\mathrm{Re}\left(\bar{I}_{2l}\right) \qquad (4.80)$$

$$\mathrm{Im}\left(\bar{I}_{2\gamma}\right) = -\mathrm{Im}\left(\bar{I}_{2l}\right). \qquad (4.81)$$

In the above equations, Re and Im denote the real and imaginary parts of the current phasor.

2. The total positive sequence current, which is source current should have desired power factor from the source, i.e.,

$$\frac{\mathrm{Im}\left(\bar{I}_{1l} + \bar{I}_{1\gamma}\right)}{\mathrm{Re}\left(\bar{I}_{1l} + \bar{I}_{1\gamma}\right)} = \tan\phi = \beta \qquad (4.82)$$

Where, ϕ is the desired phase angle between the line currents and the supply voltages. The above equation thus implies that,

$$\mathrm{Im}\left(\bar{I}_{1l} + \bar{I}_{1\gamma}\right) = \beta\,\mathrm{Re}\left(\bar{I}_{1l} + \bar{I}_{1\gamma}\right) \qquad (4.83)$$

Since $\mathrm{Re}\left(\bar{I}_{1\gamma}\right) = 0$, the above equation is rewritten as following.

$$\mathrm{Im}\left(\bar{I}_{1\gamma}\right) = -\mathrm{Im}\left(\bar{I}_{1l}\right) + \beta\,\mathrm{Re}\left(\bar{I}_{1l}\right) \qquad (4.84)$$

There are three unknown variables, i.e., the compensator susceptances, B_γ^{ab}, B_γ^{bc} and B_γ^{ca} and three conditions given by (4.80), (4.81), and (4.84). Therefore, the unknown variables can be solved as described in the following section.
Using (4.77), the current $\bar{I}_{2\gamma}$ is expressed as following.

$$
\begin{aligned}
\bar{I}_{2\gamma} &= -j(\alpha^2 B_\gamma^{ab} + B_\gamma^{bc} + \alpha B_\gamma^{ca})\sqrt{3}V \\
&= -j\left[\left(-\frac{1}{2} - j\frac{\sqrt{3}}{2}\right)B_\gamma^{ab} + B_\gamma^{bc} + \left(-\frac{1}{2} + j\frac{\sqrt{3}}{2}\right)B_\gamma^{ca}\right]\sqrt{3}V \\
&= \left[\left(-\frac{\sqrt{3}}{2}B_\gamma^{ab} + \frac{\sqrt{3}}{2}B_\gamma^{ca}\right) + j\left(\frac{1}{2}B_\gamma^{ab} - B_\gamma^{bc} + \frac{1}{2}B_\gamma^{ca}\right)\right]\sqrt{3}V \quad (4.85) \\
&= \mathrm{Re}\left(\bar{I}_{2\gamma}\right) + j\mathrm{Im}\left(\bar{I}_{2\gamma}\right)
\end{aligned}
$$

Thus, the above equation implies that,

$$\left(\frac{\sqrt{3}}{2}B_\gamma^{ab} - \frac{\sqrt{3}}{2}B_\gamma^{ca}\right) = -\frac{1}{\sqrt{3}V}\text{Re}\left(\bar{I}_{2\gamma}\right) = \frac{1}{\sqrt{3}V}\text{Re}\left(\bar{I}_{2l}\right) \tag{4.86}$$

and,

$$\left(-\frac{1}{2}B_\gamma^{ab} + B_\gamma^{bc} - \frac{1}{2}B_\gamma^{ca}\right) = -\frac{1}{\sqrt{3}V}\text{Im}\left(\bar{I}_{2\gamma}\right) = \frac{1}{\sqrt{3}V}\text{Im}\left(\bar{I}_{2l}\right) \tag{4.87}$$

or,

$$\left(-B_\gamma^{ab} + 2B_\gamma^{bc} - B_\gamma^{ca}\right) = \frac{1}{\sqrt{3}V}2\text{Im}\left(\bar{I}_{2l}\right) \tag{4.88}$$

From (4.77), $\text{Im}\left(\bar{I}_{1\gamma}\right)$ can be written as,

$$\text{Im}\left(\bar{I}_{1\gamma}\right) = \left(B_\gamma^{ab} + B_\gamma^{bc} + B_\gamma^{ca}\right)\sqrt{3}V \tag{4.89}$$

Substituting $\text{Im}\left(\bar{I}_{1\gamma}\right)$ from (4.84) into (4.89), we get the following.

$$B_\gamma^{ab} + B_\gamma^{bc} + B_\gamma^{ca} = \frac{1}{\sqrt{3}V}\text{Im}\left(\bar{I}_{1\gamma}\right) = -\frac{1}{\sqrt{3}V}\left[\text{Im}\left(\bar{I}_{1l}\right) - \beta\,\text{Re}\left(\bar{I}_{1l}\right)\right] \tag{4.90}$$

Adding (4.88) and (4.90), the following is obtained,

$$B_\gamma^{bc} = \frac{-1}{3\sqrt{3}V}\left[\text{Im}\left(\bar{I}_{1l}\right) - 2\text{Im}\left(\bar{I}_{2l}\right) - \beta\,\text{Re}\left(\bar{I}_{1l}\right)\right] \tag{4.91}$$

Now, from (4.87) we have

$$\begin{aligned}
-\frac{1}{2}B_\gamma^{ab} - \frac{1}{2}B_\gamma^{ca} &= \frac{1}{\sqrt{3}V}\text{Im}\left(\bar{I}_{2l}\right) - B_\gamma^{bc} \\
&= \frac{1}{\sqrt{3}V}\text{Im}\left(\bar{I}_{2l}\right) - \left\{\frac{-1}{3\sqrt{3}V}\left[\text{Im}\left(\bar{I}_{1l}\right) - \beta\,\text{Re}\left(\bar{I}_{1l}\right) - 2\text{Im}\left(\bar{I}_{2l}\right)\right]\right\} \\
&= \frac{1}{3\sqrt{3}V}\left[\text{Im}(\bar{I}_{1l}) + \text{Im}\left(\bar{I}_{2l}\right) - \beta\,\text{Re}\left(\bar{I}_{1l}\right)\right] \tag{4.92}
\end{aligned}$$

Reconsidering (4.92) and (4.86), we have

$$\begin{aligned}
-\frac{1}{2}B_\gamma^{ab} - \frac{1}{2}B_\gamma^{ca} &= \frac{1}{3\sqrt{3}V}\left[\text{Im}\left(\bar{I}_{1l}\right) + \text{Im}\left(\bar{I}_{2l}\right) - \beta\,\text{Re}\left(\bar{I}_{1l}\right)\right] \\
\frac{1}{2}B_\gamma^{ab} - \frac{1}{2}B_\gamma^{ca} &= \frac{1}{3\sqrt{3}V}\left[\sqrt{3}\text{Re}\left(\bar{I}_{2l}\right)\right]
\end{aligned}$$

Adding above equations, we get

$$B_\gamma^{ca} = \frac{-1}{3\sqrt{3}V}\left[\text{Im}\left(\bar{I}_{1l}\right) + \text{Im}\left(\bar{I}_{2l}\right) + \sqrt{3}\text{Re}\left(\bar{I}_{2l}\right) - \beta\,\text{Re}\left(\bar{I}_{1l}\right)\right]. \tag{4.93}$$

Therefore,

$$
\begin{aligned}
B_\gamma^{ab} &= B_\gamma^{ca} + \frac{2}{3\sqrt{3}V}[\sqrt{3}\,\mathrm{Re}\,(\bar{I}_{2l})] \\
&= \frac{-1}{3\sqrt{3}V}[\mathrm{Im}\,(\bar{I}_{1l}) + \mathrm{Im}\,(\bar{I}_{2l}) + \sqrt{3}\,\mathrm{Re}\,(\bar{I}_{2l}) - \beta\,\mathrm{Re}\,(\bar{I}_{1l}) - 2\sqrt{3}\,\mathrm{Re}\,(\bar{I}_{2l})] \\
&= \frac{-1}{3\sqrt{3}V}[\mathrm{Im}\,(\bar{I}_{1l}) + \mathrm{Im}\,(\bar{I}_{2l}) - \sqrt{3}\,\mathrm{Re}\,(\bar{I}_{2l}) - \beta\,\mathrm{Re}\,(\bar{I}_{1l})] \quad (4.94)
\end{aligned}
$$

Similarly, B_γ^{bc} can be written as in the following.

$$
B_\gamma^{bc} = -\frac{1}{3\sqrt{3}V}\left[\mathrm{Im}(\bar{I}_{1l}) - 2\,\mathrm{Im}(\bar{I}_{2l}) - \beta\,\mathrm{Re}\,(\bar{I}_{1l})\right] \quad (4.95)
$$

From the above, the compensator susceptances in terms of real and imaginary parts of the load current can be written as follows.

$$
\begin{aligned}
B_\gamma^{ab} &= \frac{-1}{3\sqrt{3}V}[\mathrm{Im}\,(\bar{I}_{1l}) + \mathrm{Im}(\bar{I}_{2l}) - \sqrt{3}\,\mathrm{Re}\,(\bar{I}_{2l}) - \beta\,\mathrm{Re}(\bar{I}_{1l})] \\
B_\gamma^{bc} &= -\frac{1}{3\sqrt{3}V}[\mathrm{Im}(\bar{I}_{1l}) - 2\,\mathrm{Im}(\bar{I}_{2l}) - \beta\,\mathrm{Re}(\bar{I}_{1l})] \quad (4.96) \\
B_\gamma^{ca} &= \frac{-1}{3\sqrt{3}V}[\mathrm{Im}\,(\bar{I}_{1l}) + \mathrm{Im}(\bar{I}_{2l}) + \sqrt{3}\,\mathrm{Re}(\bar{I}_{2l}) - \beta\,\mathrm{Re}(\bar{I}_{1l})]
\end{aligned}
$$

In the above equation, the susceptances of the compensator are expressed in terms of real and imaginary parts of symmetrical components of load currents. It is however advantageous to express these susceptances in terms of instantaneous values of voltages and currents from implementation point of view. The first step to achieve this is to express these susceptances in terms of load currents, i.e., $\bar{I}_{al}$, $\bar{I}_{bl}$, and $\bar{I}_{cl}$, which is described below. Using (4.70), the sequence components of the load currents are expressed as,

$$
\begin{aligned}
\bar{I}_{0l} &= \frac{1}{\sqrt{3}}\,(\bar{I}_{al} + \bar{I}_{bl} + \bar{I}_{cl}) \\
\bar{I}_{1l} &= \frac{1}{\sqrt{3}}\,(\bar{I}_{al} + \alpha\bar{I}_{bl} + \alpha^2\bar{I}_{cl}) \quad (4.97) \\
\bar{I}_{2l} &= \frac{1}{\sqrt{3}}\,(\bar{I}_{al} + \alpha^2\bar{I}_{1l} + \alpha\bar{I}_{cl}).
\end{aligned}
$$

Substituting these values of sequence components of load currents, in (4.96), we can obtain compensator susceptances in terms of real and imaginary components of the load currents. Let us start with B_γ^{bc}, as obtained in the following.

$$B_\gamma^{bc} = -\frac{1}{3\sqrt{3}V}\left[\text{Im}(\bar{I}_{1l}) - 2\text{Im}(\bar{I}_{2l}) - \beta\,\text{Re}(\bar{I}_{1l})\right]$$

$$= -\frac{1}{3\sqrt{3}V}\left[\text{Im}\left(\frac{\bar{I}_{al} + \alpha\bar{I}_{bl} + \alpha^2\bar{I}_{cl}}{\sqrt{3}}\right) - 2\text{Im}\left(\frac{\bar{I}_{al} + \alpha^2\bar{I}_{bl} + \alpha\bar{I}_{cl}}{\sqrt{3}}\right)\right.$$
$$\left. -\beta\,\text{Re}\left(\frac{\bar{I}_{al} + \alpha\bar{I}_{bl} + \alpha^2\bar{I}_{cl}}{\sqrt{3}}\right)\right]$$

$$= -\frac{1}{9V}\left\{\text{Im}\left[-\bar{I}_{al} + (\alpha - 2\alpha^2)\bar{I}_{bl} + (\alpha^2 - 2\alpha)\bar{I}_{cl}\right] - \beta\,\text{Re}\left(\bar{I}_{al} + \alpha\bar{I}_{bl} + \alpha^2\bar{I}_{cl}\right)\right\}$$

$$= -\frac{1}{9V}\left\{\text{Im}\left[-\bar{I}_{al} + (2 + 3\alpha)\bar{I}_{bl} + (2 + 3\alpha^2)\bar{I}_{cl}\right] - \beta\,\text{Re}\left(\bar{I}_{al} + \alpha\bar{I}_{bl} + \alpha^2\bar{I}_{cl}\right)\right\}$$

$$= -\frac{1}{9V}\left\{\text{Im}\left[-\bar{I}_{al} + 2\bar{I}_{bl} + 2\bar{I}_{cl} + 3\alpha\bar{I}_{bl} + 3\alpha^2\bar{I}_{cl}\right] - \beta\,\text{Re}\left(\bar{I}_{al} + \alpha\bar{I}_{bl} + \alpha^2\bar{I}_{cl}\right)\right\}$$

By adding and subtracting $\bar{I}_{bl}$ and $\bar{I}_{cl}$ in the above equation we get,

$$B_\gamma^{bc} = -\frac{1}{9V}\left\{\text{Im}\left[(-\bar{I}_{al} - \bar{I}_{bl} - \bar{I}_{cl}) + 3\bar{I}_{bl} + 3\bar{I}_{cl} + 3\alpha\bar{I}_{bl} + 3\alpha^2\bar{I}_{cl}\right]\right.$$
$$\left. -\beta\,\text{Re}(\bar{I}_{al} + \alpha\bar{I}_{bl} + \alpha^2\bar{I}_{cl})\right\}$$

We know that $\bar{I}_{al} + \bar{I}_{bl} + \bar{I}_{cl} = 0$, therefore $\bar{I}_{bl} + \bar{I}_{cl} = -\bar{I}_{al}$.

$$B_\gamma^{bc} = -\frac{1}{3V}\left[-\text{Im}(\bar{I}_{al}) + \text{Im}(\alpha\bar{I}_{bl}) + \text{Im}(\alpha^2\bar{I}_{cl}) - \frac{\beta}{3}\text{Re}(\bar{I}_{al} + \alpha\bar{I}_{bl} + \alpha^2\bar{I}_{cl})\right] \tag{4.98}$$

Similarly, it can be proved that,

$$B_\gamma^{ca} = -\frac{1}{3V}\left[\text{Im}(\bar{I}_{al}) - \text{Im}(\alpha\bar{I}_{bl}) + \text{Im}(\alpha^2\bar{I}_{cl}) - \frac{\beta}{3}\text{Re}(\bar{I}_{al} + \alpha\bar{I}_{bl} + \alpha^2\bar{I}_{cl})\right] \tag{4.99}$$

$$B_\gamma^{ab} = -\frac{1}{3V}\left[\text{Im}(\bar{I}_{al}) + \text{Im}(\alpha\bar{I}_{bl}) - \text{Im}(\alpha^2\bar{I}_{cl}) - \frac{\beta}{3}\text{Re}(\bar{I}_{al} + \alpha\bar{I}_{bl} + \alpha^2\bar{I}_{cl})\right] \tag{4.100}$$

The above expressions for B_γ^{ca} and B_γ^{ab} are proved below. For convenience, the last term associated with β is considered zero for unity power factor case.

$$B_\gamma^{ca} = \frac{-1}{3V}\left[\text{Im}(\bar{I}_{al}) - \text{Im}(\alpha\bar{I}_{bl}) + \text{Im}(\alpha^2\bar{I}_{cl})\right]$$

$$= -\frac{1}{3V}\left[\text{Im}\left\{\frac{(\bar{I}_{0l} + \bar{I}_{1l} + \bar{I}_{2l})}{\sqrt{3}} - \alpha\frac{(\bar{I}_{0l} + \alpha^2\bar{I}_{1l} + \alpha\bar{I}_{2l})}{\sqrt{3}}\right.\right.$$
$$\left.\left. + \alpha^2\frac{(\bar{I}_{0l} + \alpha\bar{I}_{1l} + \alpha^2\bar{I}_{2l})}{\sqrt{3}}\right\}\right]$$

Since $\bar{I}_{0l} = 0$

$$B_\gamma^{ca} = -\frac{1}{3\sqrt{3}V}\mathrm{Im}\left(\bar{I}_{1l} - 2\alpha^2\bar{I}_{2l}\right)$$

$$= -\frac{1}{3\sqrt{3}V}\mathrm{Im}\left[\bar{I}_{1l} - 2\left(-\frac{1}{2} - j\frac{\sqrt{3}}{2}\right)\bar{I}_{2l}\right]$$

$$= -\frac{1}{3\sqrt{3}V}\mathrm{Im}\left(\bar{I}_{1l} + \bar{I}_{2l} + j\sqrt{3}\bar{I}_{2l}\right)$$

$$= -\frac{1}{3\sqrt{3}V}\left[\mathrm{Im}\left(\bar{I}_{1l}\right) + \mathrm{Im}\left(\bar{I}_{2l}\right) + \sqrt{3}\mathrm{Re}\left(\bar{I}_{2l}\right)\right]$$

Note that $\mathrm{Im}(j\bar{I}_{2l}) = \mathrm{Re}(\bar{I}_{2l})$.

Adding the β term, we get the following.

$$B_\gamma^{ca} = -\frac{1}{3\sqrt{3}V}\left[\mathrm{Im}(\bar{I}_{1l}) + \mathrm{Im}(\bar{I}_{2l}) + \sqrt{3}\mathrm{Re}\left(\bar{I}_{2l}\right) - \beta\mathrm{Re}(\bar{I}_{1l})\right]$$

Similarly,

$$B_\gamma^{ab} = -\frac{1}{3V}\left[\mathrm{Im}\left(\bar{I}_{al}\right) + \mathrm{Im}\left(\alpha\bar{I}_{bl}\right) - \mathrm{Im}\left(\alpha^2\bar{I}_{cl}\right)\right]$$

$$= -\frac{1}{3V}\mathrm{Im}\left[\frac{(\bar{I}_{0l} + \bar{I}_{1l} + \bar{I}_{2l})}{\sqrt{3}} + \alpha\frac{(\bar{I}_{0l} + \alpha^2\bar{I}_{1l} + \alpha\bar{I}_{2l})}{\sqrt{3}}\right.$$

$$\left. - \alpha^2\frac{(\bar{I}_{0l} + \alpha\bar{I}_{1l} + \alpha^2\bar{I}_{2l})}{\sqrt{3}}\right]$$

$$B_\gamma^{ab} = -\frac{1}{3\sqrt{3}V}\mathrm{Im}\left(\bar{I}_{1l} - 2\alpha\bar{I}_{2l}\right)$$

$$= -\frac{1}{3\sqrt{3}V}\mathrm{Im}\left[\bar{I}_{1l} - 2\left(-\frac{1}{2} + j\frac{\sqrt{3}}{2}\right)\bar{I}_{2l}\right]$$

$$= -\frac{1}{3\sqrt{3}V}\mathrm{Im}\left(\bar{I}_{1l} + \bar{I}_{2l} - j\sqrt{3}\bar{I}_{2l}\right)$$

$$= -\frac{1}{3\sqrt{3}V}\left[\mathrm{Im}\left(\bar{I}_{1l}\right) + \mathrm{Im}\left(\bar{I}_{2l}\right) - \sqrt{3}\mathrm{Re}\left(\bar{I}_{2l}\right)\right]$$

Thus, compensator susceptances are expressed as follows.

$$B_\gamma^{ab} = -\frac{1}{3V}\left[\mathrm{Im}(\bar{I}_{al}) + \mathrm{Im}(\alpha\bar{I}_{bl}) - \mathrm{Im}(\alpha^2\bar{I}_{cl}) - \frac{\beta}{3}\mathrm{Re}(\bar{I}_{al} + \alpha\bar{I}_{bl} + \alpha^2\bar{I}_{cl})\right]$$

$$B_\gamma^{bc} = -\frac{1}{3V}\left[-\mathrm{Im}(\bar{I}_{al}) + \mathrm{Im}(\alpha\bar{I}_{bl}) + \mathrm{Im}(\alpha^2\bar{I}_{cl}) - \frac{\beta}{3}\mathrm{Re}(\bar{I}_{al} + \alpha\bar{I}_{bl} + \alpha^2\bar{I}_{cl})\right]$$

$$B_\gamma^{ca} = -\frac{1}{3V}\left[\mathrm{Im}(\bar{I}_{al}) - \mathrm{Im}(\alpha\bar{I}_{bl}) + \mathrm{Im}(\alpha^2\bar{I}_{cl}) - \frac{\beta}{3}\mathrm{Re}(\bar{I}_{al} + \alpha\bar{I}_{bl} + \alpha^2\bar{I}_{cl})\right]$$

$$(4.101)$$

An unity power factor is desired from the source. For this $\cos\phi_l = 1$, implying $\tan\phi_l = 0$ hence $\beta = 0$. Thus, we have,

$$
\begin{aligned}
B_\gamma^{ab} &= -\frac{1}{3V}\left[\operatorname{Im}\left(\bar{I}_{al}\right) + \operatorname{Im}\left(\alpha\bar{I}_{bl}\right) - \operatorname{Im}\left(\alpha^2\bar{I}_{cl}\right)\right] \\
B_\gamma^{bc} &= -\frac{1}{3V}\left[-\operatorname{Im}\left(\bar{I}_{al}\right) + \operatorname{Im}\left(\alpha\bar{I}_{bl}\right) + \operatorname{Im}\left(\alpha^2\bar{I}_{cl}\right)\right] \qquad (4.102) \\
B_\gamma^{ca} &= -\frac{1}{3V}\left[\operatorname{Im}\left(\bar{I}_{al}\right) - \operatorname{Im}\left(\alpha\bar{I}_{bl}\right) + \operatorname{Im}\left(\alpha^2\bar{I}_{cl}\right)\right]
\end{aligned}
$$

The above equations are easy to realize in order to find compensator susceptances. As mentioned above, sampling and averaging techniques will be used to convert the above equation into their time equivalents. These are described below.

4.5.1 SAMPLING METHOD

An instantaneous phase current is written as follows.

$$
i_{al}(t) = \sqrt{2}\operatorname{Im}\left(\bar{I}_{al}\,e^{j\omega t}\right) = \sqrt{2}\operatorname{Im}\left[\bar{I}_{al}(\cos\omega t + j\sin\omega t)\right] \qquad (4.103)
$$

$$
\operatorname{Im}\left(\bar{I}_{al}\right) = \frac{i_{al}(t)}{\sqrt{2}} \quad \text{at } \sin\omega t = 0,\ \cos\omega t = 1
$$

From (4.62), the phase voltages in the time domain can be expressed as below.

$$
\begin{aligned}
v_a(t) &= \sqrt{2}V\,\sin\omega t \\
v_b(t) &= \sqrt{2}V\,\sin(\omega t - 120^\circ) \qquad (4.104) \\
v_c(t) &= \sqrt{2}V\,\sin(\omega t + 120^\circ)
\end{aligned}
$$

From above voltage expressions, it is to be noted that, $\sin\omega t = 0$ and $\cos\omega t = 1$ implies that the phase-a voltage, $v_a(t)$ is going through a positive zero crossing, hence, $v_a(t) = 0$ and $\frac{dv_a(t)}{dt} = 0$. Therefore, (4.104) can be expressed as following.

$$
\operatorname{Im}(\bar{I}_{a,l}) = \frac{i_{al}(t)}{\sqrt{2}} \quad \text{when, } v_a(t) = 0, dv_a/dt > 0 \qquad (4.105)
$$

Similar to (4.103), we can express phase-b current in terms of $\operatorname{Im}(\alpha\bar{I}_{bl})$, as given below.

$$
\begin{aligned}
i_{bl}(t) &= \sqrt{2}\operatorname{Im}(\bar{I}_{bl}\,e^{j\omega t}) = \sqrt{2}\operatorname{Im}(\alpha\bar{I}_{bl}\,e^{j\omega t}\alpha^{-1}) \\
&= \sqrt{2}\operatorname{Im}\left\{\alpha\bar{I}_{bl}\left[\cos(\omega t - 120^\circ) + j\sin(\omega t - 120^\circ)\right]\right\} \qquad (4.106)
\end{aligned}
$$

From the above equation, we get the following.

$$
\operatorname{Im}\left(\alpha\bar{I}_{bl}\right) = \frac{i_{bl}(t)}{\sqrt{2}} \quad \text{at } \omega t - 120^\circ = 0,\ \text{which implies } v_b(t) = 0, dv_b/dt > 0
$$

$$
(4.107)
$$

Similarly, for phase-c, it can be proved that,

$$\text{Im}\left(\alpha^2 \bar{I}_{cl}\right) = \frac{i_{cl}(t)}{\sqrt{2}} \quad \text{when, } v_c(t) = 0, dv_c/dt > 0 \tag{4.108}$$

Substituting $\text{Im}\left(\bar{I}_{al}\right)$, $\text{Im}\left(\alpha \bar{I}_{bl}\right)$ and $\text{Im}\left(\alpha^2 \bar{I}_{cl}\right)$ from (4.105), (4.107), and (4.108), respectively, in (4.102), we get the following.

$$
\begin{aligned}
B_\gamma^{ab} &= -\frac{1}{3\sqrt{2}V}\left[i_{al}\big|_{(v_a=0, \frac{dv_a}{dt}>0)} + i_{bl}\big|_{(v_b=0, \frac{dv_b}{dt}>0)} - i_{cl}\big|_{(v_c=0, \frac{dv_c}{dt}>0)} \right] \\
B_\gamma^{bc} &= -\frac{1}{3\sqrt{2}V}\left[-i_{al}\big|_{(v_a=0, \frac{dv_a}{dt}>0)} + i_{bl}\big|_{(v_b=0, \frac{dv_b}{dt}>0)} + i_{cl}\big|_{(v_c=0, \frac{dv_c}{dt}>0)} \right] \\
B_\gamma^{ca} &= -\frac{1}{3\sqrt{2}V}\left[i_{al}\big|_{(v_a=0, \frac{dv_a}{dt}>0)} - i_{bl}\big|_{(v_b=0, \frac{dv_b}{dt}>0)} + i_{cl}\big|_{(v_c=0, \frac{dv_c}{dt}>0)} \right]
\end{aligned} \tag{4.109}
$$

Thus, the desired compensating susceptances are expressed in terms of the three line currents sampled at instants defined at positive-going zero crossings of the line to neutral voltages v_a, v_b, v_c. An artificial neutral at ground potential may be created by measuring voltages v_a, v_b, and v_c to implement above algorithm. Since the method involves computation of samples of load currents, it is called as sampling method.

4.5.2 AVERAGING METHOD

In this method, we express the compensator susceptances in terms of real and reactive power terms and finally expressed them in time domain through averaging process. The method is described below.

From (4.102), susceptance, B_γ^{ab}, can be re-written as following.

$$
\begin{aligned}
B_\gamma^{ab} &= -\frac{1}{3V}\left[\text{Im}\left(\bar{I}_{al}\right) + \text{Im}\left(\alpha \bar{I}_{bl}\right) - \text{Im}\left(\alpha^2 \bar{I}_{cl}\right) \right] \\
&= -\frac{1}{3V^2}\left[V\text{Im}\left(\bar{I}_{al}\right) + V\text{Im}\left(\alpha \bar{I}_{bl}\right) - V\text{Im}\left(\alpha^2 \bar{I}_{cl}\right) \right] \\
&= -\frac{1}{3V^2}\left[\text{Im}\left(V \bar{I}_{al}\right) + \text{Im}\left(V \alpha \bar{I}_{bl}\right) - \text{Im}\left(V \alpha^2 \bar{I}_{cl}\right) \right]
\end{aligned} \tag{4.110}
$$

Note the following property of phasors and applying it for the simplification of the above expression.

$$\text{Im}\left(\bar{V}\bar{I}\right) = -\text{Im}\left(\bar{V}\bar{I}\right)^* = -\text{Im}\left(\bar{V}^*\bar{I}^*\right) \tag{4.111}$$

Using above equation, (4.110) can be written as,

$$
\begin{aligned}
B_\gamma^{ab} &= \frac{1}{3V^2}\left[\text{Im}\left(\bar{V}_a \bar{I}_{al}\right)^* + \text{Im}\left(\alpha \bar{V}_a \bar{I}_{bl}\right)^* - \text{Im}\left(\alpha^2 \bar{V}\bar{I}_{cl}\right)^* \right] \\
&= \frac{1}{3V^2}\left[\text{Im}\left(\bar{V}_a^* \bar{I}_{al}^*\right) + \text{Im}\left(\alpha^* \bar{V}_a^* \bar{I}_{bl}^*\right) - \text{Im}\left((\alpha^2)^* \bar{V}_a^* \bar{I}_{cl}^*\right) \right] \\
&= \frac{1}{3V^2}\left[\text{Im}\left(\bar{V}_a^* \bar{I}_{al}^*\right) + \text{Im}\left(\alpha^2 \bar{V}_a^* \bar{I}_{bl}^*\right) - \text{Im}\left(\alpha \bar{V}_a^* \bar{I}_{cl}^*\right) \right]
\end{aligned}
$$

Since $\overline{V}_a = V\angle 0°$ is a reference phasor, therefore $\overline{V}_a = \overline{V}_a^* = V$, $\alpha^2 \overline{V}_a = \overline{V}_b^*$ and $\alpha \overline{V}_a = \overline{V}_c^*$. Using this, the above equation can be written as follows.

$$B_\gamma^{ab} = \frac{1}{3V^2}\left[\mathrm{Im}(\overline{V}_a\overline{I}_{al}^*) + \mathrm{Im}(\overline{V}_b\overline{I}_{bl}^*) - \mathrm{Im}(\overline{V}_c\overline{I}_{cl}^*)\right]$$

Similarly, $\quad B_\gamma^{bc} = \frac{1}{3V^2}\left[-\mathrm{Im}(\overline{V}_a\overline{I}_{al}^*) + \mathrm{Im}(\overline{V}_b\overline{I}_{bl}^*) + \mathrm{Im}(\overline{V}_c\overline{I}_{cl}^*)\right] \quad (4.112)$

$$B_\gamma^{ca} = \frac{1}{3V^2}\left[\mathrm{Im}(\overline{V}_a\overline{I}_{al}^*) - \mathrm{Im}(\overline{V}_b\overline{I}_{bl}^*) + \mathrm{Im}(\overline{V}_c\overline{I}_{cl}^*)\right]$$

It can be further proved that,

$$\mathrm{Im}(\overline{V}_a\overline{I}_a^*) = \frac{1}{T}\int_0^T v_a(t)\angle(-\pi/2)\,i_{al}(t)\,dt$$

$$\mathrm{Im}(\overline{V}_b\overline{I}_b^*) = \frac{1}{T}\int_0^T v_b(t)\angle(-\pi/2)\,i_{bl}(t)\,dt \qquad (4.113)$$

$$\mathrm{Im}(\overline{V}_c\overline{I}_c^*) = \frac{1}{T}\int_0^T v_c(t)\angle(-\pi/2)\,i_{cl}(t)\,dt$$

In (4.113), the term $v_a(t)\angle(-\pi/2)$ denotes the voltage $v_a(t)$ shifted by $-\pi/2$ radian in time domain. For fundamental balanced voltages, the following relationship between phase and line voltages are true.

$$\begin{aligned} v_a(t)\angle(-\pi/2) &= \sqrt{2}V\sin(\omega t - \pi/2) = v_{bc}(t)/\sqrt{3} \\ v_b(t)\angle(-\pi/2) &= \sqrt{2}V\sin(\omega t - 2\pi/3 - \pi/2) = v_{ca}(t)/\sqrt{3} \quad (4.114) \\ v_c(t)\angle(-\pi/2) &= \sqrt{2}V\sin(\omega t + 2\pi/3 - \pi/2) = v_{ca}(t)/\sqrt{3} \end{aligned}$$

From (4.113) and (4.114), the following can be written.

$$\mathrm{Im}(\overline{V}_a\overline{I}_a^*) = \frac{1}{\sqrt{3}\,T}\int_0^T v_{bc}(t)\,i_{al}(t)\,dt$$

$$\mathrm{Im}(\overline{V}_b\overline{I}_b^*) = \frac{1}{\sqrt{3}\,T}\int_0^T v_{ca}(t)\,i_{bl}(t)\,dt \qquad (4.115)$$

$$\mathrm{Im}(\overline{V}_c\overline{I}_c^*) = \frac{1}{\sqrt{3}\,T}\int_0^T v_{ab}(t)\,i_{cl}(t)\,dt$$

In the above, T is the time period of system voltage. Substituting the above values of $\mathrm{Im}(\overline{V}_a\overline{I}_a^*)$, $\mathrm{Im}(\overline{V}_b\overline{I}_b^*)$, and $\mathrm{Im}(\overline{V}_c\overline{I}_c^*)$ into (4.112), we get the following.

$$B_\gamma^{ab} = \frac{1}{(3\sqrt{3}V^2)}\frac{1}{T}\int_0^T (v_{bc}\,i_{al} + v_{ca}\,i_{bl} - v_{ab}\,i_{cl})\,dt$$

$$B_\gamma^{bc} = \frac{1}{(3\sqrt{3}V^2)}\frac{1}{T}\int_0^T (-v_{bc}\,i_{al} + v_{ca}\,i_{bl} + v_{ab}\,i_{cl})\,dt \qquad (4.116)$$

$$B_\gamma^{ca} = \frac{1}{(3\sqrt{3}V^2)}\frac{1}{T}\int_0^T (v_{bc}\,i_{al} - v_{ca}\,i_{bl} + v_{ab}\,i_{cl})\,dt$$

The above equations can directly be used to know the compensator susceptances by performing averaging the product of the line to line voltages and phase load currents as per (4.116). The term $\int_0^T (.)dt = \int_{t_1-T}^{t_1}(.)dt$, where t_1 is any arbitrary instant, can be implemented using moving an average of one cycle. This improves transient response by computing the average value at each instant. But in this case, the controller response, which changes the susceptance value, should match the speed of the computing algorithm.

4.6 COMPENSATOR ADMITTANCE REPRESENTED AS POSITIVE AND NEGATIVE SEQUENCE ADMITTANCE NETWORK

Recalling the following relations from equation (4.96) for unity power factor operation i.e., $\beta = 0$, we get the following.

$$
\begin{aligned}
B_\gamma^{ab} &= \frac{-1}{3\sqrt{3}V}[\text{Im }(\bar{I}_{1l}) + \text{Im }(\bar{I}_{2l}) - \sqrt{3}\text{Re }(\bar{I}_{2l})] \\
B_\gamma^{bc} &= \frac{-1}{3\sqrt{3}V}[\text{Im }(\bar{I}_{1l}) - 2\text{Im }(\bar{I}_{2l})] \\
B_\gamma^{ca} &= \frac{-1}{3\sqrt{3}V}[\text{Im }(\bar{I}_{1l}) + \text{Im }(\bar{I}_{2l}) + \sqrt{3}\text{Re }(\bar{I}_{2l})]
\end{aligned}
\tag{4.117}
$$

From these equations, it is evident that the first terms form the positive sequence susceptance as they involve $\bar{I}_{1l}$ terms. Similarly, the second and third terms in the above equation form negative sequence susceptance of the compensator, as these involve $\bar{I}_{2l}$ terms. Thus, we can write,

$$
\begin{aligned}
B_\gamma^{ab} &= B_{\gamma 1}^{ab} + B_{\gamma 2}^{ab} \\
B_\gamma^{bc} &= B_{\gamma 1}^{bc} + B_{\gamma 2}^{bc} \\
B_\gamma^{ca} &= B_{\gamma 1}^{ca} + B_{\gamma 2}^{ca}
\end{aligned}
\tag{4.118}
$$

Therefore,

$$
B_{\gamma 1}^{ab} = B_{\gamma 1}^{bc} = B_{\gamma 1}^{ca} = -\frac{1}{3\sqrt{3}V}[\text{Im}(\bar{I}_{1l})]
\tag{4.119}
$$

and,

$$
\begin{aligned}
B_{\gamma 2}^{ab} &= -\frac{1}{3\sqrt{3}V}\left[\text{Im}(\bar{I}_{2l}) - \sqrt{3}\text{Re}(\bar{I}_{2l})\right] \\
B_{\gamma 2}^{bc} &= -\frac{1}{3\sqrt{3}V}\left[-2\text{Im}(\bar{I}_{2l})\right] \\
B_{\gamma 2}^{ca} &= -\frac{1}{3\sqrt{3}V}\left[\text{Im}(\bar{I}_{2l}) + \sqrt{3}\text{Re}(\bar{I}_{2l})\right]
\end{aligned}
\tag{4.120}
$$

Earlier in (4.74), it was established that,

$$\bar{I}_{0l} = 0$$
$$\bar{I}_{1l} = \left(Y_l^{ab} + Y_l^{bc} + Y_l^{ca}\right)\sqrt{3}V$$
$$\bar{I}_{2l} = -\left(\alpha^2 Y_l^{ab} + Y_l^{bc} + \alpha Y_l^{ca}\right)\sqrt{3}V$$

Noting that,

$$Y_l^{ab} = G_l^{ab} + jB_l^{ab}$$
$$Y_l^{bc} = G_l^{bc} + jB_l^{bc}$$
$$Y_l^{ca} = G_l^{ca} + jB_l^{ca}$$

Therefore,

$$\begin{aligned} \operatorname{Im}(\bar{I}_{1l}) &= \operatorname{Im}(Y_l^{ab} + Y_l^{bc} + Y_l^{ca})\sqrt{3}V \\ &= (B_l^{ab} + B_l^{bc} + B_l^{ca})\sqrt{3}V \end{aligned} \tag{4.121}$$

Thus, (4.119) is re-written as following.

$$B_{\gamma 1}^{ab} = B_{\gamma 1}^{bc} = B_{\gamma 1}^{ca} = -\frac{1}{3}\left(B_l^{ab} + B_l^{ab} + B_l^{ca}\right) \tag{4.122}$$

Now we shall compute $B_{\gamma 2}^{ab}$, $B_{\gamma 2}^{bc}$, and $B_{\gamma 2}^{ca}$ using (4.120) as following. We know that,

$$\begin{aligned} \bar{I}_{2l} &= -\left(\alpha^2 Y_l^{ab} + Y_l^{bc} + \alpha Y_l^{ca}\right)\sqrt{3}V \\[6pt] &= -\left[\left(\frac{-1}{2} - \frac{j\sqrt{3}}{2}\right)\left(G_l^{ab} + jB_l^{ab}\right) + \left(G_l^{bc} + jB_l^{bc}\right)\right. \\[6pt] &\quad \left. + \left(\frac{-1}{2} + \frac{j\sqrt{3}}{2}\right)(G_l^{ca} + jB_l^{ca})\right]\sqrt{3}V \\[6pt] &= -\left[-\frac{G_l^{ab}}{2} + \frac{\sqrt{3}}{2}B_l^{ab} + G_l^{bc} - \frac{G_l^{ca}}{2} - \frac{\sqrt{3}}{2}B_l^{ca}\right. \\[6pt] &\quad \left. - j\left(\frac{\sqrt{3}}{2}G_l^{ab} + \frac{B_l^{ab}}{2} - B_l^{bc} - \frac{\sqrt{3}}{2}G_l^{ca} + \frac{B_l^{ca}}{2}\right)\right]\sqrt{3}V \\[6pt] &= \left[\frac{G_l^{ab}}{2} - \frac{\sqrt{3}}{2}B_l^{ab} - G_l^{bc} + \frac{G_l^{ca}}{2} + \frac{\sqrt{3}}{2}B_l^{ca}\right. \\[6pt] &\quad \left. + j\left(\frac{\sqrt{3}}{2}G_l^{ab} + \frac{B_l^{ab}}{2} - B_l^{bc} - \frac{\sqrt{3}}{2}G_l^{ca} + \frac{B_l^{ca}}{2}\right)\right]\sqrt{3}V \end{aligned} \tag{4.123}$$

The above implies that,

$$\text{Im}(\bar{I}_{2l}) = \left(\frac{\sqrt{3}}{2} G_l^{ab} + \frac{B_l^{ab}}{2} - B_l^{bc} - \frac{\sqrt{3}}{2} G_l^{ca} + \frac{B_l^{ca}}{2} \right) \sqrt{3} V$$

$$-\sqrt{3}\,\text{Re}(\bar{I}_{2l}) = \left(-\frac{\sqrt{3}}{2} G_l^{ab} + \frac{3}{2} B_l^{ab} + \sqrt{3} G_l^{bc} - \frac{\sqrt{3}}{2} G_l^{ca} - \frac{3}{2} B_l^{ca} \right) \sqrt{3} V$$

Thus, $B_{\gamma 2}^{ab}$ can be given as,

$$
\begin{aligned}
B_{\gamma 2}^{ab} &= -\frac{1}{3\sqrt{3}V} \left[\text{Im}(\bar{I}_{2l}) - \sqrt{3}\,\text{Re}(\bar{I}_{2l}) \right] \\
&= -\frac{1}{3\sqrt{3}V} \left(2B_l^{ab} - B_l^{bc} - B_l^{ca} + \sqrt{3} G_l^{bc} - \sqrt{3} G_l^{ca} \right) \sqrt{3} V \\
&= -\frac{1}{3} \left[2B_l^{ab} - B_l^{bc} - B_l^{ca} + \sqrt{3} \left(G_l^{bc} - G_l^{ca} \right) \right] \\
&= \frac{1}{\sqrt{3}} \left(G_l^{ca} - G_l^{bc} \right) + \frac{1}{3} \left(B_l^{bc} + B_l^{ca} - 2B_l^{ab} \right)
\end{aligned}
\tag{4.124}
$$

Similarly,

$$
\begin{aligned}
B_{\gamma 2}^{ca} &= -\frac{1}{3\sqrt{3}V} \left[\text{Im}(\bar{I}_{2l}) + \sqrt{3}\,\text{Re}(\bar{I}_{2l}) \right] \\
&= -\frac{1}{3\sqrt{3}V} \left(\frac{\sqrt{3}}{2} G_l^{ab} + \frac{B_l^{ab}}{2} - B_l^{bc} - \frac{\sqrt{3}}{2} G_l^{ca} + \frac{B_l^{ca}}{2} \right. \\
&\qquad\qquad \left. + \frac{\sqrt{3}}{2} G_l^{ab} - \frac{3}{2} B_l^{ab} - \sqrt{3} G_l^{bc} + \frac{\sqrt{3}}{2} G_l^{ca} + \frac{3}{2} B_l^{ca} \right) \sqrt{3} V \\
&= -\frac{1}{3} \left(\sqrt{3} G_l^{ab} - \sqrt{3} G_l^{bc} - B_l^{ab} - B_l^{bc} + 2B_l^{ca} \right) \\
&= \frac{1}{\sqrt{3}} \left(G_l^{bc} - G_l^{ab} \right) + \frac{1}{3} \left(B_l^{ab} + B_l^{bc} - 2B_l^{ca} \right)
\end{aligned}
\tag{4.125}
$$

and, $B_{\gamma 2}^{bc}$ is computed as below.

$$
\begin{aligned}
B_{\gamma 2}^{bc} &= -\frac{1}{3\sqrt{3}V} \left[-2\text{Im}(\bar{I}_{2l}) \right] \\
&= \frac{2}{3\sqrt{3}V} \left(\frac{\sqrt{3}}{2} G_l^{ab} + \frac{B_l^{ab}}{2} - B_l^{bc} - \frac{\sqrt{3}}{2} G_l^{ca} + \frac{B_l^{ca}}{2} \right) V\sqrt{3} \\
&= \frac{1}{\sqrt{3}} \left(G_l^{ab} - G_l^{ca} \right) + \frac{1}{3} \left(B_l^{ab} + B_l^{ca} - 2B_l^{bc} \right)
\end{aligned}
\tag{4.126}
$$

Using (4.119), (4.124)–(4.126), the overall compensator susceptances can be found as follows.

$$B_\gamma^{ab} = B_{\gamma 1}^{ab} + B_{\gamma 2}^{ab}$$
$$= -\frac{1}{3}\left(B_l^{ab} + B_l^{bc} + B_l^{ca}\right) + \frac{1}{\sqrt{3}}\left(G_l^{ca} - G_l^{bc}\right) + \frac{1}{3}\left(B_l^{ca} + B_l^{bc} - 2B_l^{ab}\right)$$
$$= -B_l^{ab} + \frac{1}{\sqrt{3}}\left(G_l^{ca} - G_l^{bc}\right)$$

Similarly,

$$B_\gamma^{bc} = B_{\gamma 1}^{bc} + B_{\gamma 2}^{bc}$$
$$= -\frac{1}{3}\left(B_l^{ab} + B_l^{bc} + B_l^{ca}\right) + \frac{1}{\sqrt{3}}\left(G_l^{ab} - G_l^{ca}\right) + \frac{1}{3}\left(B_l^{ab} + B_l^{ca} - 2B_l^{bc}\right)$$
$$= -B_l^{bc} + \frac{1}{\sqrt{3}}\left(G_l^{ab} - G_l^{ca}\right)$$

and,

$$B_\gamma^{ca} = B_{\gamma 1}^{ca} + B_{\gamma 2}^{ca}$$
$$= -\frac{1}{3}\left(B_l^{ab} + B_l^{bc} + B_l^{ca}\right) + \frac{1}{\sqrt{3}}\left(G_l^{bc} - G_l^{ab}\right) + \frac{1}{3}\left(B_l^{bc} + B_l^{ab} - 2B_l^{ca}\right)$$
$$= -B_l^{ca} + \frac{1}{\sqrt{3}}\left(G_l^{bc} - G_l^{ab}\right)$$

Thus, the compensator susceptances in terms of load parameters are given as follows.

$$B_\gamma^{ab} = -B_l^{ab} + \frac{1}{\sqrt{3}}\left(G_l^{ca} - G_l^{bc}\right)$$
$$B_\gamma^{bc} = -B_l^{bc} + \frac{1}{\sqrt{3}}\left(G_l^{ab} - G_l^{ca}\right) \qquad (4.127)$$
$$B_\gamma^{ca} = -B_l^{ca} + \frac{1}{\sqrt{3}}\left(G_l^{bc} - G_l^{ab}\right)$$

It is interesting to observe the above equations. The first part of the equation nullifies the effect of the load susceptances and the second part compensates the effect of unbalance in the resistive parts of the load. The two terms together as compensator susceptance, make source currents balanced and in phase with the supply voltages. The compensator's positive and negative sequence networks are shown in Fig. 4.14. The expressions in (4.127) now explain that earlier in Sub-section 4.4.4, why we have taken, $Y_\gamma^{ab} = 0$, $Y_\gamma^{bc} = jG_l^{ab}/\sqrt{3}$, and $Y_\gamma^{ca} = -jG_l^{ab}/\sqrt{3}$ for load admittances, $Y_l^{ab} = G_l^{ab}$, $Y_l^{bc} = 0$, and $Y_l^{ca} = 0$, to achieve balanced three-phase source currents.

What happens if we just use the following values of the compensator susceptances as given below?

$$B_\gamma^{ab} = -B_l^{ab}$$
$$B_\gamma^{bc} = -B_l^{bc} \qquad (4.128)$$
$$B_\gamma^{ca} = -B_l^{ca}$$

In the above case, load susceptance parts of the admittance are fully compensated. However, the source currents after compensation remain unbalanced due to unbalance resistive parts of the load. The compensator susceptances can be realized using lumped reactive components. When the load changes, their values have to be changed or switched to different values as calculated using above mentioned methods. This scheme is called an open-loop control scheme. The closed-loop control schemes can be realized using FACTs devices such as Fixed Capacitor-Thyristor Controlled Reactor (FC-TCR) [1], [12], [13], [15].

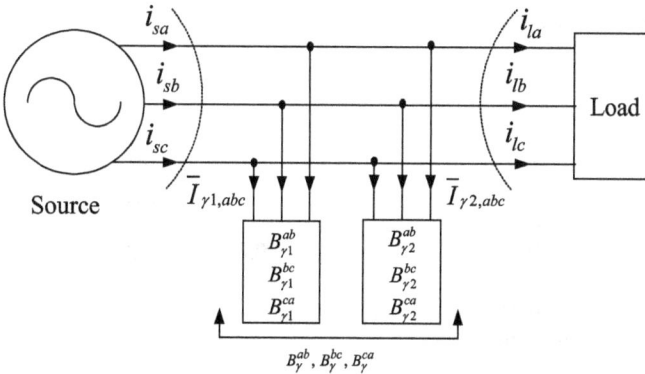

Figure 4.14 Sequence networks of the compensator

Example 4.5. For a delta connected load shown in Fig. 4.15, the load admittances are given as following,

$$Y_l^{ab} = G_l^{ab} + jB_l^{ab}$$
$$Y_l^{bc} = G_l^{bc} + jB_l^{bc}$$
$$Y_l^{ca} = G_l^{ca} + jB_l^{ca}$$

Given the load parameters:

$$Z_l^{ab} = 1/Y_l^{ab} = 5 + j12 \, \Omega$$
$$Z_l^{bc} = 1/Y_l^{bc} = 3 + j4 \, \Omega$$
$$Z_l^{ca} = 1/Y_l^{ca} = 9 - j13 \, \Omega$$

Determine compensator susceptances ($B_\gamma^{ab}, B_\gamma^{bc}, B_\gamma^{ca}$) so that the supply sees the load as balanced with unity power factor. Also, find the line currents and source active and reactive powers before and after compensation.

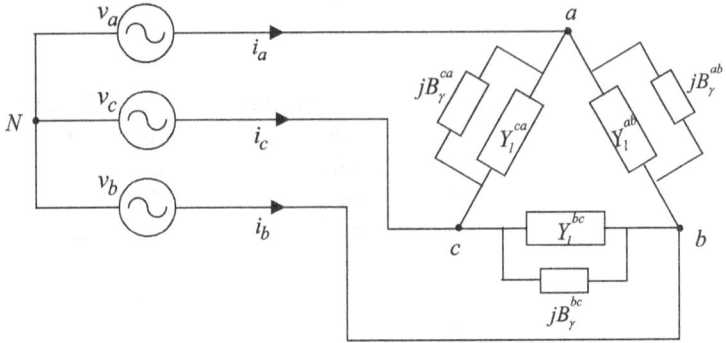

Figure 4.15 A delta connected compensated load

Solution:

$$Z_l^{ab} = 5 + j12 \ \Omega \ \Rightarrow \ Y_l^{ab} = 0.03 - j0.0710 \ \mho$$

$$Z_l^{bc} = 3 + j4 \ \Omega \ \Rightarrow \ Y_l^{bc} = 0.12 - j0.16 \ \mho$$

$$Z_l^{ca} = 9 - j12 \ \Omega \ \Rightarrow \ Y_l^{ca} = 0.04 + j0.0533 \ \mho$$

Once we know the admittances, we can compute line to line conductances and susceptances as given in the following.

$$G_l^{ab} = 0.03 \, \mho, \quad B_l^{ab} = -0.0710 \, \mho$$

$$G_l^{bc} = 0.12 \, \mho, \quad B_l^{bc} = -0.16 \, \mho$$

$$G_l^{ca} = 0.04 \, \mho, \quad B_l^{ca} = 0.0533 \, \mho$$

$$B_\gamma^{ab} = -B_l^{ab} + \frac{1}{\sqrt{3}} (G_l^{ca} - G_l^{bc}) = 0.0248 \ \mho$$

$$B_\gamma^{bc} = -B_l^{bc} + \frac{1}{\sqrt{3}} (G_l^{ab} - G_l^{ca}) = 0.1540 \ \mho$$

$$B_\gamma^{ca} = -B_l^{ca} + \frac{1}{\sqrt{3}} (G_l^{bc} - G_l^{ab}) = -0.0011 \ \mho$$

Total admittances are,

$$Y^{ab} = Y_l^{ab} + Y_\gamma^{ab} = 0.03 - j0.0462 \ \mho$$

$$Y^{bc} = Y_l^{bc} + Y_\gamma^{bc} = 0.12 - j0.006 \ \mho$$

$$Y^{ca} = Y_l^{ca} + Y_\gamma^{ca} = 0.04 + j0.0522 \ \mho$$

Knowing these total admittances, we can find line currents as below.
Current before Compensation

$$\bar{I}_{al} = \bar{I}_{abl} - \bar{I}_{cal} = \left[(1-\alpha^2)Y_l^{ab} - (\alpha-1)Y_l^{ca}\right] V = 0.2150 V \angle -9.51° A$$

$$\bar{I}_{bl} = \bar{I}_{bcl} - \bar{I}_{abl} = \left[(\alpha^2-\alpha)Y_l^{bc} - (1-\alpha^2)Y_l^{ab}\right] V = 0.4035 V \angle -161.66° A$$

$$\bar{I}_{cl} = \bar{I}_{cal} - \bar{I}_{bcl} = \left[(\alpha-1)Y_l^{ca} - (\alpha^2-\alpha)Y_l^{bc}\right] V = 0.2358 V \angle 43.54° A$$

Powers before compensation

$$
\begin{aligned}
\bar{S}_a &= \bar{V}_a(\bar{I}_{al})^* = P_a + jQ_a = V \times (0.2121 + j0.0355) \\
\bar{S}_b &= \bar{V}_b(\bar{I}_{bl})^* = P_b + jQ_b = V \times (0.3014 + j0.2682) \\
\bar{S}_c &= \bar{V}_c(\bar{I}_{cl})^* = P_c + jQ_c = V \times (0.0552 + j0.2293) \\
\text{Total real power, } P &= P_a + P_b + P_c = V \times 0.5688\,W \\
\text{Total reactive power, } Q &= Q_a + Q_b + Q_c = V \times 0.5330\,\text{VAr} \\
\text{Power factor in phase-a, } p_{f_a} &= \cos\phi_a = \cos(9.51°) = 0.9863\,\text{lag} \\
\text{Power factor in phase-b, } p_{f_b} &= \cos\phi_b = \cos(41.63°) = 0.7471\,\text{lag} \\
\text{Power factor in phase-c, } p_{f_c} &= \cos\phi_c = \cos(76.45°) = 0.2334\,\text{lag}
\end{aligned}
$$

Thus, we observe that the phases draw reactive power from the lines and currents are unbalanced in magnitude and phase angles.

Currents after compensation

$$\bar{I}_a = \bar{I}_{ab} - \bar{I}_{ca} = \left[(1-\alpha^2)Y^{ab} - (\alpha-1)Y^{ca}\right] V = 0.1896 \times V \angle 0° A$$

$$\bar{I}_b = \bar{I}_{bc} - \bar{I}_{ab} = \left[(\alpha^2-\alpha)Y^{bc} - (1-\alpha^2)Y^{ab}\right] V = 0.1896 \times V \angle -120° A$$

$$\bar{I}_a = \bar{I}_{ca} - \bar{I}_{bc} = \left[(\alpha-1)Y_l^{ca} - (\alpha^2-\alpha)Y^{bc}\right] V = 0.1896 \times V \angle 120° A$$

Powers after compensation

$$
\begin{aligned}
\bar{S}_a &= \bar{V}_a(\bar{I}_a)^* = P_a + jQ_a = V(0.1896 + j0.0) \\
\bar{S}_b &= \bar{V}_b(\bar{I}_b)^* = P_b + jQ_b = V(0.1896 + j0.0) \\
\bar{S}_c &= \bar{V}_c(\bar{I}_c)^* = P_c + jQ_c = V(0.1896 + j0.0) \\
\text{Total real power, } P &= P_a + P_b + P_c = (V \times 0.5688)W \\
\text{Total reactive power, } Q &= Q_a + Q_b + Q_c = 0\,\text{VAr} \\
\text{Power factor in phase-a, } pf_a &= \cos\phi_a = \cos(0°) = 1.0 \\
\text{Power factor in phase-b, } pf_b &= \cos\phi_b = \cos(0°) = 1.0 \\
\text{Power factor in phase-c, } pf_c &= \cos\phi_c = \cos(0°) = 1.0
\end{aligned}
$$

From the above results, we observe that after placing the compensator of suitable values as calculated above, the line currents become balanced and have a unity power factor relationship with their respective voltages.

Example 4.6. A three-phase system has delta connected load with $Z_{ab} = 10 + j15\,\Omega$, $Z_{bc} = 15 + j20\,\Omega$ and $Z_{ca} = 10 - j15\,\Omega$ supplied by balanced phase voltages with rms value of 230 V at 50 Hz.

(a) Compute these susceptances (B_γ^{ab}, B_γ^{bc}, and B_γ^{ca}) and express them in terms of physical parameters i.e., inductance or capacitance and their values.

(b) Determine the source currents after placing compensator.

(c) Determine the source, compensator load active and reactive powers.

(d) Draw the circuits as seen from the load and from the source side.

Solution:

(a) Compensator susceptances

The load admittances are computed as follows.

$$Z_l^{ab} = 10 + j15\,\Omega \Rightarrow Y_l^{ab} = 0.0308 - j0.0462\,\mho$$
$$Z_l^{bc} = 15 + j20\,\Omega \Rightarrow Y_l^{bc} = 0.0240 - j0.0320\,\mho$$
$$Z_l^{ca} = 10 - j15\,\Omega \Rightarrow Y_l^{ca} = 0.0308 + j0.0462\,\mho$$

Once we know the admittances we know,

$$G_l^{ab} = 0.0308\,\mho, \quad B_l^{ab} = -0.0462\,\mho$$
$$G_l^{bc} = 0.0240\,\mho, \quad B_l^{bc} = -0.0320\,\mho$$
$$G_l^{ca} = 0.0308\,\mho, \quad B_l^{ca} = 0.0462\,\mho$$

$$B_\gamma^{ab} = -B_l^{ab} + \frac{1}{\sqrt{3}}(G_l^{ca} - G_l^{bc}) = 0.0501\,\mho$$

$$B_\gamma^{bc} = -B_l^{bc} + \frac{1}{\sqrt{3}}(G_l^{ab} - G_l^{ca}) = 0.0320\,\mho$$

$$B_\gamma^{ca} = -B_l^{ca} + \frac{1}{\sqrt{3}}(G_l^{bc} - G_l^{ab}) = -0.0501\,\mho$$

The values of physical parameters are computed as follows.

$$B_\gamma^{ab} = \omega C_\gamma^{ab} = 0.0501 \Rightarrow C_\gamma^{ab} = \frac{0.0501}{2\pi \times 50} = 159.31\,\mu F$$

$$B_\gamma^{bc} = \omega C_\gamma^{bc} = 0.0320 \Rightarrow C_\gamma^{bc} = \frac{0.0320}{2\pi \times 50} = 101.84\,\mu F$$

$$B_\gamma^{ca} = \frac{1}{\omega L_\gamma^{ca}} = 0.0501 \Rightarrow L_\gamma^{ca} = \frac{1}{2\pi \times 50 \times 0.0501} = 63.6\,mH$$

Total admittances are:

$$Y^{ab} = Y_l^{ab} + Y_\gamma^{ab} = 0.0308 + j0.0039\,\mho$$
$$Y^{bc} = Y_l^{bc} + Y_\gamma^{bc} = 0.0240 + j0\,\mho$$
$$Y^{ca} = Y_l^{ca} + Y_\gamma^{ca} = 0.0308 - j0.0039\,\mho$$

Knowing these total admittances, we can find line currents using the following expressions.

(b) Source currents compensation
Before placing the compensator, the line currents are computed as follows.

$$\bar{I}_{la} = \bar{I}_{lab} - \bar{I}_{lca} = \left[(1 - \alpha^2)Y_l^{ab} - (\alpha - 1)Y_l^{ca}\right] = 39.6172\,V\angle 0°\,A$$

$$\bar{I}_{lb} = \bar{I}_{lbc} - \bar{I}_{lab} = \left[(\alpha^2 - \alpha)Y_l^{bc} - (1 - \alpha^2)Y_l^{ab}\right] = 32.5573\,V\angle 179.5893°\,A$$

$$\bar{I}_{lc} = \bar{I}_{lca} - \bar{I}_{lbc} = \left[(\alpha - 1)Y_l^{ca} - (\alpha^2 - \alpha)Y_l^{bc}\right] = 7.0645\,V\angle -178.1070°\,A$$

After placing the compensator, the line currents are computed as follows.

$$\bar{I}_a = \bar{I}_{ab} - \bar{I}_{ca} = \left[(1 - \alpha^2)Y^{ab} - (\alpha - 1)Y^{ca}\right]V = 19.6738\angle 0°\,A$$

$$\bar{I}_b = \bar{I}_{bc} - \bar{I}_{ab} = \left[(\alpha^2 - \alpha)Y^{bc} - (1 - \alpha^2)Y^{ab}\right]V = 19.6738\angle -120°\,A$$

$$\bar{I}_a = \bar{I}_{ca} - \bar{I}_{bc} = \left[(\alpha - 1)Y_l^{ca} - (\alpha^2 - \alpha)Y^{bc}\right]V = 19.6738\angle 120°\,A$$

Thus, we observe that the currents before compensation are unbalanced in magnitude and phase angles, and the currents after compensation are balanced in magnitude and phase angles.

(c) Source, compensator, and load powers
The load, source, compensator powers, and their power factors are computed as,

$$
\begin{aligned}
\bar{S}_{la} &= \bar{V}_a(\bar{I}_{la})^* = P_{la} + jQ_{la} = 9.1119 + j0 \text{ kVA} \\
\bar{S}_{lb} &= \bar{V}_b(\bar{I}_{lb})^* = P_{lb} + jQ_{lb} = 3.6975 + j6.5116 \text{ kVA} \\
\bar{S}_{lc} &= \bar{V}_c(\bar{I}_{lc})^* = P_{lc} + jQ_{lc} = 0.76550 - j1.4332 \text{ kVA}
\end{aligned}
$$

$$
\begin{aligned}
\text{Total real power, } P_l &= P_{la} + P_{lb} + P_{lc} = 13.575 \text{ kW} \\
\text{Total reactive power, } Q_l &= Q_{la} + Q_{lb} + Q_{lc} = 5.0784 \text{ kVAr} \\
\text{Power factor in phase-}a, pf_{la} &= \cos\phi_{la} = \cos(0°) = 1 \\
\text{Power factor in phase-}b, pf_{lb} &= \cos\phi_{lb} = \cos(59.58°) = 0.5062 \text{ (lag)} \\
\text{Power factor in phase-}c, pf_{lc} &= \cos\phi_{lc} = \cos(-58.1°) = 0.5283 \text{ (lead)}
\end{aligned}
$$

$$
\begin{aligned}
\bar{S}_a &= \bar{V}_a(\bar{I}_a)^* = P_a + jQ_a = 4.5250 + j0.0 \text{ kVA} \\
\bar{S}_b &= \bar{V}_b(\bar{I}_b)^* = P_b + jQ_b = 4.5250 + j0.0 \text{ kVA} \\
\bar{S}_c &= \bar{V}_c(\bar{I}_c)^* = P_c + jQ_c = 4.5250 + j0.0 \text{ kVA}
\end{aligned}
$$

$$\text{Total real power, } P \;=\; P_a + P_b + P_c = 13.575\,\text{kW}$$
$$\text{Total reactive power, } Q \;=\; Q_a + Q_b + Q_c = 0\,\text{kVAr}$$
$$\text{Power factor, } pf_a \;=\; pf_b = pf_c = \cos\phi_a = \cos(0°) = 1.0$$

Compensator powers are given as follows.

$$\bar{S}_{fa} \;=\; \bar{V}_a(\bar{I}_{fa})^* = \bar{V}_a(\bar{I}_{la} - \bar{I}_a)^* = 0 - j7.9448\,\text{kVA}$$
$$\bar{S}_{fb} \;=\; \bar{V}_b(\bar{I}_{fb})^* = \bar{V}_b(\bar{I}_{lb} - \bar{I}_b)^* = 0 - j5.0784\,\text{kVA}$$
$$\bar{S}_{fc} \;=\; \bar{V}_c(\bar{I}_{fc})^* = \bar{V}_c(\bar{I}_{lc} - \bar{I}_c)^* = 0 + j7.9448\,\text{kVA}$$
$$\text{Total real power, } P_f \;=\; P_{fa} + P_{fb} + P_{fc} = 0\,\text{kW}$$
$$\text{Total reactive power, } Q_f \;=\; Q_{fa} + Q_{fb} + Q_{fc} = -5.0784\,\text{kVAr}$$

From the above results, we observe that, $P = P_l = 13.575$ kW, $Q = 0$ kVAr, $P_f = 0$ kW, and $Q_f = -Q_l = -5.0784$ kVAr. This ensures that the load reactive power is fully compensated by the compensator. Also, the compensator does not consume any real power. The source is supplying only the load real power and its reactive power is zero. This makes unity power factor at the source.

(d) Equivalent circuit as seen from source and from load side is illustrated in Fig. 4.16. The value of equivalent resistance is, $R_a = R_b = R_c = \frac{230}{19.6738} = 11.71\,\Omega$ or $R_{ab} = R_{bc} = R_{ca} = 3 \times 11.69 = 35.07\,\Omega$.

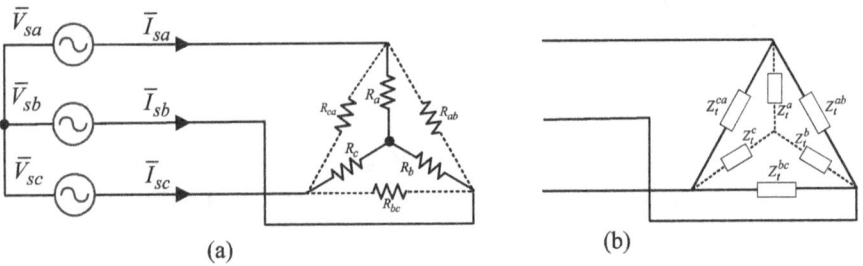

Figure 4.16 Equivalent circuit (a) from source side and (b) from load side

Example 4.7. Consider the following three-phase, three-wire system as shown in Fig. 4.17. The three-phase voltages are balanced sinusoids with rms value of 230 V at 50 Hz. The load impedances are $Z_a = 3 + j4\,\Omega$, $Z_b = 5 + j12\,\Omega$, $Z_c = 12 - j5\,\Omega$. Compute the following.

(a) The line currents $\bar{I}_{la}, \bar{I}_{lb}, \bar{I}_{lc}$.

(b) The active (P) and reactive (Q) powers of each phase.

(c) The compensator susceptance ($B_\gamma^{ab}, B_\gamma^{bc}, B_\gamma^{ca}$), so that the supply sees the load balanced and unity power factor.

(d) For case (3), compute the source, load, compensator active and reactive powers (after compensation).

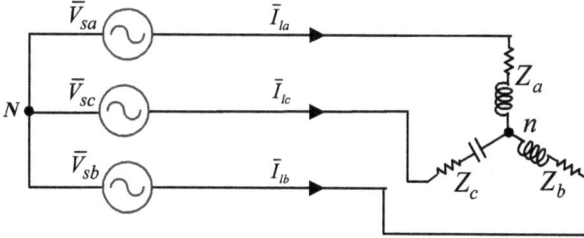

Figure 4.17   An unbalanced three-phase three-wire star connected load

Solution: Given that $Z_a = 3 + j4\,\Omega$, $Z_b = 5 + j12\,\Omega$, $Z_c = 12 - j5\,\Omega$.

(a) Line currents $\overline{I}_{la}, \overline{I}_{lb}$ and $\overline{I}_{lc}$ are found by first computing neutral voltage as given below.

$$
\begin{aligned}
\overline{V}_{nN} &= \frac{1}{\frac{1}{Z_a} + \frac{1}{Z_b} + \frac{1}{Z_c}} \left(\frac{\overline{V}_{aN}}{Z_a} + \frac{\overline{V}_{bN}}{Z_b} + \frac{\overline{V}_{cN}}{Z_c} \right) \\
&= \frac{Z_a Z_b Z_c}{Z_a Z_b + Z_b Z_c + Z_c Z_a} \left(\frac{\overline{V}_{aN}}{Z_a} + \frac{\overline{V}_{bN}}{Z_b} + \frac{\overline{V}_{cN}}{Z_c} \right) \\
&= \frac{845\angle 97.82°}{252.41\angle 55.5°} (24.12\angle -99.55°) \\
&= 43.79 - j67.04\text{V} = 80.75\angle -57.15°\,\text{V}
\end{aligned}
$$

Now the line currents are computed as below.

$$
\overline{I}_{la} = \frac{\overline{V}_{aN} - \overline{V}_{nN}}{Z_a} = \frac{230\angle 0° - 80.75\angle -57.15°}{3 + j4} = 39.63\angle -33.11°\,\text{A}
$$

$$
\overline{I}_{lb} = \frac{\overline{V}_{bN} - \overline{V}_{nN}}{Z_b} = \frac{230\angle -120° - 80.75\angle -57.15°}{5 + j12} = 15.85\angle 152.21°\,\text{A}
$$

$$
\overline{I}_{lc} = \frac{\overline{V}_{cN} - \overline{V}_{nN}}{Z_c} = \frac{230\angle 120° - 80.75\angle -57.15°}{12 - j5} = 23.89\angle 143.35°\,\text{A}
$$

(b) Active and reactive load powers
For phase-a,

$$
\begin{aligned}
P_{la} &= V_{aN} I_{la} \cos \phi_{la} = 230 \times 39.63 \times \cos(33.11°) = 7635.9\text{W} \\
Q_{la} &= V_{aN} I_{la} \sin \phi_{la} = 230 \times 39.63 \times \sin(33.11°) = 4980\text{VAr}
\end{aligned}
$$

For phase-b,

$$P_{lb} = V_{bN}I_{lb}\cos\phi_{lb} = 230 \times 15.85 \times \cos(-152.21° - 120°) = 140.92\text{W}$$
$$Q_{lb} = V_{bN}I_{lb}\sin\phi_{lb} = 230 \times 15.85 \times \sin(-152.21° - 120°) = 3643.2\text{VAr}$$

For phase-c,

$$P_{lc} = V_{cN}I_{lc}\cos\phi_{lc} = 230 \times 23.89 \times \cos(-143.35° + 120°)$$
$$= 5046.7\text{W}$$
$$Q_{lc} = V_{cN}I_{lc}\sin\phi_{lc} = 230 \times 23.89 \times \sin(-143.35° + 120°)$$
$$= -2179.3\text{VAr}$$

Total three-phase load powers are given as,

$$P_l = P_{la} + P_{lb} + P_{lc} = 12823\text{W}$$
$$Q_l = Q_{la} + Q_{lb} + Q_{lc} = 6443.8\text{VAr}$$

(c) Compensator susceptance

First, we convert the star-connected load to a delta load as given below.

$$Z_l^{ab} = \frac{Z_aZ_b + Z_bZ_c + Z_cZ_a}{Z_c} = 4 + j19 = 19.42\angle 77.11°\,\Omega$$

$$Z_l^{bc} = \frac{\Delta Z}{Z_a} = 50.44 + j2.08 = 50.42\angle 2.36°\,\Omega$$

$$Z_l^{ca} = \frac{\Delta Z}{Z_b} = 19.0 - j4.0 = 19.42\angle -11.89°\,\Omega$$

The above implies that,

$$Y_l^{ab} = 1/Z_l^{ab} = G_l^{ab} + jB_l^{ab} = 0.0106 - j0.050\,\mho$$
$$Y_l^{bc} = 1/Z_l^{bc} = G_l^{bc} + jB_l^{bc} = 0.0198 - j0.0008\,\mho$$
$$Y_l^{ca} = 1/Z_l^{ca} = G_l^{ca} + jB_l^{ca} = 0.0504 + j0.0106\,\mho$$

From the above, the compensator susceptances are computed as follows.

$$B_\gamma^{ab} = -B_l^{ab} + \frac{(G_l^{ca} - G_l^{bc})}{\sqrt{3}} = 0.0681\,\mho$$

$$B_\gamma^{bc} = -B_l^{bc} + \frac{(G_l^{ab} - G_l^{ca})}{\sqrt{3}} = -0.0222\,\mho$$

$$B_\gamma^{ca} = -B_l^{ca} + \frac{(G_l^{bc} - G_l^{ab})}{\sqrt{3}} = -0.0053\,\mho$$

The respective impedances are,

$$Z_\gamma^{ab} = -j14.69\,\Omega\,(\text{capacitance})$$
$$Z_\gamma^{bc} = j45.13\,\Omega\,(\text{inductance})$$
$$Z_\gamma^{ca} = j188.36\,\Omega\,(\text{inductance})$$

(d) After compensation

The effective impedances in the delta-connected configuration is,

$$Z_l^{ab'} = Z_l^{ab} \| Z_\gamma^{ab} = 24.97 - j41.59 = 48.52\angle - 59.01° \, \Omega$$

$$Z_l^{bc'} = Z_l^{bc} \| Z_\gamma^{bc} = 21.52 - j24.98 = 32.97\angle 49.25° \, \Omega$$

$$Z_l^{ca'} = Z_l^{ca} \| Z_\gamma^{ca} = 24.97 - j41.59 = 19.7332\angle - 6.0° \, \Omega$$

Let us convert delta-connected impedances to star-connected impedances.

$$Z_a' = \frac{Z_l^{ab'} \times Z_l^{ca'}}{Z_l^{ab'} + Z_l^{bc'} + Z_l^{ca'}} = 9.0947 - j10.55 = 13.93\angle - 49.25° \, \Omega$$

$$Z_b' = \frac{Z_l^{bc'} \times Z_l^{ab'}}{Z_l^{ab'} + Z_l^{bc'} + Z_l^{ca'}} = 23.15 + j2.43 = 23.28\angle 6.06° \, \Omega$$

$$Z_c' = \frac{Z_l^{ca'} \times Z_l^{bc'}}{Z_l^{ab'} + Z_l^{bc'} + Z_l^{ca'}} = 4.8755 + j8.12 = 9.47\angle 59.01° \, \Omega$$

The new voltage between the load and system neutral after compensation is given by,

$$\overline{V}_{nN}' = \frac{1}{(\frac{1}{Z_a'} + \frac{1}{Z_b'} + \frac{1}{Z_c'})} (\frac{\overline{V}_{aN}}{Z_a'} + \frac{\overline{V}_{bN}}{Z_b'} + \frac{\overline{V}_{cN}}{Z_c'}) = 205.157\angle 72.8° \, V$$

Based on the above, the line currents are computed as follows.

$$\overline{I}_a = \frac{\overline{V}_{aN} - \overline{V}_{nN}'}{Z_a'} = 18.584\angle 0° \, A$$

$$\overline{I}_b = \frac{\overline{V}_{bN} - \overline{V}_{nN}'}{Z_b'} = 18.584\angle - 120° \, A$$

$$\overline{I}_c = \frac{\overline{V}_{cN} - \overline{V}_{nN}'}{Z_c'} = 18.584\angle 120° \, A$$

Thus, it is seen that after compensation, the source currents are balanced and have unity power factor with respective supply voltages. Source powers after compensation are given as follows.

$$P_a = P_b = P_c = 230 \times 18.584 = 4274.32 \, W$$

$$P = 3P_a = 12822.96 \, W$$

$$Q_a = Q_b = Q_c = 0 \text{VAr}$$

$$Q = 0 \text{VAr}$$

The compensator powers are computed as following.

$$
\begin{aligned}
\overline{S}_\gamma^{ab} &= \overline{V}_{ab}\overline{I}_\gamma^{ab*} = \overline{V}_{ab}(\overline{V}_{ab}{}^{*}Y_\gamma^{ab*}) = \overline{V}_{ab}^2 Y_\gamma^{ab*} \\
&= (230 \times \sqrt{3})^2 \times (-j0.0068) = -j\,10802\,\text{VA} \\
\overline{S}_\gamma^{bc} &= V_{bc}^2 Y_\gamma^{bc*} = (230 \times \sqrt{3})^2 \times (j0.0222) = j\,3516\,\text{VA} \\
\overline{S}_\gamma^{ca} &= V_{ca}^2 Y_\gamma^{ca*} = (230 \times \sqrt{3})^2 \times (j0.0053) = j\,842\,\text{VA}
\end{aligned}
$$

The total compensator power is $\overline{S}_\gamma = \overline{S}_\gamma^{ab} + \overline{S}_\gamma^{bc} + \overline{S}_\gamma^{ca} = -j\,6444$ VA. Thus, we observe that after compensation, the source real power equals to the load power, i.e., $P_s = P_l = 12823$ W, and source reactive power is zero, i.e., $Q_s = 0$. The compensator supplies the total reactive power of the load, i.e., $Q_\gamma = -Q_l = -6443.8$ VAr, and the compensator real power is zero, i.e., $P_\gamma = 0$.

4.7 COMPENSATION OF STAR CONNECTED SYSTEM WITH GROUNDED NEUTRAL

Readers might be curious to know whether it is possible to apply these principles to compensate a three-phase four-wire system where a star-connected load is unbalanced and the neutral is connected to the ground. A three-phase four-wire supply system with an unbalanced star-connected load is shown in Fig. 4.18. Now, the question is whether we can find compensator impedances (Z_{fa}, Z_{fb}, Z_{fc}) connected between phases and the ground to balance a star-connected unbalanced load. Let the supply

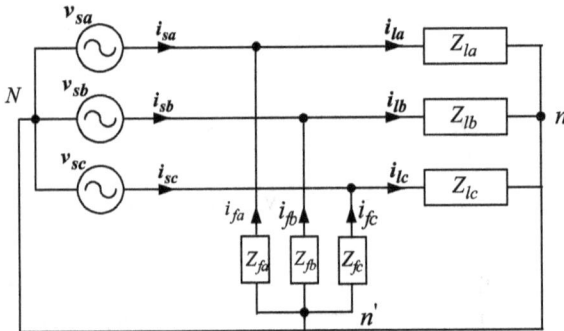

Figure 4.18 A three-phase four-wire system with unbalanced load

voltages be given as follows.

$$
\begin{aligned}
v_{sa}(t) &= \sqrt{2}V\sin(\omega t) \\
v_{sb}(t) &= \sqrt{2}V\sin(\omega t - 120°) \\
v_{sc}(t) &= \sqrt{2}V\sin(\omega t + 120°)
\end{aligned}
\tag{4.129}
$$

Before compensation, the load currents, which are also the source currents, are given as follows.

$$
\begin{aligned}
i_{la}(t) &= \sqrt{2}I_{la}\sin(\omega t - \phi_{la}) \\
i_{lb}(t) &= \sqrt{2}I_{lb}\sin(\omega t - 120° - \phi_{lb}) \\
i_{lc}(t) &= \sqrt{2}I_{lc}\sin(\omega t + 120° - \phi_{lc})
\end{aligned}
\tag{4.130}
$$

In the above equations,

$$
\begin{aligned}
I_{la} &= \frac{V}{\sqrt{R_{la}^2 + X_{la}^2}}, \quad \phi_{la} = \cos^{-1}\frac{R_{la}}{\sqrt{R_{la}^2 + X_{la}^2}} \\
I_{lb} &= \frac{V}{\sqrt{R_{lb}^2 + X_{lb}^2}}, \quad \phi_{lb} = \cos^{-1}\frac{R_{lb}}{\sqrt{R_{lb}^2 + X_{lb}^2}} \\
I_{lc} &= \frac{V}{\sqrt{R_{lc}^2 + X_{lc}^2}}, \quad \phi_{lc} = \cos^{-1}\frac{R_{lc}}{\sqrt{R_{lc}^2 + X_{lc}^2}}
\end{aligned}
\tag{4.131}
$$

For $Z_{la} = R_{la} + jX_{la}$, $Z_{lb} = R_{lb} + jX_{lb}$, and $Z_{lc} = R_{lc} + jX_{lc}$.

Let the currents after compensation be in phase with their voltages, therefore,

$$
\begin{aligned}
i_{sa}(t) &= \sqrt{2}I_s\sin(\omega t) \\
i_{sb}(t) &= \sqrt{2}I_s\sin(\omega t - 120°) \\
i_{sc}(t) &= \sqrt{2}I_s\sin(\omega t + 120°).
\end{aligned}
\tag{4.132}
$$

In phasor notations, the source voltages and currents are denoted as following.

$$
\begin{aligned}
\overline{V}_{sa}(t) &= V\angle 0 \\
\overline{V}_{sb}(t) &= V\angle -120° \\
\overline{V}_{sc}(t) &= V\angle 120°
\end{aligned}
\tag{4.133}
$$

and,

$$
\begin{aligned}
\overline{I}_{sa}(t) &= I_s\angle 0 \\
\overline{I}_{sb}(t) &= I_s\angle -120° \\
\overline{I}_{sc}(t) &= I_s\angle 120°
\end{aligned}
\tag{4.134}
$$

The total load active power is given as following.

$$
\begin{aligned}
P_l &= \text{Real}\left(\overline{V}_{sa}\overline{I}_{la}^* + \overline{V}_{sb}\overline{I}_{lb}^* + \overline{V}_{sc}\overline{I}_{lc}^*\right) \\
&= \text{Real}\left(\overline{V}_{sa}\left(\overline{V}_{sa}^*/Z_{la}^*\right) + \overline{V}_{sb}\left(\overline{V}_{sb}^*/Z_{lb}^*\right) + \overline{V}_{sc}\left(\overline{V}_{sc}^*/Z_{lc}^*\right)\right) \\
&= V^2\,\text{Real}\left(1/Z_{la} + 1/Z_{lb} + 1/Z_{lc}\right)
\end{aligned}
\tag{4.135}
$$

This load power (P_l) in the above equation must be equal to source power (P_s), which is given below.

$$P_s = P_l = 3V\,I_s = V^2\,\text{Real}(1/Z_{la} + 1/Z_{lb} + 1/Z_{lc}). \qquad (4.136)$$

The above equation implies that,

$$I_s = \frac{V}{3}\,\text{Real}(1/Z_{la} + 1/Z_{lb} + 1/Z_{lc}). \qquad (4.137)$$

From the above, $\bar{I}_{sa}$, $\bar{I}_{sb}$ and $\bar{I}_{sc}$ can be obtained using (4.134) as following.

$$\bar{I}_{sa} = \frac{\overline{V}_{sa}}{3}\,\text{Real}(1/Z_{la} + 1/Z_{lb} + 1/Z_{lc})$$

$$\bar{I}_{sb} = \frac{\overline{V}_{sb}}{3}\,\text{Real}(1/Z_{la} + 1/Z_{lb} + 1/Z_{lc}) \qquad (4.138)$$

$$\bar{I}_{sc} = \frac{\overline{V}_{sc}}{3}\,\text{Real}(1/Z_{la} + 1/Z_{lb} + 1/Z_{lc})$$

The compensator currents, $\bar{I}_{fa}$, $\bar{I}_{fb}$ and $\bar{I}_{fc}$ are given as following.

$$\begin{aligned}
\bar{I}_{fa} &= \bar{I}_{la} - \bar{I}_{sa} \\
(-\overline{V}_{sa}/Z_{fa}) &= \overline{V}_{sa}/Z_{la} - \bar{I}_{sa} \\
&= \overline{V}_{sa}/Z_{la} - \frac{\overline{V}_{sa}}{3}\,\text{Real}(1/Z_{la} + 1/Z_{lb} + 1/Z_{lc}) \qquad (4.139)
\end{aligned}$$

The above equation implies that,

$$Y_{fa} = -\frac{1}{Z_{la}} + \frac{1}{3}\,\text{Real}(1/Z_{la} + 1/Z_{lb} + 1/Z_{lc})$$

In the above equation,

$$\text{Real}(1/Z_{la} + 1/Z_{lb} + 1/Z_{lc}) = G_{la} + G_{lb} + G_{lc}$$

Therefore, the expressions for Y_{fb}, Y_{fb} and Y_{fc} can be written as following.

$$\begin{aligned}
Y_{fa} &= -Y_{la} + \frac{1}{3}(G_{la} + G_{lb} + G_{lc}) \\
Y_{fb} &= -Y_{lb} + \frac{1}{3}(G_{lb} + G_{lb} + G_{lc}) \qquad (4.140) \\
Y_{fc} &= -Y_{lc} + \frac{1}{3}(G_{la} + G_{lb} + G_{lc})
\end{aligned}$$

Or,

$$Y_{fa} = G_{fa} + jB_{fa} = \frac{1}{3}(-2G_{la} + G_{lb} + G_{lc}) - jB_{la}$$

$$Y_{fb} = G_{fb} + jB_{fb} = \frac{1}{3}(G_{la} - 2G_{lb} + G_{lc}) - jB_{lb} \qquad (4.141)$$

$$Y_{fc} = G_{fc} + jB_{fc} = \frac{1}{3}(G_{la} + G_{lb} - 2G_{lc}) - jB_{lc}$$

In the above equations,

$$Y_{la} = \frac{1}{Z_{la}} = \frac{R_{la}}{R_{la}^2 + X_{la}^2} + j\frac{(-X_{la})}{R_{la}^2 + X_{la}^2} = G_{la} + jB_{la} \qquad (4.142)$$

and

$$Y_{fa} = \frac{1}{Z_{fa}} = \frac{R_{fa}}{R_{fa}^2 + X_{fa}^2} + j\frac{(-X_{fa})}{R_{fa}^2 + X_{fa}^2} = G_{fa} + jB_{fa} \qquad (4.143)$$

Similarly, the above relations can be given for phase-b and phase-c load and filter admittances. Based on (4.141), the compensator conductances and suseptances are given as follows.

$$G_{fa} = \frac{1}{3}(-2G_{la} + G_{lb} + G_{lc})$$

$$G_{fb} = \frac{1}{3}(G_{la} - 2G_{lb} + G_{lc}) \qquad (4.144)$$

$$G_{fc} = \frac{1}{3}(G_{la} + G_{lb} - 2G_{lc})$$

and,

$$B_{fa} = -B_{la}$$

$$B_{fb} = -B_{lb} \qquad (4.145)$$

$$B_{fc} = -B_{lc}$$

Once G_{fa}, B_{fa}, G_{fb}, B_{fb}, G_{fc}, and B_{fc} are known from (4.144)–(4.145), the resistive and reactive part of the compensator network can be computed using following expressions.

$$Z_{fa} = R_{fa} + jX_{fa} = \frac{G_{fa}}{G_{fa}^2 + B_{fa}^2} + j\frac{(-B_{fa})}{G_{fa}^2 + B_{fa}^2}$$

$$Z_{fb} = R_{fb} + jX_{fb} = \frac{G_{fb}}{G_{fb}^2 + B_{fb}^2} + j\frac{(-B_{fb})}{G_{fb}^2 + B_{fb}^2} \qquad (4.146)$$

$$Z_{fc} = R_{fc} + jX_{fc} = \frac{G_{fc}}{G_{fc}^2 + B_{fc}^2} + j\frac{(-B_{fc})}{G_{fc}^2 + B_{fc}^2}$$

Comparing (4.144)–(4.145) with (4.127), we find that the compensator susceptance is negative of load susceptance in both the schemes of compensation. However, in the former scheme, the compensator network is of purely reactive parameters, and hence the scheme is practically realizable, as illustrated in Example 4.5. But in power system network where neutral is grounded, the compensator impedances have both real and reactive parts as given by (4.146). This indicates that it is not possible to compensate the unbalanced load by using purely reactive network alone. The second aspect which is important is that the resistive parts of the compensator impedances, i.e., R_{fa}, R_{fb}, and R_{fc} are not all positive. One or two of them are of negative sign, as clear from (4.144). This can be further illustrated from the losses point of view. Since total average load active power comes from source, implying that there are no losses in the compensator. That is,

$$I_{fa}^2 R_{fa} + I_{fb}^2 R_{fb} + I_{fc}^2 R_{fc} = 0 \qquad (4.147)$$

Since the terms, I_{fa}^2, I_{fa}^2, and I_{fa}^2 are positive, indicating that one or two terms out of three, i.e., R_{fa}, R_{fb}, and R_{fc} must be of negative signs. This means that there are negative resistances, which can not be realized in practice by using passive components. Negative resistances values appear in (4.146) to replenish the losses due to positive resistance. Even if all resistances are positive, making compensator impedances realizable is not a practical scheme, as irreversible losses occur. These losses have to be supplied from some active source which is not possible with passive components. Therefore, one of the possible schemes in this case, is to set the conductance part of compensator admittance to zero, i.e., $G_{fa} = G_{fb} = G_{fc} = 0$. This is given below.

$$
\begin{aligned}
Y_{fa} &= G_{fa} + jB_{fa} = jB_{fa} \\
Y_{fb} &= G_{fb} + jB_{fb} = jB_{fb} \\
Y_{fc} &= G_{fc} + jB_{fc} = jB_{fc}
\end{aligned}
$$

Using (4.145), $B_{fa} = -B_{la}$, $B_{fb} = -B_{lb}$, $B_{fc} = -B_{lc}$, therefore the overall admittance looking into the source terminal is given by,

$$
\begin{aligned}
Y_{sa} = Y_{la} + Y_{fa} &= G_{la} + jB_{la} - jB_{la} = G_{la} \\
Y_{sb} = Y_{lb} + Y_{fb} &= G_{lb} + jB_{lb} - jB_{lb} = G_{lb} \qquad (4.148) \\
Y_{sc} = Y_{lc} + Y_{fc} &= G_{lc} + jB_{lc} - jB_{lc} = G_{lc}
\end{aligned}
$$

The impedance seen by the source is given as follows.

$$
\begin{aligned}
Z_{sa} &= R_{sa} = \frac{1}{G_{la}} = \frac{R_{la}^2 + X_{la}^2}{R_{la}} \\
Z_{sb} &= R_{sb} = \frac{1}{G_{lb}} = \frac{R_{lb}^2 + X_{lb}^2}{R_{lb}} \qquad (4.149) \\
Z_{sc} &= R_{sc} = \frac{1}{G_{lc}} = \frac{R_{lc}^2 + X_{lc}^2}{R_{lc}}
\end{aligned}
$$

Thus, the source sees a three-phase resistive load as shown in Fig. 4.19. The source

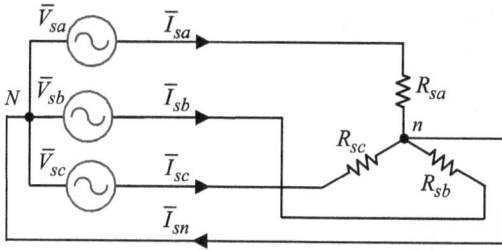

Figure 4.19 A three-phase four-wire system compensated system

currents for this scheme of compensation are given below.

$$\bar{I}_{sa} = \frac{\bar{V}_{sa}}{R_{sa}} = \bar{V}_{sa}\frac{R_{la}}{R_{la}^2+X_{la}^2} = \frac{\bar{V}_{sa}}{\sqrt{R_{la}^2+X_{la}^2}}\frac{R_{la}}{\sqrt{R_{la}^2+X_{la}^2}}$$

$$= I_{la}\angle\bar{V}_{sa}\left(\frac{R_{la}}{\sqrt{R_{la}^2+X_{la}^2}}\right) = I_{la}\cos\phi_{la}\angle\bar{V}_{sa} \qquad (4.150)$$

Similarly, source currents for the other two phases can be given by the following expressions.

$$\bar{I}_{sb} = I_{lb}\angle\bar{V}_{sb}\left(\frac{R_{lb}}{\sqrt{R_{lb}^2+X_{lb}^2}}\right) = I_{lb}\cos\phi_{lb}\angle\bar{V}_{sb} \qquad (4.151)$$

$$\bar{I}_{sc} = I_{lc}\angle\bar{V}_{sc}\left(\frac{R_{lc}}{\sqrt{R_{lc}^2+X_{lc}^2}}\right) = I_{lc}\cos\phi_{lc}\angle\bar{V}_{sc} \qquad (4.152)$$

Obviously, the source currents for this compensated system are in phase with their voltages to give a unity power factor. As seen from (4.150)–(4.152), magnitudes of the source currents are directly proportional to the power factor in respective phases, and therefore their magnitudes are reduced. Consequently, the current in the neutral wire is also reduced. Since the currents are not balanced in magnitude due to different values of resistances seen from the source, the objective of the compensation is not fully achieved. Thus, this scheme provides partial compensation. That is why in the case of the grounded system, to balance the load, active filtering is used by employing active power filters. These devices are called custom power devices [16]–[18]. The one which is used for current compensation and connected with the load in the shunt is called shunt active power filter, which is also known as distribution static compensator (DSTATCOM). These types of load compensation schemes and the devices will be discussed in the next chapter.

Example 4.8. Consider the following three-phase, four-wire system as shown in Fig. 4.18. The three-phase voltages are balanced sinusoids with rms value of 230 V at 50 Hz. The load impedances are $Z_a = 5 + j12\,\Omega$, $Z_b = 3 + j4\,\Omega$, $Z_c = 12 - j13\,\Omega$. Answer the following.

(a) The line currents, $\bar{I}_{la}, \bar{I}_{lb}, \bar{I}_{lc}, \bar{I}_{nN}$.

(b) The total active (P) and reactive (Q) load power of the load supplied by the source.

(c) Compute compensated source currents $(\bar{I}_{sa}, \bar{I}_{sb}, \bar{I}_{sc})$, compensator currents $(\bar{I}_{fa}, \bar{I}_{fb}, \bar{I}_{fc})$ and compensator impedances (Z_{fa}, Z_{fb}, Z_{fc}), so that the supply sees a balanced load with unity power factor.

(d) For case (c), compute total active and reactive source powers and compensator powers. Observe that $P_s = P_l$, $Q_s = 0$, $P_f = 0$ and $Q_f = Q_l$.

(e) Explain why this scheme is not practically realizable.

(f) Discuss the scheme which is realizable using purely reactive components in the compensator. Find out these values of compensator impedances and corresponding network parameters.

(g) For case (f), calculate source currents and neutral current and compare these values with those obtained in (c).

Solution: Given that $Z_a = 5 + j12\,\Omega$, $Z_b = 3 + j4\,\Omega$, $Z_c = 12 - j13\,\Omega$.

(a) Line currents, $\bar{I}_{la}, \bar{I}_{lb}, \bar{I}_{lc}$ are calculated as below.

$$\bar{I}_{la} = \frac{\overline{V}_{sa}}{Z_{la}} = \frac{230\angle 0°}{5 + j12} = 17.69\angle -67.38°\,\text{A}$$

$$\bar{I}_{lb} = \frac{\overline{V}_{sb}}{Z_{lb}} = \frac{230\angle -120°}{3 + j4} = 46.00\angle -173.13°\,\text{A}$$

$$\bar{I}_{lc} = \frac{\overline{V}_{sc}}{Z_{lc}} = \frac{230\angle 120°}{12 - j13} = 13.00\angle 167.29°\,\text{A}$$

The neural current, without compensation, is given as follows.

$$\bar{I}_{nN} = \bar{I}_{la} + \bar{I}_{lb} + \bar{I}_{lc} = 54.93\angle -159.79°\,\text{A}$$

(b) Active (P) and reactive (Q) powers of the load supplied by the source are given as follows.

$$\begin{aligned}
\overline{S}_{la} &= P_{la} + jQ_{la} = \overline{V}_a \bar{I}_{la}^* = 230\angle 0° \times (17.69\angle -67.38°)^* \\
&= 1565.1 + j3756.2\,\text{VA} \\
\overline{S}_{lb} &= P_{lb} + jQ_{lb} = \overline{V}_b \bar{I}_{lb}^* = 230\angle -120° \times (46.00\angle -173.13°)^* \\
&= 6348 + j8464\,\text{VA}
\end{aligned}$$

$$\overline{S}_{lc} = P_{lc} + jQ_{lc} = \overline{V}_c \overline{I}_{lc}^* = 230\angle 120° \times (13.00\angle 167.29°)^*$$
$$= 2028.1 - j2197.1 \text{ VA}$$

Total real power, $P_l = P_{la} + P_{lb} + P_{lc} = 9941.2$ W
Total reactive power, $Q_l = Q_{la} + Q_{lb} + Q_{lc} = 10023.1$ VAr

(c) **With Compensator**
The supply should see the load as a balanced and unity power factor. Therefore the source currents are computed using (4.138) and are given below.

$$\overline{I}_{sa} = \frac{\overline{V}_{sa}}{3} \text{Real}(1/Z_{la} + 1/Z_{lb} + 1/Z_{lc})$$

$$= \frac{230\angle 0°}{3} \text{Real}(\frac{1}{5+j12} + \frac{1}{3+j4} + \frac{1}{12-j13})$$

$$= \frac{230\angle 0°}{3} \text{Real}(0.1879 - j0.1895)$$

$$= \frac{230\angle 0°}{3}(0.1879)$$

$$= 14.4075\angle 0° \text{A}$$

$$\overline{I}_{sb} = \frac{\overline{V}_{sb}}{3} \text{Real}(1/Z_{la} + 1/Z_{lb} + 1/Z_{lc})$$

$$= \frac{230\angle -120°}{3} \text{Real}(\frac{1}{5+j12} + \frac{1}{3+j4} + \frac{1}{12-j13})$$

$$= \frac{230\angle -120°}{3}(0.1879)$$

$$= 14.4075\angle -120° \text{A}$$

$$\overline{I}_{sc} = \frac{\overline{V}_{sc}}{3} \text{Real}(1/Z_{la} + 1/Z_{lb} + 1/Z_{lc})$$

$$= \frac{230\angle 120°}{3} \text{Real}(\frac{1}{5+j12} + \frac{1}{3+j4} + \frac{1}{12-j13})$$

$$= \frac{230\angle 120°}{3}(0.1879)$$

$$- 14.4075\angle 120° \text{A}$$

As seen from the above calculation, the source currents are balanced and in phase with their respective phase voltages. The source neutral current is zero, as source currents are balanced. Once the source currents are known, the compensator currents are computed as follows.

$$\overline{I}_{fa} = \overline{I}_{la} - \overline{I}_{sa} = 17.69\angle -67.38° - 14.41\angle 0° = 18.01\angle -114.96° \text{A}$$
$$\overline{I}_{fb} = \overline{I}_{lb} - \overline{I}_{sb} = 46.00\angle -173.13° - 14.41\angle -120° = 39.09\angle 169.72° \text{A}$$
$$\overline{I}_{fc} = \overline{I}_{lc} - \overline{I}_{sc} = 13.00\angle 167.29° - 14.41\angle 120° = 11.07\angle -119.66° \text{A}$$

The corresponding impedances are now computed as below.

$$Z_{fa} = -\frac{\overline{V}_{sa}}{\overline{I}_{fa}} = -\frac{230\angle 0°}{18.01\angle -114.96°} = 5.38 - j11.57\,\Omega$$

$$Z_{fb} = -\frac{\overline{V}_{sb}}{\overline{I}_{fb}} = -\frac{230\angle -120°}{39.09\angle 169.72°} = -1.98 - j5.54\,\Omega$$

$$Z_{fc} = -\frac{\overline{V}_{sc}}{\overline{I}_{fc}} = -\frac{230\angle 120°}{11.07\angle -119.66°} = 10.49 + j17.93\,\Omega$$

Therefore, the compensator susceptances are given as the following.

$$Y_{fa} = \frac{1}{Z_{fa}} = 0.033 + j0.071\,\text{℧}$$

$$Y_{fb} = \frac{1}{Z_{fb}} = -0.057 + j0.16\,\text{℧}$$

$$Y_{fc} = \frac{1}{Z_{fc}} = 0.024 - j0.041\,\text{℧}$$

(d) Active (P) and reactive (Q) powers of the source after compensation

$$\overline{S}_{sa} = P_{sa} + jQ_{sa} = \overline{V}_a \overline{I}^*_{sa} = 230\angle 0° \times (14.4075\angle 0°)^*$$
$$= 3313.7\,\text{VA}$$
$$\overline{S}_{sb} = P_{sb} + jQ_{sb} = \overline{V}_b \overline{I}^*_{sb} = 230\angle -120° \times (14.4075\angle -120°)^*$$
$$= 3313.7\,\text{VA}$$
$$\overline{S}_{sc} = P_{sc} + jQ_{sc} = \overline{V}_c (\overline{I}_{sc})^* = 230\angle 120° \times (14.4075\angle 120°)^*$$
$$= 3313.7\,\text{VA}$$

Total real power, $P_s = P_{sa} + P_{sb} + P_{sc} = 9941.2$ W
Total reactive power, $Q_s = Q_{sa} + Q_{sb} + Q_{sc} = 0$ VAr

The active and reactive powers of the compensator are computed as follows.

$$\overline{S}_{fa} = P_{fa} + jQ_{fa} = \overline{V}_a \overline{I}^*_{fa} = 230\angle 0° \times (18.01\angle -114.96°)^*$$
$$= -1748.6 + j3756.2\,\text{VA}$$
$$\overline{S}_{fb} = P_{fb} + jQ_{fb} = \overline{V}_b \overline{I}^*_{fb} = 230\angle -120° \times (39.09\angle 169.72°)^*$$
$$= 3034.3 + j8464.0\,\text{VA}$$
$$\overline{S}_{fc} = P_{fc} + jQ_{fc} = \overline{V}_c \overline{I}^*_{fc} = 230\angle 120° \times (11.07\angle -119.66°)^*$$
$$= -1285.6 - j2197.1\,\text{VA}$$

Total real power, $P_f = P_{fa} + P_{fb} + P_{fc} \approx 0$ W
Total reactive power, $Q_f = Q_{fa} + Q_{fb} + Q_{fc} = 10023.1$ VAr
Thus, it is seen that $P_s = P_l$, $Q_s = 0$, $P_f = 0$ and $Q_f = Q_l$.

(e) The compensator impedances are $Z_{fa} = 5.38 - j11.57\,\Omega, Z_{fb} = -1.98 - j5.54\,\Omega, Z_{fc} = 10.49 + j17.93\,\Omega$. It can be observed that the compensator impedances have real parts which correspond to the positive and negative resistances. Due to the presence of these negative resistances, it is not possible to realize this scheme.

(f) A practical realizable compensator
As the compensator impedances have real parts which correspond to positive or negative resistances and are therefore not realizable. Thus to overcome this problem, the real parts of the compensator susceptances are set to zero. That is,

$$Z_{fa} = \frac{1}{j\,\mathrm{Imag}(Y_{fa})} = -j14.08\,\Omega$$

$$Z_{fb} = \frac{1}{j\,\mathrm{Imag}(Y_{fb})} = -j6.25\,\Omega$$

$$Z_{fc} = \frac{1}{j\,\mathrm{Imag}(Y_{fc})} = j24.07\,\Omega$$

These are realized by placing the capacitors in phases a and b, and the inductor in phase c. These values are given as follows.

$$C_{fa} = \frac{1}{2\pi \times 50 \times 14.08} = 226.02\,\mu F$$

$$C_{fb} = \frac{1}{2\pi \times 50 \times 6.25} = 509.3\,\mu F$$

$$L_{fc} = \frac{24.07}{2\pi \times 50} = 76.6\,mH$$

(g) The source currents and neutral current for this scheme of compensation are given below.

$$\bar{I}_{sa} = \overline{V}_{sa} G_{la} = \overline{V}_{sa}\frac{R_{la}}{R_{la}^2 + X_{la}^2} = 6.80\angle 0\,A$$

$$\bar{I}_{sb} = \overline{V}_{sb} G_{lb} = \overline{V}_{sb}\frac{R_{lb}}{R_{lb}^2 + X_{lb}^2} = 27.60\angle -120°\,A$$

$$\bar{I}_{sc} = \overline{V}_{sc} G_{lb} = \overline{V}_{sc}\frac{R_{lc}}{R_{lc}^2 + X_{lc}^2} = 8.82\angle 120°\,A$$

$$\bar{I}_{nN} = \bar{I}_{sa} + \bar{I}_{sb} + \bar{I}_{sc} = 19.86\angle -125.03°\,A$$

In (g), rms values of source currents are 6.8, 27.6, and 8.82 A in phase-a, b, and c, respectively. Considering the feeder of identical resistance, r, in each line, the feeder loss will be $P_{floss} = r \times (I_{sa}^2 + I_{sb}^2 + I_{sc}^2 + I_{nN}^2) = 1280.21\,r$ W. For part (c), the source currents are balanced with rms value of 14.41 A. This gives a feeder loss of $3I_s^2 r = 622.2\,r$ W. Thus, the feeder losses in (c) are approximately half that of the case in (g). This is because the source currents in (g) are partially compensated and hence remain unbalanced. The feeder losses in the uncompensated system are quite high, i.e., $P_{floss} = r \times (I_{la}^2 + I_{lb}^2 + I_{lc}^2 + I_{nN}^2) = 5615.1\,r$ W, which is approximately nine times to those in (c). This emphasizes the fact that the power system should be operated with balanced voltage and current quantities for its efficient and economic operation.

4.8 SUMMARY

In this chapter, the theory of load compensation of passive linear load is explained. It is inferred that a compensator can function independently, either as a load compensator or as a voltage regulator. Various stages of compensator currents are realized using, sampling and averaging techniques, and expressions for compensator susceptances are developed in terms of load currents at specified phase voltages. It is found that for delta-connected load or three-phase ungrounded system, it is always possible to design a compensator comprised of purely reactive elements such as inductance and capacitance to draw source currents which are balanced sinusoids at the fundamental frequency and in phase with the utility voltages. It is proved that this is not true for three-phase grounded system as it involves real power requirements from the compensator phases. This problem is addressed using active power filters, which will be discussed in the next chapter.

4.9 PROBLEMS

P 4.1 Describe various kinds of loads in the power system by giving their specific examples. What are their impacts on power system performance and characteristics?

P 4.2 Explain the term voltage regulation at the load bus in the power system. Show that the load bus voltage is a function of the load current drawn by the load. Give examples where the voltage regulation could be positive, negative, and zero.

P 4.3 Write down some of the practical aspects and characteristics of the compensator, used to correct the power factor, voltage and current unbalances in the system.

P 4.4 Explain how can locally generated reactive power by a compensator regulate the bus voltage and improve the power factor at the load bus.

P 4.5 How can the reactive power by a compensator be generated at the load bus? Which type of power system elements generates or absorbs the reactive power? Give some practical examples for applications of these elements.

P 4.6 Define the short circuit capacity and short circuit current of a bus in a power system. Develop an approximate expression for load bus voltage in terms of system voltage, load reactive power, and short circuit capacity. Explain the nature of variation of bus voltage as load reactive power varies.

P 4.7 Mention characteristics of the ideal voltage regulator and derive the expression for bus voltage (V) in terms of the supply voltage (E), compensator gain $(K\gamma)$, knee point (V_k), load reactive power (Q_l), and short circuit capacity (S_{sc}) of the bus.

P 4.8 Comment upon the neutral voltage and neutral current in three-phase, three-wire, and three-phase, four-wire electrical systems. Explain in three-phase, four-wire system, why is grounding necessary.

P 4.9 What is the relationship between zero sequence voltage (current) and neutral voltage (current) quantities?

P 4.10 Consider a three-phase supply system feeding 11 kV (L-L), 50 Hz, load with the real and reactive power demand of 40 MW and 50 MVAr, respectively. The supply bus has short-circuit level of 300 MVA and feeder with a X_s/R_s ratio of 6. Compute the following.

(a) Find the load bus voltage $(\overline{V})$ and the voltage drop $(\Delta\overline{V})$ in the supply feeder. Thus, determine system voltage $(\overline{E})$.

(b) It is required to maintain the load bus voltage to be the same as the supply bus voltage, i.e., $V_{LL} = 11$ kV. Calculate the rating of the compensator.

P 4.11 Consider a three-phase supply system at 11 kV line-to-line voltage with a short circuit level of 500 MVA and a X_s/R_s ratio of 7, supplying a star-connected inductive load with a real and reactive power demand of 50 MW and 75 MVAr, respectively.

(a) It is required to maintain the load bus voltage to be the same as the supply bus voltage i.e., $V_{LL} = 11$ kV. Calculate the rating of the compensator, compensator reactance, and the nature of its reactive element.

(b) What should be the load bus voltage and compensator current if it is required to maintain the unity power factor at the load bus?

P 4.12 A three-phase distribution feeder is designed to deliver a three-phase induction motor load of 20 MW at 11 kV (L-L), 50 Hz, and a power factor of 0.707 lagging. Now, 1 kA of extra load at a power factor of 0.5 lagging for each phase is connected to the load bus.

(a) What are the total requirement of real power (P) in MW and reactive (Q) in MVAr at the load bus?

(b) Now, this total load with the above real (P) and reactive (Q) power demands in (a), is supplied by a three-phase supply at 11 kV, 50 Hz. The supply bus has short-circuit level of 300 MVA and a feeder with a Xs/Rs ratio of 5. Compute the following.

(i) The voltage at the load bus.

(ii) The reactive power supplied by the compensator connected to the load bus to maintain the same load bus voltage (i.e., 11 kV (L-L)) and the value of compensator element per phase.

(iii) For unity power factor operation at the load bus, what is the rating of the compensator and voltage at the load bus?

P 4.13 For a delta-connected load shown in Fig. 4.15, load admittances are: $Y_l^{ab} = G_l^{ab} + jB_l^{ab}$, $Y_l^{bc} = G_l^{bc} + jB_l^{bc}$ and $Y_l^{ca} = G_l^{ca} + jB_l^{ca}$. Derive relationship for compensator susceptances, $(B_\gamma^{bc}, B_\gamma^{ab}, B_\gamma^{ca})$, such that the line currents are balanced and have unity power factor with their voltages. Assume the source voltages to be balanced and sinusoidal. Consider the load parameters as given below.

$$Z_l^{ab} = \frac{1}{Y_l^{ab}} = 5 + j12, \ Z_l^{bc} = \frac{1}{Y_l^{bc}} = 3 + j4, \ Z_l^{ca} = \frac{1}{Y_l^{ca}} = 9 - j12 \ \Omega.$$

(a) Compute line currents before the placing compensator.

(b) Compute susceptances of the compensator in order to achieve unity power factor at the source.

(c) Determine the source currents after placing the compensator.

(d) Determine the source and compensator active and reactive powers. Co-relate these powers with the load powers.

P 4.14 A three-phase system with 400 V rms (line to line) voltage, 50 Hz, is supplying a resistive load connected between phases a and b with $R = 10\,\Omega$. A purely reactive compensator is connected between the phases. Compute the line currents, active and reactive powers, power factors of each phase, and as well as three-phase. Now compute the compensator susceptances, $(B_\gamma^{ab}, B_\gamma^{bc}, B_\gamma^{ca})$ so that the supply has a balanced load with unity power factor. Co-relate the source, load, and compensator active and reactive powers. Find the rating of the compensator. Also, draw the circuit seen from the source and load side.

P 4.15 Consider the following three-phase, three-wire system as shown in Fig. 4.20. The three-phase voltages are balanced sinusoids with rms value of 230 V. The values of load impedances are given as, $Z_a = 3 + j4\,\Omega, Z_b = 5 + j12\,\Omega$ and $Z_c = 20 + j15\,\Omega$.

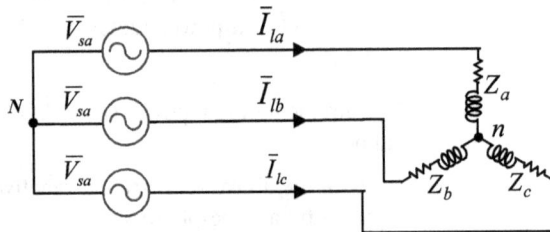

Figure 4.20 Related to problem P 4.15

Compute the following.

(a) Line currents.

(b) Total active (P) and reactive (Q) powers of the load.

(c) Compensator susceptances $(B_\gamma^{bc}, B_\gamma^{ab}, B_\gamma^{ca})$, so that the supply sees the load balanced and unity power factor.

(d) Find source currents after compensation.

(e) For case (c), compute the total source, compensator active and reactive powers (after compensation). Show that compensator supplies the reactive power and source supplies active power of the load.

(f) Draw the circuits as seen from the load and from the source.

P 4.16 Obtain and verify the compensator susceptances $(B_\gamma^{bc}, B_\gamma^{ab}, B_\gamma^{ca})$ of the above problem using Sampling Method.

P 4.17 Obtain and verify the compensator susceptances $(B_\gamma^{bc}, B_\gamma^{ab}, B_\gamma^{ca})$ of above the problem using Averaging Method.

P 4.18 Sampling and Averaging methods give you real time values of compensator susceptances to achieve a unity power factor on the source side. How would you practically realize the variable susceptances in real time?

P 4.19 A three-phase system supplies three-phase ungrounded star connected load with $Z_a = 10 + j15\,\Omega, Z_b = 15 + j20\,\Omega$, and $Z_c = \infty\,\Omega$, (phase-c open) by balanced voltages with rms value of 230 V at 50 Hz to get a unity power factor at the supply.

(a) Compute these susceptances $(B_\gamma^{ab}, B_\gamma^{bc},$ and $B_\gamma^{ca})$ and express them in terms of physical parameters, i.e., inductance or capacitance and their values.

(b) Determine the source currents after placing the compensator.

(c) Determine the source, compensator, and load active and reactive powers.

(d) Draw the circuits as seen from the load and from the source side.

P 4.20 Consider the following three-phase, four-wire system. The three-phase voltages are balanced sinusoids with rms value of 230 V at 50 Hz. The load impedances are $Z_a = 12 + j5\,\Omega, Z_b = 6 - j8\,\Omega, Z_c = 3 + j4\,\Omega$. Compute the following.

(a) Determine the total active (P) and reactive (Q) load power of the load supplied by the source.

(b) Compute the compensator currents $(\bar{I}_{fa}, \bar{I}_{fb}, \bar{I}_{fc})$ and impedances (Z_{fa}, Z_{fb}, Z_{fc}), so that the supply current see the load balanced and unity power factor.

(c) Compute the source currents before and after compensation.

    (d) With above compensator parameters, show that, $P_s = P_l, Q_s = 0, P_f = 0$ and $Q_f = Q_l$.

    (e) Comment upon the realization of calculated compensator impedances.

P 4.21 In problem P 4.20, realize the compensator with pure reactive elements and determine the following.

    (a) Compute the values of the elements of the reactances.

    (b) Determine the source and neutral currents and comment upon their values.

REFERENCES

1. A. Ghosh and G. Ledwich, *Power Quality Enhancement Using Custom Power Devices*, ser. Power Electronics and Power Systems. Springer US, 2012.

2. R. W. Erickson and D. Maksimović, *Power and Harmonics in Nonsinusoidal Systems*. Boston, MA: Springer US, 2001, pp. 589–607.

3. M. H. J. Bollen, *Understanding Power Quality Problems: Voltage Sags and Interruptions*. Wiley-IEEE Press, 1999.

4. Bhim Singh, Ambrish Chandra, and Kamal Al-Haddad, *Power Quality Problems and Mitigation Techniques*. Wiley, 2015.

5. Roger C. Dugan, Mark F. McGranaghan, Surya Santoso, and H. Wayne Beaty, *Electrical Power System Quality*. Tata McGraw-Hill, 2008.

6. M. A. Masoum and E. F. Fuchs, *Chapter 1 – Introduction to Power Quality*, M. A. Masoum and E. F. Fuchs, Eds. Academic Press, 2015.

7. M. Grady, *Understanding Power System Harmonics*. Department of Electrical and Computer Engineering, University of Texas, Austin, 2012.

8. A. Baggini, Ed., *Handbook of Power Quality*. Wiley Sons, England, 2008.

9. B. W. Kennedy, *Power Quality Primer*. Springer, 2000.

10. C. Sankaran, *Power Quality*. CRC Press, 2002.

11. G. T. Heydt, *Electric Power Quality*. Stars in a Circle Publications, 1991.

12. T. J. E. Miller, *Reactive Power Control in Electric Systems*. Wiley, 1982.

13. L. Gyugyi, "Reactive power generation and control by thyristor circuits," *IEEE Transactions on Industry Applications*, vol. IA-15, no. 5, pp. 521–532, 1979.

14. R. Otto, T. Putman, and L. Gyugyi, "Principles and applications of static, thyristor-controlled shunt compensators," *IEEE Transactions on Power Apparatus and Systems*, vol. PAS-97, no. 5, pp. 1935–1945, 1978.

15. N. Hingorani and L. Gyugyi, *Understanding FACTS: Concepts and Technology of Flexible AC Transmission Systems, 2000.* Wiley-IEEE Press, 1999.

16. M. El-Habrouk, M. K. Darwish, and P. Mehta, "Active power filters: a review," *Electric Power Applications, IEE Proceedings*, vol. 147, no. 5, pp. 403–413, 2000.

17. H. Akagi, Y. Kanazawa, and A. Nabae, "Instantaneous reactive power compensators comprising switching devices without energy storage components," *IEEE Transactions on Industry Applications*, vol. IA-20, no. 3, pp. 625–630, 1984.

18. H. Akagi, "Trends in active power line conditioners," *Power Electronics, IEEE Transactions on*, vol. 9, no. 3, pp. 263–268, 1994.

5 Control Theories for Load Compensation

5.1 INTRODUCTION

Originally, passive filters were evolved for high voltage dc transmission systems in order to reduce harmonic voltages and currents in ac power networks to acceptable levels. Later on, these were also used for balancing an unbalanced load. However, passive filters have certain disadvantages, such as increased size, lack of adaptability, inability to re-route active power in order to balance the load, etc., which are overcome by active power devices (filters). Active power filters have been proposed by eminent researchers [1]–[5]. Recent progress in voltage and current rating and switching speed of semiconductor devices such *as* IGBTs, GTO thyristors, has spurred interest in studying active power filters with a focus on practical applications. The sophisticated PWM inverter technology along with developed control methods [6]–[8] have made it possible to put active power filters into commercial installations. In the last couple of decades, researchers have developed many efficient theories for shunt compensation to control active shunt devices. The main purpose of a shunt active power compensator is to cancel the effects of poor load power factor, harmonic currents, and unbalance in the load. The dc component in load current can also be compensated by choosing a suitable compensator configuration. As a result of this compensation, source currents are balanced and sinusoidal with desired power factor. In this chapter, we shall discuss various control theories of shunt compensation, starting from sampling and averaging techniques proposed in 1970s [9], [10], to the later ones, i.e., instantaneous reactive power theory [6], and theory of instantaneous symmetrical components [11]–[13].

In the previous chapter, we studied the methods of load compensation at the fundamental frequency with a linear passive load consisting of resistance, inductance, and capacitance. These methods can eliminate only the fundamental reactive power and unbalance in the steady state for three-phase three-wire systems by employing reactive passive LC filters and thyristor controlled devices. However, when harmonics are present in the system, these methods fail to provide correct compensation. To correct load with unbalance and harmonics, instantaneous load compensation methods are used. The important theories in this context are the Instantaneous Reactive Power Theory, often known as *pq* theory, the Instantaneous Symmetrical Component Theory for load compensation and the Synchronous Reference Frame based theory (also known as *dq* theory) [3], [6], [7], [11], [14]–[21]. The first two theories will be discussed in this chapter due to their wide applications in load compensation. Their merits, demerits and applications are explored in detail.

DOI: 10.1201/9781032617305-5

5.2 INSTANTANEOUS REACTIVE POWER THEORY

To begin with pq theory, we shall first recall the $\alpha\beta 0$ transformation, which was discussed in Chapter 2. For three-phase system shown in Fig. 5.1, the $\alpha\beta 0$ transformations for voltages and currents are given below.

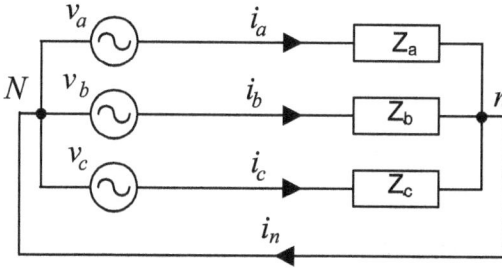

Figure 5.1   A three-phase four-wire system

$$\begin{bmatrix} v_0 \\ v_\alpha \\ v_\beta \end{bmatrix} = \sqrt{\frac{2}{3}} \begin{bmatrix} \frac{1}{\sqrt{2}} & \frac{1}{\sqrt{2}} & \frac{1}{\sqrt{2}} \\ 1 & -\frac{1}{2} & -\frac{1}{2} \\ 0 & \frac{\sqrt{3}}{2} & -\frac{\sqrt{3}}{2} \end{bmatrix} \begin{bmatrix} v_a \\ v_b \\ v_c \end{bmatrix} \tag{5.1}$$

$$\begin{bmatrix} i_0 \\ i_\alpha \\ i_\beta \end{bmatrix} = \sqrt{\frac{2}{3}} \begin{bmatrix} \frac{1}{\sqrt{2}} & \frac{1}{\sqrt{2}} & \frac{1}{\sqrt{2}} \\ 1 & -\frac{1}{2} & -\frac{1}{2} \\ 0 & \frac{\sqrt{3}}{2} & -\frac{\sqrt{3}}{2} \end{bmatrix} \begin{bmatrix} i_a \\ i_b \\ i_c \end{bmatrix} \tag{5.2}$$

The instantaneous active, $p(t)$ and reactive, $q(t)$ powers were defined in Chapter 3 through equations (3.14)–(3.15), respectively. For the sake of completeness, these are given below.

$$\begin{aligned} p_{3\phi}(t) &= v_a i_a + v_b i_b + v_c i_c \\ &= v_\alpha i_\alpha + v_\beta v_\beta + v_0 i_0 \\ &= p_\alpha + p_\beta + p_0 \\ &= p_{\alpha\beta} + p_0 \end{aligned} \tag{5.3}$$

where, $p_{\alpha\beta} = p_\alpha + p_\beta = v_\alpha i_\alpha + v_\beta i_\beta$ and $p_0 = v_0 i_0$.

In the instantaneous reactive power theory, as discussed in Chapter 3, the instantaneous reactive power, $q(t)$ was defined as,

$$
\begin{aligned}
q(t) = q_{\alpha\beta} &= v_\alpha \times i_\beta + v_\beta \times i_\alpha \\
&= v_\alpha i_\beta - v_\beta i_\alpha \\
&= -\frac{1}{\sqrt{3}}[v_{bc}i_a + v_{ca}i_b + v_{ab}i_c]
\end{aligned}
\tag{5.4}
$$

Therefore, powers p_o, $p_{\alpha\beta}$ and $q_{\alpha\beta}$ can be expressed in matrix form as given below.

$$
\begin{bmatrix} p_0 \\ p_{\alpha\beta} \\ q_{\alpha\beta} \end{bmatrix}
=
\begin{bmatrix} v_0 & 0 & 0 \\ 0 & v_\alpha & v_\beta \\ 0 & -v_\beta & v_\alpha \end{bmatrix}
\begin{bmatrix} i_0 \\ i_\alpha \\ i_\beta \end{bmatrix}
\tag{5.5}
$$

From the above equation, the currents, i_0, i_α and i_β are computed as given below.

$$
\begin{aligned}
\begin{bmatrix} i_0 \\ i_\alpha \\ i_\beta \end{bmatrix}
&=
\begin{bmatrix} v_0 & 0 & 0 \\ 0 & v_\alpha & v_\beta \\ 0 & -v_\beta & v_\alpha \end{bmatrix}^{-1}
\begin{bmatrix} p_0 \\ p_{\alpha\beta} \\ q_{\alpha\beta} \end{bmatrix} \\
&= \frac{1}{v_0(v_\alpha^2 + v_\beta^2)}
\begin{bmatrix} v_\alpha^2 + v_\beta^2 & 0 & 0 \\ 0 & v_0 v_\alpha & -v_0 v_\beta \\ 0 & v_0 v_\beta & v_0 v_\alpha \end{bmatrix}
\begin{bmatrix} p_0 \\ p_{\alpha\beta} \\ q_{\alpha\beta} \end{bmatrix}
\end{aligned}
\tag{5.6}
$$

In the above equation,

$$
i_0 = \frac{p_0(v_\alpha^2 + v_\beta^2)}{v_0(v_\alpha{}^2 + v_\beta{}^2)} = \frac{p_0}{v_0} = \frac{v_0 i_0}{v_0} = i_0
\tag{5.7}
$$

$$
\begin{aligned}
i_\alpha &= \frac{v_\alpha}{v_\alpha^2 + v_\beta^2} p_{\alpha\beta} + \frac{(-v_\beta)}{v_\alpha{}^2 + v_\beta{}^2} q_{\alpha\beta} \\
&= i_{\alpha p} + i_{\alpha q}
\end{aligned}
\tag{5.8}
$$

$$
\begin{aligned}
i_\beta &= \frac{v_\beta}{v_\alpha^2 + v_\beta^2} p_{\alpha\beta} + \frac{v_\alpha}{v_\alpha{}^2 + v_\beta{}^2} q_{\alpha\beta} \\
&= i_{\beta p} + i_{\beta q}
\end{aligned}
\tag{5.9}
$$

where,

i_0 = zero sequence instantaneous current

$i_{\alpha p}$ = α-phase instantaneous active current = $\frac{v_\alpha}{v_\alpha^2 + v_\beta^2} p_{\alpha\beta}$

$i_{\beta p}$ = β-phase instantaneous active current = $\frac{v_\beta}{v_\alpha^2 + v_\beta^2} p_{\alpha\beta}$

$i_{\alpha q}$ = α-phase instantaneous reactive current = $-\frac{v_\beta}{v_\alpha^2 + v_\beta^2} q_{\alpha\beta}$

$i_{\beta q}$ = β-phase instantaneous reactive current = $\frac{v_\alpha}{v_\alpha^2 + v_\beta^2} q_{\alpha\beta}$.

Using the above definitions of various components of currents, the three-phase instantaneous power can be expressed as,

$$
\begin{aligned}
p_{3\phi} &= v_0 i_0 + v_\alpha i_\alpha + v_\beta i_\beta \\
&= v_0 i_0 + v_\alpha (i_{\alpha p} + i_{\alpha q}) + v_\beta (i_{\beta p} + i_{\beta q}) \\
&= v_0 i_0 + v_\alpha \left[\frac{v_\alpha}{v_\alpha^2 + v_\beta^2} p_{\alpha\beta} + \frac{-v_\beta}{v_\alpha^2 + v_\beta^2} q_{\alpha\beta} \right] \\
&\quad + v_\beta \left[\frac{v_\beta}{v_\alpha^2 + v_\beta^2} p_{\alpha\beta} + \frac{v_\alpha}{v_\alpha^2 + v_\beta^2} q_{\alpha\beta} \right] \\
&= v_0 i_0 + v_\alpha i_{\alpha p} + v_\alpha i_{\alpha q} + v_\beta i_{\beta p} + v_\beta i_{\beta q} \\
&= v_0 i_0 + (p_{\alpha p} + p_{\alpha q}) + (p_{\beta p} + p_{\beta q}) \\
&= v_0 i_0 + (p_{\alpha p} + p_{\beta p})
\end{aligned}
\tag{5.10}
$$

$$
\tag{5.11}
$$

In the above equation,

$$
p_{\alpha q} + p_{\beta q} = v_\alpha i_{\alpha q} + v_\beta i_{\beta q} = 0 \tag{5.12}
$$

If referred to compensator (or filter), (5.6) can be written as,

$$
\begin{bmatrix} i_{f0} \\ i_{f\alpha} \\ i_{f\beta} \end{bmatrix} = \frac{1}{v_0(v_\alpha^2 + v_\beta^2)} \begin{bmatrix} v_\alpha^2 + v_\beta^2 & 0 & 0 \\ 0 & v_0 v_\alpha & -v_0 v_\beta \\ 0 & v_0 v_\beta & v_0 v_\alpha \end{bmatrix} \begin{bmatrix} p_{f0} \\ p_{f\alpha\beta} \\ q_{f\alpha\beta} \end{bmatrix} \tag{5.13}
$$

Since the compensator does not supply any real power, therefore,

$$
\overline{P}_{f3\phi} = \overline{P}_{f0} + \overline{P}_{f\alpha\beta} = 0 \tag{5.14}
$$

The instantaneous zero sequence power exchanges between the load and the compensator and compensator reactive power must be equal to load reactive power. Therefore, we have

$$
p_{fo} = p_{lo} = \overline{P}_{lo} + \widetilde{p}_{lo} = v_o i_{lo} \tag{5.15}
$$
$$
p_{f\alpha\beta} = \widetilde{p}_{l\alpha\beta} - \overline{P}_{lo} \tag{5.16}
$$
$$
q_{f\alpha\beta} = q_{l\alpha\beta} = v_\alpha i_{l\beta} - v_\beta i_{l\alpha} \tag{5.17}
$$

Since the overall real power from the compensator is equal to zero (the time varying active power can still be there), therefore the following should be satisfied.

$$
p_{fo} + p_{f\alpha\beta} = \widetilde{p}_{lo} + \widetilde{p}_{l\alpha\beta} \tag{5.18}
$$

and,

$$
\overline{P}_{fo} + \overline{P}_{f\alpha\beta} = 0 \tag{5.19}
$$

The power flow description is shown in Fig. 5.2. In the figure, $\Delta\bar{p} = \bar{P}_{lo} + P_{loss}$, is additional real power to be supplied from the source to maintain dc link voltage constant. The term P_{loss} represents losses in the inverter.

The zero sequence current should be circulated through the compensator, therefore,

$$i_{fo} = i_{lo} \tag{5.20}$$

Using the conditions of compensator powers as given in above, the α and β compo-

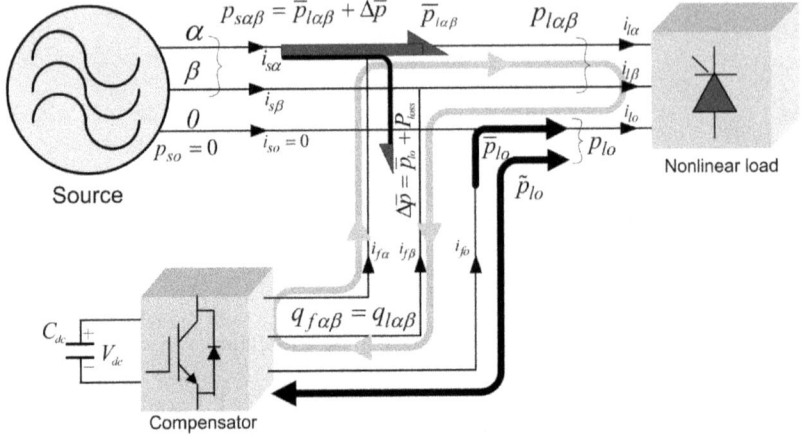

Figure 5.2 Power flow description of three-phase four-wire compensated system

nents of compensator currents can be given as follows.

$$
\begin{aligned}
i_{f\alpha} &= \frac{v_\alpha}{v_\alpha^2 + v_\beta^2} p_{f\alpha\beta} + \frac{-v_\beta}{v_\alpha^2 + v_\beta^2} q_{f\alpha\beta} \\
&= \frac{1}{v_\alpha^2 + v_\beta^2} [v_\alpha(-v_o i_{lo}) - v_\beta(v_\alpha i_{l\beta} - v_\beta i_{l\alpha})] \\
&= \frac{1}{v_\alpha^2 + v_\beta^2} [-v_\alpha v_o i_{lo} - v_\beta v_\alpha i_{l\beta} + v_\beta^2 i_{l\alpha})]
\end{aligned}
\tag{5.21}
$$

Similarly,

$$
\begin{aligned}
i_{f\beta} &= \frac{1}{v_\alpha^2 + v_\beta^2} [v_\beta(p_{f\alpha\beta}) + v_\alpha(q_{f\alpha\beta})] \\
&= \frac{1}{v_\alpha^2 + v_\beta^2} [v_\beta(-v_o i_{lo}) + v_\alpha(v_\alpha i_{l\beta} - v_\beta i_{l\alpha})] \\
&= \frac{1}{v_\alpha^2 + v_\beta^2} [-v_o v_\beta i_{lo} - v_\alpha v_\beta i_{l\alpha} + v_\alpha^2 i_{l\beta}]
\end{aligned}
\tag{5.22}
$$

The above equations are derived based on the assumption that in general, $v_o \neq 0$. If $v_o = 0$, then

$$i_{f0} = i_{lo}$$

$$i_{f\alpha} = \frac{1}{v_\alpha^2 + v_\beta^2} [v_\beta^2 i_{l\alpha} - v_\alpha v_\beta i_{l\beta}] \tag{5.23}$$

$$i_{f\beta} = \frac{1}{v_\alpha^2 + v_\beta^2} [-v_\alpha v_\beta i_{l\alpha} + v_\alpha^2 i_{l\beta}]$$

Once the compensator currents, i_{fo}, $i_{f\alpha}$ and $i_{f\beta}$ are known, they are transformed back to the *abc* frame in order to implement in real time. This transformation is given below.

$$\begin{bmatrix} i_{fa}^* \\ i_{fb}^* \\ i_{fc}^* \end{bmatrix} = \sqrt{\frac{2}{3}} \begin{bmatrix} \frac{1}{\sqrt{2}} & 1 & 0 \\ \frac{1}{\sqrt{2}} & -\frac{1}{2} & \frac{\sqrt{3}}{2} \\ \frac{1}{\sqrt{2}} & -\frac{1}{2} & -\frac{\sqrt{3}}{2} \end{bmatrix} \begin{bmatrix} i_{fo} \\ i_{f\alpha} \\ i_{f\beta} \end{bmatrix} \tag{5.24}$$

These reference currents are shown in Fig. 5.3. Once reference compensator currents are known, these are tracked using a voltage source inverter (VSI). The other details of the scheme are given as follows.

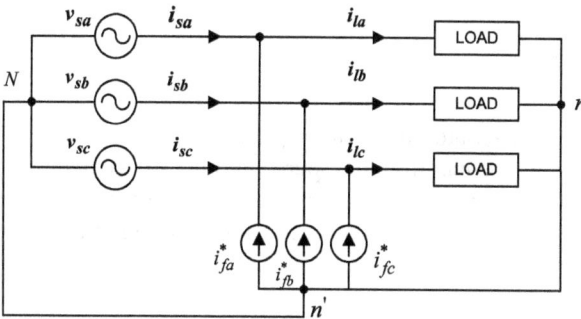

Figure 5.3 A three-phase four-wire compensated system with ideal compensator

The power components and various components of currents are related as follows.

$$\begin{bmatrix} p_\alpha \\ p_\beta \end{bmatrix} = \begin{bmatrix} v_\alpha i_\alpha \\ v_\beta i_\beta \end{bmatrix}$$

$$= \begin{bmatrix} v_\alpha(i_{\alpha p} + i_{\alpha q}) \\ v_\beta(i_{\beta p} + i_{\beta q}) \end{bmatrix}$$

$$= \begin{bmatrix} v_\alpha i_{\alpha p} \\ v_\beta i_{\beta p} \end{bmatrix} + \begin{bmatrix} v_\alpha i_{\alpha q} \\ v_\beta i_{\beta q} \end{bmatrix}$$

$$= \begin{bmatrix} p_{\alpha p} \\ p_{\beta p} \end{bmatrix} + \begin{bmatrix} p_{\alpha q} \\ p_{\beta q} \end{bmatrix} \tag{5.25}$$

The following quantities are defined.

α - axis instantaneous active power $= p_{\alpha p} = v_\alpha \, i_{\alpha p}$

α - axis instantaneous reactive power $= p_{\alpha q} = v_\alpha \, i_{\alpha q}$

β - axis instantaneous active power $= p_{\beta p} = v_\beta \, i_{\beta p}$

β - axis instantaneous reactive power $= p_{\beta q} = v_\beta \, i_{\beta q}$

It is seen that,

$$
\begin{aligned}
p_{\alpha p} + p_{\beta p} &= v_\alpha i_{\alpha p} + v_\beta i_{\beta p} = v_\alpha \frac{v_\alpha}{v_\alpha^2 + v_\beta^2} p_{\alpha \beta} + v_\beta \frac{v_\beta}{v_\alpha^2 + v_\beta^2} p_{\alpha \beta} \\
&= \frac{v_\alpha^2 + v_\beta^2}{v_\alpha^2 + v_\beta^2} p_{\alpha \beta} = p_{\alpha \beta} \quad\quad (5.26)
\end{aligned}
$$

and,

$$
p_{\alpha q} + p_{\beta q} = v_\alpha i_{\alpha q} + v_\beta i_{\beta q} = v_\alpha \frac{-v_\beta}{v_{\alpha^2} + v_{\beta^2}} q_{\alpha \beta} + v_\beta \frac{v_\alpha}{v_{\alpha^2} + v_{\beta^2}} q_{\alpha \beta} = 0 \quad (5.27)
$$

Thus, it can be observed that the sum of $p_{\alpha p}$ and $p_{\beta p}$ is equal to total instantaneous active power $p_{\alpha \beta}$, and the sum of $p_{\alpha q}$ and $p_{\beta q}$ is equal to zero. Therefore,

$$
\begin{aligned}
p_{3\phi} &= p_{\alpha \beta} + p_o \\
&= p_\alpha + p_\beta + p_o \\
&= p_{\alpha p} + p_{\beta p} + p_o \quad\quad (5.28)
\end{aligned}
$$

For a practical compensator, the switching and ohmic losses should be considered. These losses should be met from the source in order to maintain the dc link voltage constant. Let these losses be denoted by P_{loss}, and the average power that must be supplied to the compensator be $\Delta \bar{p}$, as illustrated in Fig. 5.2, then $\Delta \bar{p}$ is given as following.

$$
\Delta \bar{p} = \bar{P}_{lo} + P_{loss} \quad\quad (5.29)
$$

Now, the compensator powers can be expressed as,

$$
\begin{aligned}
p_{f0} &= p_{l0} \\
p_{f\alpha \beta} &= \tilde{p}_{l\alpha \beta} - \Delta \bar{p} \quad\quad (5.30) \\
q_{f\alpha \beta} &= q_{l\alpha \beta}
\end{aligned}
$$

With the above compensator powers, the reference compensator currents are computed as follows.

$$
\begin{bmatrix} i_{fo} \\ i_{f\alpha} \\ i_{f\beta} \end{bmatrix} = \frac{1}{v_0(v_\alpha^2 + v_\beta^2)} \begin{bmatrix} v_\alpha^2 + v_\beta^2 & 0 & 0 \\ 0 & v_o v_\alpha & -v_o v_\beta \\ 0 & v_o v_\beta & v_o v_\alpha \end{bmatrix} \begin{bmatrix} p_{lo} \\ \tilde{p}_{l\alpha \beta} - \Delta \bar{p} \\ q_{l\alpha \beta} \end{bmatrix} \quad (5.31)
$$

Knowing these currents, we can obtain compensator currents in the *abc* frame using the following equation.

$$
\begin{bmatrix} i_{fa}^* \\ i_{fb}^* \\ i_{fc}^* \end{bmatrix} = \sqrt{\frac{2}{3}} \begin{bmatrix} \frac{1}{\sqrt{2}} & 1 & 0 \\ \frac{1}{\sqrt{2}} & -\frac{1}{2} & \frac{\sqrt{3}}{2} \\ \frac{1}{\sqrt{2}} & -\frac{1}{2} & -\frac{\sqrt{3}}{2} \end{bmatrix} \begin{bmatrix} i_{fo} \\ i_{f\alpha} \\ i_{f\beta} \end{bmatrix}
\tag{5.32}
$$

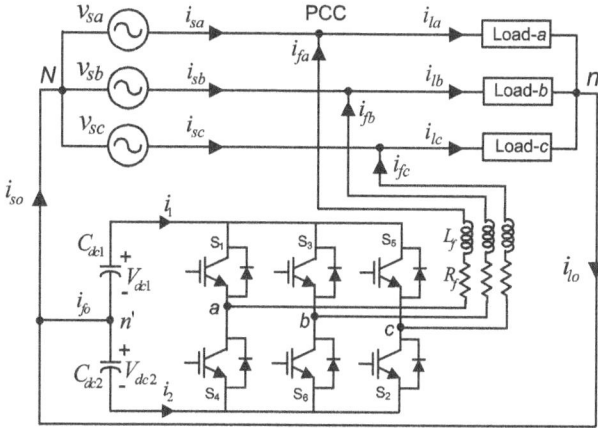

Figure 5.4 A neutral clamped voltage source inverter

These reference currents are realized using a voltage source inverter (VSI). One of the common VSI topologies, which is used to track the reference currents, is illustrated in Fig. 5.4. This VSI topology is known as neutral clamped inverter. The topology has two dc storage capacitors (C_{dc1}, C_{dc2}) with voltages, V_{dc1} and V_{dc2}, respectively. Together these capacitors form the dc link with voltage, $V_{dc} = V_{dc1} + V_{dc2}$ for the inverter. It consists of six switches (S_1 to S_6) with anti-parallel diodes (D_1 to D_6) to provide a freewheeling path for the respective phase currents, as shown in the figure. The switches shown in the figure are realized using Insulated Gate Bipolar Transistors (IGBTs) or Metal Oxide Field Effect Transistors (MOSFETs) depending on their voltage, current, switching frequency and other important parameters. The voltage source inverter is connected to the point of common coupling (PCC) through interface inductors ($L_f - R_f$) in each phase [22]. A schematic diagram of a three-phase four-wire compensated system is shown in Fig. 5.5. As explained above, the reference compensator currents, i.e., i_{fa}^*, i_{fb}^*, i_{fc}^* are generated using an active filter controller. The actual compensator currents, i.e., i_{fa}, i_{fb}, i_{fc} are measured using current transducers. These currents are compared with respective phase reference currents within the defined hysteresis band. This is indicated by dynamic hysteresis current control block, which generates switching signals for the voltage source inverter. While realizing the compensator using a voltage source inverter, there are switching and other losses in the inverter circuit. Therefore, a voltage regulator is

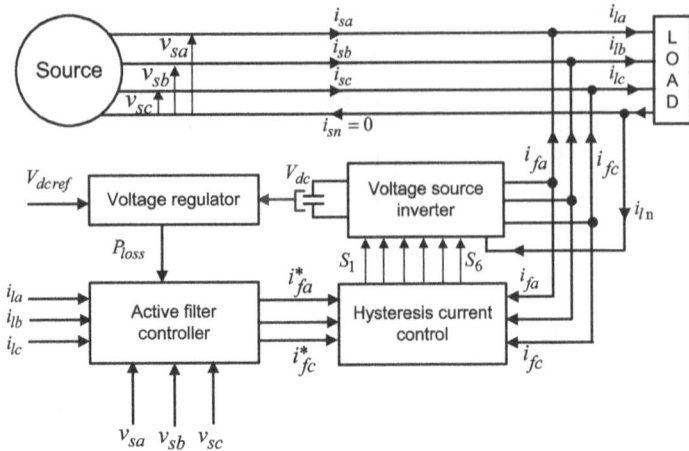

Figure 5.5   Control scheme for three-phase four-wire compensated system

used to maintain the dc link voltage to its reference value. This voltage regulator is basically a proportional integral (PI) controller, which generates P_{loss} term to account for the losses in the inverter circuit.

Example 5.1. Consider a three-phase four-wire system as shown in Fig. 5.3. The supply voltages are balanced with rms value of 230 V at 50 Hz. The load impedances are: $Z_{la} = 5 + j8\,\Omega$, $Z_{lb} = j4\,\Omega$ and $Z_{lc} = -j12\,\Omega$. Using instantaneous reactive power theory, calculate the following.

(a) $\alpha\beta0$ components of three-phase supply voltages ($v_{s\alpha}$, $v_{s\beta}$, v_{s0},) and load currents ($i_{l\alpha}$, $i_{l\beta}$, i_{l0}).

(b) The various components of instantaneous and average powers i.e., p_α, p_β, p_0, $q_{\alpha\beta}$.

(c) Verify these powers on the basis of real power values, i.e., $P_{abc} = P_\alpha + P_\beta + P_0 = P_{\alpha\beta0}$ and reactive power, $Q_{abc} = Q_{\alpha\beta}$.

(d) Reference compensator currents in $\alpha\beta0$ frame i.e., ($i_{f\alpha}$, $i_{f\beta}$, i_{f0}) and abc frame i.e., (i_{fa}, i_{fb}, i_{fc}) so that source supplies balanced currents from the source with unity power factor relationship with their voltages.

(e) Check whether the source currents are balanced and in phase with their voltages.

(f) Check whether $p_s(t) = P_l$, $q_s(t) = 0$, $p_f(t) = 0$ and $q_f(t) = q_l$.

Solution:

(a) The three-phase voltages and currents are expressed as follows.

$$v_{sa}(t) = \sqrt{2}V\sin(\omega t)$$
$$v_{sb}(t) = \sqrt{2}V\sin(\omega t - 120°)$$
$$v_{sc}(t) = \sqrt{2}V\sin(\omega t + 120°)$$

$$i_{la}(t) = \sqrt{2}I_{la}\sin(\omega t - \phi_a)$$
$$i_{lb}(t) = \sqrt{2}I_{lb}\sin(\omega t - 120° - \phi_b)$$
$$i_{lc}(t) = \sqrt{2}I_{lc}\sin(\omega t + 120° - \phi_c)$$

In phasor form, for $V = 230$ V, the voltages and currents are given as follows.

$$\overline{V}_{sa} = 230\angle0° \text{ V}$$
$$\overline{V}_{sb} = 230\angle-120° \text{ V}$$
$$\overline{V}_{sc} = 230\angle120° \text{ V}$$

$$\overline{I}_{la} = \frac{\overline{V}_{sa}}{5+j8} = 24.38\angle-58° \text{ A}$$

$$\overline{I}_{lb} = \frac{\overline{V}_{sb}}{j4} = 57.5\angle-210° \text{ A}$$

$$\overline{I}_{lc} = \frac{\overline{V}_{sc}}{-j12} = 19.16\angle210° \text{ A}$$

Based on the above current phasors, the rms values of currents (I_{la}, I_{lb}, I_{lc}) and their phase angles (ϕ_a, ϕ_b, ϕ_c) are given as following.

$I_{la} = 24.38$ A, $I_{lb} = 57.5$ A, $I_{lc} = 19.16$ A, $\phi_a = -58°$, $\phi_b = 90°$, $\phi_c = -90°$.

Since the voltages and currents have only fundamental components, the calculation of $\alpha\beta0$ components can be made using phasors. Based on the phasor values, we can write time expressions. These are given in the following.

$$\overline{V}_{s\alpha} = \sqrt{\frac{2}{3}}\left(\overline{V}_{sa} - \frac{\overline{V}_{sb}}{2} - \frac{\overline{V}_{sc}}{2}\right) = \sqrt{\frac{3}{2}}V\angle0° = 281.7\angle0° \text{ V}$$

$$\overline{V}_{s\beta} = \sqrt{\frac{2}{3}}\frac{\sqrt{3}}{2}(\overline{V}_{sb} - \overline{V}_{sc}) = \sqrt{\frac{3}{2}}V\angle-90° = 281.7\angle-90° \text{ V}$$

$$\overline{V}_{s0} = \frac{1}{\sqrt{3}}(\overline{V}_{sa} + \overline{V}_{sb} + \overline{V}_{sc}) = 0\text{ V}$$

Similarly,

$$\bar{I}_{l\alpha} = \sqrt{\frac{2}{3}} \left(\bar{I}_{la} - \frac{\bar{I}_{lb}}{2} - \frac{\bar{I}_{lc}}{2} \right) = 45.04\angle -33.27^\circ \text{ A}$$

$$\bar{I}_{l\beta} = \sqrt{\frac{2}{3}} \frac{\sqrt{3}}{2} (\bar{I}_{lb} - \bar{I}_{lc}) = 35.85\angle 130.89^\circ \text{ A}$$

$$\bar{I}_{l0} = \frac{1}{\sqrt{3}} (\bar{I}_{la} + \bar{I}_{lb} + \bar{I}_{lc}) = 30.85\angle -178.38^\circ \text{ A}$$

Based on the above equations, time expressions of $\alpha\beta 0$ components of three-phase supply voltages are given as follows.

$$v_{s\alpha} = \sqrt{\frac{2}{3}} \left(v_{sa} - \frac{v_{sb}}{2} - \frac{v_{sc}}{2} \right) = 281.7\sqrt{2} \sin \omega t$$

$$v_{s\beta} = \sqrt{\frac{2}{3}} \frac{\sqrt{3}}{2} (v_{sb} - v_{sc}) = 281.7\sqrt{2} \sin(\omega t - 90^\circ) \qquad (5.33)$$

$$v_{s0} = \frac{1}{\sqrt{3}} (v_{sa} + v_{sb} + v_{sc}) = 0$$

Similarly, the time expressions of $\alpha\beta 0$ components of three-phase load currents are given as follows.

$$i_{l\alpha} = \sqrt{\frac{2}{3}} \left(i_{la} - \frac{i_{lb}}{2} - \frac{i_{lc}}{2} \right) = 45.04\sqrt{2} \sin(\omega t - 33.27^\circ)$$

$$i_{l\beta} = \sqrt{\frac{2}{3}} \frac{\sqrt{3}}{2} (i_{lb} - i_{lc}) = 35.85\sqrt{2} \sin(\omega t + 130.89^\circ)$$

$$i_{l0} = \frac{1}{\sqrt{3}} (i_{la} + i_{lb} + i_{lc}) = 30.85\sqrt{2} \sin(\omega t - 178.38^\circ)$$

Therefore, we have the following voltages and currents in $\alpha\beta 0$ phasor form,

$$\bar{V}_{s\alpha} = 281.7\angle 0^\circ \text{ V}$$
$$\bar{V}_{s\beta} = 281.7\angle -90^\circ \text{ V}$$
$$\bar{V}_{s0} = 0 \text{ V}$$

and the currents are as follows.

$$\bar{I}_{l\alpha} = 45.04\angle -33.27^\circ \text{ A}$$
$$\bar{I}_{l\beta} = 35.85\angle 130.89^\circ \text{ A}$$
$$\bar{I}_{l0} = 30.85\angle -178.38^\circ \text{ A}$$

(b) Power components $p_{l\alpha}$, $p_{l\beta}$, p_{l0}, $q_{l\alpha\beta}$

$$
\begin{aligned}
p_{l\alpha} &= v_{s\alpha}\, i_{l\alpha} = 281.7\sqrt{2}\,\sin\omega t \; 45.04\sqrt{2}\,\sin(\omega t - 33.27^\circ)\\
&= 12.68\,\{\cos 33.27^\circ - \cos(2\omega t - 33.27^\circ)\}\\
&= 10.61 - 12.68\,\cos(2\omega t - 33.27^\circ)\,\mathrm{kW}
\end{aligned}
$$

Therefore, the average value of $p_{l\alpha} = P_{l\alpha} = 10.61\ \mathrm{kW}$
Oscillating value of $p_{l\alpha} = \widetilde{p}_{l\alpha} = -12.68\,\cos(2\omega t - 33.27^\circ)\ \mathrm{kW}$.

$$
\begin{aligned}
p_{l\beta} &= v_{s\beta}\, i_{l\beta} = 281.7\sqrt{2}\,\sin(\omega t - 90^\circ)\; 35.85\sqrt{2}\,\sin(\omega t + 130.89^\circ)\\
&= 10.09\,\{\cos 220.89^\circ - \cos(2\omega t + 40.89^\circ)\}\\
&= -7.64 - 10.09\,\cos(2\omega t + 40.89^\circ)\,\mathrm{kW}
\end{aligned}
$$

Average value of $p_{l\beta} = P_{l\beta} = -7.63\ \mathrm{kW}$
Oscillating value of $p_{l\beta} = \widetilde{p}_{l\beta} = -10.09\,\cos(2\omega t + 40.89^\circ)\ \mathrm{kW}$.

Therefore, the total active power in $\alpha\beta$ frame i.e., $p_{l\alpha\beta}$ is given as,

$$
\begin{aligned}
p_{l\alpha\beta} &= p_{l\alpha} + p_{l\beta}\\
&= 2.97 + 18.24\,\cos(2\omega t + 178.86^\circ)\,\mathrm{kW}
\end{aligned}
\tag{5.34}
$$

Average value of $p_{l\alpha\beta} = P_{l\alpha\beta} = 2.97\ \mathrm{kW}$.
Oscillating value of $p_{l\alpha\beta} = \widetilde{p}_{l\alpha\beta} = 18.24\,\cos(2\omega t + 178.86^\circ)\ \mathrm{kW}$.
The zero sequence power is zero as $v_{s0} = 0$.
Therefore,

$$
p_{l0} = v_{s0}\, i_{l0} = 0
$$

The instantaneous reactive power in $\alpha\beta 0$ frame is given by,

$$
\begin{aligned}
q_{l\alpha\beta} &= v_{s\alpha}\, i_{l\beta} - v_{s\beta}\, i_{l\alpha}\\
&= 281.7\sqrt{2}\,\sin(\omega t)\; 35.85\sqrt{2}\,\sin(\omega t + 130.89^\circ)\\
&\quad -281.7\sqrt{2}\,\sin(\omega t - 90^\circ)\; 45.04\sqrt{2}\,\sin(\omega t - 33.27^\circ)\\
&= 10.09\,\{\cos 130.89^\circ - \cos(2\omega t + 130.89^\circ)\}\\
&\quad -12.69\,\{\cos 56.73^\circ - \cos(2\omega t - 123.27^\circ)\}\\
&= -6.61 - 10.09\,\cos(2\omega t + 130.89^\circ) - 6.96\\
&\quad +12.69\,\cos(2\omega t - 123.27^\circ)\\
&= -13.57 - 18.24\,\cos(2\omega t + 88.89^\circ)\,\mathrm{kVAr}
\end{aligned}
$$

From above expressions, the average value of these terms is given below.

$$
\begin{aligned}
P_{3\phi} = P_{l\alpha} + P_{l\beta} + P_0 &= 2.97\,\mathrm{kW}\\
Q_{l\alpha\beta} &= -13.57\,\mathrm{kVAr}
\end{aligned}
$$

(c) The three-phase average real and reactive powers of the load are given by,

$$P_{3\phi} = V_{sa}I_{la}\cos\phi_a + V_{sb}I_{lb}\cos\phi_b + V_{sc}I_{lc}\cos\phi_c = 2.97 + 0 + 0 = 2.97\,\text{kW}$$
$$Q_{3\phi} = -(V_{sa}I_{la}\sin\phi_a + V_{sb}I_{lb}\sin\phi_b + V_{sc}I_{lc}\sin\phi_c)$$
$$= 4.75 + 13.23 - 4.41 = 13.57\,\text{kVAr}$$

Thus, we observe that,

$$
\begin{aligned}
P_{3\phi} &= P_{abc} = P_{\alpha\beta0} = 2.97\,\text{kW} \\
Q_{3\phi} &= Q_{abc} = -Q_{\alpha\beta} = 13.57\,\text{kVAr}
\end{aligned}
$$

(d) The required filter powers, p_{f0}, $p_{f\alpha\beta}$ and $q_{f\alpha\beta}$ in terms of load powers are given below.

$$
\begin{aligned}
p_{fo} &= p_{lo} = 0 \\
p_{f\alpha\beta} &= \tilde{p}_{l\alpha\beta} = 18.24\cos(2\omega t + 178.86°)\,\text{kW} \\
q_{f\alpha\beta} &= q_{l\alpha\beta} = -13.57 - 18.24\cos(2\omega t + 88.89°)\,\text{kVAr}
\end{aligned}
$$

Now the filter currents can be found by using (5.13) and substituting the above filter powers. These are expressed below.

$$
\begin{aligned}
i_{f0} &= i_{l0} = 43.63\sin(\omega t - 178.38°)\,\text{A} \\
i_{f\alpha} &= \frac{v_\alpha}{v_\alpha^2 + v_\beta^2}p_{f\alpha\beta} + \frac{-v_\beta}{v_\alpha^2 + v_\beta^2}q_{f\alpha\beta} = 57.59\sin(\omega t - 37.36°)\,\text{A} \\
i_{f\beta} &= \frac{v_\beta}{v_\alpha^2 + v_\beta^2}p_{f\alpha\beta} + \frac{v_\alpha}{v_\alpha^2 + v_\beta^2}q_{f\alpha\beta} = 56.52\sin(\omega t + 126.93°)\,\text{A}
\end{aligned}
$$

Substituting these values in equation (5.32), we get filter currents in *abc* frame. These are expressed below.

$$
\begin{aligned}
i_{fa} &= \sqrt{2} \times 22.39\sin(\omega t - 67.38°)\,\text{A} \\
i_{fb} &= \sqrt{2} \times 57.66\sin(\omega t + 145.72°)\,\text{A} \\
i_{fc} &= \sqrt{2} \times 19.65\sin(\omega t - 137.33°)\,\text{A}
\end{aligned}
$$

(e) Therefore, the source currents i_{sa}, i_{sb}, i_{sc} after compensation are given below.

$$
\begin{aligned}
i_{sa} &= i_{la} - i_{fa} = \sqrt{2} \times 4.3\sin\omega t\,\text{A} \\
i_{sb} &= i_{lb} - i_{fb} = \sqrt{2} \times 4.3\sin(\omega t - 120°)\,\text{A} \\
i_{sc} &= i_{lc} - i_{fc} = \sqrt{2} \times 4.3\sin(\omega t + 120°)\,\text{A}
\end{aligned}
$$

It is seen that the above source currents are balanced and in phase with the respective voltages.

(f) From the above values of voltage and currents of source, load and compensator, the powers are given below.

$$p_s = v_{sa}i_{sa} + v_{sb}i_{sb} + v_{sc}i_{sc} = 2.97\,\text{kW} = P_l$$

$$q_s = -\frac{1}{\sqrt{3}}(v_{sbc}i_{sa} + v_{sca}i_{sb} + v_{sab}i_{sc}) = 0\,\text{kVAr}$$

$$p_f = v_{sa}i_{fa} + v_{sb}i_{fb} + v_{sc}i_{fc} = 0\,\text{kW}$$

$$q_f = -\frac{1}{\sqrt{3}}(v_{sbc}i_{fa} + v_{sca}i_{fb} + v_{sab}i_{fc}) \tag{5.35}$$

$$= -13.57 - 18.24\cos(2\omega t + 88.89°)\,\text{kVAr} = q_l$$

5.2.1 STATE SPACE MODELING OF THE COMPENSATOR

There are various VSI topologies that can be used to realize DSTATCOM [23]. The most commonly used is neutral clamped inverter topology as shown in Fig. 5.4. Since this is a three-phase four-wire system each phase can be considered independently. Therefore, to analyze the above circuit, only one phase is considered. When switch S_1 is closed and S_4 is open, as shown in Fig. 5.6(a), then Kirchhoff's Voltage Law (KVL) is given in the following equation.

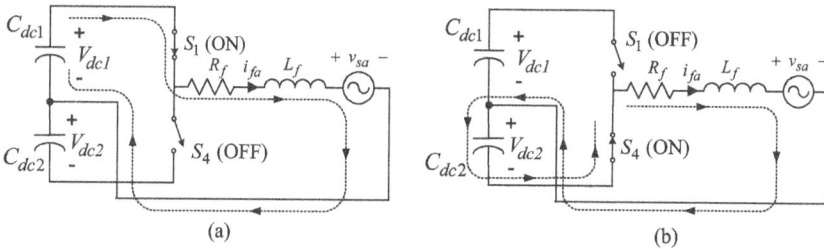

(a)   (b)

Figure 5.6 Equivalent circuit (a) S_1 (ON), S_4 (OFF) and (b) S_1 (OFF), S_4 (ON)

$$L_f\frac{di_{fa}}{dt} + R_f i_{fa} + v_{sa} - V_{dc1} = 0 \tag{5.36}$$

From the above equation,

$$\frac{di_{fa}}{dt} = -\frac{R_f}{L_f}i_{fa} - \frac{v_{sa}}{L_f} + \frac{V_{dc1}}{L_f} \tag{5.37}$$

Similarly, when S_1 is open and S_4 is closed as shown in Fig. 5.6(b),

$$\frac{di_{fa}}{dt} = -\frac{R_f}{L_f}i_{fa} - \frac{v_{sa}}{L_f} - \frac{V_{dc2}}{L_f} \tag{5.38}$$

The above two equations can be combined into one equation by using switching signals $S_a = S_1$, $\overline{S}_a = S_4$, (S_1 and S_4 are complementary binary signals for switches in leg-*a* of the inverter), as given below.

$$\frac{di_{fa}}{dt} = -\frac{R_f}{L_f}i_{fa} + S_a\frac{V_{dc1}}{L_f} - \overline{S}_a\frac{V_{dc2}}{L_f} - \frac{v_{sa}}{L_f} \tag{5.39}$$

Similarly, for phases *b* and *c*, the first order derivative of filter currents can be written as follows.

$$\frac{di_{fb}}{dt} = -\frac{R_f}{L_f}i_{fb} + S_b\frac{V_{dc1}}{L_f} - \overline{S}_b\frac{V_{dc2}}{L_f} - \frac{v_{sb}}{L_f} \tag{5.40}$$

$$\frac{di_{fc}}{dt} = -\frac{R_f}{L_f}i_{fc} + S_c\frac{V_{dc1}}{L_f} - \overline{S}_c\frac{V_{dc2}}{L_f} - \frac{v_{sc}}{L_f} \tag{5.41}$$

where, $S_a = 0$ and $\overline{S}_a = 1$ implies that the top switch is open and bottom switch is closed and, $S_a = 1$ and $\overline{S}_a = 0$ implies that the top switch is closed and bottom switch is open. The two logic signals S_a and $\overline{S}_a$ are complementary to each other. This logic also holds for the other two phases.

The inverter currents i_1 and i_2, as shown in Fig. 5.4, can be expressed in terms of filter currents and switching signals. These are given below.

$$\begin{aligned} i_1 &= S_a i_{fa} + S_b i_{fb} + S_c i_{fc} \\ i_2 &= \overline{S}_a i_{fa} + \overline{S}_b i_{fb} + \overline{S}_c i_{fc} \end{aligned} \tag{5.42}$$

The relationship between dc capacitor voltages V_{dc1}, V_{dc2} and inverter currents i_1 and i_2 are given as below.

$$C_{dc1}\frac{dV_{dc1}}{dt} = -i_1$$

$$C_{dc2}\frac{dV_{dc2}}{dt} = i_2 \tag{5.43}$$

Considering $C_{dc1} = C_{dc2} = C_{dc}$ and substituting i_1 and i_2 from (5.42), the above equations can be written as,

$$\frac{dV_{dc1}}{dt} = -\frac{S_a}{C_{dc}}i_{fa} - \frac{S_b}{C_{dc}}i_{fb} - \frac{S_c}{C_{dc}}i_{fc} \tag{5.44}$$

$$\frac{dV_{dc2}}{dt} = \frac{\overline{S}_a}{C_{dc}}i_{fa} + \frac{\overline{S}_b}{C_{dc}}i_{fb} + \frac{\overline{S}_c}{C_{dc}}i_{fc} \tag{5.45}$$

The equations (5.39), (5.40), (5.41), (5.44) and (5.45) can be represented in state space form as given below.

$$\frac{d}{dt}\begin{bmatrix} i_{fa} \\ i_{fb} \\ i_{fc} \\ V_{dc1} \\ V_{dc2} \end{bmatrix} = \begin{bmatrix} -\frac{R_f}{L_f} & 0 & 0 & \frac{S_a}{L_f} & -\frac{\overline{S}_a}{L_f} \\ 0 & -\frac{R_f}{L_f} & 0 & \frac{S_b}{L_f} & -\frac{\overline{S}_b}{L_f} \\ 0 & 0 & -\frac{R_f}{L_f} & \frac{S_c}{L_f} & -\frac{\overline{S}_c}{L_f} \\ -\frac{S_a}{C_{dc}} & -\frac{S_b}{C_{dc}} & -\frac{S_c}{C_{dc}} & 0 & 0 \\ \frac{\overline{S}_a}{C_{dc}} & \frac{\overline{S}_b}{C_{dc}} & \frac{\overline{S}_c}{C_{dc}} & 0 & 0 \end{bmatrix}\begin{bmatrix} i_{fa} \\ i_{fb} \\ i_{fc} \\ V_{dc1} \\ V_{dc2} \end{bmatrix} + \begin{bmatrix} -\frac{1}{L_f} & 0 & 0 \\ 0 & -\frac{1}{L_f} & 0 \\ 0 & 0 & -\frac{1}{L_f} \\ 0 & 0 & 0 \\ 0 & 0 & 0 \end{bmatrix}\begin{bmatrix} v_{sa} \\ v_{sb} \\ v_{sc} \end{bmatrix}$$

$$\tag{5.46}$$

The above equation is of the form,

$$\dot{x} = Ax + Bu \tag{5.47}$$

where, x is a state vector, A is system matrix, B is input matrix and u is input vector. This state space equation can be solved using MATLAB to implement the compensator for the simulation study.

5.2.2 SWITCHING CONTROL OF THE VSI

In (5.46), the switching signals S_a, $\overline{S}_a$, S_b, $\overline{S}_b$, S_c, and $\overline{S}_c$ are generated using a hysteresis band current control. This is described as follows.

The upper and lower bands of the reference filter current (say phase-a) are formed using hysteresis h, i.e., $i^*_{fa} + h$ and $i^*_{fa} - h$. The following logic is used to generate switching signals.

$$\text{If } i_{fa} \geq (i^*_{fa} + h)$$
$$S_a = 0, \text{ and } \overline{S}_a = 1$$
$$\text{else if } i_{fa} \leq (i^*_{fa} - h)$$
$$S_a = 1, \text{ and } \overline{S}_a = 0$$
$$\text{else if } (i^*_{fa} - h) < i_{fa} < (i^*_{fa} + h)$$
$$\text{Retain the current status of the switches}$$
$$\text{end}$$

The first-order derivative of state variables can be easily solved using "c2d" (continuous to discrete) command in MATLAB, as given below.

$$[A_d \; B_d] = c2d(A, B, T_d) \tag{5.48}$$

The value of the state vector is updated using the following equation.

$$x[(k+1)T_d] = A_d x[kT_d] + B_d u[kT_d] \tag{5.49}$$

where $x(k+1)$ refers the value of the state vector at $(k+1)^{th}$ sample or at the instant $t = (k+1)T_d$ for T_d as the time duration between two consecutive samples. The solution of state equation given by (5.47) is given as follows [24].

$$x(t) = e^{A(t-t_0)} x(t_0) + \int_{t_o}^{t} e^{A(t-\tau)} Bu(\tau) d\tau \tag{5.50}$$

where t_o represents initial time and t represents current time.

Writing the above equation for a small time interval, $kT_d \le t \le (k+1)T_d$ with $t_o = kT_d$ and $t = (k+1)T_d$,

$$x[(k+1)T_d] = e^{AT_d} x[kT_d] + \int_{kT_d}^{(k+1)T_d} e^{A\{(k+1)T_d - \tau\}} B u[kT_d] d\tau$$

$$= e^{AT_d} x[kT_d] + \int_{kT_d}^{(k+1)T_d} \left\{ e^{A\{(k+1)T_d - \tau\}} B d\tau \right\} u[kT_d] \quad (5.51)$$

Comparing (5.49) and (5.51), the discrete matrices A_d and B_d computed by "c2d" MATLAB function can be written as follows.

$$A_d = e^{AT_d}$$

$$B_d = \int_{kT_d}^{(k+1)T_d} e^{A\{(k+1)T_d - \tau\}} B d\tau \quad (5.52)$$

5.2.3 GENERATION OF P_{LOSS} TO MAINTAIN DC CAPACITOR VOLTAGE

The next step is to determine P_{loss} in order to maintain the dc link voltage close to its reference value. In compensation, the average voltage variation of dc link is an indicator of P_{loss} in the inverter. If losses are more than the power supplied by the source through inverter, the dc link voltage, i.e., $V_{dc} = V_{dc1} + V_{dc2}$, will decrease toward zero and vice-versa. For proper operation of the compensator, we need to maintain the dc capacitor voltage to two times of the reference value of each capacitor voltage, i.e., $V_{dc1} + V_{dc2} = V_{dc} = 2V_{dcref}$. The value of V_{dcref} is chosen well above (approximately 1.5 times) the peak value of the system voltage [25]. Thus, we have to replenish losses in the inverter and sustain the dc capacitor voltage to $2V_{dcref}$ with each capacitor voltage at V_{dcref}. This is achieved by using proportional integral (PI) controller described below [7]. Let us define an error signal as follows.

$$e_{Vdc} = 2V_{dcref} - (V_{dc1} + V_{dc2}) = 2V_{dcref} - V_{dc}$$

Then, the term P_{loss} is computed as follows.

$$P_{loss} = K_p e_{Vdc} + K_i \int_0^{T_d} e_{Vdc} dt$$

This dc link voltage control loop need not be very fast and may be updated once in a voltage cycle, preferably at the positive zero crossings of phase-a voltage and generate P_{loss} term at these points. The above controller can be implemented using the digital domain as follows.

$$P_{loss}(k) = K_p e_{Vdc}(k) + K_i \sum_{j=0}^{k} e_{Vdc}(j) T_d. \quad (5.53)$$

In the above equation, k represents the kth sample of error, $e_{V_{dc}}$. For $k = 1$, the above equation can be written as,

$$P_{loss}(1) = K_p e_{V_{dc}}(1) + K_i \sum_{j=0}^{1} e_{V_{dc}}(j) T_d$$
$$= K_p e_{V_{dc}}(1) + K_i[e_{V_{dc}}(0) + e_{V_{dc}}(1)] T_d \qquad (5.54)$$

Similarly for $k = 2$, we can write,

$$P_{loss}(2) = K_p e_{V_{dc}}(1) + K_i \left[e_{V_{dc}}(0) + e_{V_{dc}}(1) + e_{V_{dc}}(2) \right] T_d \qquad (5.55)$$

Replacing $K_i[e_{V_{dc}}(0) + e_{V_{dc}}(1)]$ from (5.54), we get,

$$K_i \left[e_{V_{dc}}(0) + e_{V_{dc}}(1) \right] T_d = P_{loss}(1) - K_p e_{V_{dc}}(1) \qquad (5.56)$$

Substituting above value in (5.55), we obtain the following.

$$P_{loss}(2) = K_p e_{V_{dc}}(2) + P_{loss}(1) - K_p e_{V_{dc}}(1) + K_i e_{V_{dc}}(2) T_d$$
$$= P_{loss}(1) + K_p \left[e_{V_{dc}}(2) - e_{V_{dc}}(1) \right] + K_i e_{V_{dc}}(2) T_d \qquad (5.57)$$

In general, for k^{th} sample of P_{loss},

$$P_{loss}(k) = P_{loss}(k-1) + K_p \left[e_{V_{dc}}(k) - e_{V_{dc}}(k-1) \right] + K_i e_{V_{dc}}(k) T_d \qquad (5.58)$$

The above algorithm can be used to implement PI controller to generate P_{loss}. The control action of the dc link voltage control loop can be updated once at every positive zero crossing of phase-a voltage.

5.2.4 COMPUTATION OF LOAD AVERAGE POWER

In reference current expressions, the average load power (P_{lavg}) is required to be computed. Although a low pass filter can be used to find this, its dynamic response is slow, which may not be suitable for effective compensation. The dynamic performance of computation of P_{lavg} plays a significant role in compensation. For this reason, a moving average algorithm can be used to compute average power, which is described below.

$$P_{lavg} = \frac{1}{T} \int_0^T (v_a i_{la} + v_b i_{lb} + v_c i_{lc}) dt \qquad (5.59)$$

The above equation can be written as an integration operation from $t_1 - T$ to t_1 as given in the following.

$$P_{lavg} = \frac{1}{T} \int_{t_1-T}^{t_1} (v_a i_{la} + v_b i_{lb} + v_c i_{lc}) dt \qquad (5.60)$$

This is known as moving average filter (MAF). Any change in variables instantly reflects within the settling time of one cycle. In the above equation, T represents the time period of fundamental component of voltage or current. The time t_1 is any arbitrary time instant.

5.3 SOME MISCONCEPTIONS IN INSTANTANEOUS REACTIVE POWER THEORY

The instantaneous reactive power theory has evolved from Fortescue, Clarke and Park Transformations of voltage and current specified in phases-a, b and c coordinates [26]. In general, for three-phase, four-wire system,

$$\begin{bmatrix} v_o \\ v_\alpha \\ v_\beta \end{bmatrix} = \sqrt{\frac{2}{3}} \begin{bmatrix} \frac{1}{\sqrt{2}} & \frac{1}{\sqrt{2}} & \frac{1}{\sqrt{2}} \\ 1 & -\frac{1}{2} & -\frac{1}{2} \\ 0 & \frac{\sqrt{3}}{2} & -\frac{\sqrt{3}}{2} \end{bmatrix} \begin{bmatrix} v_a \\ v_b \\ v_c \end{bmatrix} \tag{5.61}$$

Similarly, for currents, the $\alpha\beta 0$ components are given as follows.

$$\begin{bmatrix} i_o \\ i_\alpha \\ i_\beta \end{bmatrix} = \sqrt{\frac{2}{3}} \begin{bmatrix} \frac{1}{\sqrt{2}} & \frac{1}{\sqrt{2}} & \frac{1}{\sqrt{2}} \\ 1 & -\frac{1}{2} & -\frac{1}{2} \\ 0 & \frac{\sqrt{3}}{2} & -\frac{\sqrt{3}}{2} \end{bmatrix} \begin{bmatrix} i_a \\ i_b \\ i_c \end{bmatrix} \tag{5.62}$$

For balanced system $v_0 = (v_a + v_b + v_c)/\sqrt{3} = 0$. For three-phase three-wire system, $i_a + i_b + i_c = 0$, which implies that $i_0 = 0$. Using these details, the above transformations in (5.61) and (5.62) result in the following.

$$\begin{bmatrix} v_\alpha \\ v_\beta \end{bmatrix} = \sqrt{\frac{2}{3}} \begin{bmatrix} 1 & -\frac{1}{2} & -\frac{1}{2} \\ 0 & \frac{\sqrt{3}}{2} & -\frac{\sqrt{3}}{2} \end{bmatrix} \begin{bmatrix} v_a \\ v_b \\ v_c \end{bmatrix} \tag{5.63}$$

$$\begin{bmatrix} i_\alpha \\ i_\beta \end{bmatrix} = \sqrt{\frac{2}{3}} \begin{bmatrix} 1 & -\frac{1}{2} & -\frac{1}{2} \\ 0 & \frac{\sqrt{3}}{2} & -\frac{\sqrt{3}}{2} \end{bmatrix} \begin{bmatrix} i_a \\ i_b \\ i_c \end{bmatrix} \tag{5.64}$$

From equation (5.63), and using $v_a + v_b + v_c = 0$, we get the following.

$$\begin{aligned} v_\alpha &= \sqrt{\frac{2}{3}} \left[v_a - \frac{v_b + v_c}{2} \right] = \sqrt{\frac{2}{3}} \left[v_a - \frac{v_b}{2} - \frac{v_c}{2} \right] \\ &= \sqrt{\frac{2}{3}} \left[v_a - \frac{v_b}{2} - \left(-\frac{v_a}{2} - \frac{v_b}{2} \right) \right] \\ &= \sqrt{\frac{3}{2}} v_a \end{aligned} \tag{5.65}$$

$$v_\beta = \sqrt{\frac{2}{3}} \left[\frac{\sqrt{3}}{2} v_b - \frac{\sqrt{3}}{2} v_c \right] = \sqrt{\frac{2}{3}} \left[\frac{\sqrt{3}}{2} v_b - \frac{\sqrt{3}}{2} (-v_a - v_b) \right]$$

$$= \sqrt{\frac{2}{3}} \left[\frac{\sqrt{3}}{2} v_b + \frac{\sqrt{3}}{2} v_a + \frac{\sqrt{3}}{2} v_b \right]$$

$$= \frac{1}{\sqrt{2}} v_a + \sqrt{2} v_b \qquad (5.66)$$

Writing (5.65) and (5.66) in matrix form we get,

$$\begin{bmatrix} v_\alpha \\ v_\beta \end{bmatrix} = \begin{bmatrix} \sqrt{\frac{3}{2}} & 0 \\ \frac{1}{\sqrt{2}} & \sqrt{2} \end{bmatrix} \begin{bmatrix} v_a \\ v_b \end{bmatrix} \qquad (5.67)$$

Similary, using $i_a + i_b + i_c = 0$ the following can be written.

$$\begin{bmatrix} i_\alpha \\ i_\beta \end{bmatrix} = \begin{bmatrix} \sqrt{\frac{3}{2}} & 0 \\ \frac{1}{\sqrt{2}} & \sqrt{2} \end{bmatrix} \begin{bmatrix} i_a \\ i_b \end{bmatrix} \qquad (5.68)$$

According to the *pq* theory, the *abc* components of voltages and currents are transformed to the α and β coordinates and the instantaneous powers p and q of the load can be expressed as follows.

$$p = v_\alpha i_\alpha + v_\beta i_\beta \qquad (5.69)$$
$$q = v_\alpha i_\beta - v_\beta i_\alpha \qquad (5.70)$$

The above equations representing instantaneous active and reactive powers can be expressed in the matrix form as following [6].

$$\begin{bmatrix} p \\ q \end{bmatrix} = \begin{bmatrix} v_\alpha & v_\beta \\ -v_\beta & v_\alpha \end{bmatrix} \begin{bmatrix} i_\alpha \\ i_\beta \end{bmatrix} \qquad (5.71)$$

Therefore, from the above equation (5.71), the α β components of currents can be expressed as follows.

$$\begin{bmatrix} i_\alpha \\ i_\beta \end{bmatrix} = \begin{bmatrix} v_\alpha & v_\beta \\ -v_\beta & v_\alpha \end{bmatrix}^{-1} \begin{bmatrix} p \\ q \end{bmatrix}$$

The matrix, $\begin{bmatrix} v_\alpha & v_\beta \\ -v_\beta & v_\alpha \end{bmatrix}^{-1}$ is given as following.

$$\begin{bmatrix} v_\alpha & v_\beta \\ -v_\beta & v_\alpha \end{bmatrix}^{-1} = \frac{1}{v_\alpha^2 + v_\beta^2} \begin{bmatrix} v_\alpha & -v_\beta \\ v_\beta & v_\alpha \end{bmatrix} \qquad (5.72)$$

From the above equations,

$$i_\alpha = \frac{v_\alpha}{v_\alpha^2 + v_\beta^2} p - \frac{v_\beta}{v_\alpha^2 + v_\beta^2} q \qquad (5.73)$$

$$i_\beta = \frac{v_\beta}{v_\alpha^2 + v_\beta^2} p + \frac{v_\alpha}{v_\alpha^2 + v_\beta^2} q \qquad (5.74)$$

Which can further be written as,

$$i_\alpha = i_{\alpha p} + i_{\alpha q} \qquad (5.75)$$

$$i_\beta = i_{\beta p} + i_{\beta q} \qquad (5.76)$$

In the above equation,

$$i_{\alpha p} = \frac{v_\alpha}{v_\alpha^2 + v_\beta^2} p \qquad (5.77)$$

$$i_{\alpha q} = -\frac{v_\beta}{v_\alpha^2 + v_\beta^2} q \qquad (5.78)$$

$$i_{\beta p} = \frac{v_\beta}{v_\alpha^2 + v_\beta^2} p \qquad (5.79)$$

$$i_{\beta q} = \frac{v_\alpha}{v_\alpha^2 + v_\beta^2} q \qquad (5.80)$$

The instantaneous active and reactive components of currents in supplying line can be calculated from the α and β components of the current as given in the following.

$$\begin{bmatrix} i_a \\ i_b \end{bmatrix} = \begin{bmatrix} \sqrt{\frac{3}{2}} & 0 \\ \frac{1}{\sqrt{2}} & \sqrt{2} \end{bmatrix}^{-1} \begin{bmatrix} i_\alpha \\ i_\beta \end{bmatrix} = \begin{bmatrix} \sqrt{\frac{2}{3}} & 0 \\ -\frac{1}{\sqrt{6}} & \frac{1}{\sqrt{2}} \end{bmatrix} \begin{bmatrix} i_{\alpha p} + i_{\alpha q} \\ i_{\beta p} + i_{\beta q} \end{bmatrix}$$

$$= \begin{bmatrix} \sqrt{\frac{2}{3}} & 0 \\ -\frac{1}{\sqrt{6}} & \frac{1}{\sqrt{2}} \end{bmatrix} \begin{bmatrix} i_{\alpha p} \\ i_{\beta p} \end{bmatrix} + \begin{bmatrix} \sqrt{\frac{2}{3}} & 0 \\ -\frac{1}{\sqrt{6}} & \frac{1}{\sqrt{2}} \end{bmatrix} \begin{bmatrix} i_{\alpha q} \\ i_{\beta q} \end{bmatrix}$$

$$\begin{bmatrix} i_{ap} \\ i_{bp} \end{bmatrix} = \begin{bmatrix} \sqrt{\frac{2}{3}} & 0 \\ -\frac{1}{\sqrt{6}} & \frac{1}{\sqrt{2}} \end{bmatrix} \begin{bmatrix} i_{\alpha p} \\ i_{\beta p} \end{bmatrix} \qquad (5.81)$$

and,

$$\begin{bmatrix} i_{aq} \\ i_{bq} \end{bmatrix} = \begin{bmatrix} \sqrt{\frac{2}{3}} & 0 \\ -\frac{1}{\sqrt{6}} & \frac{1}{\sqrt{2}} \end{bmatrix} \begin{bmatrix} i_{\alpha q} \\ i_{\beta q} \end{bmatrix} \qquad (5.82)$$

The active and reactive components of the line currents must be consistent to the basic definitions. However, these components of currents have little in common with the reactive power of the load as defined in [26]. This is shown in the following illustration.

Example 5.2. Consider a resistive load connected as shown in Fig. 5.7. It is supplied from a three-phase balanced sinusoidal voltage source with $v_a = \sqrt{2}V \sin \omega t$, with $V = 230$ V. Express the voltage and currents for primary and secondary side of the transformer. Express the active and reactive component of the currents, powers and discuss about them.

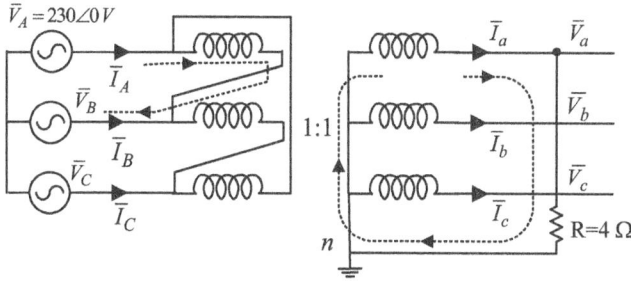

Figure 5.7 An unbalanced resistive load supplied by three-phase delta-star connected transformer

Solution: With the above given values, the primary side phase voltages with respect to virtual ground could be expressed as the following.

$$
\begin{aligned}
v_A &= \sqrt{2}V \sin \omega t = 230\sqrt{2}\sin \omega t \\
v_B &= \sqrt{2}V \sin(\omega t - 120°) = 230\sqrt{2}\sin(\omega t - 120°) \\
v_C &= \sqrt{2}V \sin(\omega t + 120°) = 230\sqrt{2}\sin(\omega t + 120°)
\end{aligned} \tag{5.83}
$$

In phasor form,

$$
\begin{aligned}
\overline{V}_A &= 230\angle 0° \text{ V} \\
\overline{V}_B &= 230\angle -120° \text{ V} \\
\overline{V}_C &= 230\angle 120° \text{ V}
\end{aligned}
$$

Therefore, the primary side line-to-line voltages are expressed as follows.

$$
\begin{aligned}
v_{AB} &= \sqrt{2}\sqrt{3}V \sin(\omega t + 30°) \\
v_{BC} &= \sqrt{2}\sqrt{3}V \sin(\omega t - 90°) \\
v_{CA} &= \sqrt{2}\sqrt{3}V \sin(\omega t + 150°)
\end{aligned} \tag{5.84}
$$

In phasor form,

$$
\begin{aligned}
\overline{V}_{AB} &= 398.37\angle 30° \text{ V} \\
\overline{V}_{BC} &= 398.37\angle -90° \text{ V} \\
\overline{V}_{CA} &= 398.37\angle 150° \text{ V}
\end{aligned}
$$

These voltages are transformed into the secondaries and are expressed below.

$$
\begin{aligned}
v_a &= \sqrt{2}\sqrt{3}V\sin(\omega t + 30°) = 398.37\sqrt{2}\sin(\omega t + 30°) \\
v_b &= \sqrt{2}\sqrt{3}V\sin(\omega t - 90°) = 398.37\sqrt{2}\sin(\omega t - 90°) \quad\quad (5.85) \\
v_c &= \sqrt{2}\sqrt{3}V\sin(\omega t + 150°) = 398.37\sqrt{2}\sin(\omega t + 150°)
\end{aligned}
$$

In phasor form,

$$
\begin{aligned}
\overline{V}_a &= 398.37\angle 30°\ \text{V} \\
\overline{V}_b &= 398.37\angle -90°\ \text{V} \\
\overline{V}_c &= 398.37\angle 150°\ \text{V}
\end{aligned}
$$

Therefore, the currents on the secondary side are given below.

$$
\begin{aligned}
i_a &= \frac{\sqrt{2}\sqrt{3}V}{R}\sin(\omega t + 30) \\
i_b &= 0 \quad\quad\quad\quad\quad\quad\quad\quad\quad\quad (5.86) \\
i_c &= 0
\end{aligned}
$$

Taking $V = 230$ V and $R = 4\ \Omega$, the currents on the secondary side of the transformer are given as follows.

$$
\begin{aligned}
i_a &= \frac{v_a}{R} = \frac{\sqrt{2}\sqrt{3}V}{4}\sin(\omega t + 30°) \\
&= 99.59\sqrt{2}\sin(\omega t + 30°) = \sqrt{2}I\sin(\omega t + 30°) \\
i_b &= 0 \quad\quad\quad\quad\quad\quad\quad\quad\quad\quad\quad\quad\quad (5.87) \\
i_c &= 0
\end{aligned}
$$

In phasor form, the above can be expressed as,

$$
\begin{aligned}
\overline{I}_a &= 99.59\angle 30°\ \text{A} \\
\overline{I}_b &= 0\,\text{A} \quad\quad\quad\quad\quad (5.88) \\
\overline{I}_c &= 0\,\text{A}
\end{aligned}
$$

This phase-*a* current, i_a in the secondary side of the transformer, is transformed to the primary of the delta-connected winding, therefore the currents on the primary side of the transformer are given as follows.

$$
\begin{aligned}
i_A &= \sqrt{2}I\sin(\omega t + 30°) \\
i_B &= -i_A = -\sqrt{2}I\sin(\omega t + 30°) \quad\quad (5.89) \\
i_C &= 0
\end{aligned}
$$

The above can be written in phasor form as given below.

$$
\begin{aligned}
\bar{I}_A &= 99.59\angle 30° = I\angle 30°\,\text{A} \\
\bar{I}_B &= -\bar{I}_A = 1\angle -180° \times 99.59\angle 30° \\
&= 99.59\angle -150° = I\angle -150°\,\text{A} \\
\bar{I}_C &= 0\,\text{A}
\end{aligned}
\tag{5.90}
$$

After, knowing the voltages and currents of the primary side of the transformer, their α and β components are expressed as.

$$
\begin{bmatrix} v_\alpha \\ v_\beta \end{bmatrix} = \begin{bmatrix} \sqrt{\tfrac{3}{2}} & 0 \\ \tfrac{1}{\sqrt{2}} & \sqrt{2} \end{bmatrix} \begin{bmatrix} v_A \\ v_B \end{bmatrix}
\tag{5.91}
$$

Substituting v_A and v_B from (5.83) in the above equation, we get the following.

$$
\begin{bmatrix} v_\alpha \\ v_\beta \end{bmatrix} = \begin{bmatrix} \sqrt{\tfrac{3}{2}} & 0 \\ \tfrac{1}{\sqrt{2}} & \sqrt{2} \end{bmatrix} \begin{bmatrix} \sqrt{2}V\sin\omega t \\ \sqrt{2}V\sin(\omega t - 120°) \end{bmatrix} = \begin{bmatrix} \sqrt{3}V\sin\omega t \\ -\sqrt{3}V\cos\omega t \end{bmatrix}
\tag{5.92}
$$

and,

$$
\begin{aligned}
\begin{bmatrix} i_\alpha \\ i_\beta \end{bmatrix} &= \begin{bmatrix} \sqrt{\tfrac{3}{2}} & 0 \\ \tfrac{1}{\sqrt{2}} & \sqrt{2} \end{bmatrix} \begin{bmatrix} i_A \\ i_B \end{bmatrix} = \begin{bmatrix} \sqrt{\tfrac{3}{2}} & 0 \\ \tfrac{1}{\sqrt{2}} & \sqrt{2} \end{bmatrix} \begin{bmatrix} \sqrt{2}I\sin(\omega t + 30°) \\ -\sqrt{2}I\sin(\omega t + 30°) \end{bmatrix} \\
&= \begin{bmatrix} \sqrt{3}I\sin(\omega t + 30°) \\ -I\sin(\omega t + 30°) \end{bmatrix}
\end{aligned}
\tag{5.93}
$$

Based on the above transformation matrix, the active and reactive powers are computed as,

$$
\begin{aligned}
p(t) &= v_\alpha i_\alpha + v_\beta i_\beta \\
&= \sqrt{3}V\sin\omega t\,\sqrt{3}I\sin(\omega t + 30°) - \sqrt{3}V\cos\omega t\,[-I\sin(\omega t + 30°)] \\
&= 2\sqrt{3}V I\sin(\omega t + 30°)\left(\frac{\sqrt{3}}{2}\sin\omega t + \frac{1}{2}\cos\omega t\right) \\
&- \sqrt{3}V I\left[2\sin^2(\omega t + 30°)\right] \\
p(t) &= \sqrt{3}V I[1 - \cos 2(\omega t + 30°)]
\end{aligned}
\tag{5.94}
$$

$$
\begin{aligned}
q(t) &= v_\alpha i_\beta - v_\beta i_\alpha \\
&= \sqrt{3}V\sin\omega t\,[-I\sin(\omega t + 30°)] - (-\sqrt{3}V\cos\omega t)\sqrt{3}I\sin(\omega t + 30°) \\
&= -2\sqrt{3}V I\sin(\omega t + 30°)\left(\frac{1}{2}\sin\omega t - \frac{\sqrt{3}}{2}\cos\omega t\right) \\
&= -2\sqrt{3}V I\sin(\omega t + 30°)(-\cos(\omega t + 30°)) \\
q(t) &= \sqrt{3}V I\sin 2(\omega t + 30°)
\end{aligned}
\tag{5.95}
$$

Based on the above values of p and q powers, the α and β components of active and reactive components are given below.

$$\begin{bmatrix} i_\alpha \\ i_\beta \end{bmatrix} = \begin{bmatrix} i_{\alpha p} + i_{\alpha q} \\ i_{\beta p} + i_{\beta q} \end{bmatrix}$$

where,

$$
\begin{aligned}
i_{\alpha p} &= \frac{v_\alpha}{v_\alpha^2 + v_\beta^2} p \\
&= \frac{\sqrt{3}V \sin \omega t}{(\sqrt{3}V \sin \omega t)^2 + (-\sqrt{3}V \cos \omega t)^2} p \\
&= \frac{1}{\sqrt{3}V} \sin \omega t \; \sqrt{3}VI [1 - \cos 2(\omega t + 30°)] \\
&= I \sin \omega t \, [1 - \cos 2(\omega t + 30°)] \quad\quad (5.96)
\end{aligned}
$$

Similarly,

$$
\begin{aligned}
i_{\beta p} &= \frac{v_\beta}{v_\alpha^2 + v_\beta^2} p \\
&= \frac{-\sqrt{3}V \cos \omega t}{(\sqrt{3}VI \sin \omega t)^2 + (-\sqrt{3}V \cos \omega t)^2} p \\
&= \frac{-\sqrt{3}V \cos \omega t}{3V^2} \sqrt{3}VI [1 - \cos 2(\omega t + 30°)] \\
&= -I \cos \omega t \, [1 - \cos 2(\omega t + 30°)] \quad\quad (5.97)
\end{aligned}
$$

$$
\begin{aligned}
i_{\alpha q} &= \frac{-v_\beta}{v_\alpha^2 + v_\beta^2} q \\
&= \frac{-(-\sqrt{3}V \cos \omega t)}{3V^2} \sqrt{3}VI \sin 2(\omega t + 30°) \\
&= I \cos \omega t \sin 2(\omega t + 30°) \quad\quad (5.98)
\end{aligned}
$$

$$
\begin{aligned}
i_{\beta q} &= \frac{v_\alpha}{v_\alpha^2 + v_\beta^2} q \\
&= \frac{\sqrt{3}V \sin \omega t}{3V^2} \sqrt{3}VI \sin 2(\omega t + 30°) \\
&= I \sin \omega t \sin 2(\omega t + 30°) \quad\quad (5.99)
\end{aligned}
$$

Thus, knowing $i_{\alpha p}, i_{\alpha q}, i_{\beta p}$ and $i_{\beta q}$, we can determine active and reactive components of currents on the source side as given below.

$$\begin{bmatrix} i_{Ap} \\ i_{Bp} \end{bmatrix} = [C]^{-1} \begin{bmatrix} i_{\alpha p} \\ i_{\beta p} \end{bmatrix}$$

where,

$$[C]^{-1} = \begin{bmatrix} \sqrt{\frac{3}{2}} & 0 \\ \frac{1}{\sqrt{2}} & \sqrt{2} \end{bmatrix}^{-1} = \begin{bmatrix} \sqrt{\frac{2}{3}} & 0 \\ \frac{-1}{\sqrt{6}} & \frac{1}{\sqrt{2}} \end{bmatrix}$$

Using the above equation, we can find out the active and reactive components of the current, as given below.

$$\begin{bmatrix} i_{Ap} \\ i_{Bp} \end{bmatrix} = \begin{bmatrix} \sqrt{\frac{2}{3}} & 0 \\ \frac{-1}{\sqrt{6}} & \frac{1}{\sqrt{2}} \end{bmatrix} \begin{bmatrix} I \sin \omega t \, \{1 - \cos 2(\omega t + 30°)\} \\ -I \cos \omega t \, \{1 - \cos 2(\omega t + 30°)\} \end{bmatrix}$$

From the above,

$$i_{Ap} = \sqrt{\frac{2}{3}} I \sin \omega t \, \{1 - \cos 2(\omega t + 30°)\}$$

$$= \sqrt{\frac{2}{3}} \frac{I}{2} \{2 \sin \omega t - 2 \sin \omega t \cos 2(\omega t + 30°)\}$$

$$= \frac{I}{\sqrt{6}} \{2 \sin \omega t - \sin(3\omega t + 60°) - \sin(-\omega t - 60°)\}$$

$$= \frac{I}{\sqrt{6}} \{2 \sin \omega t + \sin(\omega t + 60°) - \sin(3\omega t + 60°)\}$$

$$i_{Bp} = -\frac{1}{\sqrt{6}} i_{\alpha p} + \frac{1}{\sqrt{2}} i_{\beta p}$$

$$= -\frac{1}{\sqrt{6}} I \sin \omega t \, \{1 - \cos 2(\omega t + 30°)\}$$

$$\quad + \frac{1}{\sqrt{2}} (-I \cos \omega t) \{1 - \cos 2(\omega t + 30°)\}$$

$$= -\frac{2I}{\sqrt{6}} \{1 - \cos 2(\omega t + 30°)\} \left(\frac{1}{2} \sin \omega t + \frac{\sqrt{3}}{2} \cos \omega t \right)$$

$$= -\frac{2I}{\sqrt{6}} \{1 - \cos 2(\omega t + 30°)\} \sin(\omega t + 60°)$$

$$= -\frac{2I}{\sqrt{6}} \{\sin(\omega t + 60°) - \sin(\omega t + 60°) \cos 2(\omega t + 30°)\}$$

$$= -\frac{2I}{\sqrt{6}} \left\{ \sin(\omega t + 60°) + \frac{1}{2} \sin(\omega t) - \frac{1}{2} \sin(3\omega t + 120°) \right\}$$

$$= -\frac{I}{\sqrt{6}} \{\sin \omega t + 2 \sin(\omega t + 60°) - \sin(3\omega t + 120°)\}$$

$$i_{Aq} = \sqrt{\frac{2}{3}} i_{\alpha q} = \sqrt{\frac{2}{3}} \frac{I}{2} \{2 \sin 2(\omega t + 30°) \cos \omega t\}$$

$$= \frac{I}{\sqrt{6}} \{\sin(\omega t + 60°) + \sin(3\omega t + 60°)\}$$

$$i_{Bq} = -\frac{1}{\sqrt{6}} i_{\alpha q} + \frac{1}{\sqrt{2}} i_{\beta q}$$

$$= -\frac{1}{\sqrt{6}} I \cos \omega t \sin 2(\omega t + 30°) + \frac{1}{\sqrt{2}} I \sin \omega t \sin 2(\omega t + 30°)$$

$$= -\frac{I}{\sqrt{6}} 2 \sin 2(\omega t + 30°) \left(-\frac{1}{2} \cos \omega t + \frac{\sqrt{3}}{2} \sin \omega t \right)$$

$$= \frac{I}{\sqrt{6}} 2 \sin 2(\omega t + 30°) \sin(\omega t - 30°)$$

$$= \frac{I}{\sqrt{6}} (\cos(\omega t + 90°) - \cos(3\omega t + 30°))$$

$$= \frac{I}{\sqrt{6}} \{-\sin \omega t - \cos(3\omega t + 30°)\}$$

Thus, we have,

$$i_{Ap} = \frac{I}{\sqrt{6}} \{2 \sin \omega t + \sin(\omega t + 60°) - \sin(3\omega t + 60°)\} \tag{5.100}$$

$$i_{Bp} = -\frac{I}{\sqrt{6}} \{\sin(\omega t) + 2 \sin(\omega t + 60°) - \sin(3\omega t + 120°)\} \tag{5.101}$$

$$i_{Aq} = \frac{I}{\sqrt{6}} \{\sin(\omega t + 60°) + \sin(3\omega t + 60°)\} \tag{5.102}$$

$$i_{Bq} = \frac{I}{\sqrt{6}} \{-\sin \omega t - \cos(3\omega t + 30°)\}. \tag{5.103}$$

From the above equations, the terms instantaneous active current and instantaneous reactive current given in the pq theory do not have commonality with the notion of active and reactive currents used in electrical engineering. Also, the reactive current i_q occurs in the supply lines of the load in spite of the absence of any reactive element in the load. Furthermore, the nature of the load is linear, and harmonics are absent. From (5.100) to (5.103), it is observed that the net currents in phase-A and B are obtained by adding the active and reactive terms, i.e., $i_A = i_{Ap} + i_{Aq}$ and $i_B = i_{Bp} + i_{Bq}$, as given in (5.89). These currents do not have any harmonics, however the resolutions of active and reactive components of the currents ($i_{Ap}, i_{Aq}, i_{Bp}, i_{Bq}$), based on pq theory have harmonics. For example, in the above discussion,

$$i_{Ap} = \frac{I}{\sqrt{6}} \{2 \sin \omega t + \sin(\omega t + 60°) - \sin(3\omega t + 60°)\} \tag{5.104}$$

is the active current component in the phase-A, and it contains the third order harmonic. This contradicts the basic notion of the active current that was introduced to electrical engineering by Fryze [26]. Thus, it seems that there is a misconception of electrical phenomenon in three-phase circuits with balanced sinusoidal voltages for linear loads that do not have harmonics. Moreover, the active and reactive currents, $i_{Ap}, i_{Aq}, i_{Bp}, i_{Bq}$ which result from the pq theory, are not the current that should remain

in the supply lines after the load is compensated to unity power factor as defined by
Fryze [26].

Also, it is evident that the instantaneous reactive power $q(t)$ as defined by the pq
theory, does not really identify the power properties of load instantaneously. For example, for the above discussion, the active and reactive powers are given as follows.

$$p(t) = \sqrt{3}VI\{1 - \cos 2(\omega t + 30°)\}$$
$$q(t) = \sqrt{3}VI\sin 2(\omega t + 30°)$$

Hence, the following points are noted.

1. The active components of currents (i_{Ap}, i_{Bp}, i_{Cp}) and reactive components of
   currents, (i_{Aq}, i_{Bq}, i_{Cq}) contain third harmonic, which is not possible for a linear
   load as discussed above.

2. The instantaneous reactive power $q(t)$ defined by pq theory does not identify
   with the power properties of the load instantaneously. Both powers $p(t)$ and
   $q(t)$ are time varying quantities, so that a pair of their values at any single
   point of time does not identify with the power properties of the load. The
   possibility of instantaneous identification of active and reactive power $p(t)$ and
   $q(t)$ does not mean that power properties of load are identified instantaneously.
   For example,

$$\text{at } (\omega t + 30°) = 90°, \left\{ \begin{array}{l} p(t) = 2\sqrt{3}VI \\ q(t) = 0 \end{array} \right\}$$

The above implies that as if it is a resistive load.

$$\text{Similarly at } (\omega t + 30°) = 0°, \left\{ \begin{array}{l} p(t) = 0 \\ q(t) = 0 \end{array} \right\}$$

which implies as there is no load.

$$\text{When } (\omega t + 30°) = 105°, \left\{ \begin{array}{l} p(t) = \sqrt{3}VI(1 + \sqrt{\frac{3}{2}}) \\ q(t) = -\sqrt{3}VI(\frac{1}{2}) \end{array} \right\}$$

implies that it is capacitive load.

$$\text{Similarly when } (\omega t + 30°) = 75°, \left\{ \begin{array}{l} p(t) = \sqrt{3}VI(1 + \sqrt{\frac{3}{2}}) \\ q(t) = \sqrt{3}VI(\frac{1}{2}) \end{array} \right\}$$

implies as if the load is inductive.

We, therefore, conclude that power properties cannot be identified without monitoring $p(t)$ and $q(t)$ powers over the entire cycle period. For example, in the above case,
the instantaneous reactive power $q(t)$ has occurred, not because of the load reactive
elements, but because of load currents unbalance. This unbalance nature of load can

not be identified by instantaneous reactive power $q(t)$. Therefore, pq theory does not offer an advantage with respect to the time interval needed to identify the nature of load and its property over the power theories based on the time domain or frequency domain approaches that require the system to be monitored over one time period.

Power computations

The secondary side powers are given as following.

$$
\begin{aligned}
\overline{S}_a &= P_a + jQ_a = \overline{V}_a \overline{I}_a^* = 398.37\angle 30°\, 99.59\angle -30° = 39675\,\text{VA} \\
P_a &= 39675\,\text{W},\, Q_a = 0\,\text{VAr} \\
\overline{S}_b &= P_b + jQ_b = \overline{V}_b \overline{I}_b^* = 398.37\angle -90° \times 0 = 0\,\text{VA} \\
P_b &= 0\,\text{W},\, Q_b = 0\,\text{VAr} \\
\overline{S}_c &= P_c + jQ_c = \overline{V}_c \overline{I}_c^* = 398.37\angle 150° \times 0 = 0\,\text{VA} \\
P_c &= 0\,\text{W},\, Q_c = 0\,\text{VAr}
\end{aligned}
$$

The total active and reactive powers on the secondary side are given as following.

$$
\begin{aligned}
P &= P_a + P_b + P_c = 39675\,\text{W} \\
Q &= Q_a + Q_b + Q_c = 0\,\text{VAr} \\
S_{vect} &= S_{arith} = P = 39675\,\text{VA} \\
pf_{vect} &= pf_{arith} = P/S = 1.0
\end{aligned}
$$

The primary side powers are given as following.

$$
\begin{aligned}
\overline{S}_A &= P_A + jQ_A = \overline{V}_A \overline{I}_A^* = 230\angle 0° \times 99.59\angle -30° \\
&= 22905.7\angle -30° = 19837.50 - j11453.16\,\text{VA} \\
P_A &= 19837.50\,\text{W},\, Q_A = -11453.160\,\text{VAr} \\
\overline{S}_B &= P_B + jQ_B = \overline{V}_B \overline{I}_B^* = 230\angle -120° \times (-99.59\angle 30°)^* \\
&= 22905.7\angle 30° = 19837.50 + j11453.16\,\text{VA} \\
P_B &= 19837.50\,\text{W},\, Q_B = 11453.160\,\text{VAr} \\
\overline{S}_C &= P_C + jQ_C = \overline{V}_C \overline{I}_C^* = 230\angle 120° \times 0 \\
P_C &= 0\,\text{W},\, Q_C = 0\,\text{VAr}
\end{aligned}
$$

The total active and reactive powers on the primary side are given as follows.

$$
\begin{aligned}
P &= P_A + P_B + P_C = 39675\,\text{W} \\
Q &= Q_A + Q_B + Q_C = 0\,\text{VAr} \\
S_{vect} &= |\overline{S}_A + \overline{S}_B + \overline{S}_C| = P = 39675\,\text{VA} \\
S_{Arith} &= |\overline{S}_A| + |\overline{S}_B| + |\overline{S}_C| = 22906 + 22906 + 0 = 45813\,\text{VA} \\
S_{eff} &= 3V_e Ie = 3*230*81.32 = 56109\,\text{VA} \\
pf_{vect} &= P/S_{vect} = 1.0 \\
pf_{Arith} &= P/S_{Arith} = 39675/45813 = 0.866 \\
pf_e &= P/S_e = 39675/45813 = 0.707
\end{aligned}
$$

Example 5.3. Consider an inductive load connected as shown in Fig. 5.7. It is supplied from a symmetrical source of a sinusoidal balanced voltage with $v_a = \sqrt{2}V \sin \omega t$, with V = 230 Volts, 50 Hz. Express the voltage and currents for the primary and secondary side of the transformer. Express the active and reactive components of the currents, powers and discuss them.

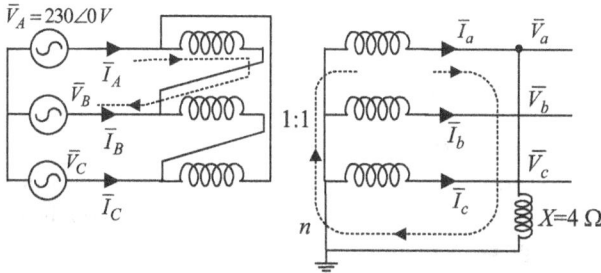

Figure 5.8 An unbalanced reactive load supplied by three-phase delta-star connected transformer

Solution: With the above given values, the primary side phase voltages with respect to the virtual ground could be expressed as the following.

$$
\begin{aligned}
v_A &= \sqrt{2}V \sin \omega t = 230\sqrt{2}\sin \omega t \\
v_B &= \sqrt{2}V \sin(\omega t - 120°) = 230\sqrt{2}\sin(\omega t - 120°) \quad\quad (5.105) \\
v_C &= \sqrt{2}V \sin(\omega t + 120°) = 230\sqrt{2}\sin(\omega t + 120°)
\end{aligned}
$$

Therefore, the primary side line-to-line voltages are expressed as follows.

$$
\begin{aligned}
v_{AB} &= \sqrt{2}\sqrt{3}V \sin(\omega t + 30°) \\
v_{BC} &= \sqrt{2}\sqrt{3}V \sin(\omega t - 90°) \\
v_{CA} &= \sqrt{2}\sqrt{3}V \sin(\omega t + 150°)
\end{aligned}
$$

These voltages are transformed to the secondaries and are expressed below.

$$
\begin{aligned}
v_a &= \sqrt{2}\sqrt{3}V \sin \omega t = 398.37\sqrt{2}\sin(\omega t + 30°) \\
v_b &= \sqrt{2}\sqrt{3}V \sin(\omega t - 120°) = 398.37\sqrt{2}\sin(\omega t - 90°) \\
v_c &= \sqrt{2}\sqrt{3}V \sin(\omega t + 120°) = 398.37\sqrt{2}\sin(\omega t + 150°)
\end{aligned}
$$

In phasor form, the above voltages are expressed below.

$$
\begin{aligned}
\overline{V}_a &= \sqrt{3}V \angle 30° \text{ V} \\
\overline{V}_b &= \sqrt{3}V \angle -90° \text{ V} \\
\overline{V}_c &= \sqrt{3}V \angle 150° \text{ V}
\end{aligned}
$$

Currents on the secondary side are calculated as.

$$i_a = \frac{\sqrt{2}\sqrt{3}V}{X}\sin(\omega t - 60°) = 99.59\sqrt{2}\sin(\omega t - 60°)$$

$$i_b = 0$$

$$i_c = 0$$

In phasor form, the above can be expressed as,

$$\bar{I}_a = 99.59\angle - 60° \text{ A}.$$

The above phase-*a* current (i_a) is transformed to the primary of the delta-connected winding. Since the currents should have a 90° phase shift with respect to the voltages across the windings as given by (5.84), therefore the currents on the primary side of the transformer are given as follows.

$$i_A = i_{AB} = i_a = \sqrt{2}I\sin(\omega t - 60°)$$

$$i_B = -i_A = -\sqrt{2}I\sin(\omega t - 60°)$$

$$i_C = 0$$

In phasor form, the above can be expressed as,

$$\bar{I}_A = I\angle - 60° = 99.59\angle - 60° \text{ A}$$

$$\bar{I}_B = -I\angle - 60° = -99.59\angle - 60° \text{ A}$$

$$\bar{I}_C = 0\text{A}.$$

After knowing the voltages and currents of the primary side of the transformer, their α and β components are expressed as.

$$\begin{bmatrix} v_\alpha \\ v_\beta \end{bmatrix} = \begin{bmatrix} \sqrt{\frac{3}{2}} & 0 \\ \frac{1}{\sqrt{2}} & \sqrt{2} \end{bmatrix} \begin{bmatrix} v_A \\ v_B \end{bmatrix}$$

Substituting v_A and v_B from (5.105) in the above equation, we get the following.

$$\begin{bmatrix} v_\alpha \\ v_\beta \end{bmatrix} = \begin{bmatrix} \sqrt{\frac{3}{2}} & 0 \\ \frac{1}{\sqrt{2}} & \sqrt{2} \end{bmatrix} \begin{bmatrix} \sqrt{2}V\sin\omega t \\ \sqrt{2}V\sin(\omega t - 120°) \end{bmatrix} = \begin{bmatrix} \sqrt{3}V\sin\omega t \\ -\sqrt{3}V\cos(\omega t) \end{bmatrix} \qquad (5.106)$$

and,

$$\begin{bmatrix} i_\alpha \\ i_\beta \end{bmatrix} = \begin{bmatrix} \sqrt{\frac{3}{2}} & 0 \\ \frac{1}{\sqrt{2}} & \sqrt{2} \end{bmatrix} \begin{bmatrix} i_A \\ i_B \end{bmatrix} = \begin{bmatrix} \sqrt{\frac{3}{2}} & 0 \\ \frac{1}{\sqrt{2}} & \sqrt{2} \end{bmatrix} \begin{bmatrix} \sqrt{2}I\sin(\omega t - 60°) \\ -\sqrt{2}I\sin(\omega t - 60°) \end{bmatrix}$$

$$= \begin{bmatrix} \sqrt{3}I\sin(\omega t - 60°) \\ -I\sin(\omega t - 60°) \end{bmatrix} \qquad (5.107)$$

Based on the above, the active and reactive powers are computed as.

$$
\begin{aligned}
p(t) = p_{\alpha\beta} &= v_\alpha i_\alpha + v_\beta i_\beta \\
&= \sqrt{3}V \sin(\omega t)\sqrt{3}I \sin(\omega t - 60°) \\
&\quad + (-\sqrt{3}V \cos \omega t)(-I \sin(\omega t - 60°)) \\
&= 2\sqrt{3}V I \sin(\omega t - 60°)\left[\frac{\sqrt{3}}{2}\sin \omega t + \frac{1}{2}\cos \omega t\right] \\
&= \sqrt{3}V I [2\sin(\omega t - 60°)\cos(\omega t - 60°)] \\
&= \sqrt{3}V I \sin 2(\omega t - 60°)
\end{aligned}
\tag{5.108}
$$

$$
\begin{aligned}
q(t) &= v_\alpha i_\beta - v_\beta i_\alpha \\
&= \sqrt{3}V \sin \omega t \{-I \sin(\omega t - 60°)\} - (-\sqrt{3}V \cos \omega t)\sqrt{3}I \sin(\omega t - 60°) \\
&= -\sqrt{3}VI2\sin(\omega t - 60°)\left[\frac{1}{2}\sin \omega t - \frac{\sqrt{3}}{2}\cos \omega t\right] \\
&= -\sqrt{3}VI2\sin(\omega t - 60°)(\sin(\omega t - 60°)) \\
&= -\sqrt{3}VI2\sin^2(\omega t - 60°) \\
&= -\sqrt{3}V I \{1 - \cos 2(\omega t - 60°)\}
\end{aligned}
\tag{5.109}
$$

Based on above values of p and q powers, the α and β components of active and reactive components are given below.

$$
\begin{bmatrix} i_\alpha \\ i_\beta \end{bmatrix} = \begin{bmatrix} i_{\alpha p} + i_{\alpha q} \\ i_{\beta p} + i_{\beta q} \end{bmatrix}
$$

where,

$$
\begin{aligned}
i_{\alpha p} &= \frac{v_\alpha}{v_\alpha^2 + v_\beta^2} p \\
&= \frac{\sqrt{3}V \sin \omega t}{(\sqrt{3}V \sin \omega t)^2 + (-\sqrt{3}V \cos \omega t)^2} p \\
&= \frac{1}{\sqrt{3}V}\sin \omega t \sqrt{3}VI\{\sin 2(\omega t - 60°)\} \\
&= I\sin \omega t \sin 2(\omega t - 60°)
\end{aligned}
\tag{5.110}
$$

Similarly,

$$
\begin{aligned}
i_{\beta p} &= \frac{v_\beta}{v_\alpha^2 + v_\beta^2}\, p \\
&= \frac{-\sqrt{3}V \cos\omega t}{(\sqrt{3}VI\sin\omega t)^2 + (-\sqrt{3}V\cos\omega t)^2}\, p \\
&= \frac{-\sqrt{3}V\cos\omega t}{3V^2}\,\sqrt{3}VI\,\sin 2(\omega t - 60^\circ) \\
&= -I\cos\omega t\,\sin 2(\omega t - 60^\circ)
\end{aligned}
\tag{5.111}
$$

$$
\begin{aligned}
i_{\alpha q} &= \frac{-v_\beta}{v_\alpha^2 + v_\beta^2}\, q \\
&= \frac{-(-\sqrt{3}V\cos\omega t)}{3V^2}\left\{-\sqrt{3}V I\,[1 - \cos 2(\omega t - 60^\circ)]\right\} \\
&= -I\cos\omega t\,\left\{1 - \cos 2(\omega t - 60^\circ)\right\}
\end{aligned}
\tag{5.112}
$$

$$
\begin{aligned}
i_{\beta q} &= \frac{v_\alpha}{v_\alpha^2 + v_\beta^2}\, q \\
&= \frac{\sqrt{3}V\sin\omega t}{3V^2}\left\{-\sqrt{3}V I\,[1 - \cos 2(\omega t - 60^\circ)]\right\} \\
&= -I\sin\omega t\,\left\{1 - \cos 2(\omega t - 60^\circ)\right\}
\end{aligned}
\tag{5.113}
$$

Thus, knowing $i_{\alpha p}, i_{\alpha q}, i_{\beta p}$ and $i_{\beta q}$, we can determine active and reactive components of currents on the source side as given below.

$$
\begin{bmatrix} i_{Ap} \\ i_{Bp} \end{bmatrix} = [C]^{-1} \begin{bmatrix} i_{\alpha p} \\ i_{\beta p} \end{bmatrix}
$$

where,

$$
[C]^{-1} = \begin{bmatrix} \sqrt{\frac{3}{2}} & 0 \\ \frac{1}{\sqrt{2}} & \sqrt{2} \end{bmatrix}^{-1} = \begin{bmatrix} \sqrt{\frac{2}{3}} & 0 \\ \frac{-1}{\sqrt{6}} & \frac{1}{\sqrt{2}} \end{bmatrix}
\tag{5.114}
$$

Using the above equation, we can find out the active and reactive components of the current, as given below.

$$
\begin{bmatrix} i_{Ap} \\ i_{Bp} \end{bmatrix} = \begin{bmatrix} \sqrt{\frac{2}{3}} & 0 \\ \frac{-1}{\sqrt{6}} & \frac{1}{\sqrt{2}} \end{bmatrix} \begin{bmatrix} I\sin\omega t\,\sin 2(\omega t - 60^\circ) \\ -I\cos\omega t\,\sin 2(\omega t - 60^\circ) \end{bmatrix}
$$

From the above,

$$i_{Ap} = \sqrt{\frac{2}{3}} I \sin \omega t \sin 2(\omega t - 60°)$$

$$= \sqrt{\frac{2}{3}} \frac{I}{2} \{2 \sin \omega t \sin 2(\omega t - 60°)\}$$

$$= \frac{I}{\sqrt{6}} \{\cos(\omega t - 120°) - \cos(3\omega t - 120°)\}$$

$$i_{Bp} = -\frac{1}{\sqrt{6}} i_{\alpha p} + \frac{1}{\sqrt{2}} i_{\beta p}$$

$$= -\frac{1}{\sqrt{6}} I \sin \omega t \{\sin 2(\omega t - 60°)\} + \frac{1}{\sqrt{2}} (-I \cos \omega t) \{\sin 2(\omega t - 60°)\}$$

$$= -\frac{I}{\sqrt{6}} \{2 \sin 2(\omega t - 60°)\} \left(\frac{1}{2} \sin \omega t + \frac{\sqrt{3}}{2} \cos \omega t\right)$$

$$= \frac{I}{\sqrt{6}} \{\cos \omega t + \cos(3\omega t - 60°)\}$$

$$i_{Aq} = \sqrt{\frac{2}{3}} i_{\alpha q}$$

$$= -\sqrt{\frac{2}{3}} I \{1 - \cos 2(\omega t - 60°)\} \cos \omega t$$

$$= -\frac{I}{\sqrt{6}} \{2 \cos \omega t - 2 \cos \omega t \cos 2(\omega t - 60°)\}$$

$$= -\frac{I}{\sqrt{6}} \{2 \cos \omega t - \cos(3\omega t - 120°) - \cos(\omega t - 120°)\}$$

$$= \frac{I}{\sqrt{6}} \{-2 \cos \omega t + \cos(\omega t - 120°) + \cos(3\omega t - 120°)\}$$

$$i_{Bq} = -\frac{1}{\sqrt{6}} i_{\alpha q} + \frac{1}{\sqrt{2}} i_{\beta q}$$

$$= -\frac{1}{\sqrt{6}} \{-I \cos \omega t [1 - \cos 2(\omega t - 60°)]\}$$

$$+ \frac{1}{\sqrt{2}} \{-I \sin \omega t [1 - \cos 2(\omega t - 60°)]\}$$

$$= \frac{2I}{\sqrt{6}} \{1 - \cos 2(\omega t - 60°)\} \left(\frac{1}{2} \cos \omega t - \frac{\sqrt{3}}{2} \sin \omega t\right)$$

$$= \frac{2I}{\sqrt{6}} \{1 - \cos 2(\omega t - 60°)\} \cos(\omega t + 60°)$$

$$= \frac{I}{\sqrt{6}} \{2 \cos(\omega t + 60°) - 2 \cos(\omega t + 60°) \cos 2(\omega t - 60°)\}$$

$$= \frac{I}{\sqrt{6}} \{2 \cos(\omega t + 60°) + \cos \omega t - \cos(3\omega t - 60°)\}$$

Thus, we have,

$$i_{Ap} = \frac{I}{\sqrt{6}}\{\cos(\omega t - 120°) - \cos(3\omega t - 120°)\} \tag{5.115}$$

$$i_{Bp} = \frac{I}{\sqrt{6}}\{\cos\omega t + \cos(3\omega t - 60°)\} \tag{5.116}$$

$$i_{Aq} = \frac{I}{\sqrt{6}}\{-2\cos\omega t + \cos(\omega t - 120°) + \cos(3\omega t - 120°)\} \tag{5.117}$$

$$i_{Bq} = \frac{I}{\sqrt{6}}\{\cos\omega t + 2\cos(\omega t + 60°) - \cos(3\omega t - 60°)\} \tag{5.118}$$

From the above equations, it is clear that there exist active components of currents (i_{Ap}, i_{Bp}), even though there is a purely reactive load. Additionally, both active and reactive components of currents $(i_{Ap}, i_{Aq}, i_{Bp}, i_{Bq})$, have third harmonics, although there is purely linear inductive load. Thus, this does not go in line with the definitions of active and reactive components of currents proposed by Fryze given in [26].

Powers computation
The secondary side powers are given as follows.

$$
\begin{aligned}
\overline{S}_a &= P_a + jQ_a = \overline{V}_a \overline{I}_a^* = 398.37\angle 30° \, 99.59\angle 60° = 39673.8361\angle 90° \text{ VA} \\
P_a &= 0\,\text{W}, Q_a = 39673.8361 \text{ VAr} \\
\overline{S}_b &= P_b + jQ_b = \overline{V}_b \overline{I}_b^* = 398.37\angle -90° \times 0 = 0\,\text{VA} \\
P_b &= 0\,\text{W}, Q_b = 0\,\text{VAr} \\
\overline{S}_c &= P_c + jQ_c = \overline{V}_c \overline{I}_c^* = 398.37\angle 150° \times 0 = 0\,\text{VA} \\
P_c &= 0\,\text{W}, Q_c = 0\,\text{VAr}
\end{aligned}
$$

The total active and reactive powers on the secondary side are given as follows.

$$
\begin{aligned}
P &= P_a + P_b + P_c = 0\,\text{W} \\
Q &= Q_a + Q_b + Q_c = 39673.83 \text{ VAr} \\
S_{vect} &= S_{arith} = Q = 39673.83 \text{ VA} \\
pf_{vect} &= pf_{arith} = P/S = 0
\end{aligned}
$$

The primary side powers are given as follows.

$$
\begin{aligned}
\overline{S}_A &= P_A + jQ_A = \overline{V}_a \overline{I}_a^* = 230\angle 0° \, 99.59\angle 60° = 11452.85 + j19836.91 \text{ VA} \\
P_A &= 11452.85\,\text{W}, Q_A = 19836.91 \text{ VAr} \\
\overline{S}_B &= P_B + jQ_B = \overline{V}_b \overline{I}_b^* = 230\angle -120° \, (-99.59\angle -120°)^* = \\
P_B &= -11452.85\,\text{W}, Q_B = 19836.91 \text{ VAr} \\
\overline{S}_C &= P_C + jQ_C = \overline{V}_c \overline{I}_C^* = 230\angle 120° \times 0 \\
P_C &= 0\,\text{W}, Q_C = 0\,\text{VAr}
\end{aligned}
$$

The total active and reactive powers on the primary side are given as follows.

$$
\begin{aligned}
P &= P_A + P_B + P_C = 0\,\text{W} \\
Q &= Q_A + Q_B + Q_C = 39673.82\,\text{VAr} \\
S_{vect} &= |\bar{S}_A + \bar{S}_B + \bar{S}_C| = 39673.82\,\text{VA} \\
S_{Arith} &= |\bar{S}_A| + |\bar{S}_B| + |\bar{S}_C| = 22906 + 22906 + 0 = 45811.4\,\text{VA} \\
S_e &= 3V_e I_e = 3*230*81.32 = 56109\,\text{VA} \\
pf_{vect} &= P/S_{vect} = 0 \\
pf_{Arith} &= P/S_{Arith} = 0 \\
pf_e &= P/S_e = 39675/45813 = 0
\end{aligned}
$$

Example 5.4. Consider the star-delta connected ideal transformer with 1:1 turns ratio, as shown in Fig. 5.9. The secondary side of transformer, a load of 3 ohms is connected between the phase-a and b. The primary is star connected and neutral is grounded. Compute the following.

(a) Draw the circuit diagram and compute time domain expressions of currents in each phase on both primary and secondary side.

(b) Does the load require reactive power from the source? If any, find its value. Also, compute the reactive power in each phase on either sides of the transformer.

(c) Also, determine active powers in each phase and the overall active power on either sides of the transformer.

(d) If you have a similar arrangement with balanced load with same output power, comment upon the rating of the line conductors and the transformer.

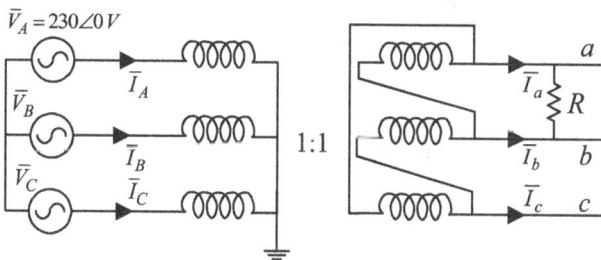

Figure 5.9 An unbalanced load supplied by three-phase star-delta connected transformer

Solution: In this example, we have a three-phase star-delta connected transformer of turns ratio 1:1 with the star side connected to a three-phase balanced voltage source and the neutral connected to the ground. Thus, in this three-phase star-delta connected transformer, the delta side phase voltages equal the star side line voltages.

The primary side instantaneous phase voltages are given by,

$$
\begin{aligned}
v_A &= 230\sqrt{2}\,\sin\omega t = 325.27\,\sin\omega t \\
v_B &= 230\sqrt{2}\,\sin(\omega t - 120°) = 325.27\,\sin(\omega t - 120°) \\
v_C &= 230\sqrt{2}\,\sin(\omega t + 120°) = 325.27\,\sin(\omega t + 120°)
\end{aligned}
$$

Therefore, the instantaneous line to line voltages on the delta side (secondary) are given by,

$$
\begin{aligned}
v_{ab} &= 230\sqrt{2}\,\sin\omega t = 325.27\,\sin\omega t \\
v_{bc} &= 230\sqrt{2}\,\sin(\omega t - 120°) = 325.27\,\sin(\omega t - 120°) \\
v_{ca} &= 230\sqrt{2}\,\sin(\omega t + 120°) = 325.27\,\sin(\omega t + 120°)
\end{aligned}
$$

Therefore, the instantaneous phase voltages with respect to the ground are given as follows.

$$
\begin{aligned}
v_a &= \frac{230\sqrt{2}}{\sqrt{3}}\,\sin(\omega t - 30°) = 132.79\sqrt{2}\,\sin(\omega t - 30°) \\[2mm]
v_b &= \frac{230\sqrt{2}}{\sqrt{3}}\,\sin(\omega t - 150°) = 132.79\sqrt{2}\,\sin(\omega t - 150°) \\[2mm]
v_c &= \frac{230\sqrt{2}}{\sqrt{3}}\,\sin(\omega t + 90°) = 132.79\sqrt{2}\,\sin(\omega t + 90°)
\end{aligned}
$$

(a) On the delta side, we have a resistive load of $R = 3\,\Omega$ connected between the terminals a and b. Thus, the expression for instantaneous currents flowing out of terminals a, b, and c of the transformer are given by,

$$
\begin{aligned}
i_a &= \frac{v_{ab}}{R} = \frac{230\sqrt{2}}{3}\,\sin\omega t = 76.67\sqrt{2}\,\sin\omega t \\[2mm]
i_b &= -i_a = -\frac{v_{ab}}{R} = -\frac{230\sqrt{2}}{3}\,\sin\omega t = -76.67\sqrt{2}\,\sin\omega t \\[2mm]
i_c &= 0
\end{aligned}
$$

Therefore, the winding currents on the secondary side are given as.

$$
\begin{aligned}
i_{ab} &= 76.67\sqrt{2}\,\sin\omega t \\
i_{bc} &= 0 \\
i_{ca} &= 0
\end{aligned}
$$

These currents are transformed to the primary windings. Thus, the time domain expressions of these currents are given as,

$$
\begin{aligned}
i_A &= 76.67\sqrt{2}\,\sin\omega t \\
i_B &= 0 \\
i_C &= 0
\end{aligned}
$$

(b) Load does not require any reactive power from the source because it is a purely resistive load. This fact can also be verified by looking at the expressions for instantaneous phase voltages and currents on the star side.

(c) Similarly, no current (and hence power) is drawn by the load from phases-B and C. Thus, various powers on the primary side are as follows.

For phase-A,

$$\begin{aligned}
\overline{S}_A &= P_A + jQ_A = \overline{V}_A \overline{I}_A^* = 230\angle 0° \times 76.67\angle 0° = 17634.1\,\text{VA}\\
P_A &= 17634.1\,\text{W}, Q_A = 0\,\text{VAr}
\end{aligned}$$

For phase-B,

$$\begin{aligned}
\overline{S}_B &= P_B + jQ_B = \overline{V}_B \overline{I}_B^* = 230\angle -120° \times 0 = 0\,\text{VA}\\
P_B &= 0\,\text{W}, Q_B = 0\,\text{VAr}
\end{aligned}$$

For phase-C,

$$\begin{aligned}
\overline{S}_C &= P_C + jQ_C = \overline{V}_C \overline{I}_C^* = 230\angle 120° \times 0 = 0\,\text{VA}\\
P_C &= 0\,\text{W}, Q_C = 0\,\text{VAr}
\end{aligned}$$

Thus, various powers on the secondary side are as follows.

For phase-A,

$$\begin{aligned}
\overline{S}_a &= P_a + jQ_a = \overline{V}_a \overline{I}_a^* = 132.79\angle -30° \times 76.67\angle 0°\\
&= 8817.05 - j5090.53\,\text{VA}\\
P_a &= 8817.05\,\text{W}, Q_a = -5090.53\,\text{VAr}
\end{aligned}$$

For phase-B,

$$\begin{aligned}
\overline{S}_b &= P_b + jQ_b = \overline{V}_b \overline{I}_b^* = 132.79\angle -150° \times (-76.67\angle 0°)\\
&= 8817.05 + j5090.53\,\text{VA}\\
P_b &= 8817.05\,\text{W}, Q_b = 5090.53\,\text{VAr}
\end{aligned}$$

For phase-C,

$$\begin{aligned}
\overline{S}_c &= P_c + jQ_c = \overline{V}_c \overline{I}_c^* = 132.79\angle 90° \times 0 = 0\,\text{VA}\\
P_c &= 0\,\text{W}, Q_c = 0\,\text{VAr}
\end{aligned}$$

For the above analysis, it is observed that the total active power, $P = P_A + P_B + P_C = P_a + P_b + P_c = 17634.1$ W and the total reactive power $Q = Q_A + Q_B + Q_C = Q_a + Q_b + Q_c = 0$ VAr. However, due to the unbalanced load, on the delta side of the transformer phase-a and phase-b experience reactive powers as calculated above. This causes a non-unity power factor in phase-a and b. The vector and arithmetic apparent powers are: $S_{vect} = |\overline{S}_a + \overline{S}_b + \overline{S}_c| = 17634.1$ VA and $S_{Arith} = |\overline{S}_a| + |\overline{S}_b| + |\overline{S}_c| = 17634.1$ VA, and $S_e = 3V_e I_e = 3*230*62.59 = 41393$ VA. Accordingly the vector, arithmetic, and effective power factors are 1.0, 1.0, and 0.4082, respectively.

(d) Earlier, the load which was getting power from one phase on the delta side, will now be shared equally among the three phases. Thus, the current rating of the line conductors and transformer will reduce.

5.4   THEORY OF INSTANTANEOUS SYMMETRICAL COMPONENTS

The theory of instantaneous symmetrical components can be used for the purpose of load balancing, harmonic suppression, and power factor correction [7]. The control algorithms based on instantaneous symmetrical component theory can practically compensate any kind of unbalance and harmonics in the load, provided we have a high bandwidth power converter to track the filter reference currents. These algorithms are discussed in this section. For any set of three-phase instantaneous voltages, the instantaneous symmetrical components are given as.

$$\begin{bmatrix} \overline{v}_{a0} \\ \overline{v}_{a+} \\ \overline{v}_{a-} \end{bmatrix} = \frac{1}{3} \begin{bmatrix} 1 & 1 & 1 \\ 1 & a & a^2 \\ 1 & a^2 & a \end{bmatrix} \begin{bmatrix} v_a \\ v_b \\ v_c \end{bmatrix} \tag{5.119}$$

Similarly, three-phase instantaneous currents can be expressed in their instantaneous symmetrical components, as given in the following.

$$\begin{bmatrix} \overline{i}_{a0} \\ \overline{i}_{a+} \\ \overline{i}_{a-} \end{bmatrix} = \frac{1}{3} \begin{bmatrix} 1 & 1 & 1 \\ 1 & a & a^2 \\ 1 & a^2 & a \end{bmatrix} \begin{bmatrix} i_a \\ i_b \\ i_c \end{bmatrix} \tag{5.120}$$

In the above equations, a is a complex operator and it is given by $a = e^{j2\pi/3}$ and $a^2 = e^{j4\pi/3}$. It is to be noted that the instantaneous components of currents, $\overline{i}_{a+}$ and $\overline{i}_{a-}$ are complex time varying quantities and also they are complex conjugates of each other. The same is true for $\overline{v}_{a+}$ and $\overline{v}_{a-}$ quantities. The terms $\overline{i}_{a0}$ and $\overline{v}_{a0}$ are real quantities, however (-) has been used as upper script for the sake of uniformity of notation. These instantaneous symmetrical components are used to formulate equations for load compensation. First, a three-phase, four-wire system supplying star-connected load is considered.

5.4.1   COMPENSATING STAR CONNECTED LOAD

A three-phase four-wire compensated system is shown in Fig. 5.10. In the figure, three-phase load currents (i_{la}, i_{lb} and i_{lc}), can be unbalanced and nonlinear load. The objective in either three or four-wire compensation system is to provide a balanced supply current such that its zero sequence component is zero. We therefore have,

$$i_{sa} + i_{sb} + i_{sc} = 0 \tag{5.121}$$

Using equations (5.119) and (5.120), instantaneous positive sequence voltage ($\overline{v}_{a+}$) and current ($\overline{i}_{a+}$) are computed from instantaneous values of v_{sa}, v_{sb}, v_{sc} and i_{sa}, i_{sb}, i_{sc}, respectively. To have a predefined power factor from the source, the relationship between the angle of $\overline{v}_{a+}$ and $\overline{i}_{a+}$ is given as follows.

$$\angle \overline{v}_{sa+} = \angle \overline{i}_{sa+} + \phi_+ \tag{5.122}$$

where ϕ_+ is desired phase angle between $\overline{v}_{a+}$ and $\overline{i}_{a+}$.
The above equation is rewritten as follows.

$$\angle \left[\frac{1}{3} (v_{sa} + a v_{sb} + a^2 v_{sc}) \right] = \angle \left[\frac{1}{3} (i_{sa} + a i_{sb} + a^2 i_{sc}) \right] + \phi_+$$

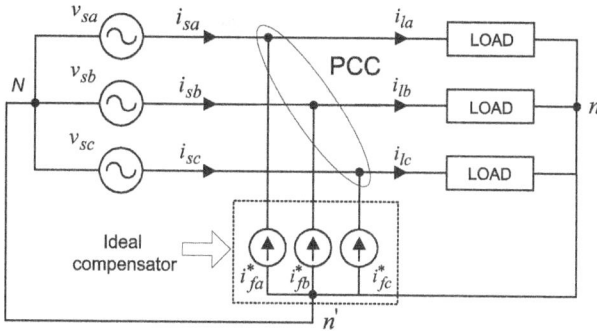

Figure 5.10 A three-phase four-wire compensated system

L.H.S. of the above equation is expressed as below.

$$
\begin{aligned}
\text{L.H.S.} &= \angle\left\{\frac{1}{3}\left[v_{sa} + \left(-\frac{1}{2} + j\frac{\sqrt{3}}{2}\right)v_{sb} + \left(-\frac{1}{2} - j\frac{\sqrt{3}}{2}\right)v_{sc}\right]\right\} \\
&= \angle\left\{\frac{1}{3}\left[\left(v_{sa} - \frac{v_{sb}}{2} - \frac{v_{sc}}{2}\right) + j\frac{\sqrt{3}}{2}(v_{sb} - v_{sc})\right]\right\} \\
&= \tan^{-1}\frac{(\sqrt{3}/2)(v_{sb} - v_{sc})}{(v_{sa} - v_{sb}/2 - v_{sc}/2)} \\
&= \tan^{-1}\frac{K_1}{K_2}
\end{aligned}
\tag{5.123}
$$

where, $K_1 = (\sqrt{3}/2)(v_{sb} - v_{sc})$ and $K_2 = (v_{sa} - v_{sb}/2 - v_{sc}/2)$.
Similarly, R.H.S. of the equation is expanded as below.

$$
\begin{aligned}
\text{R.H.S.} &= \angle\left\{\frac{1}{3}\left[i_{sa} + \left(-\frac{1}{2} + j\frac{\sqrt{3}}{2}\right)i_{sb} + \left(-\frac{1}{2} - j\frac{\sqrt{3}}{2}\right)i_{sc}\right]\right\} + \phi_+ \\
&= \angle\left\{\frac{1}{3}\left[\left(i_{sa} - \frac{i_{sb}}{2} - \frac{i_{sc}}{2}\right) + j\frac{\sqrt{3}}{2}(i_{sb} - i_{sc})\right]\right\} + \phi_+ \\
&= \tan^{-1}\frac{(\sqrt{3}/2)(i_{sb} - i_{sc})}{(i_{sa} - i_{sb}/2 - i_{sc}/2)} + \phi_+ \\
&= \tan^{-1}\frac{K_3}{K_4} + \phi_+
\end{aligned}
\tag{5.124}
$$

where, $K_3 = (\sqrt{3}/2)(i_{sb} - i_{sc})$ and $K_4 = (i_{sa} - i_{sb}/2 - i_{sc}/2)$.
Equating (5.123) and (5.124), we get the following.

$$
\tan^{-1}\frac{K_1}{K_2} = \tan^{-1}\frac{K_3}{K_4} + \phi_+
$$

Taking tangent on both sides, the following is obtained.

$$\tan\left(\tan^{-1}\frac{K_1}{K_2}\right) = \tan\left(\tan^{-1}\frac{K_3}{K_4} + \phi_+\right)$$

Therefore,
$$\frac{K_1}{K_2} = \frac{(K_3/K_4) + \tan\phi_+}{1 - (K_3/K_4)\times\tan\phi_+}$$

$$\frac{K_1}{K_2} = \frac{K_3 + K_4\tan\phi_+}{K_4 - K_3\times\tan\phi_+}$$

The above equation implies that,

$$K_1 K_4 - K_1 K_3\tan\phi_+ - K_2 K_3 - K_2 K_4\tan\phi_+ = 0$$

Substituting the values of K_1, K_2, K_3, K_4 in the above equation, the following expression is obtained.

$$\frac{\sqrt{3}}{2}(v_{sb} - v_{sc})\left(i_{sa} - \frac{i_{sb}}{2} - \frac{i_{sc}}{2}\right) - \frac{3}{4}(v_{sb} - v_{sc})(i_{sb} - i_{sc})\tan\phi_+$$

$$-\frac{\sqrt{3}}{2}\left(v_{sa} - \frac{v_{sb}}{2} - \frac{v_{sc}}{2}\right)(i_{sb} - i_{sc})$$

$$-\left(v_{sa} - \frac{v_{sb}}{2} - \frac{v_{sc}}{2}\right)\left(i_{sa} - \frac{i_{sb}}{2} - \frac{i_{sc}}{2}\right)\tan\phi_+ = 0$$

The above equation can be arranged according to the terms associated with i_{sa}, i_{sb}, and i_{sc} as given below.

$$\left\{\frac{\sqrt{3}}{2}(v_{sb} - v_{sc}) + \frac{\tan\phi_+}{2}(v_{sb} + v_{sc} - 2v_{sa})\right\}i_{sa}$$

$$+\left\{\frac{\sqrt{3}}{2}(v_{sc} - v_{sa}) + \frac{\tan\phi_+}{2}(v_{sc} + v_{sa} - 2v_{sb})\right\}i_{sb}$$

$$+\left\{\frac{\sqrt{3}}{2}(v_{sa} - v_{sb}) + \frac{\tan\phi_+}{2}(v_{sa} + v_{sb} - 2v_{sc})\right\}i_{sc} = 0$$

Dividing above equation by $\frac{\sqrt{3}}{2}$, it can be written as follows.

$$\left\{(v_{sb} - v_{sc}) + \frac{\tan\phi_+}{\sqrt{3}}(v_{sb} + v_{sc} - 2v_{sa})\right\}i_{sa}$$

$$+\left\{(v_{sc} - v_{sa}) + \frac{\tan\phi_+}{\sqrt{3}}(v_{sc} + v_{sa} - 2v_{sb})\right\}i_{sb}$$

$$+\left\{(v_{sa} - v_{sb}) + \frac{\tan\phi_+}{\sqrt{3}}(v_{sa} + v_{sb} - 2v_{sc})\right\}i_{sc} = 0$$

Let $\tan\phi_+/\sqrt{3} = \beta$, the above equation is further simplified to,

$$\{(v_{sb} - v_{sc}) + \beta(v_{sb} + v_{sc} - 2v_{sa})\}\, i_{sa}$$
$$+ \{(v_{sc} - v_{sa}) + \beta(v_{sc} + v_{sa} - 2v_{sb})\}\, i_{sb}$$
$$+ \{(v_{sa} - v_{sb}) + \beta(v_{sa} + v_{sb} - 2v_{sc})\}\, i_{sc} = 0. \qquad (5.125)$$

The third objective of compensation is that the power supplied from the source (p_s) must be equal to the average load power (P_{lavg}). Therefore the following holds true.

$$p_s = v_{sa} i_{sa} + v_{sb} i_{sb} + v_{sc} i_{sc} = P_{lavg} \qquad (5.126)$$

The above equation has important implications. For example, when supply voltages are balanced, the equation is satisfied for balanced source currents. However, if supply voltages are unbalanced and distorted, the equation gives a set of currents that are also not balanced and sinusoidal in order to supply constant power.

Equations (5.121), (5.125), and (5.126), can be written in matrix form as given below.

$$\begin{bmatrix} 1 & 1 & 1 \\ (v_{sb} - v_{sc}) & (v_{sc} - v_{sa}) & (v_{sa} - v_{sb}) \\ +\beta(v_{sb} + v_{sc} - 2v_{sa}) & +\beta(v_{sc} + v_{sa} - 2v_{sb}) & +\beta(v_{sa} + v_{sb} - 2v_{sc}) \\ v_{sa} & v_{sb} & v_{sc} \end{bmatrix} \begin{bmatrix} i_{sa} \\ i_{sb} \\ i_{sc} \end{bmatrix}$$
$$= \begin{bmatrix} 0 \\ 0 \\ P_{lavg} \end{bmatrix}$$

which can be further written as,

$$[A]\,[i_{sabc}] = [P_{lavg}] \qquad (5.127)$$

Therefore,

$$[i_{sabc}] = [A^{-1}]\,[P_{lavg}] = \frac{1}{|A|}\begin{bmatrix} a_{c11} & a_{c12} & a_{c13} \\ a_{c21} & a_{c22} & a_{c23} \\ a_{c31} & a_{c32} & a_{c33} \end{bmatrix}^{T} \begin{bmatrix} 0 \\ 0 \\ P_{lavg} \end{bmatrix}$$

$$= \frac{1}{|A|}\begin{bmatrix} a_{c11} & a_{c21} & a_{c31} \\ a_{c12} & a_{c22} & a_{c32} \\ a_{c13} & a_{c23} & a_{c33} \end{bmatrix} \begin{bmatrix} 0 \\ 0 \\ P_{lavg} \end{bmatrix}$$

where a_{cij} is the cofactor of ith row and j^{th} column element of matrix A in (5.127) and $|A|$ is the determinant of matrix A. Due to the presence of zero elements in the first two rows of the column matrix with power elements, the cofactors in the first two columns need not be computed. These are indicated by dots in the following matrix.

$$[i_{sabc}] = \begin{bmatrix} i_{sa} \\ i_{sb} \\ i_{sc} \end{bmatrix} = \frac{1}{|A|}\begin{bmatrix} \cdot & \cdot & a_{c31} \\ \cdot & \cdot & a_{c32} \\ \cdot & \cdot & a_{c33} \end{bmatrix} \begin{bmatrix} 0 \\ 0 \\ P_{lavg} \end{bmatrix} = \frac{1}{|A|}\begin{bmatrix} a_{c31} \\ a_{c32} \\ a_{c33} \end{bmatrix} P_{lavg}$$

The determinant of matrix A is computed as below.

$$
\begin{aligned}
|A| &= [(v_{sc}-v_{sa})+\beta(v_{sc}+v_{sa}-2v_{sb})]v_{sc} \\
&\quad -[(v_{sa}-v_{sb})+\beta(v_{sa}+v_{sb}-2v_{sc})]v_{sb} \\
&\quad -[(v_{sb}-v_{sc})+\beta(v_{sb}+v_{sc}-2v_{sa})]v_{sc} \\
&\quad +[(v_{sa}-v_{sb})+\beta(v_{sa}+v_{sb}-2v_{sc})]v_{sa} \\
&\quad +[(v_{sb}-v_{sc})+\beta(v_{sb}+v_{sc}-2v_{sa})]v_{sb} \\
&\quad -[(v_{sc}-v_{sa})+\beta(v_{sc}+v_{sa}-2v_{sb})]v_{sa} \\
&= \beta[v_{sc}^2+v_{sa}v_{sc}-2v_{sb}v_{sc}-v_{sa}v_{sb}-v_{sb}^2+2v_{sb}v_{sc}-v_{sb}v_{sc}-v_{sc}^2+2v_{sa}v_{sc} \\
&\quad +v_{sa}^2+v_{sb}v_{sa}-2v_{sc}v_{sa}+v_{sb}^2+v_{sc}v_{sb}-2v_{sa}v_{sb}-v_{sc}v_{sa}-v_{sa}^2+2v_{sb}v_{sa}] \\
&\quad +(v_{sc}-v_{sa})v_{sc}-(v_{sa}-v_{sb})v_{sb}-(v_{sb}-v_{sc})v_{sc} \\
&\quad +(v_{sa}-v_{sb})v_{sa}+(v_{sb}-v_{sc})v_{sb}-(v_{sc}-v_{sa})v_{sa}
\end{aligned}
$$

The above equation can be further simplified to,

$$
\begin{aligned}
|A| &= \beta\cdot 0+v_{sc}^2-v_{sa}v_{sc}-v_{sb}v_{sa}+v_{sb}^2-v_{sb}v_{sc}+v_{sc}^2+v_{sa}^2 \\
&\quad -v_{sa}v_{sb}+v_{sb}^2-v_{sc}v_{sb}-v_{sa}v_{sc}+v_{sa}^2 \\
&= 2v_{sa}^2+2v_{sb}^2+2v_{sc}^2-2v_{sa}v_{sb}-2v_{sb}v_{sc}-2v_{sc}v_{sa} \\
&= v_{sa}^2+v_{sb}^2-2v_{sa}v_{sb}+v_{sb}^2+v_{sc}^2-2v_{sb}v_{sc}+v_{sc}^2+v_{sa}^2-2v_{sc}v_{sa} \\
&= (v_{sa}-v_{sb})^2+(v_{sb}-v_{sc})^2+(v_{sc}-v_{sa})^2 \\
&= (v_{sab}^2+v_{sbc}^2+v_{sca}^2)
\end{aligned}
$$

Further adding and subtracting, $v_{sa}^2+v_{sb}^2+v_{sc}^2$ in the above equation, we get the following.

$$
|A| = 3(v_{sa}^2+v_{sb}^2+v_{sc}^2)-(v_{sa}^2+v_{sb}^2+v_{sc}^2+2v_{sa}v_{sb}+2v_{sb}v_{sc}+2v_{sc}v_{sa}) \quad (5.128)
$$

Further,

$$
(v_{sa}+v_{sb}+v_{sc})^2 = v_{sa}^2+v_{sb}^2+v_{sc}^2+2v_{sa}v_{sb}+2v_{sb}v_{sc}+2v_{sc}v_{sa} \quad (5.129)
$$

Using equations (5.128) and (5.129), we obtain the following.

$$
\begin{aligned}
|A| &= 3(v_{sa}^2+v_{sb}^2+v_{sc}^2)-(v_{sa}+v_{sb}+v_{sc})^2 \\
&= 3\sum_{j=a,b,c} v_{sj}^2-9v_{so}^2 \\
&= 3\left(\sum_{j=a,b,c} v_{sj}^2-3v_{so}^2\right) \quad (5.130)
\end{aligned}
$$

In above equation, the term v_{so} is the instantaneous zero sequence component of the source voltage and it is given as following.

$$
v_{so} = \frac{(v_{sa}+v_{sb}+v_{sc})}{3} \quad (5.131)
$$

The determinant of matrix A, using (5.130), can also be expressed as,

$$
\begin{aligned}
|A| &= 3\left[v_{sa}^2 + v_{sb}^2 + v_{sc}^2 - 3v_{so}^2\right] \\
&= 3\left[v_{sa}^2 + v_{sb}^2 + v_{sc}^2 + 3v_{so}^2 - 2 \times 3v_{so}^2\right] \\
&= 3\left[v_{sa}^2 + v_{sb}^2 + v_{sc}^2 + v_{so}^2 + v_{so}^2 + v_{so}^2 - 2v_{so}(v_{sa} + v_{sb} + v_{sc})\right] \\
&= 3\left[(v_{sa}^2 + v_{so}^2 - 2v_{sa}v_{so}) + (v_{sb}^2 + v_{so}^2 - 2v_{sb}v_{so}) + (v_{sc}^2 + v_{so}^2 - 2v_{sc}v_{so})\right] \\
&= 3\left[(v_{sa} - v_{so}^2) + (v_{sb} - v_{so}^2) + (v_{sc} - v_{so}^2)\right]
\end{aligned}
$$

The cofactors a_{c31}, a_{c32}, and a_{c33} are computed as below.

$$
\begin{aligned}
a_{c31} &= \left[-(v_{sc} - v_{sa}) - \beta(v_{sc} + v_{sa} - 2v_{sb}) + (v_{sa} - v_{sb}) + \beta(v_{sa} + v_{sb} - 2v_{sc})\right] \\
&= v_{sa} - v_{sb} - v_{sc} + v_{sa} + \beta(-v_{sc} - v_{sa} + 2v_{sb} + v_{sa} + v_{sb} - 2v_{sc}) \\
&= (2v_{sa} - v_{sb} - v_{sc}) + 3\beta(v_{sb} - v_{sc}) \\
&= (2v_{sa} + v_{sa} - v_{sa} - v_{sb} - v_{sc}) + 3\beta(v_{sb} - v_{sc}) \\
&= 3(v_{sa} - v_{s0}) + 3\beta(v_{sb} - v_{sc})
\end{aligned}
$$

Similarly,

$$
\begin{aligned}
a_{c32} &= \left[-(v_{sa} - v_{sb}) - \beta(v_{sa} + v_{sb} - 2v_{sc}) + (v_{sb} - v_{sc}) + \beta(v_{sb} + v_{sc} - 2v_{sa})\right] \\
&= (-v_{sa} + v_{sb} + v_{sb} - v_{sc}) + \beta(-v_{sa} - v_{sb} + 2v_{sc} + v_{sb} + v_{sc} - 2v_{sa}) \\
&= (2v_{sb} - v_{sc} - v_{sa}) + 3\beta(v_{sc} - v_{sa}) \\
&= 3(v_{sb} - v_{so}) + 3\beta(v_{sc} - v_{sa})
\end{aligned}
$$

and,

$$
\begin{aligned}
a_{c33} &= \left[(v_{sc} - v_{sa}) + \beta(v_{sc} + v_{sa} - 2v_{sb}) - (v_{sb} - v_{sc}) - \beta(v_{sb} + v_{sc} - 2v_{sa})\right] \\
&= (2v_{sc} - v_{sa} - v_{sb}) + 3\beta(v_{sa} - v_{sb}) \\
&= 3(v_{sc} - v_{so}) + 3\beta(v_{sa} - v_{sb})
\end{aligned}
$$

Knowing the value of cofactors, we now have,

$$
\begin{aligned}
\begin{bmatrix} i_{sa} \\ i_{sb} \\ i_{sc} \end{bmatrix} &= \frac{1}{3\left(\sum_{j=a,b,c} v_{sj}^2 - 3v_{so}^2\right)} \begin{bmatrix} 3(v_{sa} - v_{s0}) + 3\beta(v_{sb} - v_{sc}) \\ 3(v_{sb} - v_{s0}) + 3\beta(v_{sc} - v_{sa}) \\ 3(v_{sc} - v_{s0}) + 3\beta(v_{sa} - v_{sb}) \end{bmatrix} \left[P_{lavg}\right] \\
&= \frac{1}{\left(\sum_{j=a,b,c} v_{sj}^2 - 3v_{so}^2\right)} \begin{bmatrix} (v_{sa} - v_{so}) + \beta(v_{sb} - v_{sc}) \\ (v_{sb} - v_{so}) + \beta(v_{sc} - v_{sa}) \\ (v_{sc} - v_{so}) + \beta(v_{sa} - v_{sb}) \end{bmatrix} \left[P_{lavg}\right] \quad (5.132)
\end{aligned}
$$

From the above equation, the desired source currents can be written as following.

$$
\begin{aligned}
i_{sa} &= \frac{(v_{sa}-v_{so})+\beta(v_{sb}-v_{sc})}{\sum_{j=a,b,c}v_{sj}^2-3v_{so}^2}P_{lavg} \\[2mm]
i_{sb} &= \frac{(v_{sb}-v_{so})+\beta(v_{sc}-v_{sa})}{\sum_{j=a,b,c}v_{sj}^2-3v_{so}^2}P_{lavg} \\[2mm]
i_{sc} &= \frac{(v_{sc}-v_{so})+\beta(v_{sa}-v_{sb})}{\sum_{j=a,b,c}v_{sj}^2-3v_{so}^2}P_{lavg}
\end{aligned}
\tag{5.133}
$$

Applying Kirchoff's current law at the point of common coupling (PCC), we have,

$$
\begin{aligned}
i_{fa}^* &= i_{la}-i_{sa} \\
i_{fb}^* &= i_{lb}-i_{sb} \\
i_{fc}^* &= i_{lc}-i_{sc}
\end{aligned}
\tag{5.134}
$$

Replacing i_{sa}, i_{sb}, and i_{sc} from equations (5.133)–(5.134), we obtain the reference filter currents as given in the following.

$$
\begin{aligned}
i_{fa}^* &= i_{la}-i_{sa}=i_{la}-\frac{(v_{sa}-v_{so})+\beta(v_{sb}-v_{sc})}{\sum_{j=a,b,c}v_{sj}^2-3v_{so}^2}P_{lavg} \\[2mm]
i_{fb}^* &= i_{lb}-i_{sb}=i_{lb}-\frac{(v_{sb}-v_{so})+\beta(v_{sc}-v_{sa})}{\sum_{j=a,b,c}v_{sj}^2-3v_{so}^2}P_{lavg} \\[2mm]
i_{fc}^* &= i_{lc}-i_{sc}=i_{lc}-\frac{(v_{sc}-v_{so})+\beta(v_{sa}-v_{sb})}{\sum_{j=a,b,c}v_{sj}^2-3v_{so}^2}P_{lavg}
\end{aligned}
\tag{5.135}
$$

For balanced supply voltages, $v_{s0}=0$. For the unity power factor, $\beta=0$. Therefore, expressions for the reference filter currents are simplified to the following.

$$
\begin{aligned}
i_{fa}^* &= i_{la}-i_{sa}=i_{la}-\frac{v_{sa}}{\sum_{j=a,b,c}v_{sj}^2}P_{lavg} \\[2mm]
i_{fb}^* &= i_{lb}-i_{sb}=i_{lb}-\frac{v_{sb}}{\sum_{j=a,b,c}v_{sj}^2}P_{lavg} \\[2mm]
i_{fc}^* &= i_{lc}-i_{sc}=i_{lc}-\frac{v_{sc}}{\sum_{j=a,b,c}v_{sj}^2}P_{lavg}
\end{aligned}
\tag{5.136}
$$

Example 5.5. Consider a three-phase four-wire compensated system shown in Fig. 5.10. The three-phase voltages and load impedances as follows.

$$
\begin{aligned}
v_{sa}(t) &= 230\sqrt{2}\sin\omega t \\
v_{sb}(t) &= 230\sqrt{2}\sin(\omega t-120°) \\
v_{sc}(t) &= 230\sqrt{2}\sin(\omega t+120°)
\end{aligned}
$$

Load impedances: $Z_{la}=5+j12\,\Omega$, $Z_{lb}=3+j4\,\Omega$, $Z_{lc}=5-j12\,\Omega$. Using instantaneous symmetrical component theory,

(a) Find reference filter currents (i_{fa}, i_{fb}, i_{fc}) of an ideal compensator such that source currents are balanced sinusoids and in phase with their respective source voltages. Also, find the active and reactive powers of the compensator.

(b) Based on these reference filter currents, identify the nature and values of Z_{fa}, Z_{fb}, Z_{fc}. Are these practically realizable using passive impedances?

(c) Find out the time expressions for the source currents based on the above filter currents. Check whether these are balanced and in phase with their respective voltages.

(d) Compute the source powers and check if $P_s = P_l$, $Q_s = 0$, $P_f = 0$, and $Q_f = Q_l$.

Solution:

(a) Given the source voltage and load impedances, the load currents are computed as following.

$$\bar{I}_{la} = \frac{\bar{V}_{sa}}{Z_{la}} = \frac{230\angle 0°}{5 + j12} = 17.69\angle -67.38° \text{ A}$$

$$\bar{I}_{lb} = \frac{\bar{V}_{sb}}{Z_{lb}} = \frac{230\angle -120°}{3 + j4} = 46.0\angle -173.13° \text{ A}$$

$$\bar{I}_{lc} = \frac{\bar{V}_{sc}}{Z_{lc}} = \frac{230\angle 120°}{5 - j12} = 17.69\angle 187.38° \text{ A}$$

For the given source voltages, the value of $v_{sa}^2 + v_{sb}^2 + v_{sc}^2$ is given below.

$$v_{sa}^2 + v_{sb}^2 + v_{sc}^2 = 3V_{rms}^2 = 3 \times 230^2 = 158700 \text{ Volt}^2$$

The load real and reactive powers are computed as following.

$$\bar{S}_l = \bar{V}_{sa}\bar{I}_{la}^* + \bar{V}_{sb}\bar{I}_{lb}^* + \bar{V}_{sc}\bar{I}_{lc}^* = 9478.2 + j8464 \text{ VA}$$

This gives, $P_l = P_{lavg} = 9478.2$ W, and $Q_l = 8464$ VAr.

The reference filter currents are computed using (5.136). Since the voltage and current quantities involved here have only fundamental components, the reference filter currents can be computed using phasor equations equivalent to (5.136). These are expressed below.

$$\bar{I}_{fa} = \bar{I}_{la} - \bar{V}_{sa}\frac{P_{lavg}}{v_{sa}^2 + v_{sb}^2 + v_{sc}^2}$$

$$\bar{I}_{fb} = \bar{I}_{lb} - \bar{V}_{sb}\frac{P_{lavg}}{v_{sa}^2 + v_{sb}^2 + v_{sc}^2}$$

$$\bar{I}_{fc} = \bar{I}_{lc} - \bar{V}_{sc}\frac{P_{lavg}}{v_{sa}^2 + v_{sb}^2 + v_{sc}^2}$$

Substituting the values of source voltages, load currents, P_{lavg}, and $v_{sa}^2 + v_{sb}^2 + v_{sc}^2$ as computed above in (5.137), we get the following.

$$\bar{I}_{fa} = 17.69\angle -67.38° - 230\angle 0 \frac{9478.2}{158700} = 17.74\angle -113°\,\mathrm{A}$$

$$\bar{I}_{fb} = 46.0\angle -173.13° - 230\angle -120° \frac{9478.2}{158700} = 39.32\angle 170.64°\,\mathrm{A}$$

$$\bar{I}_{fc} = 17.69\angle 187.38° - 230\angle 120° \frac{9478.2}{158700} = 17.74\angle -127°\,\mathrm{A}$$

Therefore, the time expressions for filter currents are given as following.

$$
\begin{aligned}
i_{fa}(t) &= 17.74\sqrt{2}\sin(\omega t - 113°)\,\mathrm{A} \\
i_{fb}(t) &= 39.32\sqrt{2}\sin(\omega t + 170.64°)\,\mathrm{A} \\
i_{fc}(t) &= 17.74\sqrt{2}\sin(\omega t - 127°)\,\mathrm{A}
\end{aligned}
$$

Based on the above reference filter currents, the compensator powers are computed as below.

$$
\begin{aligned}
P_f &= \mathrm{Real}\left(\bar{V}_{sa}\bar{I}_{fa}^* + \bar{V}_{sb}\bar{I}_{fb}^* + \bar{V}_{sc}\bar{I}_{fc}^*\right) = 0\,\mathrm{W} \\
Q_f &= \mathrm{Imag}\left(\bar{V}_{sa}\bar{I}_{fa}^* + \bar{V}_{sb}\bar{I}_{fb}^* + \bar{V}_{sc}\bar{I}_{fc}^*\right) = 8464\,\mathrm{VAr} = Q_l
\end{aligned}
$$

From the values of compensator real and reactive powers, it can be observed that the compensator does not exchange any real power with the load or source, and it supplies the required reactive power to the load. Thus, the source is relieved of supplying reactive power of the load and maintains a unity power factor.

(b) Based on the voltages at the load point and filter currents, the compensator impedances Z_{fa}, Z_{fb}, Z_{fc} are computed as following.

$$\bar{Z}_{fa} = \frac{\bar{V}_{sa}}{-\bar{I}_{fa}} = \frac{230\angle 0°}{-17.74\angle -113°} = 5.06 - j11.93\,\Omega$$

$$\bar{Z}_{fb} = \frac{\bar{V}_{sb}}{-\bar{I}_{fb}} = \frac{230\angle -120°}{-39.32\angle 170.64°} = -2.06 - j5.47\,\Omega$$

$$\bar{Z}_{fc} = \frac{\bar{V}_{sc}}{-\bar{I}_{fc}} = \frac{230\angle 120°}{-17.74\angle -127°} = 5.06 + j11.93\,\Omega$$

It can be seen that the compensator impedances in leg-a and c can be realized as capacitive and inductive loads, respectively. However, leg-b has a negative real part in the impedance and hence can not be realized using a passive component. Thus, the scheme can not be implemented using a passive filter. That is why the idea of active filtering is employed to realize such compensation using a voltage source inverter.

(c) The compensated source currents are calculated below.

$$\begin{aligned}
\overline{I}_{sa} &= \overline{I}_{la} - \overline{I}_{fa} = 13.74\angle 0\,\text{A} \\
\overline{I}_{sb} &= \overline{I}_{lb} - \overline{I}_{fb} = 13.74\angle -120°\,\text{A} \\
\overline{I}_{sc} &= \overline{I}_{lc} - \overline{I}_{fc} = 13.74\angle 120°\,\text{A}
\end{aligned}$$

Therefore, the time expressions of the source currents are given as following.

$$\begin{aligned}
i_{sa}(t) &= 13.74\sqrt{2}\sin\omega t\,\text{A} \\
i_{sb}(t) &= 13.74\sqrt{2}\sin(\omega t - 120°)\,\text{A} \\
i_{sc}(t) &= 13.74\sqrt{2}\sin(\omega t + 120°)\,\text{A}
\end{aligned}$$

From the above, it can be observed that the compensated source currents are balanced and in phase with their respective source voltages.

(d) The source and compensator powers are given as below.

$$\begin{aligned}
P_s &= P_l = \text{Real}\left(\overline{V}_{sa}\overline{I}_{sa}^* + \overline{V}_{sb}\overline{I}_{sb}^* + \overline{V}_{sc}\overline{I}_{sc}^*\right) = 9478.2\,\text{W} \\
Q_s &= \text{Imag}\left(\overline{V}_{sa}\overline{I}_{sa}^* + \overline{V}_{sb}\overline{I}_{sb}^* + \overline{V}_{sc}\overline{I}_{sc}^*\right) = 0\,\text{VAr} \\
P_f &= 0\,\text{W} \\
Q_f &= Q_l = 8464\,\text{VAr}
\end{aligned}$$

It can be observed that the source only supplies the real power of the load, and the compensator supplies the total reactive power of the load.

5.4.2 COMPENSATING DELTA CONNECTED LOAD

The balancing of an unbalanced Δ-connected load is a generic problem and the theory of instantaneous symmetrical components can be used to balance the load. The schematic diagram of this compensated scheme is shown in Fig. 5.11.

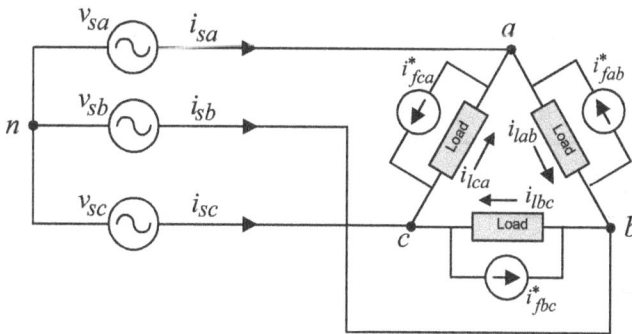

Figure 5.11 A compensation for delta connected load

This compensator is connected between the phases of the load. The aim is to generate the three reference current waveforms denoted by i^*_{fab}, i^*_{fbc}, and i^*_{fca}, respectively based on the measurement of system voltages and load currents such that the supply sees a balanced load. As there is no fourth or neutral wire, the following equation is valid.

$$i_{sa} + i_{sb} + i_{sc} = 0 \qquad (5.137)$$

Applying Kirchhoff's current law at nodes, we can express i_{sa}, i_{sb}, and i_{sc}, respectively as follows.

$$
\begin{aligned}
i_{sa} &= (i_{lab} - i^*_{fab}) - (i_{lca} - i^*_{fca}) \\
i_{sb} &= (i_{lbc} - i^*_{fbc}) - (i_{lab} - i^*_{fab}) \\
i_{sc} &= (i_{lca} - i^*_{fca}) - (i_{lbc} - i^*_{fbc})
\end{aligned}
\qquad (5.138)
$$

As can be seen from the above equations, (5.137) is satisfied. Since, in a balanced Δ-connected load after compensation, the zero sequence current within the delta cannot flow, therefore,

$$(i_{lab} - i^*_{fab}) + (i_{lbc} - i^*_{fbc}) + (i_{lca} - i^*_{fca}) = 0 \qquad (5.139)$$

The source supplies the average load power, P_{lavg} and the following equation is satisfied.

$$v_{sa}i_{sa} + v_{sb}i_{sb} + v_{sc}i_{sc} = P_{lavg} \qquad (5.140)$$

From the power factor condition between the source voltages and currents, as stated in (5.122), the following equation was derived.

$$
\begin{aligned}
&\{(v_{sb} - v_{sc}) + \beta (v_{sb} + v_{sc} - 2v_{sa})\} \, i_{sa} \\
&+ \{(v_{sc} - v_{sa}) + \beta (v_{sc} + v_{sa} - 2v_{sb})\} \, i_{sb} \\
&+ \{(v_{sa} - v_{sb}) + \beta (v_{sa} + v_{sb} - 2v_{sc})\} \, i_{sc} = 0.
\end{aligned}
\qquad (5.141)
$$

Replacing i_{sa}, i_{sb} and i_{sc} from (5.138) in the above equation, the following equation is obtained.

$$\underbrace{\{[(v_{sb} - v_{sc}) + \beta(v_{sb} + v_{sc} - 2v_{sa})] - [(v_{sc} - v_{sa}) + \beta(v_{sc} + v_{sa} - 2v_{sb})]\} (i_{lab} - i^*_{fab})}_{\text{I}}$$

$$+ \underbrace{\{[(v_{sc} - v_{sa}) + \beta(v_{sc} + v_{sa} - 2v_{sb})] - [(v_{sa} - v_{sb}) + \beta(v_{sa} + v_{sb} - 2v_{sc})]\} (i_{lbc} - i^*_{fbc})}_{\text{II}}$$

$$+ \underbrace{\{[(v_{sa} - v_{sb}) + \beta(v_{sa} + v_{sb} - 2v_{sc})] - [(v_{sb} - v_{sc}) + \beta(v_{sb} + v_{sc} - 2v_{sa})]\} (i_{lca} - i^*_{fca})}_{\text{III}} = 0$$

The first term, I is simplified, as given below.

$$
\begin{aligned}
\text{I} &= \{(v_{sb}-v_{sc}-v_{sc}+v_{sa})+\beta(v_{sb}+v_{sc}-2v_{sa}-v_{sc}-v_{sa}+2v_{sb})\}\,(i_{lab}-i^*_{fab})\\
&= \{(v_{sa}+v_{sb}-2v_{sc})-3\beta(v_{sa}-v_{sb})\}\,(i_{lab}-i^*_{fab})\\
&= -3\{(v_{sc}-v_{s0})+\beta(v_{sa}-v_{sb})\}\,(i_{lab}-i^*_{fab})
\end{aligned}
\tag{5.142}
$$

Similarly, the second and third terms are given as below.

$$
\begin{aligned}
\text{II} &= \{(v_{sc}-v_{sa}-v_{sa}+v_{sb})+\beta(v_{sc}+v_{sa}-2v_{sb}-v_{sa}-v_{sb}+2v_{sc})\}\,(i_{lbc}-i^*_{fbc})\\
&= \{(v_{sb}+v_{sc}-2v_{sa})-3\beta(v_{sb}-v_{sc})\}\,(i_{lbc}-i^*_{fbc})\\
&= -3\{(v_{sa}-v_{s0})+\beta(v_{sb}-v_{sc})\}\,(i_{lbc}-i^*_{fbc})
\end{aligned}
\tag{5.143}
$$

$$
\begin{aligned}
\text{III} &= \{(v_{sa}-v_{sb}-v_{sb}+v_{sc})+\beta(v_{sa}+v_{sb}-2v_{sc}-v_{sb}-v_{sc}+2v_{sa})\}\,(i_{lca}-i^*_{fca})\\
&= \{(v_{sc}+v_{sa}-2v_{sb})-3\beta(v_{sc}-v_{sa})\}\,(i_{lca}-i^*_{fca})\\
&= -3\{(v_{sb}-v_{s0})+\beta(v_{sc}-v_{sa})\}\,(i_{lca}-i^*_{fca})
\end{aligned}
\tag{5.144}
$$

Summing the above three terms and simplifying we get,

$$
\begin{aligned}
&\{(v_{sc}-v_{s0})+\beta(v_{sa}-v_{sb})\}\,(i_{lab}-i^*_{fab})\\
&+\{(v_{sa}-v_{s0})+\beta(v_{sb}-v_{sc})\}\,(i_{lbc}-i^*_{fbc})\\
&+\{(v_{sb}-v_{s0})+\beta(v_{sc}-v_{sa})\}\,(i_{lca}-i^*_{fca})=0
\end{aligned}
\tag{5.145}
$$

The third condition for load compensation ensures that the average load power should be supplied from the sources. Therefore,

$$
v_{sa}i_{sa}+v_{sb}i_{sb}+v_{sc}i_{sc}=P_{lavg}
\tag{5.146}
$$

The terms i_{sa}, i_{sb}, and i_{sc} are substituted from (5.138) in the above equation, and the modified equation is given below.

$$
\begin{aligned}
v_{sa}\{(i_{lab}-i^*_{fab})-(i_{lca}-i^*_{fca})\}+v_{sb}\{(i_{lbc}-i^*_{fbc})-(i_{lab}-i^*_{fab})\}\\
+v_{sc}\{(i_{lca}-i^*_{fca})-(i_{lbc}-i^*_{fbc})\}=P_{lavg}
\end{aligned}
\tag{5.147}
$$

The above is simplified to,

$$
(v_{sa}-v_{sb})(i_{lab}-i^*_{fab})+(v_{sb}-v_{sc})(i_{lbc}-i^*_{fbc})+(v_{sc}-v_{sa})(i_{lca}-i^*_{fca})=P_{lavg}
\tag{5.148}
$$

Equations (5.139), (5.145), (5.148) can be written in the matrix form as given below.

$$
\begin{bmatrix}
1 & 1 & 1\\
(v_{sc}-v_{s0}) & (v_{sa}-v_{s0}) & (v_{sb}-v_{s0})\\
+\beta(v_{sa}-v_{sb}) & +\beta(v_{sb}-v_{sc}) & +\beta(v_{sc}-v_{sa})\\
v_{sa}-v_{sb} & v_{sb}-v_{sc} & v_{sc}-v_{sa}
\end{bmatrix}
\begin{bmatrix}
i_{lab}-i^*_{fab}\\
i_{lbc}-i^*_{fbc}\\
i_{lca}-i^*_{fca}
\end{bmatrix}
=
\begin{bmatrix}
0\\
0\\
P_{lavg}
\end{bmatrix}
\tag{5.149}
$$

The above equation can be written in the following form.

$$[A_\Delta] \begin{bmatrix} i_{lab} - i^*_{fab} \\ i_{lbc} - i^*_{fbc} \\ i_{lca} - i^*_{fca} \end{bmatrix} = \begin{bmatrix} 0 \\ 0 \\ P_{lavg} \end{bmatrix} \tag{5.150}$$

Therefore,

$$\begin{bmatrix} i_{lab} - i^*_{fab} \\ i_{lbc} - i^*_{fbc} \\ i_{lca} - i^*_{fca} \end{bmatrix} = [A_\Delta]^{-1} \begin{bmatrix} 0 \\ 0 \\ P_{lavg} \end{bmatrix} \tag{5.151}$$

The above equation is solved by finding the determinate of A_Δ and the cofactors transpose as given below.

$$\begin{bmatrix} i_{lab} - i^*_{fab} \\ i_{lbc} - i^*_{fbc} \\ i_{lca} - i^*_{fca} \end{bmatrix} = \frac{1}{|A_\Delta|} \begin{bmatrix} a_{c11} & a_{c12} & a_{c13} \\ a_{c21} & a_{c22} & a_{c23} \\ a_{c31} & a_{c32} & a_{c33} \end{bmatrix}^T \begin{bmatrix} 0 \\ 0 \\ P_{lavg} \end{bmatrix}$$

$$= \frac{1}{|A_\Delta|} \begin{bmatrix} a_{c11} & a_{c21} & a_{c31} \\ a_{c12} & a_{c22} & a_{c32} \\ a_{c13} & a_{c23} & a_{c33} \end{bmatrix} \begin{bmatrix} 0 \\ 0 \\ P_{lavg} \end{bmatrix} \tag{5.152}$$

The determinant $|A_\Delta|$ and the cofactors in the above equation are calculated below.

$$|A_\Delta| =$$
$$[(v_{sa} - v_{so}) + \beta(v_{sb} - v_{sc})](v_{sc} - v_{sa}) - [(v_{sb} - v_{so}) + \beta(v_{sc} - v_{sa})](v_{sb} - v_{sc})$$
$$- [(v_{sc} - v_{so}) + \beta(v_{sa} - v_{sb})](v_{sc} - v_{sa}) + [(v_{sb} - v_{so}) + \beta(v_{sc} - v_{sa})](v_{sa} - v_{sb})$$
$$+ [(v_{sc} - v_{so}) + \beta(v_{sa} - v_{sb})](v_{sb} - v_{sc}) - [(v_{sa} - v_{so}) + \beta(v_{sb} - v_{sc})](v_{sa} - v_{sb})$$

Separating all the terms containing β and rearranging the above equation, we get,

$$|A_\Delta| = (v_{sa} - v_{so})(v_{sc} - v_{sa}) - (v_{sb} - v_{so})(v_{sb} - v_{sc}) - (v_{sc} - v_{so})(v_{sc} - v_{sa})$$
$$+ (v_{sb} - v_{so})(v_{sa} - v_{sb}) + (v_{sc} - v_{so})(v_{sb} - v_{sc}) - (v_{sa} - v_{so})(v_{sa} - v_{sb})$$
$$+ \beta[(v_{sb} - v_{sc})(v_{sc} - v_{sa}) - (v_{sc} - v_{sa})(v_{sb} - v_{sc}) - (v_{sa} - v_{sb})(v_{sc} - v_{sa})$$
$$+ (v_{sc} - v_{sa})(v_{sa} - v_{sb}) + (v_{sa} - v_{sb})(v_{sb} - v_{sc}) - (v_{sb} - v_{sc})(v_{sa} - v_{sb})]$$

It is seen that the terms containing β cancel out each other and give zero. Thus, the above equation becomes,

$$|A_\Delta| = (v_{sa} - v_{so})(v_{sc} - v_{sa} - v_{sa} + v_{sb})$$
$$+ (v_{sb} - v_{so})(-v_{sb} + v_{sc} + v_{sa} - v_{sb})$$
$$+ (v_{sc} - v_{so})(-v_{sc} + v_{sa} + v_{sb} - v_{sc}) + \beta \times 0 \tag{5.153}$$

In the above, using $(v_{sa} + v_{sb} + v_{sc}) = 3v_{s0}$,

$$
\begin{aligned}
v_{sc} - v_{sa} - v_{sa} + v_{sb} &= v_{sa} + v_{sb} + v_{sc} - 3v_{sa} = -3(v_{sa} - v_{so}) \\
-v_{sb} + v_{sc} + v_{sa} - v_{sb} &= v_{sa} + v_{sb} + v_{sc} - 3v_{sb} = -3(v_{sb} - v_{so}) \\
-v_{sc} + v_{sa} + v_{sb} - v_{sc} &= v_{sa} + v_{sb} + v_{sc} - 3v_{sc} = -3(v_{sc} - v_{so}).
\end{aligned}
$$

Replacing the above term in (5.153), we get the following.

$$
\begin{aligned}
A_\Delta &= -3\left[(v_{sa} - v_{s0})(v_{sa} - v_{s0}) + (v_{sb} - v_{s0})(v_{sb} - v_{s0}) + (v_{sc} - v_{s0})(v_{sc} - v_{s0})\right] \\
&= -3\left[(v_{sa} - v_{s0})^2 + (v_{sb} - v_{s0})^2 + (v_{sc} - v_{s0})^2\right] \\
&= -3 \sum_{j=a,b,c} (v_{sj} - v_{s0})^2
\end{aligned}
\tag{5.154}
$$

The above equation can also be written as,

$$
\begin{aligned}
|A_\Delta| &= -3\left[v_{sa}^2 + v_{sb}^2 + v_{sc}^2 + 3v_{s0}^2 - 2v_{s0}(v_{sa} + v_{sb} + v_{sc})\right] \\
&= -3\left[v_{sa}^2 + v_{sb}^2 + v_{sc}^2 + 3v_{s0}^2 - 6v_{s0}^2\right] \\
&= -3\left[v_{sa}^2 + v_{sb}^2 + v_{sc}^2 - 3v_{s0}^2\right] \\
&= -3\left(\sum_{j=a,b,c} v_{sj}^2 - 3v_{s0}^2\right)
\end{aligned}
\tag{5.155}
$$

Calculation of cofactors of A_Δ

We need to calculate a_{c31}, a_{c32}, and a_{c33}. These are computed as following.

$$
\begin{aligned}
a_{c31} &= \left[(v_{sb} - v_{so}) + \beta(v_{sc} - v_{sa}) - (v_{sa} - v_{so}) - \beta(v_{sb} - v_{sc})\right] \\
&= \left[(v_{sb} - v_{sa}) - \beta(v_{sa} + v_{sb} - 2v_{sc})\right] \\
&= \left[(v_{sb} - v_{sa}) - \beta(v_{sa} + v_{sb} + v_{sc} - 3v_{sc})\right] \\
&= \left[(v_{sb} - v_{sa}) - \beta(3v_{s0} - 3v_{sc})\right] \\
&= -\left[v_{sab} - 3\beta(v_{sc} - v_{s0})\right]
\end{aligned}
$$

Similarly,

$$
\begin{aligned}
a_{c32} &= -\left[(v_{sb} - v_{so}) + \beta(v_{sc} - v_{sa}) - (v_{sc} - v_{so}) + \beta(v_{sa} - v_{sb})\right] \\
&= -\left[(v_{sb} - v_{sc}) + \beta(v_{sc} + v_{sb} - 2v_{sa})\right] \\
&= -\left[(v_{sb} - v_{sc}) + \beta(v_{sa} + v_{sb} + v_{sc} - 3v_{sa})\right] \\
&= -\left[(v_{sb} - v_{sc}) + \beta(3v_{s0} - 3v_{sa})\right] \\
&= -\left[v_{sbc} - 3\beta(v_{sa} - v_{s0})\right]
\end{aligned}
$$

and

$$
\begin{aligned}
a_{c33} &= [(v_{sa} - v_{so}) + \beta(v_{sb} - v_{sc}) - (v_{sc} - v_{so}) + \beta(v_{sa} - v_{sb})] \\
&= [(v_{sa} - v_{sc}) - \beta(v_{sa} + v_{sc} - 2v_{sb})] \\
&= [(v_{sa} - v_{sc}) - \beta(v_{sa} + v_{sb} + v_{sc} - 3v_{sb})] \\
&= [(v_{sa} - v_{sc}) - \beta(3v_{s0} - 3v_{sb})] \\
&= -[v_{sca} - 3\beta(v_{sb} - v_{s0})]
\end{aligned}
$$

Therefore, the solution of the equation is given by,

$$
\begin{aligned}
\begin{bmatrix} i_{lab} - i^*_{fab} \\ i_{lbc} - i^*_{fbc} \\ i_{lca} - i^*_{fca} \end{bmatrix} &= \frac{1}{A_\Delta} \begin{bmatrix} a_{c11} & a_{c12} & a_{c13} \\ a_{c21} & a_{c22} & a_{c23} \\ a_{c31} & a_{c32} & a_{c33} \end{bmatrix}^T \begin{bmatrix} 0 \\ 0 \\ P_{lavg} \end{bmatrix} \\
&= \frac{1}{A_\Delta} \begin{bmatrix} a_{c11} & a_{c21} & a_{c31} \\ a_{c12} & a_{c22} & a_{c32} \\ a_{c13} & a_{c23} & a_{c33} \end{bmatrix} \begin{bmatrix} 0 \\ 0 \\ P_{lavg} \end{bmatrix}
\end{aligned} \tag{5.156}
$$

From the above equation and substituting the values of cofactors obtained above, we get the following.

$$
\begin{aligned}
i_{lab} - i^*_{fab} &= \frac{a_{c31}}{|A_\Delta|} P_{lavg} \\
&= \frac{-[v_{sab} - 3\beta(v_{sc} - v_{s0})]}{-3\left[\sum_{j=a,b,c} v^2_{sj} - 3v^2_{s0}\right]} P_{lavg} \\
&= \frac{[v_{sab}/3 - \beta(v_{sc} - v_{s0})]}{\sum_{j=a,b,c} v^2_{sj} - 3v^2_{s0}} P_{lavg}
\end{aligned}
$$

From the above equation, the reference compensator current (i^*_{fab}) can be given as follows.

$$
i^*_{fab} = i_{lab} - \frac{v_{sab}/3 - \beta(v_{sc} - v_{s0})}{\sum_{j=a,b,c} v^2_{sj} - 3v^2_{s0}} P_{lavg} \tag{5.157}
$$

Similarly,

$$
i^*_{fbc} = i_{lbc} - \frac{v_{sbc}/3 - \beta(v_{sa} - v_{s0})}{\sum_{j=a,b,c} v^2_{sj} - 3v^2_{s0}} P_{lavg} \tag{5.158}
$$

and

$$
i^*_{fca} = i_{lca} - \frac{v_{sca}/3 - \beta(v_{sb} - v_{s0})}{\sum_{j=a,b,c} v^2_{sj} - 3v^2_{s0}} P_{lavg} \tag{5.159}
$$

When the source power factor is unity, $\beta = 0$ and for balanced source voltages (fundamental), $v_{s0} = 0$. Substituting these values in the above equations, we get,

$$
\begin{aligned}
i^*_{fab} &= i_{lab} - \frac{v_{sab}}{3\sum_{j=a,b,c} v^2_{sj}} P_{lavg} \\
i^*_{fbc} &= i_{lbc} - \frac{v_{sbc}}{3\sum_{j=a,b,c} v^2_{sj}} P_{lavg} \\
i^*_{fca} &= i_{lca} - \frac{v_{sca}}{3\sum_{j=a,b,c} v^2_{sj}} P_{lavg}
\end{aligned}
\tag{5.160}
$$

Earlier, it was established that,

$$
v^2_{sab} + v^2_{sbc} + v^2_{sca} = 3\left(\sum_{j=a,b,c} v^2_{sj} - 3v^2_{s0} \right)
\tag{5.161}
$$

Replacing $3\left(\sum_{j=a,b,c} v^2_{sj} - 3v^2_{s0}\right)$ in (5.157), (5.158) and (5.159), we get the following.

$$
\begin{aligned}
i^*_{fab} &= i_{lab} - \frac{v_{sab} - 3\beta\,(v_{sc} - v_{s0})}{\left(v^2_{sab} + v^2_{sbc} + v^2_{sca}\right)} P_{lavg} \\
i^*_{fbc} &= i_{lbc} - \frac{v_{sbc} - 3\beta\,(v_{sa} - v_{s0})}{\left(v^2_{sab} + v^2_{sbc} + v^2_{sca}\right)} P_{lavg} \\
i^*_{fca} &= i_{lca} - \frac{v_{sca} - 3\beta\,(v_{sb} - v_{s0})}{\left(v^2_{sab} + v^2_{sbc} + v^2_{sca}\right)} P_{lavg}
\end{aligned}
\tag{5.162}
$$

For unity power factor and balanced source voltages (fundamental), the reference compensator currents are given as follows.

$$
\begin{aligned}
i^*_{fab} &= i_{lab} - \frac{v_{sab}}{9V^2} P_{lavg} \\
i^*_{fbc} &= i_{lbc} - \frac{v_{sbc}}{9V^2} P_{lavg} \\
i^*_{fca} &= i_{lca} - \frac{v_{sca}}{9V^2} P_{lavg}
\end{aligned}
\tag{5.163}
$$

5.4.3 COMPENSATION FOR THREE-PHASE THREE-WIRE SYSTEM SUPPLYING A PASSIVE LOAD

If we simplify the above for passive load (resistance, inductance and capacitance), with balanced three-phase supply voltage and current phasors at the fundamental frequency, we can find out the nature of source currents and the equivalent impedance seen by the source. This is illustrated below.

$$
\begin{aligned}
\bar{I}_{sab} &= \bar{I}_{lab} - \bar{I}_{fab} = \frac{\bar{V}_{sab} P_{lavg}}{9V^2} \\
\bar{I}_{sbc} &= \bar{I}_{lbc} - \bar{I}_{fbc} = \frac{\bar{V}_{sbc} P_{lavg}}{9V^2} \\
\bar{I}_{sca} &= \bar{I}_{lca} - \bar{I}_{fca} = \frac{\bar{V}_{sca} P_{lavg}}{9V^2}
\end{aligned}
\tag{5.164}
$$

From (5.164) we can write,

$$
\begin{aligned}
\bar{I}_{sab} &= \frac{\bar{V}_{sab} P_{lavg}}{9V^2} = \frac{\bar{V}_{sab}}{9V^2} [\mathrm{Real}(\bar{V}_{sab}\bar{I}_{lab}^* + \bar{V}_{sbc}\bar{I}_{lbc}^* + \bar{V}_{sca}\bar{I}_{lca}^*)] \\
&= \frac{\bar{V}_{sab}}{9V^2} \left\{ \mathrm{Real}[\bar{V}_{sab}\bar{V}_{sab}^* (Y_l^{ab})^* + \bar{V}_{sbc}\bar{V}_{sbc}^* (Y_l^{bc})^* + \bar{V}_{sca}\bar{V}_{sca}^* (Y_l^{ca})^*] \right\} \\
&= \frac{\bar{V}_{sab}}{9V^2} 3V^2 \mathrm{Real}[(G_l^{ab} - jB_l^{ab}) + (G_l^{bc} - jB_l^{bc}) + (G_l^{ca} - jB_l^{ca})] \\
&= \frac{\bar{V}_{sab}}{3} (G_l^{ab} + G_l^{bc} + G_l^{ca})
\end{aligned}
$$

Similarly,

$$
\begin{aligned}
\bar{I}_{sbc} &= \frac{\bar{V}_{sbc}}{3} (G_l^{ab} + G_l^{bc} + G_l^{ca}) \\
\bar{I}_{sca} &= \frac{\bar{V}_{sca}}{3} (G_l^{ab} + G_l^{bc} + G_l^{ca})
\end{aligned}
\tag{5.165}
$$

The above equation also implies that,

$$
\begin{aligned}
\bar{I}_{sa} &= \bar{V}_{sa} (G_l^{ab} + G_l^{bc} + G_l^{ca}) \\
\bar{I}_{sb} &= \bar{V}_{sb} (G_l^{ab} + G_l^{bc} + G_l^{ca}) \\
\bar{I}_{sc} &= \bar{V}_{sc} (G_l^{ab} + G_l^{bc} + G_l^{ca})
\end{aligned}
\tag{5.166}
$$

From (5.165), it is clear that the three-phase source sees a balanced delta connected load with the resistance values, $R_{ab} = R_{bc} = R_{ca} = \frac{3}{G_{lab}+G_{lbc}+G_{lca}}$, as shown in Fig. 5.12(a). This represents an equivalent star-connected load with the resistance values of $R_a = R_b = R_c = \frac{1}{G_{lab}+G_{lbc}+G_{lca}}$ as given in (5.166) and illustrated in Fig. 5.12(b). This ensures that the compensated source currents are balanced and in phase with the respective phase voltage.

5.4.4 EQUIVALENCE BETWEEN THE PASSIVE AND ACTIVE LOAD COMPENSATION FOR THREE-PHASE THREE-WIRE SYSTEM

At this point, a query may arise in the reader's mind, what is the equivalence between the passive load compensation for a three-phase three-wire system (discussed

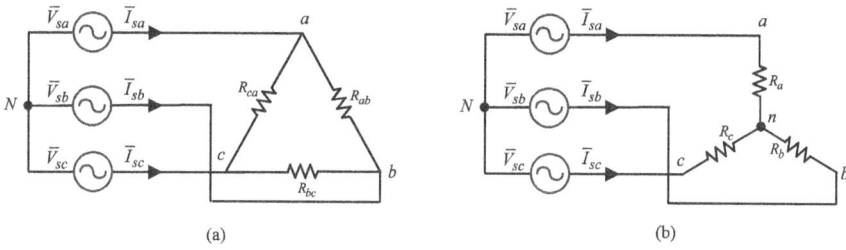

Figure 5.12 Three-phase three-wire compensated system (a) delta equivalent and (b) star equivalent

in Section 4.5 of the previous chapter) and the active load compensation scheme based on symmetrical component theory (discussed above)? While the end results are the same, i.e., the compensated source currents are balanced and in phase with the source voltages, what is their impact on the compensator rating and the losses in the system? Which scheme should be called better? These points need investigation. The two compensation schemes are shown in Fig. 5.13. Fig. 5.13(a) shows passive compensation using a purely reactive network as discussed in Section 4.6 and Fig. 5.13(b) illustrates the compensation scheme using an active filter using symmetrical component theory, discussed in the above section. These schemes thus can be referred to as passive and active compensation for three-phase three-wire system. The three-phase delta-connected load consists of unbalanced passive load, i.e., resistance, inductance, and capacitance. The three-phase supply voltages are balanced and sinusoidal.

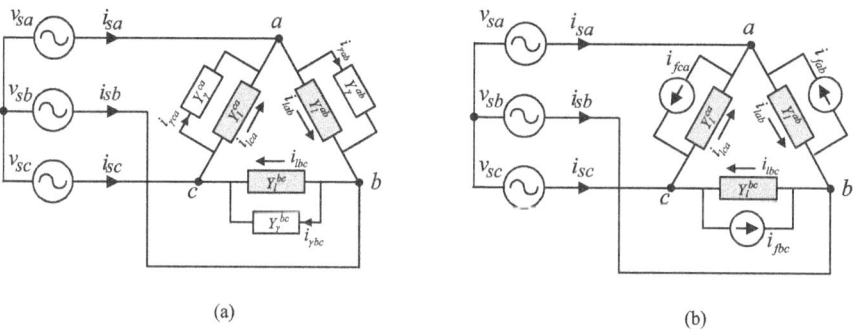

Figure 5.13 Compensation schemes (a) passive compensation and (b) active compensation

To compare the currents for the compensator, we shall denote $i'_{fab} = -i_{fab}$, $i'_{fbc} = -i_{fbc}$, $i'_{fca} = -i_{fca}$. Since only fundamental frequency components are present, the currents can be expressed in phasor form to simplify the analysis. Using passive compensation, the compensated source currents are given as follows.

$$
\begin{aligned}
\bar{I}_{sa\,passive}(t) &= (\bar{I}_{lab} + \bar{I}_{\gamma ab}) - (\bar{I}_{lca} + \bar{I}_{\gamma ca}) = (\bar{I}_{lab} - \bar{I}_{lca}) + (\bar{I}_{\gamma ab} - \bar{I}_{\gamma ca}) \\
\bar{I}_{sb\,passive}(t) &= (\bar{I}_{lbc} + \bar{I}_{\gamma bc}) - (\bar{I}_{lab} + \bar{I}_{\gamma ab}) = (\bar{I}_{lbc} - \bar{I}_{lab}) + (\bar{I}_{\gamma bc} - \bar{I}_{\gamma ab}) \\
\bar{I}_{sc\,passive}(t) &= (\bar{I}_{lca} + \bar{I}_{\gamma ca}) - (\bar{I}_{lbc} + \bar{I}_{\gamma bc}) = (\bar{I}_{lca} - \bar{I}_{lbc}) + (\bar{I}_{\gamma ca} - \bar{I}_{\gamma bc})
\end{aligned}
$$

$$(5.167)$$

Similarly for the active compensation scheme,

$$
\begin{aligned}
\bar{I}_{sa\,active}(t) &= (\bar{I}_{lab} + \bar{I}'_{fab}) - (\bar{I}_{lca} + \bar{I}'_{fca}) = (\bar{I}_{lab} - \bar{I}_{lca}) + (\bar{I}'_{fab} - \bar{I}'_{fca}) \\
\bar{I}_{sb\,active}(t) &= (\bar{I}_{lbc} + \bar{I}'_{fbc}) - (\bar{I}_{lab} + \bar{I}'_{fab}) = (\bar{I}_{lbc} - \bar{I}_{lab}) + (\bar{I}'_{fbc} - \bar{I}'_{fab}) \\
\bar{I}_{sc\,active}(t) &= (\bar{I}_{lca} + \bar{I}'_{fca}) - (\bar{I}_{lbc} + \bar{I}'_{fbc}) = (\bar{I}_{lca} - \bar{I}_{lbc}) + (\bar{I}'_{fca} - \bar{I}'_{fbc})
\end{aligned}
$$

$$(5.168)$$

In the previous section, we have proved that the source currents after compensation, using active filter are balanced and sinusoidal, i.e., $\bar{I}_{sa} = \bar{I}_{sa\,active}, \bar{I}_{sb} = \bar{I}_{sb\,active}$, and $\bar{I}_{sc} = \bar{I}_{sc\,active}$, through (5.165), illustrated in Fig. 5.12.

Let us verify the nature of source current for load compensation using purely reactive elements (passive compensation) in (5.167).

First consider the expression of $\bar{I}_{sa\,passive}$.

$$
\begin{aligned}
\bar{I}_{sa\,passive}(t) &= (\bar{I}_{lab} + \bar{I}_{\gamma ab}) - (\bar{I}_{lca} + \bar{I}_{\gamma ca}) = (\bar{I}_{lab} - \bar{I}_{lca}) + (\bar{I}_{\gamma ab} - \bar{I}_{\gamma ca}) \\
&= \bar{V}_{ab}(Y_l^{ab} + Y_\gamma^{ab}) - \bar{V}_{ca}(Y_l^{ca} + Y_\gamma^{ca})
\end{aligned}
$$

$$(5.169)$$

In the above expression,

$$
\begin{aligned}
\bar{V}_{ab} &= \sqrt{3}V\angle 30° = \frac{\sqrt{3}V}{2}(\sqrt{3} + j) \\
\bar{V}_{bc} &= \sqrt{3}V\angle -90° = -j\sqrt{3}V \\
\bar{V}_{ca} &= \sqrt{3}V\angle 150° = \frac{\sqrt{3}V}{2}(-\sqrt{3} + j) \\
Y_l^{ab} &= G_l^{ab} + jB_l^{ab} \\
Y_l^{bc} &= G_l^{bc} + jB_l^{bc} \\
Y_l^{ca} &= G_l^{ca} + jB_l^{ca} \\
Y_\gamma^{ab} &= jB_\gamma^{ab} = j\left(-B_l^{ab} + \frac{G_l^{ca} - G_l^{bc}}{\sqrt{3}}\right) \\
Y_\gamma^{bc} &= jB_\gamma^{bc} = j\left(-B_l^{bc} + \frac{G_l^{ab} - G_l^{ca}}{\sqrt{3}}\right) \\
Y_\gamma^{ca} &= jB_\gamma^{ca} = j\left(-B_l^{ca} + \frac{G_l^{bc} - G_l^{ab}}{\sqrt{3}}\right)
\end{aligned}
$$

Substituting the above values of $\overline{V}_{ab}$, $\overline{V}_{bc}$, $\overline{V}_{ca}$, Y_l^{ab}, Y_l^{bc}, Y_l^{ca}, Y_γ^{ab}, Y_γ^{bc}, and Y_γ^{ca} in (5.169) simplifies to,

$$
\begin{aligned}
\overline{I}_{sa\ passive} &= \overline{V}_{sa}(G_l^{ab} + G_l^{bc} + G_l^{ca}) \\
\overline{I}_{sb\ passive} &= \overline{V}_{sb}(G_l^{ab} + G_l^{bc} + G_l^{ca}) \\
\overline{I}_{sc\ passive} &= \overline{V}_{sc}(G_l^{ab} + G_l^{bc} + G_l^{ca})
\end{aligned}
\tag{5.170}
$$

Thus, the final compensated source currents using passive compensation, i.e., $\overline{I}_{sa\ passive}$, $\overline{I}_{sb\ passive}$, and $\overline{I}_{sc\ passive}$ given in (5.170) are same as the compensated source currents using active compensation, i.e., $\overline{I}_{sa\ active}$, $\overline{I}_{sb\ active}$, and $\overline{I}_{sc\ active}$, given in (5.166). Since the final compensated currents are same, does it mean that, the line compensator currents are also same for active ($\overline{I}'_{fa}$, $\overline{I}'_{fb}$, and $\overline{I}'_{fc}$) and passive ($\overline{I}_{\gamma a}$, $\overline{I}_{\gamma b}$, and $\overline{I}_{\gamma c}$) compensation? Let us find out. Comparing (5.167) and (5.168), we can write,

$$
\begin{aligned}
\overline{I}'_{fab} - \overline{I}'_{fca} &= \overline{I}_{\gamma ab} - \overline{I}_{\gamma ca} \implies \overline{I}'_{fa} = \overline{I}_{\gamma a} \\
\overline{I}'_{fbc} - \overline{I}'_{fab} &= \overline{I}_{\gamma bc} - \overline{I}_{\gamma ab} \implies \overline{I}'_{fb} = \overline{I}_{\gamma b} \\
\overline{I}'_{fca} - \overline{I}'_{fbc} &= \overline{I}_{\gamma ca} - \overline{I}_{\gamma bc} \implies \overline{I}'_{fc} = \overline{I}_{\gamma c}
\end{aligned}
\tag{5.171}
$$

The above equation (5.171) indicates that the net compensator currents, as seen from star connected voltage supply, are the same for active and passive compensation. Let us find out $\overline{I}'_{fab}, \overline{I}_{\gamma ab}, \overline{I}'_{fbc}, \overline{I}_{\gamma bc}$, and $\overline{I}'_{fca}, \overline{I}_{\gamma ca}$ and compare their values. Using (5.164), the current, $\overline{I}'_{fab}$ is given by,

$$
\overline{I}'_{fab} = \overline{I}_{sab} - \overline{I}_{lab}
\tag{5.172}
$$

From (5.165), $\overline{I}_{sab} = \frac{V_{sab}}{3}(G_l^{ab} + G_l^{bc} + G_l^{ca})$ and $\overline{I}_{lab} = \overline{V}_{sab}Y_l^{ab} = \overline{V}_{sab}(G_l^{ab} + jB_l^{ab})$. Thus from (5.172), $\overline{I}'_{fab}$ is simplified as following.

$$
\begin{aligned}
\overline{I}'_{fab} &= \overline{V}_{sab}\left(\frac{G_l^{ab} + G_l^{bc} + G_l^{ca}}{3} - G_l^{ab} - jB_l^{ab}\right) \\
&= \overline{V}_{sab}\left(\frac{-2G_l^{ab} + G_l^{bc} + G_l^{ca}}{3} - jB_l^{ab}\right)
\end{aligned}
\tag{5.173}
$$

Similarly,

$$
\overline{I}'_{fbc} = \overline{V}_{sbc}\left(\frac{G_l^{ab} - 2G_l^{bc} + G_l^{ca}}{3} - jB_l^{bc}\right)
$$

and,

$$
\overline{I}'_{fca} = \overline{V}_{sbc}\left(\frac{G_l^{ab} + G_l^{bc} - 2G_l^{ca}}{3} - jB_l^{ca}\right)
$$

Now we compute, $\bar{I}_{\gamma ab}, \bar{I}_{\gamma bc}$, and $\bar{I}_{\gamma ca}$, as given below.

$$\bar{I}_{\gamma ab} = \bar{V}_{sab} Y_{\gamma}^{ab} = \bar{V}_{sab} jB_{\gamma}^{ab} = \bar{V}_{sab} \left(-jB_l^{ab} + j\frac{G_l^{ca} - G_l^{bc}}{\sqrt{3}} \right) \qquad (5.174)$$

Similarly,

$$\bar{I}_{\gamma bc} = \bar{V}_{sbc} Y_{\gamma}^{bc} = \bar{V}_{sbc} jB_{\gamma}^{bc} = \bar{V}_{sbc} \left(-jB_l^{bc} + j\frac{G_l^{ab} - G_l^{ca}}{\sqrt{3}} \right)$$

and,

$$\bar{I}_{\gamma ca} = \bar{V}_{sca} Y_{\gamma}^{ca} = \bar{V}_{sca} jB_{\gamma}^{ca} = \bar{V}_{sca} \left(-jB_l^{ca} + j\frac{G_l^{bc} - G_l^{ab}}{\sqrt{3}} \right)$$

From (5.173) and (5.174), it is clear that,

$$\begin{aligned} \bar{I}'_{fab} &\neq \bar{I}_{\gamma ab} \\ \bar{I}'_{fbc} &\neq \bar{I}_{\gamma bc} \\ \bar{I}'_{fca} &\neq \bar{I}_{\gamma ca} \end{aligned} \qquad (5.175)$$

Nevertheless from (5.171),

$$\bar{I}_{\gamma ab} - \bar{I}'_{fab} = \bar{I}_{\gamma bc} - \bar{I}'_{fbc} = \bar{I}_{\gamma ca} - \bar{I}'_{fca} = K \text{ (say)} \qquad (5.176)$$

Then the relationship between the active phase filter currents $(\bar{I}'_{fab}, \bar{I}'_{fbc}, \bar{I}'_{fca})$ and passive filter phase currents $(\bar{I}_{\gamma ab}, \bar{I}_{\gamma bc}, \bar{I}_{\gamma ca})$ is given as,

$$\begin{aligned} \bar{I}'_{fab} &= \bar{I}_{\gamma ab} - K \\ \bar{I}'_{fbc} &= \bar{I}_{\gamma bc} - K \\ \bar{I}'_{fca} &= \bar{I}_{\gamma ca} - K \end{aligned} \qquad (5.177)$$

From (5.173) and (5.174), it can be found that,

$$K = \sqrt{3}V \left[\frac{G_l^{ab} - G_l^{ca}}{\sqrt{3}} + j\frac{G_l^{ab} - 2G_l^{bc} + G_l^{ca}}{3} \right] \qquad (5.178)$$

From load currents, $\bar{I}_{lab}, \bar{I}_{lbc}, \bar{I}_{lca}$ and passive compensator currents, $\bar{I}_{\gamma ab}, \bar{I}_{\gamma bc}, \bar{I}_{\gamma ca}$, it is found that,

$$\frac{\bar{I}_{lab} + \bar{I}_{lbc} + \bar{I}_{lca}}{3} + \frac{\bar{I}_{\gamma ab} + \bar{I}_{\gamma bc} + \bar{I}_{\gamma ca}}{3} = \bar{I}_{lab0} + \bar{I}_{\gamma ab0} = \bar{I}_{sab0} = K \qquad (5.179)$$

From (5.177) and (5.179),

$$\frac{\bar{I}_{lab} + \bar{I}_{lbc} + \bar{I}_{lca}}{3} + \frac{\bar{I}'_{fab} + \bar{I}'_{fbc} + \bar{I}'_{fca}}{3} = 0 \qquad (5.180)$$

In (5.179), the terms, $\bar{I}_{lab0}, \bar{I}_{\gamma ab0}$ are load and compensator circulating currents. Their sum, i.e., $\bar{I}_{sab0} = K$, is the net circulating current in the compensated delta network. This actually means that in the passive compensation scheme, the net circulating current exists in the compensated system. While from (5.180), it is evident that there is no circulating current in the active compensation scheme. This is the main difference between the active and passive compensation schemes for three-phase three-wire systems, although the source currents after compensation are the same for both schemes. For the requirement of zero circulating current after compensation, the active compensation scheme should be chosen.

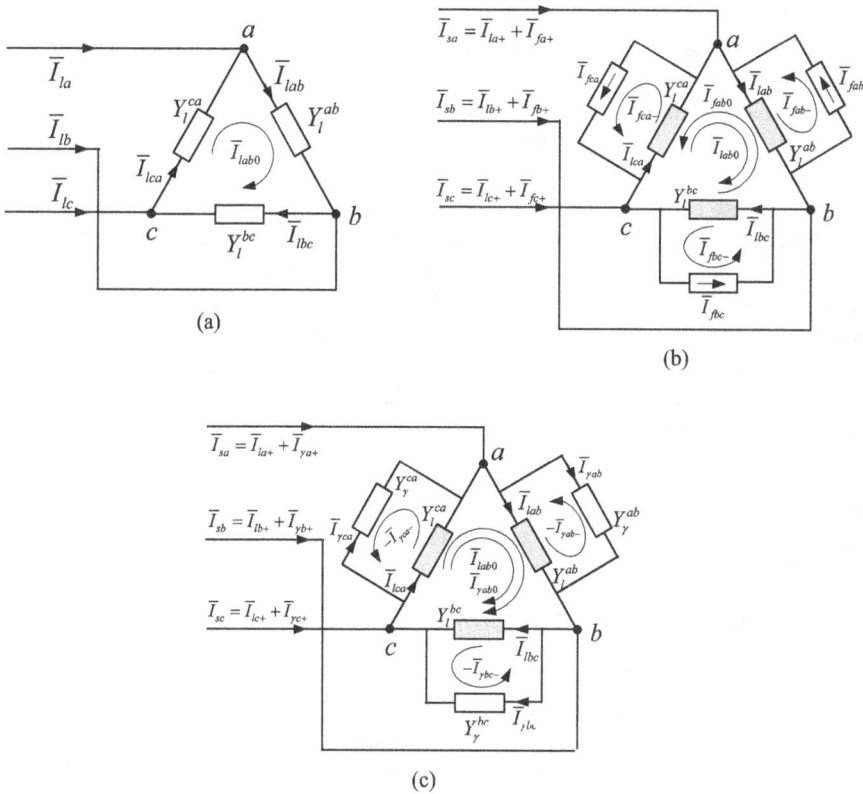

Figure 5.14 Compensation of delta connected load (a) without compensation, (b) active compensation, and (c) passive compensation

These situations are illustrated through circuits shown in Fig. 5.14(a)–(c). Fig. 5.14(a) shows a delta-connected load without compensation. Since the load is unbalanced, the phase currents, $i_{lab}, i_{lbc}, i_{lca}$ have positive, negative, and zero sequence currents, as expressed below.

$$\begin{aligned}
i_{lab} &= i_{lab+} + i_{lab-} + i_{lab0} \\
i_{lbc} &= i_{lbc+} + i_{lbc-} + i_{lbc0} \\
i_{lca} &= i_{lca+} + i_{lca-} + i_{lca0}
\end{aligned} \qquad (5.181)$$

The zero sequence current, $i_{lab0} = i_{lbc0} = i_{lca0}$, that flows in the delta structure, is also known as circulating current, as shown in Fig. 5.14(a). In the active compensation scheme shown in Fig. 5.14(b), the compensator generates an equal circulating current in the opposite direction, thus leading to zero circulating current in the delta. The active compensator also compensates negative sequence currents for each phase. The resultant phase current components are the summation of positive sequence components of load and the compensator, which makes line currents balanced with the unity power factor. The source sees the compensated load as it is, i.e., balanced resistive load.

In the case of the passive load compensation scheme, as illustrated in 5.14(c), the net circulating current is not zero. In fact, the load circulating current and passive compensator circulating current are not related and these will add up to some non-zero finite value. However, negative sequence components of the load currents are canceled by the compensator negative sequence currents. Due to this, the actual compensated load consists of three different impedance elements, however, when looked at from the source, it is seen as a balanced resistive load, similar to that of active compensation.

Example 5.6. Consider a three-phase balanced system shown in Fig. 5.11. The supply voltages are: $\overline{V}_a = 230\angle 0°$, $\overline{V}_b = 230\angle -120°$, $\overline{V}_c = 230\angle 120°$. The impedances of delta connected load are; $Z_{ab} = 5 + j12\,\Omega$, $Z_{bc} = 3 + j4\,\Omega$ and $Z_{ca} = 12 - j13\,\Omega$. Using instantaneous symmetrical component theory,

(a) Find the line currents (i_{la}, i_{lb}, i_{lc}) for the above delta connected load.

(b) Find reference filter currents ($i_{fab}, i_{fbc}, i_{fca}$) of an ideal compensator such that the source currents are balanced sinusoids and in phase with their respective phase voltages. Also, find out the time expressions for the source currents and check whether these are balanced and in phase with their respective voltages.

(c) Based on these reference filter currents, identify the nature and values of compensator emulated impedances, i.e., Z_{fab}, Z_{fbc}, Z_{fca}. Compare their values with those obtained using the theory load compensation technique discussed in Section 4.5.

(d) Compute the active and reactive powers of the load, compensator, and source. Verify that, $P_s = P_l$, $Q_s = 0$, $P_f = 0$, and $Q_f = Q_l$.

(e) Compare the above results with those of passive compensation (purely reactive network) for the system.

Solution:

(a) To compute the source currents, first line to line currents are calculated. These are given below.

$$\bar{I}_{lab} = \frac{\bar{V}_{sa} - \bar{V}_{sb}}{Z_{lab}} = \frac{\bar{V}_{sab}}{Z_{lab}} = \frac{398.37\angle 30°}{5 + j12} = 30.64\angle -37.38° \text{ A}$$

$$\bar{I}_{lbc} = \frac{\bar{V}_{sb} - \bar{V}_{sc}}{Z_{lbc}} = \frac{\bar{V}_{sbc}}{Z_{lbc}} = \frac{398.37\angle -90°}{3 + j4} = 79.67\angle -143.13° \text{ A}$$

$$\bar{I}_{lca} = \frac{\bar{V}_{sc} - \bar{V}_{sa}}{Z_{lca}} = \frac{\bar{V}_{sca}}{Z_{lca}} = \frac{398.37\angle 150°}{12 - j13} = 22.52\angle -162.71° \text{ A}$$

By applying KCL at load points, the line currents are calculated as below.

$$\bar{I}_{la} = \bar{I}_{lab} - \bar{I}_{lca} = 47.37\angle -14.56° \text{ A}$$
$$\bar{I}_{lb} = \bar{I}_{lbc} - \bar{I}_{lab} = 92.80\angle -161.66° \text{ A}$$
$$\bar{I}_{lc} = \bar{I}_{lca} - \bar{I}_{lbc} = 58.94\angle 44.22° \text{ A}$$

Therefore, the time expressions for the line currents are given as follows.

$$i_{la}(t) = 47.37\sqrt{2}\sin(\omega t - 14.56°) \text{ A}$$
$$i_{lb}(t) = 92.80\sqrt{2}\sin(\omega t - 161.66°) \text{ A}$$
$$i_{lc}(t) = 58.94\sqrt{2}\sin(\omega t + 44.22°) \text{ A}$$

It can be seen that the source currents are not balanced in magnitude as well as in phase angles and hence cause various reactive powers and power factors in different phases.

(b) Now to obtain balanced source currents at unity power factor, the filter currents $(i_{fab}, i_{fbc}, i_{fca})$ are computed using (5.163), as given below.

$$i_{fab}^* = i_{lab} - \frac{v_{sab}}{9V^2}P_{lavg}$$

$$i_{fbc}^* = i_{lbc} - \frac{v_{sbc}}{9V^2}P_{lavg}$$

$$i_{fca}^* = i_{lca} - \frac{v_{sca}}{9V^2}P_{lavg}$$

Since only fundamental frequency components are involved in computation, the above equation can be used in the phasor domain. Since the line currents and line to line voltages are known, the remaining term, P_{lavg} is calculated as follows.

$$P_{lavg} = \text{Real}\left(\bar{V}_{sab}\bar{I}_{lab}^* + \bar{V}_{sbc}\bar{I}_{lbc}^* + \bar{V}_{sca}\bar{I}_{lca}^*\right) = 29824 \text{ W}$$

Finally, reference filter currents are computed as follows.

$$\bar{I}_{fab} = \bar{I}_{lab} - \frac{\bar{V}_{sab}}{9V^2}P_{lavg} = 30.64\angle - 30.38° - \frac{398.37\angle 30° \times 29824}{9 \times 230^2}$$

$$= 31.2\angle -84.96°\,\text{A}$$

$$\bar{I}_{fbc} = \bar{I}_{lbc} - \frac{\bar{V}_{sbc}}{9V^2}P_{lavg} = 79.67\angle - 143.13° - \frac{398.37\angle - 90° \times 29824}{9 \times 230^2}$$

$$= 67.71\angle -160.28°\,\text{A}$$

$$\bar{I}_{fca} = \bar{I}_{lca} - \frac{\bar{V}_{sca}}{9V^2}P_{lavg} = 22.52\angle - 162.71° - \frac{398.37\angle 150° \times 29824}{9 \times 230^2}$$

$$= 19.17\angle -89.67°\,\text{A}$$

Therefore, the time expressions for the compensator currents are given as follows.

$$\begin{aligned}
i_{fab}(t) &= 31.2\sqrt{2}\sin(\omega t - 84.96°)\,\text{A}\\
i_{fbc}(t) &= 67.71\sqrt{2}\sin(\omega t - 160.28°)\,\text{A}\\
i_{fca}(t) &= 19.17\sqrt{2}\sin(\omega t - 89.67°)\,\text{A}
\end{aligned}$$

After compensation, the source currents are computed as follows.

$$\begin{aligned}
\bar{I}_{sa} &= (\bar{I}_{lab} - \bar{I}_{fab}) - (\bar{I}_{lca} - \bar{I}_{fca}) = 43.22\angle 0°\,\text{A}\\
\bar{I}_{sb} &= (\bar{I}_{lbc} - \bar{I}_{fbc}) - (\bar{I}_{lab} - \bar{I}_{fab}) = 43.22\angle - 120°\,\text{A}\\
\bar{I}_{sc} &= (\bar{I}_{lca} - \bar{I}_{fca}) - (\bar{I}_{lbc} - \bar{I}_{fbc}) = 43.22\angle 120°\,\text{A}
\end{aligned}$$

Therefore, the time expressions of source currents after compensation are given as follows.

$$\begin{aligned}
i_{sa}(t) &= 43.22\sqrt{2}\sin(\omega t)\,\text{A}\\
i_{sb}(t) &= 43.22\sqrt{2}\sin(\omega t - 120°)\,\text{A}\\
i_{sc}(t) &= 43.22\sqrt{2}\sin(\omega t + 120°)\,\text{A}
\end{aligned}$$

Thus, it is seen that the source currents after compensation becomes balanced and are in phase with their respective source voltages.

(c) Based on the line to line voltages and the filter currents, the compensator impedances Z_{fab}, Z_{fbc}, Z_{fca} are computed as follows.

$$\bar{Z}_{fab} = \frac{\bar{V}_{sab}}{-\bar{I}_{fab}} = \frac{398.37\angle 30°}{-31.2\angle - 84.96°} = 5.3885 - j11.5748\,\Omega$$

$$\bar{Z}_{fbc} = \frac{\bar{V}_{sbc}}{-\bar{I}_{fbc}} = \frac{398.37\angle - 90°}{-67.71\angle - 160.28°} = -1.9854 - j5.5382\,\Omega \quad (5.182)$$

$$\bar{Z}_{fca} = \frac{\bar{V}_{sca}}{-\bar{I}_{fca}} = \frac{398.37\angle 150°}{-19.17\angle - 89.67°} = 10.495 + j17.9359\,\Omega$$

As seen from the expressions, the compensator impedance Z_{fab} is a combination of a resistance and a capacitive reactance, Z_{fbc} has a negative resistance and a capacitive reactance, and Z_{fca} has a resistance and an inductive reactance. This negative resistance is realized using active power flow between the phases of the compensator.

Since this is a three-phase three-wire system, the three-phase unbalanced delta load can also be compensated using purely reactive components, as discussed in Section 4.6. Based on this, the compensator admittances are computed as follows.

$$Y_\gamma^{ab} = jB_\gamma^{ab} = j\left(-B_l^{ab} + \frac{G_l^{ca} - G_l^{bc}}{\sqrt{3}}\right) = j0.0238\, \mho$$

$$Y_\gamma^{bc} = jB_\gamma^{bc} = j\left(-B_l^{bc} + \frac{G_l^{ab} - G_l^{ca}}{\sqrt{3}}\right) = j0.1549\, \mho \qquad (5.183)$$

$$Y_\gamma^{ca} = jB_\gamma^{ca} = j\left(-B_l^{ca} + \frac{G_l^{bc} - G_l^{ab}}{\sqrt{3}}\right) = j0.0107\, \mho$$

Thus, the compensator impedances are as follows.

$$Z_{\gamma ab} = \frac{1}{Y_\gamma^{ab}} = -j41.91\,\Omega$$

$$Z_{\gamma bc} = \frac{1}{Y_\gamma^{bc}} = -j6.45\,\Omega \qquad (5.184)$$

$$Z_{\gamma ca} = \frac{1}{Y_\gamma^{ca}} = -j93.7\,\Omega$$

As it can be seen from (5.182) and (5.184), the nature of compensator impedances are not the same. In active filtering, the compensator emulates negative resistance such that the net real power of the active power filter is zero, nevertheless, real power flows in the phases of the active filter. In the case of the passive filter, the passive compensator impedances are purely reactive. However, in both cases, the compensated source currents are the same, i.e., balanced and in phase with their phase voltages.

(d) The active and reactive powers of the load, compensator, and source can be found as given below.

$$P_l = \text{Real}\left(\overline{V}_{sab}\overline{I}_{lab}^* + \overline{V}_{sbc}\overline{I}_{lbc}^* + \overline{V}_{sc}\overline{I}_{lca}^*\right) = 29824\,\text{W}$$

$$Q_l = \text{Imag}\left(\overline{V}_{sab}\overline{I}_{lab}^* + \overline{V}_{sbc}\overline{I}_{lbc}^* + \overline{V}_{sc}\overline{I}_{lca}^*\right) = 30069\,\text{VAr}$$

$$P_f = \text{Real}\left(\overline{V}_{sab}\overline{I}_{fab}^* + \overline{V}_{sbc}\overline{I}_{fbc}^* + \overline{V}_{sca}\overline{I}_{fca}^*\right) = 0\,\text{W}$$

$$Q_f = \text{Imag}\left(\overline{V}_{sab}\overline{I}_{fab}^* + \overline{V}_{sbc}\overline{I}_{fbc}^* + \overline{V}_{sca}\overline{I}_{fca}^*\right) = -30069\,\text{VAr}$$

$$P_s = \text{Real} \left(\overline{V}_{sa} \overline{I}_{sa}^* + \overline{V}_{sb} \overline{I}_{sb}^* + \overline{V}_{sc} \overline{I}_{sc}^* \right) = 29824 \, W$$

$$Q_s = \text{Imag} \left(\overline{V}_{sa} \overline{I}_{sa}^* + \overline{V}_{sb} \overline{I}_{sb}^* + \overline{V}_{sc} \overline{I}_{sc}^* \right) = 0 \, VAr$$

From the above, it is observed that after compensation, $P_s = P_l = 29824$ W, $Q_s = 0$ VAr, $P_f = 0$ W, $Q_f = Q_l = 30069$ VAr. This ensures that the source only supplies average load power while the compensator supplies reactive power of the load.

(e) Passive compensation method

Using (5.174), the passive compensator currents are computed as follows:

$$\overline{I}_{\gamma ab} = \overline{V}_{sab} Y_\gamma^{ab} = \overline{V}_{sab} \, jB_\gamma^{ab} = \overline{V}_{sab} \left(-jB_l^{ab} + j\frac{G_l^{ca} - G_l^{bc}}{\sqrt{3}} \right) = 9.5\angle 120° \, A$$

$$\overline{I}_{\gamma bc} = \overline{V}_{sbc} Y_\gamma^{bc} = \overline{V}_{sbc} \, jB_\gamma^{bc} = \overline{V}_{sbc} \left(-jB_l^{bc} + j\frac{G_l^{ab} - G_l^{ca}}{\sqrt{3}} \right) = 61.73\angle 0° \, A$$

$$\overline{I}_{\gamma ca} = \overline{V}_{sca} Y_\gamma^{ca} = \overline{V}_{sca} \, jB_\gamma^{ca} = \overline{V}_{sca} \left(-jB_l^{ca} + j\frac{G_l^{bc} - G_l^{ab}}{\sqrt{3}} \right) = 4.25\angle - 120° \, A$$

Therefore, the source currents are given as,

$$\overline{I}_{sa} = (\overline{I}_{lab} - \overline{I}_{\gamma ab}) - (\overline{I}_{lca} - \overline{I}_{\gamma ca}) = 43.22\angle 0° \, A$$

$$\overline{I}_{sb} = (\overline{I}_{lbc} - \overline{I}_{\gamma bc}) - (\overline{I}_{lab} - \overline{I}_{\gamma ab}) = 43.22\angle - 120° \, A$$

$$\overline{I}_{sc} = (\overline{I}_{lca} - \overline{I}_{\gamma ca}) - (\overline{I}_{lab} - \overline{I}_{\gamma ab}) = 43.22\angle 120° \, A$$

As can be seen from the above equation, the source currents are the same as that in the case of the active compensator. Nevertheless, the individual phase compensator currents are different. As discussed earlier, the difference between the two schemes accounts for the circulating current in the compensated delta circuit. The net circulating current after passive compensation is:

$$\frac{\overline{I}_{lab} + \overline{I}_{lbc} + \overline{I}_{lca}}{3} + \frac{\overline{I}_{\gamma ab} + \overline{I}_{\gamma bc} + \overline{I}_{\gamma ca}}{3} = 22.94\angle - 95.03° \, A$$

And the net circulating current for the active compensation method is zero as given in the following.

$$\frac{\overline{I}_{lab} + \overline{I}_{lbc} + \overline{I}_{lca}}{3} + \frac{\overline{I}'_{fab} + \overline{I}'_{fbc} + \overline{I}'_{fca}}{3} = 0$$

Further, it can be seen that,

$$\overline{I}_{\gamma ab} - \overline{I}'_{fab} = \overline{I}_{\gamma bc} - \overline{I}'_{fbc} = \overline{I}_{\gamma ca} - \overline{I}'_{fca} = 22.94\angle - 95.03° \, A$$

Therefore, the difference between the compensator currents for each phase is the difference between the circulating currents for the two compensation schemes. The passive compensator powers are:

$$P_\gamma = \text{Real} \left(\overline{V}_{sab} \overline{I}^*_{\gamma ab} + \overline{V}_{sbc} \overline{I}^*_{\gamma bc} + \overline{V}_{sca} \overline{I}^*_{\gamma ca} \right) = 0\,\text{W}$$

$$Q_\gamma = \text{Imag} \left(\overline{V}_{sab} \overline{I}^*_{\gamma ab} + \overline{V}_{sbc} \overline{I}^*_{\gamma bc} + \overline{V}_{sca} \overline{I}^*_{\gamma ca} \right) = 30069\,\text{VAr}$$

Due to the opposite direction of active filter currents to that of passive filter currents, $Q_\gamma = -Q_f$. The phase wise ratings of the active and passive compensators are:

$$S_{\gamma ab} = Q_{\gamma ab} = 3.78\,\text{kVA, capacitive}$$

$$S_{\gamma bc} = Q_{\gamma bc} = 24.59\,\text{kVA, capacitive}$$

$$S_{\gamma ca} = Q_{\gamma bc} = 1.69\,\text{kVA, capacitive}$$

and,

$$S_{fab} = 12.43\,\text{kVA}$$

$$S_{fbc} = 26.97\,\text{kVA}$$

$$S_{fca} = 7.64\,\text{kVA}$$

As evident from these values, the phase wise ratings of the active compensator are more as compared to those of the passive compensator. It is due to the additional circulating current that is compensated by the active compensator. However, this may not always be the case, as it depends upon the nature of the load.

5.5 SUMMARY

This chapter has primarily focused on control theories for the compensation of a three-phase load with unbalanced harmonic components. Two theories, i.e., instantaneous reactive power theory and instantaneous symmetrical component theory have been discussed in detail with illustrations. Some conceptual anomalies of the instantaneous reactive power theory are also discussed. A generic, neutral clamped voltage source converter topology is discussed and its state space model is developed to simulate three-phase compensated systems with active power filters on a digital platform. Various components of the three-phase compensated system, such as the computation of losses in the converter using dc bus voltage regulation, computation of average load power, and other power components, are described. A comparison between active and passive compensation schemes is brought out with mathematical and circuit descriptions for a three-phase three-wire system. A number of examples are presented to explain the concepts, followed by subjective and numerical problems.

5.6 PROBLEMS

P 5.1 Explain the concept of instantaneous reactive power (pq) theory. Based on this concept:

(a) Define instantaneous active, $p(t)$ and instantaneous reactive, $q(t)$ powers for a given three-phase four-wire system using $\alpha\beta0$ transformation.

(b) Show that power is constant in both abc and $\alpha\beta0$ frames of reference.

(c) Explain how you will extract reference compensator currents based on this theory.

(d) Knowing reference currents, how do you realize these currents using a voltage source inverter?

Assume terms and notations for three-phase system quantities and circuits to illustrate the above.

P 5.2 With the help of a figure, illustrate the power flows between the source, load, and compensator in $\alpha\beta0$ frame of reference.

P 5.3 Explain how you will extract the compensator reference currents using instantaneous reactive power theory (pq theory) with the help of equations and a schematic diagram. Also, derive the expression for these currents.

P 5.4 Develop state space model for the three-phase, four-wire compensated system. The compensator is realized using a neutral clamped voltage source inverter. How this state space model is solved using MATLAB program?

P 5.5 Explain how the dc link voltage supporting the voltage source inverter operation is regulated to reference value.

P 5.6 Discuss how the extracted reference compensator currents are synthesized using a neutral clamped voltage source inverter with a hysteresis current control scheme.

P 5.7 Describe the instantaneous symmetrical component theory to compensate for an unbalanced nonlinear load for the star as well as delta connected load.

P 5.8 Show that the locus of real versus imaginary part of instantaneous positive (or negative) sequence symmetrical components of balanced voltages at the fundamental frequency, is a circle. Discuss the applications of this in identifying the nature of voltage and current waveforms.

P 5.9 Consider a three-phase four-wire compensated system in the figure below. The three-phase voltages and load impedances are as follows.

$$
\begin{aligned}
v_{sa}(t) &= 230\sqrt{2}\sin\omega t \text{ V} \\
v_{sb}(t) &= 230\sqrt{2}\sin(\omega t - 120°)\text{ V} \\
v_{sc}(t) &= 230\sqrt{2}\sin(\omega t + 120°)\text{ V}
\end{aligned}
$$

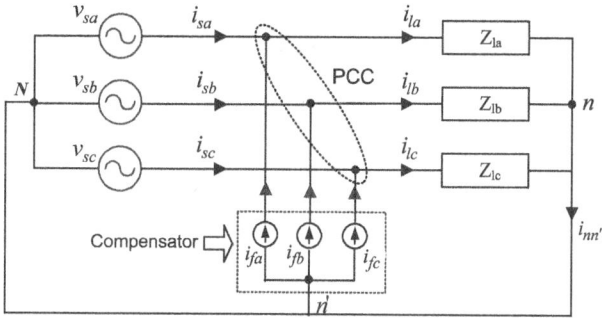

Figure 5.15 Related to problem P 5.9

Load impedances: $Z_{la} = 3 + j4\,\Omega$, $Z_{lb} = 5 - j12\,\Omega$, $Z_{lc} = 7 + j9\,\Omega$. Using instantaneous symmetrical component theory,

(a) Find reference filter currents (i_{fa}, i_{fb}, i_{fc}) of an ideal compensator such that source currents are balanced sinusoids and in phase with their respective phase voltages. Also, find the active and reactive powers of the compensator.

(b) Based on these reference filter currents, identify the nature and values of Z_{fa}, Z_{fb}, Z_{fc}. Are these practically realizable using passive impedances?

(c) If you realize the compensator using a voltage source inverter, explain for the actual (not ideal) compensator how expressions of reference filter currents are modified to maintain dc link voltage supported by dc storage capacitors to a constant reference value.

(d) While generating reference currents, which parameters play significant roles in deciding the dynamic performance of the compensator?

P 5.10 Refer to the above question to answer the following using instantaneous reactive power (pq) theory and verify answers.

(a) α, β and 0 components of three-phase supply voltages $(v_{s\alpha}, v_{s\beta}, v_{s0},)$ and load currents $(i_{l\alpha}, i_{l\beta}, i_{l0})$.

(b) The various components of instantaneous and average powers i.e., p_α, p_β, p_0, $q_{\alpha\beta}$.

(c) Verify these powers on the basis of their average values, i.e., $P_{abc} = P_\alpha + P_\beta + P_0 = P_{\alpha\beta 0}$ and $Q_{abc} = -Q_{\alpha\beta}$.

(d) Reference compensator currents in $\alpha\beta 0$ frame, i.e., $(i_{f\alpha}, i_{f\beta}, i_{f0})$ and abc frame, i.e., (i_{fa}, i_{fb}, i_{fc}) so that source supplies balanced currents from the source with unity power factor relationship with their voltages.

(e) Check whether the source currents are balanced and sinusoidal and in phase with their voltages.

(f) Check whether $p_s(t) = P_l$, $q_s(t) = 0$, $p_f(t) = 0$ and $q_f(t) = q_l$?

P 5.11 Consider a three-phase four-wire compensated system. The three-phase source voltages and load impedances are as follows.

$$
\begin{aligned}
v_{sa}(t) &= 230\sqrt{2}\sin\omega t \text{ V} \\
v_{sb}(t) &= 230\sqrt{2}\sin(\omega t - 2\pi/3) \text{ V} \\
v_{sc}(t) &= 230\sqrt{2}\sin(\omega t + 2\pi/3) \text{ V}
\end{aligned}
$$

Load impedances: $Z_{la} = 6 + j8\,\Omega$, $Z_{lb} = 3 - j4\,\Omega$, $Z_{lc} = 12 + j5\,\Omega$.

(a) Resolve system voltages and load currents into α, β and 0 components.

(b) Based on α, β, and 0 components of voltages and currents, compute the following.

   (i) Instantaneous α-axis (p_α), β-axis (p_β), $p_{\alpha\beta}$, 0-axis (p_0) instantaneous powers.

   (ii) Instantaneous reactive power, $q_{\alpha\beta}(t)$.

   (iii) Then, compute real power (P_l) and reactive power (Q_l) of the load.

(c) If an active power filter is employed, what are compensator currents in $\alpha\beta0$ and abc frames of reference?

(d) Compute the source currents (i_{sa}, i_{sb}, i_{sc}), after compensation.

P 5.12 Verify answers to parts (c) and (d) of the above problem, using instantaneous symmetrical component theory.

P 5.13 Consider a three-phase three-wire compensated system. The three-phase source voltages are balanced with a phase voltage of 230 V rms, 50 Hz and supplies three-phase unbalanced delta load with $Z_{lab} = 10 + j8\,\Omega$, $Z_{lbc} = 12 + j5\,\Omega$, $Z_{lca} = 12 - j8\,\Omega$.

(a) Draw the schematic of the above system.

(b) Compute reference compensator phase currents, $i_{fab}, i_{fbc}, i_{fca}$, using instantaneous symmetrical component theory, considering an ideal compensator such that source currents are balanced sinusoid and in phase with their respective phase voltages.

(c) Compute source currents after compensation.

(d) Find active and reactive powers of the compensator and also source active and reactive, and load active and reactive powers. Establish a relationship between these powers.

(e) Compare it with a compensation scheme with a purely reactive network.

P 5.14 Consider the delta-star connected ideal transformer with 1:1 turn ratio shown in Fig. 5.16. On the secondary side, a load of R ohms is connected between the phase-a and neutral n.

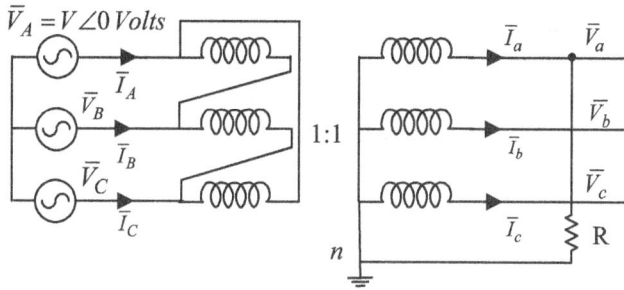

Figure 5.16 Related to problem P 5.14

Assuming balanced source voltages, compute the following.

(a) Time domain expressions of currents in each phase on both primary and secondary sides.

(b) Does the load require reactive power from the source? If any, find its value. Also, compute the reactive power on each phase of either side of the transformer.

(c) Also, determine active and reactive powers on each phase and overall active power on either side of the transformer.

(d) If you have a similar arrangement with a balanced load and the same output power, comment upon the rating of line conductors, and transformers.

P 5.15 In the above question, the primary and secondaries are interchanged and the load is connected between phase-a and b, as shown in Fig. 5.9. Answer points (a)–(d) in the question.

P 5.16 Consider a three-phase balanced system shown in Fig. 5.11. The supply voltages are: $\overline{V}_a = 230\angle0°$, $\overline{V}_b = 230\angle-120°$, $\overline{V}_c = 230\angle120°$. The impedances of delta connected load are, $Z_{ab} = 3 + j4\,\Omega$, $Z_{bc} = 5 + j10\,\Omega$ and $Z_{ca} = 5 - j13\,\Omega$. Using instantaneous symmetrical component theory,

(a) Find the line currents (i_{la}, i_{lb}, i_{lc}) for the above delta connected load.

(b) Find reference filter currents (i_{fab}, i_{fbc}, i_{fca}) of an ideal compensator such that source currents are balanced sinusoids and in phase with their respective phase voltages. Also, find out the time expressions of the source currents after compensation. Check whether these are balanced and in phase with their respective voltages.

(c) Based on these reference filter currents, identify the nature and values of Z_{fab}, Z_{fbc}, Z_{fca}. Verify these values obtained using the theory of load compensation technique discussed in Section 4.5 and comment upon the values.

(d) Compute the active and reactive powers of the load, compensator, and source. Verify that, $P_s = P_l$, $Q_s = 0$, $P_f = 0$, and $Q_f = Q_l$.

(e) Compare the above results with those of passive compensation (purely reactive network) for the system.

P 5.17 Consider a three-phase four-wire compensated system as shown in Fig. 5.15. The three-phase source voltages and load impedances are as follows.

Supply voltages: $v_{sa} = 220\sqrt{2}\sin\omega t$, $v_{sb} = 150\sqrt{2}\sin(\omega t - 150°)$
$$v_{sc} = 300\sqrt{2}\sin(\omega t + 100°) \text{ V}$$
Load impedances: $Z_{la} = 12 + j5\,\Omega$, $Z_{lb} = 3 - j4\,\Omega$, $Z_{lc} = 25 - j7\,\Omega$.

(a) Using instantaneous symmetrical component theory, find reference filter currents $(i_{fa}^*, i_{fb}^*, i_{fc}^*)$ in time domain considering an ideal compensator such that source currents are balanced sinusoids and in phase with their respective positive sequence fundamental phase voltages, while source supplies the same load power.

(b) Find active (P_f) and reactive (Q_f) powers of the compensator and also source active (P_s) and reactive (Q_s) and load active (P_l) and reactive (Q_l) powers. Check whether, $P_f = 0$ and $Q_f = -Q_l$ and $P_s = P_l$, $Q_s = 0$.

[Hint: Since the three-phase voltages are unbalanced at the fundamental frequency, the compensated source currents cannot be balanced to meet the condition, $v_{sa}i_{sa} + v_{sb}i_{sb} + v_{sc}i_{sc} = P_{lavg}$ (a constant value). To meet this condition with balanced fundamental source currents, fundamental positive sequence voltages should be computed to be used in the algorithm to find reference filter currents].

P 5.18 Consider a three-phase four-wire compensated system as shown (refer Fig. 5.15). Due to some abnormal condition, phase-c supply voltage becomes zero. Thus, the three-phase unbalanced voltages are expressed as follows.

Supply voltages: $v_{sa} = 220\sqrt{2}\sin\omega t$, $v_{sb} = 220\sqrt{2}\sin(\omega t - 150°)$, and $v_{sc} = 0$ V.
The load impedances are: $Z_{la} = 9 + j3$, $Z_{lb} = 6 - j8$, $Z_{lc} = 5 + j12\,\Omega$.

(a) Find reference filter currents (i_{fa}, i_{fb}, i_{fc}) of an ideal compensator such that source currents (i_{sa}, i_{sb}, i_{sc}) are balanced sinusoids and in phase with their respective fundamental positive sequence phase voltages, while source supplies the same load power. Also, compute compensated source currents.

(b) With the above compensation scheme, compute load, compensator, and source active (P) and reactive (Q) powers comment upon the values of these powers. Will the instantaneous power from the supply be constant (i.e., P_{lavg})?

REFERENCES

1. N. Mohan, H. A. Peterson, G. R. D. W. F. Long, and J. J. Vithayathil, "Active filters for ac harmonic suppression," *IEEE/PES Winter Meeting, New York City, USA*, pp. 529–535, Jan./Feb. 1977.

2. L. Gyugyi and E. C. Strycula, "Active AC Power Filters," *IEEE-IAS Annual Meeting Record*, pp. 529–535, 1976.

3. S. Bhattacharya and D. Divan, "Active filter for utility interface of industrial load," *IEEE Transactions on Industry Application*, vol. 32, no. 6, pp. 1312–1322, Nov./Dec. 1996.

4. S. Bhattacharya and D. Divan, "Active filter for utility interface of industrial load," *Proceedings IEEE Conf. PEDES '98, New Delhi*, pp. 1079–1084, Jan. 1998.

5. H. Sasaki and T. Machida, "A new method to eliminate ac harmonic currents by magnetic flux compensation considerations on basic design," *IEEE Transactions on Power Apparatus and Systems*, vol. PAS-90, no. 5, pp. 2009–2019, 1971.

6. H. Akagi, Y. Kanazawa, and A. Nabae, "Instantaneous reactive power compensators comprising switching devices without energy storage components," *IEEE Transactions on Industry Applications*, vol. IA-20, no. 3, pp. 625–630, 1984.

7. A. Ghosh and A. Joshi, "A new approach to load balancing and power factor correction in power distribution system," *IEEE Transactions on Power Delivery*, vol. 15, no. 1, pp. 417–422, Jan. 2000.

8. A. Ghosh and A. Joshi, "The use of instantaneous symmetrical components for balancing a delta connected load and power factor correction," *Electric Power System Research*, vol. 54, no. 1, pp. 67–74, April 2000.

9. R. A. Otto, T. H. Putman, and L. Gyugyi, "Principles and applications of static, thyristor-controlled shunt compensators," *IEEE Transactions on Power Apparatus and Systems*, vol. PAS-97, no. 5, pp. 1935–1945, 1978.

10. L. Gyugyi, "Reactive power generation and control by thyristor circuits," *IEEE Transactions on Industry Applications*, vol. IA-15, no. 5, pp. 521–532, 1979.

11. A. Ghosh and A. Joshi, "A new approach to load balancing and power factor correction in power distribution system," *IEEE Transactions on Power Delivery*, vol. 15, no. 1, pp. 417–422, 2000.

12. A. Ghosh and A. Joshi, "A new method for load balancing and power factor correction using instantaneous symmetrical components," *IEEE Power Engineering Review*, vol. 18, no. 9, pp. 60–61, 1998.

13. A. Ghosh and A. Joshi, "The use of instantaneous symmetrical components for balancing a delta connected load and power factor correction," *Electric Power Systems Research*, vol. 54, no. 1, pp. 67–74, 2000.

14. A. Ghosh and G. Ledwich, *Power Quality Enhancement Using Custom Power Devices*, ser. Power Electronics and Power Systems. Springer US, 2012.

15. E. Watanabe, R. Stephan, and M. Aredes, "New concepts of instantaneous active and reactive powers in electrical systems with generic loads," *IEEE Transactions on Power Delivery*, vol. 8, no. 2, pp. 697–703, 1993.

16. M. Aredes and E. Watanabe, "New control algorithms for series and shunt three-phase four-wire active power filters," *IEEE Transactions on Power Delivery*, vol. 10, no. 3, pp. 1649–1656, 1995.

17. T. Furuhashi, S. Okuma, and Y. Uchikawa, "A study on the theory of instantaneous reactive power," *IEEE Transactions on Industrial Electronics*, vol. 37, no. 1, pp. 86–90, 1990.

18. M. Aredes, J. Hafner, and K. Heumann, "Three-phase four-wire shunt active filter control strategies," *IEEE Transactions on Power Electronics*, vol. 12, no. 2, pp. 311–318, 1997.

19. F. Peng, H. Akagi, and A. Nabae, "A new approach to harmonic compensation in power systems-a combined system of shunt passive and series active filters," *IEEE Transactions on Industry Applications*, vol. 26, no. 6, pp. 983–990, 1990.

20. H. Akagi, "Trends in active power line conditioners," *IEEE Transactions on Power Electronics*, vol. 9, no. 3, pp. 263–268, 1994.

21. H. Fujita and H. Akagi, "A practical approach to harmonic compensation in power systems-series connection of passive and active filters," *IEEE Transactions on Industry Applications*, vol. 27, no. 6, pp. 1020–1025, 1991.

22. Mahesh K. Mishra, A. Ghosh, and A. Joshi, "A new STATCOM topology to compensate loads containing ac and dc components," in *IEEE Power Engineering Society Winter Meeting. Conference Proceedings*, vol. 4, 2000, pp. 2636–2641.

23. V. George and Mahesh K. Mishra, "DSTATCOM topologies for three phase high power applications," *International Journal of Power Electronics*, vol. 2, no. 2, pp. 107–124, 2010.

24. I. J. Nagrath and M. Gopal, *Control Systems Engineering*. New Age International Pvt., New Delhi, 2005.

25. Mahesh K. Mishra and K. Karthikeyan, "An investigation on design and switching dynamics of a voltage source inverter to compensate unbalanced and non-linear loads," *IEEE Transactions on Industrial Electronics*, vol. 56, no. 8, pp. 2802–2810, 2009.

26. L. Czarnecki, "On some misinterpretations of the instantaneous reactive power pq theory," *IEEE Transactions on Power Electronics*, vol. 19, no. 3, pp. 828–836, 2004.

6 Voltage Compensation Using Dynamic Voltage Restorer

6.1 INTRODUCTION

Power system should ensure good quality of electric power supply, which means voltage and current waveforms should be balanced and sinusoidal. Furthermore, the voltage levels on the system should be within reasonable limits, such as $\pm 10\%$ of their rated value and the frequency of grid supply should be normally within limits of $\pm 1\%$ of the nominal value, depending upon the grid codes of the country [1]–[5]. If the voltage and frequency are more or less than their pre-specified value, the performance of the equipment is compromised. In case of low voltages, picture on television starts rolling, the torque of induction motor reduces to the square of voltage and therefore there is a need for voltage compensation. The frequency level control is achieved through governor dynamics of the synchronous generators in the power system. However, grid frequency control is not part of discussion in this book. Voltage related disturbances and variations are more common in power distribution network, which arise due to various reasons such as voltage drops in the power network, different types of faults in the network, energization of large loads and capacitors bank, switching off heavy loads, loose connections in wiring, switching transients, etc. These variations in the voltage are termed and quantified as voltage sag, swell, interruptions, fluctuations, unbalance, waveform distortion, etc. The detailed descriptions of these terms were given in the Chapter 1.

One study reveals that approximately 60–70% of disturbances in power system are due to single-line to ground faults, 11% due to line-line faults and 6% due to three-phase symmetrical faults. Due to finite impedance of power lines, distribution feeders and transformers, these faults get translated into voltage sag, swell or interruption at the point of common coupling, where various loads are connected [6]–[9].

To compensate voltage, two broad categories of voltage regulating devices such as passive and active are used. Each of these devices has either series or shunt configuration. In this chapter we shall focus on series active power filter, which is termed as Dynamic Voltage Restorer (DVR). This is one of the variants of voltage source converter based compensating device. Various aspects of DVR such as topology, analysis and design of series active voltage compensator are described in [10]–[14]. In this chapter, we shall mainly focus on voltage variation, specially short term variations, such as voltage sag, swell and their mitigation using a series active power filter, i.e., DVR.

DOI: 10.1201/9781032617305-6

6.2   METHODS TO REGULATE VOLTAGE

In order to regulate the load bus voltage within allowable limits, various compensating devices such as listed below can be used.

1. Shunt capacitor

2. Series capacitor

3. Synchronous capacitor

4. Tap changing transformer

5. Booster transformer

6. Static synchronous series capacitor

7. Dynamic voltage restorer

The first six methods are normally employed at transmission level while the DVR is used in power distribution network to protect any voltage variation at the load bus connected to the sensitive and critical electrical units. The DVR is a series connected custom power device used to mitigate the voltage unbalance, sags, swells, harmonics and any abrupt changes due to abnormal conditions in the system. In the following section, dynamic voltage restorer will be described in detail.

6.3   DYNAMIC VOLTAGE RESTORER (DVR)

A dynamic voltage restorer is a solid-state inverter based on injection of voltage in series with a power distribution system [15]–[18]. A single-line diagram of a DVR connected to a power distribution system is shown in Fig. 6.1. The dc side of the DVR is connected to an energy source or an energy storage device, while its ac side is connected to the distribution feeder by an interfacing transformer. In the figure, $v_s(t)$ represents the supply voltage, $v_t(t)$ represents the terminal voltage and $v_l(t)$ represents the load voltage. The inductance, L_s in series with resistance, R_s represents the feeder impedance of the line. Since DVR is a series connected device, the source current, $i_s(t)$ is the same as the load current, $i_l(t)$. Also, note in the figure, $v_f(t)$ is

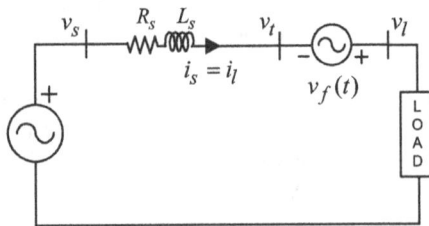

Figure 6.1   A single-line diagram of DVR compensated system

DVR injected voltage in series with the line such that the load voltage is maintained sinusoidal at the nominal value. The three-phase DVR compensated system is shown in Fig. 6.2. It is assumed that the transmission line has the same impedance in all three phases. A DVR unit which is represented in Figs. 6.1 and 6.2, have following components, as shown in Fig. 6.3 [19]– [21].

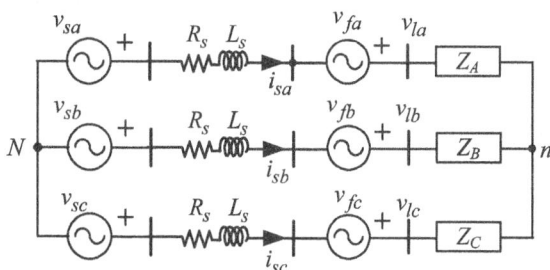

Figure 6.2   A three-phase DVR compensated system

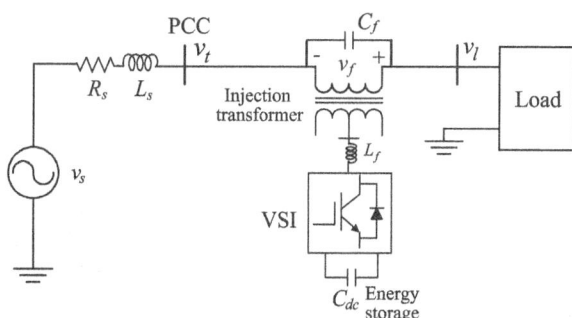

Figure 6.3   Schematic diagram of a DVR based compensation in a distribution system

1. Voltage source inverter (VSI)

2. Filter capacitors and inductors

3. Injection transformer

4. DC storage system

Some other important issues, i.e., how much voltage should be injected in series using the appropriate algorithm, choice of suitable power converter topology to synthesize voltage, and design of filter capacitor and inductor components, have to be addressed while designing the DVR unit.

6.4 OPERATING PRINCIPLE OF DVR

Consider a DVR compensated single-phase system as shown in Fig. 6.1. Let us assume that the source voltage is 1.0 p.u. and we want to regulate the load voltage to 1.0 p.u. Let us denote the phase angle between $\overline{V}_s$ and $\overline{V}_l$ as δ. In this analysis, harmonics are not considered. Further, during DVR operation, minimal real power is required to counter losses in the inverter and the filter components. These losses, for the time being, are considered to be zero. Also, in this analysis, only reactive power exchange is considered for the DVR operation. This condition implies that the phase difference between $\overline{V}_f$ and $\overline{I}_s$ should be $90°$. Let us first consider a general case to understand the concept.

The DVR equivalent circuit with fundamental voltages and current is shown in Fig. 6.1. Applying Kirchhoff's voltage law in the circuit,

$$\begin{aligned} \overline{V}_s + \overline{V}_f &= \overline{I}_s(R_s + jX_s) + \overline{V}_l \\ \overline{V}_s + \overline{V}_f &= \overline{I}_s Z_s + \overline{V}_l. \end{aligned} \tag{6.1}$$

Note that in above circuit $\overline{I}_s = \overline{I}_l = \overline{I}$. The load voltage $\overline{V}_l$ can be written in terms of load current and load impedance as given below.

$$\overline{V}_s + \overline{V}_f = \overline{I}(Z_s + Z_l) \tag{6.2}$$

In the above equation, $\overline{V}_l = \overline{I}Z_l = \overline{I}(R_l + jX_l)$. Using (6.1), the source voltage can be expressed as in the following.

$$\overline{V}_s = \overline{V}_l + \overline{I}R_s - (\overline{V}_f - j\overline{I}X_s) \tag{6.3}$$

In the above equation, the DVR voltage, $\overline{V}_f$ is associated with the term $j\overline{I}X_s$ to indicate that the $\overline{V}_f$ is along phasor $j\overline{I}X_s$. This is required for the DVR operation without involvement of real power. Substituting $\overline{I} = \frac{\overline{V}_l}{Z_l}$ in (6.2), the relationship between load, source, and DVR voltages can be expressed as below.

$$\overline{V}_l = \left(\frac{\overline{V}_s + \overline{V}_f}{Z_s + Z_l}\right) Z_l \tag{6.4}$$

The equation indicates that the load voltage can be controlled using appropriate DVR voltage, $\overline{V}_f$. The following example illustrates the DVR compensation under various conditions of load and feeder impedances.

Example 6.1. Let us apply the condition to maintain load voltage same as source voltage, i.e., $\overline{V}_l = \overline{V}_s$ and discuss the feasibility of the DVR voltage in series with the line as shown in Fig. 6.1. Consider the following cases.

(a) Line resistance is negligible with $Z_s = j0.25$ p.u. and $Z_l = 0.5 + j0.25$ p.u.

(b) When the line resistance is not negligible, i.e., $Z_s = 0.45 + j0.25$ p.u. and $Z_l = 0.5 + j0.25$ p.u.

Solution:

(a) When line resistance is negligible

The above condition implies that, $R_s = 0$. Without DVR, the load terminal voltage $\overline{V}_l$ can be given as follows.

$$\overline{I}_l = \frac{\overline{V}_s}{Z_l + Z_s} = \frac{1.0 \angle 0°}{0.5 + j0.5} = 1.0 - j1.0 = 1.4142 \angle -45° \text{ p.u.}$$

Therefore the load voltage is given as following. $\overline{V}_l = Z_l \overline{I}_l = 0.5590 \angle 26.56° \times 1.4142 \angle -45° = 0.7906 \angle -18.43°$ p.u. This is illustrated in Fig. 6.4(a). Thus, the load voltage has reduced by 21%. Now it is desired to maintain the load voltage same as the supply voltage in magnitude and phase angle. Thus, substituting $\overline{V}_s = \overline{V}_l$ in equation (6.1), we get,

$$\overline{V}_s + \overline{V}_f = \overline{I}(R_s + jX_s) + \overline{V}_l$$
$$\Rightarrow \quad \overline{V}_f = \overline{I}(R_s + jX_s)$$
$$\overline{V}_f = \frac{jX_s}{Z_l} \overline{V}_l, \quad \text{since,} \quad R_s = 0 \quad \text{and} \quad \overline{I} = \overline{V}_l / Z_l$$

Neglecting the resistance part of the feeder impedance, $Z_s = j0.25$, the DVR voltage can be computed as above, i.e.,

$$\overline{V}_f = \frac{j0.25}{0.5 + j0.25} 1.0 \angle 0° \quad \text{for } \overline{V}_l = 1.0 \angle 0°$$
$$= 0.4472 \angle 63.43° \text{ p.u.}$$

From the above, the line current is computed as follows.

$$\overline{I}_s = \frac{\overline{V}_l}{Z_l} = \frac{1.0 \angle 0°}{0.5590 \angle 26.56°} = 1.7889 \angle -26.56° \text{ p.u..}$$

It is to be noted that, although $\overline{V}_s = \overline{V}_l = 1.0 \angle 0°$ p.u., it does not imply that power does not flow from source to the load. In fact the total effective source voltage is $\overline{V}_s' = \overline{V}_s + \overline{V}_f = 1.2649 \angle 18.8°$ p.u. Therefore, the effective source voltage is leading the load voltage by an angle of 18.43°. This ensures the power flow from the source to the load. This is illustrated by drawing a phasor diagram in Fig. 6.4(b).

(b) When line resistance is not negligible

For this case, $Z_l = R_l + jX_l = 0.5 + j0.25$ p.u. Substituting $\overline{V}_s = \overline{V}_l$ in (6.1), we get the following.

$$\overline{V}_f = (R_s + jX_s)\overline{I} = R_s \overline{I}_s + jX_s \overline{I}_s.$$

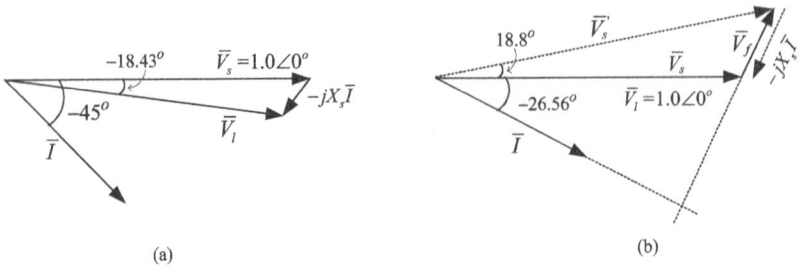

Figure 6.4 Load voltage (a) without DVR and (b) with DVR

From the above equation, it is indicated that DVR voltage has two components, one is in phase with $\bar{I}_s$ and the other is in phase quadrature with $\bar{I}_s$. This implies that for a finite value of feeder resistance, it is not possible to maintain $\bar{V}_l = \bar{V}_s$ without active power supplied from the DVR to the load. This is due to the presence of in-phase component of the DVR voltage in the above equation. This is illustrated in the phasor diagram given in Fig. 6.5. It is to be noted that $\bar{V}_l = \bar{V}_s$ is not a necessary and useful condition to regulate the load voltage. If the amplitude of load voltage equals to the supply voltage, i.e., $V_l = V_s$, the objective of the voltage compensation is achieved without real power requirement by the DVR, even if the feeder resistance is non-zero. These aspects of the DVR will be discussed in the following section.

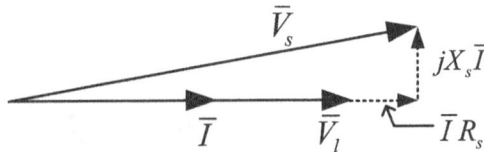

Figure 6.5 When the load is purely resistive

6.4.1 GENERAL CASE

In general, it is desired to maintain the magnitude of the load voltage equal to the source voltage i.e., 1.0 p.u. The voltage equation, in general, relating the source, load and DVR has been expressed in (6.3) and is given below.

$$\bar{V}_s = \bar{V}_l + \bar{I}R_s - (\bar{V}_f - j\bar{I}X_s)$$

The above equation is illustrated using the phasor diagram description in Fig. 6.6 given below. Three cases of voltage compensation are discussed below.

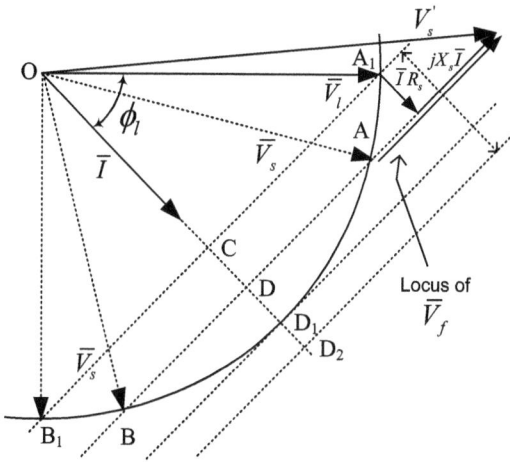

Figure 6.6 Compensation using DVR

Case 1: When $IR_s(=CD) < CD_1$

For this case, it is always possible to maintain the load voltage same as the source voltage, i.e., $V_l = V_s$. The DVR is expected to supply a wide range of reactive power to meet this condition. When IR_s is quite smaller than CD_1, the above condition can be met by supplying less reactive power from the DVR. For this condition, there are two solutions. Graphically, these solutions are represented by points A and B in Fig. 6.6.

Case 2: When $IR_s(=CD) > CD_1$

For this condition, it is not possible to meet $V_l = V_s$. This is shown by lines passing through points between D_1 and D_2. This may take place due to the higher feeder resistance or high current, thus making the product of IR_s relatively large.

Case 3: When $IR_s(=CD) = CD_1$

This is a limiting case of compensation to obtain $V_s = V_l$. This condition is now satisfied at only one point when $CD_1 = IR_s$. This is indicated by point D_1 in Fig. 6.6. Now let us set the following objective for the load compensation.

$$V_l = V_s = V = 1.0 \text{p.u.} \tag{6.5}$$

From Fig. 6.6, $OC = V \cos \phi_l = \cos \phi_l$. Therefore, $CD_1 = OD_1 - OC = V(1 - \cos \phi_l) = (1 - \cos \phi_l)$ p.u.. In order to meet the condition given by (6.5), the following must be satisfied.

$$IR_s \leq V(1 - \cos \phi_l) \tag{6.6}$$

The above implies that

$$R_s \leq \frac{V(1-\cos\phi_l)}{I} \tag{6.7}$$

$$\text{or} \quad I \leq \frac{V(1-\cos\phi_l)}{R_s}. \tag{6.8}$$

Thus, it is observed that for a given power factor, the DVR characteristics can be obtained by varying R_s and computing I or vice-versa for a given value of ϕ_l, as described below. Let us consider three values $R_s = 0.04$ p.u., $R_s = 0.1$ p.u. and $R_s = 0.4$ p.u. For these values of feeder resistance, the line currents are expressed as follows using (6.8).

$$
\begin{aligned}
I &= 25(1-\cos\phi_l)\,\text{p.u.} \quad \text{for } R_s = 0.04\,\text{p.u.}\\
I &= 10(1-\cos\phi_l)\,\text{p.u.} \quad \text{for } R_s = 0.1\,\text{p.u.}\\
I &= 2.5(1-\cos\phi_l)\,\text{p.u.} \quad \text{for } R_s = 0.4\,\text{p.u.}
\end{aligned}
$$

The above currents are plotted as a function of load power factor and are shown in Fig. 6.7. Since $IR_s = V_l(1-\cos\phi_l)$, when R_s increases, I has to decrease to make $V_l(1-\cos\phi_l)$ to be a constant for a given power factor and load voltage. Thus, if the load requires more current than the permissible value, the DVR will not be able to regulate the load voltage at the nominal value, i.e., 1.0 p.u. However, we can regulate bus voltage less than 1.0 p.u. For regulating the load voltage less than 1.0 p.u., the current drawing capacity of the load increases.

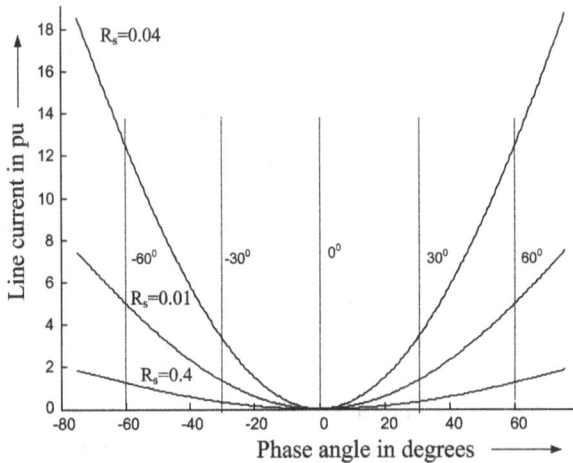

Figure 6.7 DVR characteristics for different load power factor and feeder resistance

6.5 MATHEMATICAL DESCRIPTION TO COMPUTE DVR VOLTAGE

The previous section explains DVR characteristics and describes the feasibility of realizing DVR voltage graphically under different operating conditions. In this section,
a feasible solution for the DVR voltage is presented with a mathematical description.
This plays a significant role while implementing DVR on real time basis. Re-writing
(6.1),

$$\overline{V}_s + \overline{V}_f = \overline{V}_l + (R_s + jX_s)\overline{I}$$

Denoting, $(R_s + jX_s)\overline{I} = a_2 + jb_2$ and $\overline{V}_f = V_f \angle \overline{V}_f = V_f(a_1 + jb_1)$, the above equation can be written as following.

$$\begin{aligned}
\overline{V}_s &= \overline{V}_l + (a_2 + jb_2) - \overline{V}_f \\
&= \overline{V}_l + (a_2 + jb_2) - V_f(a_1 + jb_1) \\
&= 1.0\angle 0° + (a_2 + jb_2) - V_f(a_1 + jb_1)
\end{aligned}$$

Since, source voltage and load voltage have to be maintained at nominal value, i.e.,
1.0 p.u., therefore $\overline{V}_s = V_s \angle \delta = 1.0\angle \delta$. Substituting this value of $\overline{V}_s$ in above equation, we get,

$$\overline{V}_s = 1.0\angle \delta = \cos\delta + j\sin\delta = \left[(1 + a_2) - V_f a_1\right] + j(b_2 - V_f b_1)$$

Squaring and adding the real and imaginary parts from both the sides of the above
equation, we get,

$$(1 + a_2)^2 + V_f^2 a_1^2 - 2(1 + a_2)a_1 V_f + b_2^2 + V_f^2 b_1^2 - 2b_1 b_2 V_f - V_s^2 = 0$$

Since $a_1^2 + b_1^2 = 1$, therefore summation of underlines terms, $V_f^2(a_1^2 + b_1^2) = V_f^2$. Using
this and rearranging above equation in the power of V_f, with $V_s = 1.0$ p.u., we get
the following.

$$V_f^2 - 2\left[(1 + a_2)a_1 + b_1 b_2\right]V_f + (1 + a_2)^2 + b_2^2 - 1.0 = 0 \tag{6.9}$$

The above equation gives two solutions for V_f. These are equivalent to two points A
and B shown in Fig. 6.6. However, the feasible value of the voltage is chosen on the
basis of the rating of the DVR.

In general, the load voltage may not always be $1.0\angle 0°$ p.u. Hence, the load voltage is expressed as $\overline{V}_l = V_l \angle \delta_l$ p.u. Substituting this value of the load voltage in
(6.9), the following is obtained.

$$\begin{aligned}
\overline{V}_s &= V_l \angle \delta_l + a_2 + jb_2 - V_f(a_1 + jb_1) \\
&= (V_l \cos\delta_l + a_2 - a_1 V_f) + j(V_l \sin\delta_l + b_2 - b_1 V_f) \tag{6.10}
\end{aligned}$$

Denoting $V_l \cos\delta_l + a_2 = a_2'$ and $V_l \sin\delta_l + b_2 = b_2'$, the above equation can be written
as following.

$$V_s \cos\delta + jV_s \sin\delta = (a_2' - a_1 V_f) + j(b_2' - b_1 V_f)$$

Squaring and adding the real and imaginary terms in (6.10), the following is obtained.

$$V_f^2 - 2(a_1 a_2' + b_1 b_2')V_f + a_2'^2 + b_2'^2 - V_s^2 = 0 \tag{6.11}$$

To solve DVR voltages for three-phase system, the nominal load voltages with their respective phase angles should be considered in (6.10). Therefore,

$$\overline{V}_{sk} = (V_{lk} \cos \delta_{lk} + a_{2k} - a_{1k} V_{fk}) + j(V_{lk} \sin \delta_{lk} + b_{2k} - b_{1k} V_{fk}) \tag{6.12}$$

In the above equation, k denotes phases a, b, c. Denoting $V_{lk} \cos \delta_{lk} + a_{2k} = a_{2k}'$ and $V_{lk} \sin \delta_{lk} + b_{2k} = b_{2k}'$, the above equation can be written as following.

$$V_{sk} \cos \delta + j V_{sk} \sin \delta \;\; = \;\; (a_{2k}' - a_{1k} V_{fk}) + j(b_{2k}' - b_{1k} V_{fk})$$

Squaring and adding the real and imaginary terms in (6.12), the following equation is obtained.

$$V_{fk}^2 - 2(a_{1k} a_{2k}' + b_{1k} b_{2k}')V_{fk} + a_{2k}'^2 + b_{2k}'^2 - V_{sk}^2 = 0 \tag{6.13}$$

Example 6.2. Consider a system with supply voltage 230 V = 1.0 p.u., 50 Hz as shown in Fig. 6.8. Consider feeder impedance as $Z_s = 0.05 + j0.3$ p.u. and load impedance $Z_l = 0.5 + j0.3$ p.u.

(a) Compute the load voltage without DVR.

(b) Compute the current and DVR voltage such that $V_l = V_s$.

(c) Compute the effective source voltage including DVR. Explain the power flow in the circuit.

(d) Compute the terminal voltage with DVR compensation.

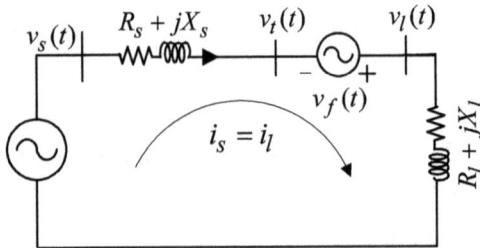

Figure 6.8   A DVR compensated system

Solution:

(a) When DVR is not connected

The system parameters are given as follows. The supply voltage $V_s = 1.0\angle0°$ p.u., $Z_s = R_s + jX_s = 0.05 + j0.3$ and $Z_l = R_l + jX_l = 0.5 + j0.3$ p.u. The current in the circuit is given by,

$$\bar{I}_s = \bar{I} = \bar{I}_l = \frac{\bar{V}_s}{Z_s + Z_l} = \frac{1.0\angle0°}{0.55 + j0.6}$$
$$= 0.83 - j0.91 = 1.2286\angle - 47.49° \, \text{p.u.}$$

The load voltage is therefore given by,

$$\bar{V}_l = Z_l \bar{I}_s = 1.2286\angle - 47.49° \times (0.5 + j0.3)$$
$$= 0.7164\angle - 16.53° \, \text{p.u.}$$

Thus, we observe that the load voltage is 71% of the rated value. Due to a reduction in the load voltage, the load may not perform to the expected level.

(b) When DVR is connected

It is desired to maintain $V_l = V_s$ by connecting the DVR. Taking $\bar{V}_l$ as reference phasor, i.e., $\bar{V}_l = 1.0\angle0°$, The line current is computed as below.

$$\bar{I} = \frac{1.0\angle0}{0.5 + j0.3} = 1.47 - j0.88 = 1.715\angle - 30.96° \, \text{p.u.}$$

Writing KVL for the circuit shown in Fig. 6.8,

$$\bar{V}_s + \bar{V}_f = \bar{V}_l + (R_s + jX_s)\bar{I}.$$

The DVR voltage $\bar{V}_f$ can be expressed as following.

$$\bar{V}_f = V_f\angle(\angle\bar{I}_s + 90°)$$

The angle of $\bar{V}_f$ is taken as $(\angle\bar{I}_s + 90°)$ so that DVR does not exchange any real power with the system.

$$\bar{V}_f = V_f\angle(\angle\bar{I}_s + 90°)$$
$$= V_f\angle(-30.96° + 90°)$$
$$= V_f\angle59.04° = V_f(0.51 + j0.86) = V_f(a_1 + jb_1)\text{p.u.}$$

The above equation implies that $a_1 = 0.51, b_1 = 0.86$.
Let us now compute $(R_s + jX_s)\bar{I}$.

$$(R_s + jX_s)\bar{I} = (0.05 + j0.3)(1.47 - j0.88)$$
$$= 0.3041\angle80.54° \times 1.715\angle - 30.96°$$
$$= 0.5215\angle49.58°$$
$$= 0.3382 + j0.3971 \, \text{p.u.}$$

This gives, $a_2 = 0.3382$ and $b_2 = 0.3971$. As discussed in previous section, the equation $\overline{V}_s + \overline{V}_f = \overline{V}_l + (R_s + jX_s)\overline{I}$ can be written in following form.

$$V_f^2 - 2\left[(1+a_2)a_1 + b_1 b_2\right]V_f + (1+a_2)^2 + b_2^2 - 1.0 = 0.$$

Substituting a_1, b_1, a_2, b_2 in the above equation, we get the following quadratic equation for the DVR.

$$V_f^2 - 2.058 V_f + 0.9485 = 0$$

Solving the above equation, we get $V_f = 0.6969, 1.3611$ p.u. as two values of the DVR voltage. These two values correspond to points A and B, respectively in Fig. 6.6. However, the feasible solution is $V_f = 0.6969$ p.u., as it ensures less rating of the DVR.

Therefore,

$$
\begin{aligned}
\overline{V}_f &= 0.6969\angle 59.04° \\
&= 0.3585 + j0.5976 \,\text{p.u.}
\end{aligned}
$$

The source voltage can be computed using the following equation.

$$
\begin{aligned}
\overline{V}_s &= \overline{V}_l + (R_s + jX_s)\overline{I} - \overline{V}_f. \\
&= 1.0\angle 0° + (0.05 + j0.3)\,1.715\angle -30.96° - 0.6969\angle 59.04° \\
&= 0.9797 - j0.2005 = 1.0\angle -11.56°\,\text{p.u.}
\end{aligned}
$$

(c) Effective source voltage

It is seen that the magnitude of $\overline{V}_s$ is 1.0 p.u. which satisfies the condition $V_s = V_l$. However, the angle of $\overline{V}_s$ is $\angle -11.56°$, which implies that power is flowing from the load to the source. This is not true because the effective source voltage is now $\overline{V}_s' = \overline{V}_s + \overline{V}_f$. This is computed below.

$$
\begin{aligned}
\overline{V}_s + \overline{V}_f = \overline{V}_s' &= 0.9797 - j0.2005 + 0.3585 + j0.5976 \\
&= 1.3382 + j0.3971 \\
&= 1.3959\angle 16.52°\,\text{p.u.}
\end{aligned}
$$

From the above, it is evident that the effective source voltage has a magnitude of 1.3959 p.u. and an angle of $\angle 16.52°$ which ensures that power flows from the source to the load. For this, the equivalent circuit is shown in Fig. 6.9.

(d) Terminal voltage with DVR compensation

The terminal voltage can also be computed as in the following.

$$
\begin{aligned}
\overline{V}_t = \overline{V}_s - Z_s\overline{I} &= \overline{V}_l - \overline{V}_f \\
&= 1.0\angle 0° - 0.6969\angle 59.04° \\
&= 0.8767\angle -42.97°\,\text{p.u.}
\end{aligned}
$$

This indicates that for the rated current flowing in the load, the terminal voltage is less than 1.0 p.u. After compensation, the load voltage is 1.0 p.u. as shown in Fig. 6.9.

$$\overline{V}_s' = \overline{V}_s + \overline{V}_f$$
$$= 1.39\angle 16.52^o \quad R_s + jX_s \qquad \overline{V}_l = 1.0\angle 0^o$$

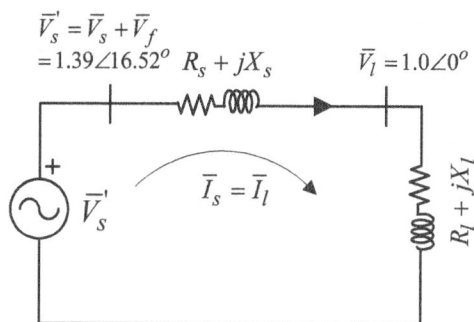

Figure 6.9 A DVR compensated system

6.6 TRANSIENT OPERATION OF THE DVR

In the previous section, the operation of the DVR in the steady state was discussed with the assumption that complete system information is available. While implementing the DVR compensation scheme, the above discussed method should be implemented on a real-time basis. For the single-phase DVR operation, the following steps are required.

1. Define a reference quantity such as the load voltage $v_l(t)$, and then other quantities are synchronized to it.

2. The phase angle of the DVR voltage is obtained by computing the phase angle of the line current with respect to the reference quantity, as given below.

$$\angle \overline{V}_f = \angle \overline{I}_l + 90^\circ = \angle \overline{I}_s + 90^\circ$$

3. Then DVR voltage is computed using (6.11), which is given below.

$$V_f^2 - 2(a_1 a_2' + b_1 b_2')V_f + a_2'^2 + b_2'^2 - V_s^2 = 0$$

4. DVR voltage v_f is calculated using magnitude V_f and phase angle obtained above in time domain form as given below.

$$v_f(t) = \sqrt{2}V_f \sin(\omega t + \angle \overline{V}_f)$$

5. The DVR voltage $v_f(t)$ is then synthesized using appropriate switching control of the voltage source inverter.

The above method can be referred to as Type 1 control [15]. The method assumes that all circuit parameters are known along with the information of the source impedance. However, this may not be feasible in all circumstances. To solve this problem, Type 2 control is suggested. In Type 2 control, only local quantities are required to compute the DVR voltage. The method is described below.

The terminal voltage, which is the local quantity to the DVR as shown in Fig. 6.1, can be expressed as follows.

$$
\begin{aligned}
\overline{V}_t &= \overline{V}_l - \overline{V}_f \\
&= V_l \angle \delta_l - V_f(a_1 + jb_1) = V_l \cos \delta_l + jV_l \sin \delta_l - a_1 V_f - jb_1 V_f \\
&= (V_l \cos \delta_l - a_1 V_f) + j(V_l \sin \delta_l - b_1 V_f)
\end{aligned}
\tag{6.14}
$$

Since, $\overline{V}_t = V_t \angle \delta_t = V_t \cos \delta_t + jV_t \sin \delta_t$, the above equation is written as following.

$$
V_t \cos \delta_t + jV_t \sin \delta_t = (V_l \cos \delta_l - a_1 V_f) + j(V_l \sin \delta_l - b_1 V_f)
\tag{6.15}
$$

Squaring and adding real and imaginary parts on both sides, we get the following.

$$
\begin{aligned}
V_t^2 &= (V_l \cos \delta_l - a_1 V_f)^2 + (V_l \cos \delta_l - b_1 V_f)^2 \\
&= V_l^2 + V_f^2 - 2(a_1 \cos \delta_l + b_1 \sin \delta_l) V_l V_f
\end{aligned}
\tag{6.16}
$$

The above equation can be arranged in the powers of the DVR voltage as given below.

$$
V_f^2 - 2(a_1 \cos \delta_l + b_1 \sin \delta_l) V_l V_f + V_l^2 - V_t^2 = 0
\tag{6.17}
$$

For $\overline{V}_l$ as the reference voltage, $\delta_l = 0$, the above equation is written as follows.

$$
V_f^2 - 2a_1 V_l V_f + V_l^2 - V_t^2 = 0
\tag{6.18}
$$

We can see that the equation does not need the information of voltage drop across the feeder impedance, hence a_2 and b_2 are not present. The above equation is quadratic in V_f and gives a positive value as a feasible solution for V_f. If the roots are complex, it is impossible to compensate for the voltage without real power support.

6.6.1   OPERATION OF THREE PHASE DVR WITH UNBALANCE VOLTAGES WITHOUT HARMONICS

To implement DVR for the unbalanced three-phase system without harmonics, angles of DVR voltages are found by shifting current angles by 90°, i.e.,

$$
\begin{aligned}
\angle \overline{V}_{fa} &= \angle \overline{I}_a + 90° \\
\angle \overline{V}_{fb} &= \angle \overline{I}_b + 90° \\
\angle \overline{V}_{fc} &= \angle \overline{I}_c + 90°
\end{aligned}
\tag{6.19}
$$

The magnitude of DVR voltage can be found using (6.9) and (6.17) for Type 1 and Type 2 control, respectively. Based on the above, the DVR voltages v_{fa}, v_{fb}, v_{fc} can be expressed in the time domain as given below.

$$
\begin{aligned}
v_{fa} &= \sqrt{2} V_{fa} \sin(\omega t + \angle \overline{V}_{fa}) \\
v_{fb} &= \sqrt{2} V_{fb} \sin(\omega t + \angle \overline{V}_{fb}) \\
v_{fc} &= \sqrt{2} V_{fc} \sin(\omega t + \angle \overline{V}_{fc})
\end{aligned}
\tag{6.20}
$$

6.6.2 OPERATION OF THREE PHASE DVR WITH UNBALANCE VOLTAGES WITH HARMONICS

In the previous analysis, it was assumed that the supply voltages are unbalanced without harmonics. In this section, the operation of the DVR with harmonics will be discussed. The terminal voltages (v_{ta}, v_{tb} and v_{tc}) are resolved into their fundamental voltages and the rest part, as given below.

$$
\begin{aligned}
v_{ta} &= v_{ta1} + v_{ta\,rest} \\
v_{tb} &= v_{tb1} + v_{tb\,rest} \\
v_{tc} &= v_{tc1} + v_{tc\,rest}
\end{aligned}
\tag{6.21}
$$

The angles of fundamental DVR voltages ($\angle \overline{V}_{fa1}, \angle \overline{V}_{fb1}$, and $\angle \overline{V}_{fc1}$) are extracted as following.

$$
\begin{aligned}
\angle \overline{V}_{fa1} &= \angle \overline{I}_{a1} + 90° \\
\angle \overline{V}_{fb1} &= \angle \overline{I}_{b1} + 90° \\
\angle \overline{V}_{fc1} &= \angle \overline{I}_{c1} + 90°
\end{aligned}
\tag{6.22}
$$

The angles of currents, i.e., $\angle \overline{I}_{a1}, \angle \overline{I}_{b1}, \angle \overline{I}_{c1}$, are computed by extracting fundamental of the three-phase load currents using Fourier transform. The magnitudes of the fundamental DVR voltages (V_{fa1}, V_{fb1} and V_{fc1}) can be computed using equations (6.11) and (6.18) for Type 1 and Type 2 control, respectively. For example, using Type 2 control, the fundamental phase-a DVR voltage is computed as per the following equation.

$$
V_{fa1}^2 - 2 a_{a1} V_l V_{fa1} + V_l^2 - V_{ta1}^2 = 0
\tag{6.23}
$$

In (6.23), $a_{a1} + jb_{a1} = \angle \overline{V}_{fa1}$ and V_{ta1} is fundamental phase-a terminal voltage as given above in (6.21). A similar expression can be written for phase-b and phase-c. This equation gives a solution only for the fundamental component of the DVR voltage. The rest of the DVR voltages consist of harmonics, which are equal and opposite to that of the terminal voltage harmonics, i.e., $v_{ta\,rest}, v_{tb\,rest}$ and $v_{tc\,rest}$. Therefore, these can be given using the following equations.

$$
\begin{aligned}
v_{fa\,rest} &= -v_{ta\,rest} \\
v_{fb\,rest} &= -v_{tb\,rest} \\
v_{fc\,rest} &= -v_{tc\,rest}
\end{aligned}
\tag{6.24}
$$

Thus, the total DVR voltage to be injected can be given as follows.

$$
\begin{aligned}
v_{fa} &= v_{fa1} + v_{fa\,rest} \\
v_{fa} &= v_{fb1} + v_{fb\,rest} \\
v_{fa} &= v_{fc1} + v_{fc\,rest}
\end{aligned}
\tag{6.25}
$$

In the above equation, v_{fa1}, v_{fb1}, and v_{fc1} are constructed using (6.20). Once v_{fa}, v_{fb}, and v_{fc} are known, these voltages are synthesized using a suitable power electronic circuit. This is discussed in the following section.

6.7 REALIZATION OF DVR VOLTAGE USING VOLTAGE SOURCE INVERTER

In the previous section, a reference voltage of DVR was extracted using discussed control algorithms. This DVR voltage, however should be realized at the power level. This is achieved with the help of a power electronic converter which is also known as a voltage source inverter. Various components of the DVR are shown in detail in Fig. 6.10. The transformer injects the required voltage in series with the line to maintain the load bus voltage at the nominal value. The transformer not only reduces the voltage requirement but also provides isolation between the inverters and the line. The filter components of the DVR, such as external inductance (L_t), which also includes the leakage of the transformer on the primary side and ac filter capacitor on the secondary side, play a significant role in the performance of the DVR [15]. The

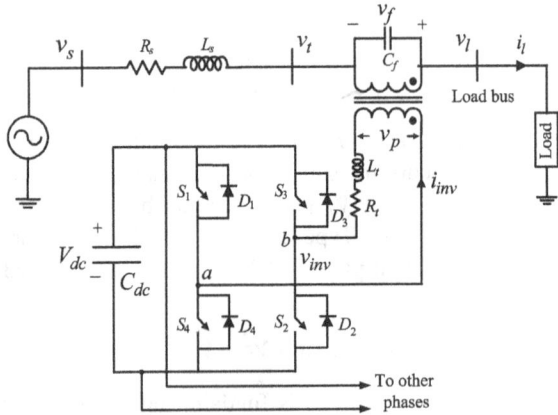

Figure 6.10   A DVR compensated system

same dc link can be extended to other phases as shown in Fig. 6.10. The single-phase equivalent of the DVR is shown in Fig. 6.11. In Figs. 6.10 and 6.11, v_{inv} denotes

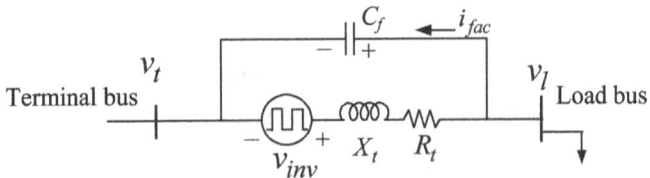

Figure 6.11   Equivalent circuit of the DVR

the switched voltage generated at the inverter output terminals. The resistance, R_t, models the switching losses of the inverter and the copper loss of the connected transformer and filter inductance. The voltage source inverter (VSI) is operated in a switching band voltage control mode to track the reference voltages generated using control logic as discussed below.

Let V_f^* be the reference voltage of a phase that DVR needs to inject in series with the line with the help of the VSI explained above. We form a voltage hysteresis band of $\pm h$ over the reference value. Thus, the upper and lower limits within which the DVR has to track the voltage can be given as follows.

$$
\begin{aligned}
v_{fup} &= v_f^* + h \\
v_{fdn} &= v_f^* - h
\end{aligned}
\tag{6.26}
$$

The following switching logic is used to synthesize the reference DVR voltage.

> If $v_f \geq v_{fup}$
> > S_1, S_2 OFF, and S_3, S_4 ON ('-1' state)
> > else if $v_f \leq v_{fdn}$
> > > S_1, S_2 ON, and S_3, S_4 OFF ('+1' state)
> > end

It is to be noted that switches status $S_1 - S_2$ ON and $S_3 - S_4$ OFF is denoted by '+1' state and it gives $v_{inv} = +V_{dc}$. The switches status $S_1 - S_2$ OFF and $S_3 - S_4$ ON corresponds to '-1' state providing $v_{inv} = -V_{dc}$ as shown in Fig. 6.10. The above switching logic is very basic and has a scope to be refined. For example '0' state of the switches of the VSI as shown in Fig. 6.10, can also be used to have smooth switching and to minimize switching losses. In the zero state, $v_{inv} = 0$ and refers switches status as $S_1 D_3$ or $S_2 D_4$ for positive inverter current ($i_{inv} > 0$). Similarly, for negative inverter current ($i_{inv} < 0$), '0' state is obtained through $S_3 D_1$ or $S_4 D_2$. With the addition of '0' state, the modified switching logic becomes as follows.

> If $v_f^* \geq 0$
> > if $v_f \geq v_{fup}$
> > > '0' state
> > > else if $v_f \leq v_{fdn}$
> > > > '+1' state
> > > end
> > else if $v_f^* < 0$
> > > if $v_f \geq v_{fup}$
> > > > '-1' state
> > > > else if $v_f \leq v_{fdn}$
> > > > > '0' state
> > > > end
> > end

In order to improve the switching performance, one more term is added in the above equation based on the feedback of filter capacitor current, i_{fac}.

$$v_{fup} = v_f^* + h + \alpha i_{fac}$$
$$v_{fdn} = v_f^* - h + \alpha i_{fac} \tag{6.27}$$

Where α is a proportional gain given to smoothen and stabilize the switching performance of the VSI [16]. The dimension of α is Ohm and is equivalent to virtual resistance, whose effect is to damp out and smoothen the DVR voltage trajectory resulting from the switching of the inverter [20]. The value of the hysteresis band (h) should be chosen in such a way that it limits switching frequency within the specified maximum value. This kind of voltage control using VSI is called as switching band control. The actual DVR voltage is compared with these upper and lower bands of the voltage (v_{fup}, v_{fdn}), and accordingly switching commands to the power switches are generated. The switching control logic is described in Table 6.1. To minimize the switching frequency losses of the VSI, three level logic has been used. For this, an additional check of the polarity of the reference voltage has been taken into consideration. Based on this switching status, the inverter supplies $+V_{dc}$, 0, and $-V_{dc}$ levels of voltage corresponding to the 1, 0, and -1 given in the table in order to synthesize the reference DVR voltage.

Table 6.1
Three level switching logic for the VSI

Conditions		Switching value
$v_f^* \geq 0$	$v_f > v_{fup}$	0
$v_f^* \geq 0$	$v_f < v_{fdn}$	1
$v_f^* < 0$	$v_f > v_{fup}$	−1
$v_f^* < 0$	$v_f < v_{fdn}$	0

In addition to switching band control, an additional loop is required to correct the voltage in the dc storage capacitor against losses in the inverter and transformer. During transients, the dc capacitor voltage may rise or fall from the reference value due to real power flow for a short duration. To correct this voltage deviation, a small amount of real power must be drawn from the source to replenish the losses. To accomplish this, a simple proportional-plus-integral controller (PI) is used. The signal, u_c, is generated from this PI controller as given below.

$$u_c = K_p e_{Vdc} + K_i \int e_{Vdc} dt \tag{6.28}$$

Where, $e_{Vdc} = V_{dcref} - V_{dc}$. This control loop need not to be as fast as the switching control logic. It may be updated once in an ac cycle preferably synchronized to positive zero crossing of phase-a voltage. Based on this information, the variable u_c will be included in the generation of the fundamental DVR voltage as given below.

$$\overline{V}_{f1} = V_{f1} \angle (\angle \overline{I}_s + 90° - u_c) = V_{f1} (\tilde{a}_1 + j\tilde{b}_1) \qquad (6.29)$$

Then, (6.17) is modified to the following.

$$V_{f1}^2 - 2\tilde{a}_1 V_l V_{f1} + V_l^2 - V_{t1}^2 = 0 \qquad (6.30)$$

The above equation is used to find the DVR voltage with dc link voltage regulation. It can be found that the phase difference between line current and DVR voltage differs slightly from $90°$ (say $87°$) in order to account for the losses in the inverter.

6.8   MAXIMUM COMPENSATION CAPACITY OF THE DVR WITHOUT REAL POWER SUPPORT FROM THE DC LINK

There is a direct relationship between the terminal voltage, the power factor of the load, and the maximum possible achievable load voltage, with the assumption that no real power is required from the dc bus. Referring to the quadratic equation in (6.30), for a given value of V_{t1} and a target load bus voltage V_l, the equation gives two real values of V_{f1} for a feasible solution. In case, the solution is not feasible, the equation gives two complex conjugate roots. This concludes that the maximum voltage that DVR can compensate corresponds to the single solution in the limiting case of the above equation. This solution corresponds to point D_1 in Fig. 6.6.

$$V_{f1} = \frac{2\tilde{a}_1 V_l \pm \sqrt{(2\tilde{a}_1 V_l)^2 - 4(V_l^2 - V_{t1}^2)}}{2} \qquad (6.31)$$

Since voltage should not be a complex number, the value of the terms within the square root must not be negative. Therefore,

$$(2\tilde{a}_1 V_l)^2 \geq 4(V_l^2 - V_{t1}^2) \qquad (6.32)$$

The above equation implies that,

$$V_l \leq \frac{V_{t1}}{\sqrt{1 - \tilde{a}_1^2}}. \qquad (6.33)$$

Therefore, the DVR voltage is given by the following equation.

$$V_{f1} = \tilde{a}_1 V_l \qquad (6.34)$$

With no losses in the VSI, $u_c = 0$,

$$V_l \leq \frac{V_{t1}}{\sqrt{1 - \tilde{a}_1^2}} = \frac{V_{t1}}{\sqrt{1 - a_1^2}} \qquad (6.35)$$

Since, $\tilde{a}_1 + j\tilde{b}_1 = 1\angle(90° + \phi_l) = \cos(90° + \phi_l) + j\sin(90° + \phi_l) = -\sin\phi_l + j\cos\phi_l$. This implies $\tilde{a}_1 = -\sin\phi_l$, therefore $\sqrt{1 - \tilde{a}_1^2} = \sqrt{1 - (-\sin\phi_l)^2} = \cos\phi_l$. Using this relation, the above equation can be written in the following.

$$V_l \leq \frac{V_{t1}}{\cos\phi_l} \tag{6.36}$$

For this condition, the DVR voltage V_f is as follows.

$$V_f = \tilde{a}_1 V_l \tag{6.37}$$

Example 6.3. A DVR is shown in Fig. 6.12. The feeder impedance of the line $0.1 + j0.5$ p.u. Assume a constant load current, i_l drawn from the supply and is represented by square waveform approximated by the following expression.

$$i_l = 1.0\sqrt{2}\sin(\omega t - 30°) + 0.3\sqrt{2}\sin(3\omega t - 90°)\,\text{p.u.}$$

In absence of the DVR the load current is given with respect to the source voltage, $\overline{V}_s = 1\angle 0°$ p.u., as if there is no effect of the feeder impedance. With the DVR, the given load current is the same but with respect to the load voltage. Although, there is a finite feeder impedance given in the problem, this assumption is required to simplify the analysis.

(a) Find the load voltage $v_l(t)$ with respect to source voltage without DVR compensation, i.e., $v_f = 0$.

(b) Is it possible to maintain load voltage, V_l to be 1.0 p.u. with sinusoidal waveform? If yes what is the DVR voltage, $v_f(t)$?

(c) If no, how much maximum voltage can be maintained at the load terminal with the DVR without taking any real power from the dc bus?

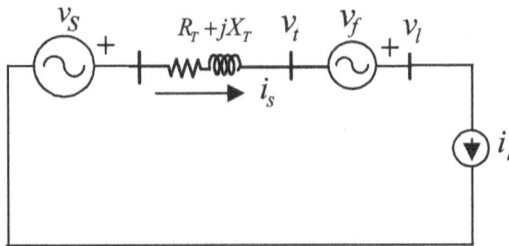

Figure 6.12   A DVR compensated system

Solution:

(a) Without DVR, $v_f = 0$

$$
\begin{aligned}
v_t &= v_s - R_s\left(i_{l1} + i_{l3}\right) - L_s \frac{d\left(i_{l1} + i_{l3}\right)}{dt} \\
&= v_s - \left(R_s\, i_{l1} + L_s \frac{di_{l1}}{dt}\right) - \left(R_s\, i_{l3} + L_s \frac{di_{l3}}{dt}\right) \\
&= v_s - v_{zs1} - v_{zs3}
\end{aligned}
$$

where, $v_{zs1} = R_s\, i_{l1} + L_s \frac{di_{l1}}{dt}$ and $v_{zs3} = R_s\, i_{l3} + L_s \frac{di_{l3}}{dt}$. In the frequency domain, the impedance at the fundamental frequency, $Z_{s1} = 0.1 + j0.5 = 0.51\angle 78.7°$ p.u. The impedance at third harmonic, $Z_{s3} = 0.1 + j1.5 = 1.50\angle 86.18°$ p.u. Therefore the voltage drop due to the fundamental component of the current,

$$
\begin{aligned}
\overline{V}_{zs1} &= (0.1 + j0.5) \times 1\angle{-30°} \\
&= 0.51\angle 78.69° \times 1\angle{-30°} \\
&= 0.51\angle 48.69 \,\text{p.u.}
\end{aligned}
$$

The voltage drop due to the third harmonic component of the current,

$$
\begin{aligned}
\overline{V}_{zs3} &= (0.1 + j1.5) \times 0.3\angle{-90°} \\
&= 0.45\angle{-3.81°} \,\text{p.u..}
\end{aligned}
$$

Thus, the load voltage in the time domain can be expressed as,

$$
\begin{aligned}
v_l = v_t = v_s - v_{zs1} - v_{zs3} \\
= \underbrace{\sqrt{2}\sin\omega t - 0.51\sqrt{2}\sin(\omega t + 48.69°)} - 0.45\sqrt{2}\sin(3\,\omega t - 3.81°) \\
= 0.7659\sqrt{2}\sin(\omega t - 30°) - 0.45\sqrt{2}\sin(3\,\omega t - 3.81°)\,\text{p.u.} \\
= v_{t1}(t) + v_{th}(t).
\end{aligned}
$$

Implying that,

$$
\overline{V}_{t1} = 0.7659\angle{-30°}\,\text{p.u.}
$$

(b) With DVR

From the above equation, $V_{t1} = 0.7659$ p.u. With load voltage $v_l = \sqrt{2}\sin\omega t$, the DVR voltage V_{f1} can be solved using the quadratic equation as mentioned in Type 2 control. Further,

$$
\begin{aligned}
\overline{V}_{f1} = V_{f1}\angle(\angle\overline{I}_{s1} + 90°) = V_{f1}\angle(-30° + 90°) = V_{f1}\angle 60° \\
= V_{f1}\left(\cos 60° + j\sin 60°\right) = V_{f1}\left(0.5 + j0.8666\right) = V_{f1}\left(a_1 + jb_1\right)\text{p.u.}
\end{aligned}
$$

The above implies $a_1 = 0.5$, $b_1 = 0.866$. Knowing this, we can solve V_{f1} using the following quadratic equation.

$$
V_{f1}^2 - 2a_1 V_l V_{f1} + V_l^2 - V_{t1}^2 = 0
$$

In the above, $V_{t1} = 0.7659$ p.u. is considered as computed from (a), for further computations.

$$\begin{aligned} V_{f1} &= a_1 V_l \pm \sqrt{a_1^2 V_l^2 - (V_l^2 - V_{t1}^2)} \\ &= 0.5 \pm \sqrt{(0.5)^2 - \{1^2 - 0.7659^2\}} \\ &= 0.5 \pm \sqrt{0.25 - 0.4134} \end{aligned}$$

The above solution is a complex quantity, which implies that it is not possible to maintain load voltage at $\sqrt{2} \sin \omega t$.

(c) Maximum possible load voltage

The maximum load voltage that can be obtained with the DVR, without any real power from the dc bus can be given as follows.

$$V_l = \frac{V_{t1}}{\sqrt{1 - a_1^2}} = \frac{0.7659}{\sqrt{1 - 0.5^2}} = 0.8845 \, \text{p.u.}$$

In the time domain the load voltage $v_l = v_{l1} = \sqrt{2} \times 0.8845 \sin \omega t$. For this load voltage, the DVR voltage is given as following.

$$V_{f1} = a_1 V_l = 0.5 \times 0.8845 = 0.4423 \, \text{p.u.}$$

This implies

$$\overline{V}_{f1} = 0.4423 \angle 60° \, \text{p.u.}$$

The time domain expression for the fundamental DVR voltage is given as,

$$v_{f1}(t) = 0.4423 \sqrt{2} \sin(\omega t + 60°) \, \text{p.u.}$$

The harmonic voltage that DVR compensates is as follows.

$$v_{fh}(t) = -v_{th} = 0.45 \sqrt{2} \sin(3\omega t - 3.81°) \, \text{p.u.}$$

The total DVR voltage is given as below.

$$\begin{aligned} v_f(t) &= v_{f1}(t) + v_{fh}(t) \\ &= 0.4423 \sqrt{2} \sin(\omega t + 60°) + 0.45 \sqrt{2} \sin(3\omega t - 3.81°) \, \text{p.u.} \end{aligned}$$

Example 6.4. A DVR compensated single-phase system is shown in Fig. 6.13. Determine V_{D1} and V_{D2} such that the voltage remains 1.0 p.u. at all three buses. Also, determine the value of δ_1 and δ to ensure power flow from source to load. The value of relevant parameters, all in per unit, are given as follows.
$Z_{s1} = 0.02 + j0.5$, $Z_{s2} = 0.05 + j0.3$, $Z_{l1} = 0.25 + j0.4$ and $Z_{l2} = 0.3 + j0.4$.

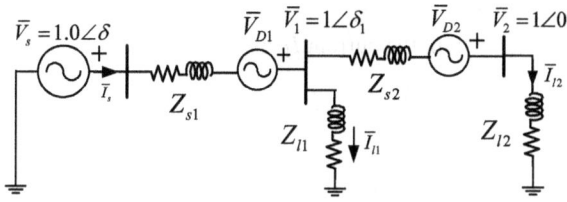

Figure 6.13 A DVR compensated single-phase system

Solution: The DVR2 is operated at nominal voltage, i.e., $\overline{V}_2 = 1\angle 0$. Therefore, the current $\overline{I}_{l2}$ is given as follows.

$$\overline{I}_{l2} = \frac{\overline{V}_2}{Z_{l2}} = \frac{1\angle 0}{0.3 + j0.4} = 2.0\angle -53.13° \, \text{p.u.}$$

Since, the DVR2 injects a voltage with an angle of $90°$ with respect to the current $\overline{I}_{l2}$. Therefore,

$$\angle \overline{V}_{D2} = 90° - 53.13° = 36.87°.$$

The values of a_{1l} and b_{1l} is computed as follows:

$$a_{1l} + jb_{1l} = 1.0\angle \overline{V}_{D2} = 0.8 + j0.6 \, \text{p.u.}$$

Voltage across the feeder 2 (Z_{s2}) is given as,

$$\Delta \overline{V}_{f2} = Z_{s2}\overline{I}_{l2} = (0.05 + j0.3) \times 2.0\angle -53.13° = 0.54 + j0.28 \, \text{p.u.} = a_{2l} + jb_{2l}.$$

The above equation implies that $a_{2l} = 0.54$ and $b_{2l} = 0.28$. Now, the voltage V_{D2} is computed by solving following equation.

$$V_{D2}^2 - 2\left[(1 + a_{2l})a_{1l} + b_{1l}b_{2l}\right] V_{D2} + (1 + a_{2l})^2 + b_{2l}^2 - 1.0 = 0.$$

Substituting variables, the following quadratic equation is used to determine the DVR voltage.

$$V_{D2}^2 - 2.8 V_{D2} + 1.45 = 0.$$

Solving the equation, $V_{D2} = 2.11$ p.u. and 0.68 p.u. are obtained. However, $V_{D2} = 0.68$ p.u. is the feasible solution.
Now, the voltage $\overline{V}_1$ is computed as follows.

$$\begin{aligned}
\overline{V}_1 &= Z_{s2}\overline{I}_{l2} + \overline{V}_2 - \overline{V}_{D2} \\
&= 0.54 + j0.28 + 1\angle 0 - 0.68\angle 36.87° \\
&= 1\angle -7.55° \, \text{p.u.}
\end{aligned}$$

The above implies that $\delta_1 = -7.55°$.

Once value of $\overline{V}_1$ is known, the current $\overline{I}_{l1}$ is given by

$$\overline{I}_{l1} = \frac{\overline{V}_1}{Z_{l1}} = \frac{1\angle -7.55°}{0.25 + j0.4} = 2.12\angle -65.55° \text{ p.u.}$$

Using KCL, source current is given as

$$\overline{I}_s = \overline{I}_{l1} + \overline{I}_{l2} = 2.0\angle -53.13° + 2.12\angle -65.55° = 4.09\angle -59.52° \text{ p.u.}$$

Again, DVR1 will inject a voltage perpendicular to current $\overline{I}_s$. Therefore,

$$\angle \overline{V}_{D1} = -59.52° + 90° = 30.48°.$$

$$a_{1s} + jb_{1s} = 1.0\angle \overline{V}_{D1} = 1.0\angle 30.48° = 0.86 + j0.51 \text{ p.u.}$$

Thus, voltage across the feeder 1 (Z_{s1}) is given as

$$\Delta \overline{V}_{f1} = Z_{s1}\overline{I}_s = (0.02 + j0.5) \times 4.09\angle -59.52° = 1.8065 + j0.968 \text{ p.u.}$$
$$= a_{2s} + jb_{2s}.$$

Using Kirchoff's Voltage Law (KVL), the voltage equation is as follows.

$$\overline{V}_s = \overline{V}_1 + Z_{s1}\overline{I}_s - \overline{V}_{D1}$$

Using steps given by (6.10) to (6.11), the following is obtained.

$$V_{D1}^2 - 2(a'_{2s}a_{1s} + b'_{2s}b_{1s})V_{D1} + a'_{2s}{}^2 + b'_{2s}{}^2 - V_s^2 = 0$$

In above equation, $a'_{2s} = V_1 \cos \delta_1 + a_{2s} = 2.7978$ and $b'_{2s} = V_1 \sin \delta_1 + b_{2s} = 0.8366$. Now substituting these values in (6.38), the voltage, V_{D1} is computed by solving the following equation.

$$V_{D1}^2 - 5.6711 V_{D1} + 7.5277 = 0.$$

Solving the equation, $V_{D2} = 2.1195$ p.u. and 3.5516 p.u. are obtained. However, $V_{D2} = 2.1195$ p.u. is the feasible solution. Now, the source voltage is given in the following.

$$\begin{aligned}
\overline{V}_s &= Z_{s1}\overline{I}_s + \overline{V}_1 - \overline{V}_{D1} \\
&= 2.049\angle 28.18° + 1\angle -7.55° - 2.1195\angle 30.48° \\
&= 1\angle -13.79° \text{ p.u.}
\end{aligned}$$

Hence, $\delta = -13.79°$.

Comments on the Results

Although it seems that the angle of the source voltage is negative and, therefore, power should flow from load to source. However, it is not true due to the presence of DVR voltages $\overline{V}_{D1}$ and $\overline{V}_{D2}$. This can be seen by looking into the effective or equivalent source voltage i.e.,

$$\overline{V}_s' = \overline{V}_s + \overline{V}_{D1} = 1\angle -13.79° + 2.1195\angle 30.48°$$
$$= 2.902\angle 16.64° \text{ p.u.}$$

Similarly,

$$\overline{V}_{s1}' = \overline{V}_1 + \overline{V}_{D2} = 1\angle -7.55° + 0.68\angle 36.87°$$
$$= 1.5652\angle 10.3° \text{ p.u.}$$

The positive angles of the effective source voltages indicate that the power flows from source to the load.

Example 6.5. For the circuit shown in Fig. 6.14, what will be the load voltages without DVR compensation? Analyze whether DVR compensation is possible or not by satisfying the condition $|V_l| = |V_s| = 1.0$ p.u. in all three phases. If yes, then find the voltages injected by the DVR in all three phases. Also, calculate the terminal voltages (v_{ta}, v_{tb}, and v_{tc}), source voltages (v_{sa}, v_{sb}, and v_{sc}) and effective source voltages (v_{sa}', v_{sb}', and v_{sc}') after compensation. Solve the above problem for the following conditions.

Case 1: $V_n = V_N$, i.e., for three-phase four-wire system.

Case 2: $V_n \neq V_N$, i.e., for three-phase three-wire system.

Parameters of the circuit are given as follows:
Source voltage: 230 V rms per phase (1 p.u.), 50 Hz.
Feeder impedance (Z_s): $0.01 + j0.1$ p.u. in all three phases.
Load impedance: $Z_{la} = 0.3 + j0.4$ p.u., $Z_{lb} = 1.2 + j1.3$ p.u. and $Z_{lc} = \infty$.

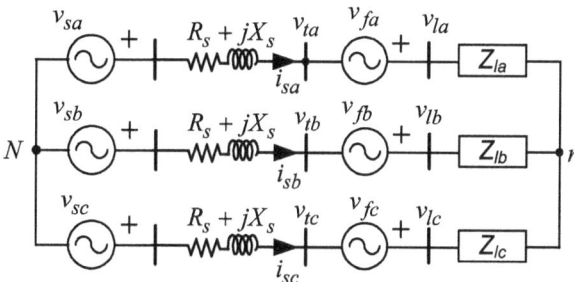

Figure 6.14 A three-phase DVR compensated system

Solution:

Case 1: Three-phase four-wire system ($\overline{V}_n = \overline{V}_N$)

The circuit diagram of the system is shown in Fig. 6.15. Here, all three phases behave as individual phases as their neutral points are connected to the supply neutral.

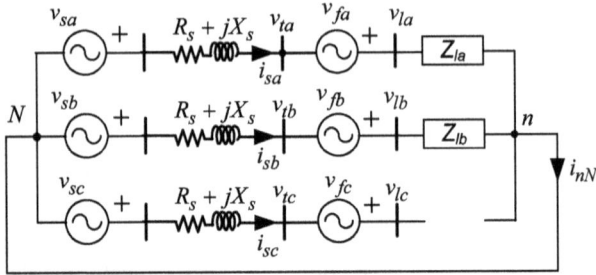

Figure 6.15 A three-phase four-wire DVR compensated system

First, phase-a is considered. The load voltage $\overline{V}_{la}$ without compensation is computed as follows.

$$
\begin{aligned}
\overline{V}_{la} &= \overline{V}_{sa} \frac{Z_{la}}{Z_{la} + Z_s} \\
&= 1\angle 0 \frac{0.3 + j0.4}{0.3 + j0.4 + 0.01 + j0.1} \\
&= 0.85\angle -5.07° \text{ p.u.}
\end{aligned}
$$

Now, the DVR is operated to maintain load voltage at nominal voltage. Therefore, the load current $\overline{I}_{la}$ is given as follows.

$$
\overline{I}_{la} = \frac{\overline{V}_{la}}{Z_{la}} = \frac{1\angle 0°}{0.3 + j0.4} = 2.0\angle -53.13° \text{ p.u.}
$$

The DVR will inject a voltage with an angle of 90° with respect to the current $\overline{I}_{la}$. Therefore,

$$
\angle \overline{V}_{fa} = 90° - 53.13° = 36.87°.
$$

The values of a_{1a} and b_{1a} is computed as follows:

$$
a_{1a} + jb_{1a} = 1.0\angle \overline{V}_{fa} = 0.8 + j0.6 \text{ p.u.}
$$

Above equation implies that $a_{1a} = 0.8$ and $b_{1a} = 0.6$. Also, voltage across the feeder (Z_{sa}) is given as

$$
Z_{sa}\overline{I}_{la} = (0.01 + j0.1) \times 2.0\angle -53.13° = 0.172 + j0.104 \text{ p.u.} = a_{2a} + jb_{2a}.
$$

From the above equation, $a_{2a} = 0.172$ and $b_{2a} = 0.104$. Now, the voltage V_{fa} is computed by solving the following equation.

$$V_{fa}^2 - 2\left[(1 + a_{2a})a_{1a} + b_{1a}b_{2a}\right]V_{fa} + (1 + a_{2a})^2 + b_{2a}^2 - 1.0 = 0.$$

Substituting variables, the following quadratic equation is used to find out the DVR voltage.

$$V_{fa}^2 - 2V_{fa} + 0.3844 = 0.$$

Solving the equation, $V_{fa} = 1.7846$ p.u. and 0.2153 p.u. are obtained. However, $V_{fa} = 0.2153$ p.u. is the feasible magnitude. Hence, $\overline{V}_{fa} = 0.2153\angle 36.87°$ p.u.

Terminal voltage after compensation:

$$
\begin{aligned}
\overline{V}_{ta} &= -\overline{V}_{fa} + \overline{V}_{la} \\
&= -0.2153\angle 36.87° + 1\angle 0° \\
&= 0.837\angle - 8.87° \text{ p.u.}
\end{aligned}
$$

Source voltage after compensation:

$$
\begin{aligned}
\overline{V}_{sa} &= \overline{V}_{ta} + \overline{I}_{la}Z_s \\
&= 0.837\angle - 8.87° + \left[(2.0\angle - 53.13°) \times (0.01 + j0.1)\right] \\
&= 1.0\angle - 1.43° \text{ p.u.}
\end{aligned}
$$

Effective source voltage after compensation:

$$
\begin{aligned}
\overline{V}_{sa}' &= \overline{V}_{sa} + \overline{V}_{fa} \\
&= 1.0\angle - 1.43° + 0.2153\angle 36.87° \\
&= 1.17\angle 5.1° \text{ p.u.}
\end{aligned}
$$

Now, phase-b is considered. The load voltage $\overline{V}_{lb}$ without compensation is computed as follows.

$$
\begin{aligned}
\overline{V}_{lb} &= \overline{V}_{sb}\frac{Z_{lb}}{Z_{lb} + Z_s} \\
&= 1\angle - 120°\frac{1.2 + j1.3}{1.2 + j1.3 + 0.01 + j0.1} \\
&= 0.956\angle - 121.87° \text{ p.u.}
\end{aligned}
$$

The DVR voltage for phase-b is computed as follows. The load current $\overline{I}_{lb}$ is given as below.

$$\overline{I}_{lb} = \frac{\overline{V}_{lb}}{Z_{lb}} = \frac{1\angle - 120°}{1.2 + j1.3} = 0.565\angle - 167.29° \text{ p.u.}$$

The DVR will inject a voltage with an angle of 90° with respect to the current $\bar{I}_{lb}$. Therefore,

$$\angle \bar{V}_{fb} = -167.29° + 90° = -77.29°.$$

The values of a_{1b} and b_{1b} is computed as follows:

$$a_{1b} + jb_{1b} = 1.0\angle \bar{V}_{fb} = 0.22 - j0.9755 \text{ p.u.}$$

Above equation implies that $a_{1b} = 0.22$ and $b_{1b} = -0.9755$. Also, voltage across the feeder (Z_{sb}) is given as

$$Z_{sb}\bar{I}_{lb} = (0.01 + j0.1) \times 0.565\angle -167.29° = 0.0069 - j0.0564 \text{ p.u.}$$
$$= a_{2b} + jb_{2b}.$$

Equating both sides of above equation, $a_{2b} = 0.0069$ and $b_{2b} = -0.0564$. Now, the voltage V_{fb} is computed by solving the following equation.

$$V_{fb}^2 - 2\{a'_{2b}a_{1b} + b'_{2b}b_{1b}\}V_{fb} + a'^{\,2}_{2b} + b'^{\,2}_{2b} - V_{sb}^2 = 0. \tag{6.38}$$

In above equation, $a'_{2b} = V_{lb}\cos(-120°) + a_{2b} = -0.4931$, $b'_{2b} = V_{lb}\sin(-120°) + b_{2b} = -0.9224$ and $V_{sb} = 1.0$. Substituting these values, the following quadratic equation is used to find out DVR voltage.

$$V_{fb}^2 - 1.5827 V_{fb} + 0.094 = 0.$$

Solving the equation, $V_{fb} = 1.5209$ p.u. and 0.0618 p.u. are obtained. However, $V_{fb} = 0.0618$ p.u. is the feasible magnitude. Hence, $\bar{V}_{fb} = 0.0618\angle -77.29°$ p.u.
Terminal voltage after compensation is given as follow.

$$\begin{aligned}
\bar{V}_{tb} &= -\bar{V}_{fb} + \bar{V}_{lb} \\
&= -0.0618\angle -77.29° + 1\angle -120° \\
&= 0.955\angle -122.52° \text{ p.u.}
\end{aligned}$$

Source voltage after compensation is given below.

$$\begin{aligned}
\bar{V}_{sb} &= \bar{V}_{tb} + \bar{I}_{lb}Z_s \\
&= 0.955\angle -122.52° + [(0.565\angle -167.29°) \times (0.01 + j0.1)] \\
&= 1.0\angle -120.45° \text{ p.u.}
\end{aligned}$$

Effective source voltage after compensation is computed as below.

$$\begin{aligned}
\bar{V}'_{sb} &= \bar{V}_{sb} + \bar{V}_{fb} \\
&= 1.0\angle -120.45° + 0.0618\angle -77.3° \\
&= 1.0459\angle -118.13° \text{ p.u.}
\end{aligned}$$

In phase-c, the load is not connected. Therefore, the current in this phase will be zero. Hence, the load voltage will be the same as the source voltage, i.e., $V_{lc} = 1.0\angle 120°$ p.u. Here, the voltage drop in the feeder is zero. Thus, DVR in phase-c does not inject any voltage, i.e., $V_{fc} = 0$.

Case 2: Three-phase three-wire system $(\overline{V}_n \neq \overline{V}_N)$

(a) **Without DVR**

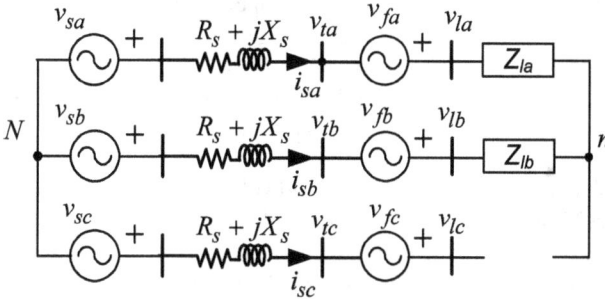

Figure 6.16 A three-phase three-wire DVR compensated system

The circuit diagram of the three-phase three-wire system is shown in Fig. 6.16. In this system, different phases are dependent on each other. Therefore, it is required to find out neutral voltage $\overline{V}_{nN}$ without DVR, using the impedances, $Z_a = Z_{la} + Z_s, Z_b = Z_{lb} + Z_s, Z_c = Z_{lc} + Z_s$. The neutral voltage, $\overline{V}_{nN}$ is computed as follows.

$$\overline{V}_{nN} = \frac{1}{\frac{1}{Z_a} + \frac{1}{Z_b} + \frac{1}{Z_c}} \left(\frac{\overline{V}_{saN}}{Z_a} + \frac{\overline{V}_{sbN}}{Z_b} + \frac{\overline{V}_{scN}}{Z_c} \right)$$

$$= \frac{1}{\frac{1}{0.31+j0.5} + \frac{1}{1.21+j1.4} + \frac{1}{\infty}} \left(\frac{1\angle 0°}{0.31 + j0.5} + \frac{1\angle -120°}{1.21 + j1.4} + \frac{1\angle 120°}{\infty} \right)$$

$$= 0.665 - j0.2512 \,\text{p.u.} = 0.7108\angle -20.69° \,\text{p.u.}$$

Now, consider the phase-a. The load voltage $\overline{V}_{lan}$ without compensation is computed as follows.

$$\overline{V}_{lan} = (\overline{V}_{saN} - \overline{V}_{nN}) \frac{Z_{la}}{Z_{la} + Z_s}$$

$$= [(1\angle 0°) - (0.7108\angle -20.69°)] \frac{0.3 + j0.4}{0.3 + j0.4 + 0.01 + j0.1}$$

$$= 0.356\angle 31.79° \,\text{p.u.}$$

Similarly, the load voltage $\overline{V}_{lbn}$, without compensation is computed as follows.

$$\overline{V}_{lbn} = \left(\overline{V}_{sbN} - \overline{V}_{nN}\right) \frac{Z_{lb}}{Z_{lb} + Z_s}$$

$$= [(1\angle - 120°) - (0.7108\angle - 20.69°)] \frac{1.2 + j1.3}{1.2 + j1.3 + 0.01 + j0.1}$$

$$= 1.2594\angle - 154.05° \text{ p.u.}$$

The load voltage $\overline{V}_{lcn}$, without compensation is computed as given below.

$$
\begin{aligned}
\overline{V}_{lcn} &= \left(\overline{V}_{scN} - \overline{V}_{nN}\right) \\
&= [(1\angle 120°) - (0.7108\angle - 20.69°)] \\
&= 1.6141\angle 136.2° \text{ p.u.}
\end{aligned}
$$

We can observe here that the load voltages with respect to the load neutral are highly unbalanced in magnitude and in phase angles. Now, these load voltages have to be maintained balanced with 1.0 p.u. as rms magnitude. For this, the DVR voltages are calculated as below.

(b) **With DVR**

Let the new load voltages with respect to the load neutral, are balanced with rms value of 1 p.u., as expressed below.

$$
\begin{aligned}
\overline{V}_{lan} &= 1.0\angle 0° \text{ p.u.} \\
\overline{V}_{lbn} &= 1.0\angle - 120° \text{ p.u.} \qquad\qquad (6.39) \\
\overline{V}_{lbn} &= 1.0\angle 120° \text{ p.u.}
\end{aligned}
$$

Therefore,

$$\overline{V}_{lan} + \overline{V}_{lbn} + \overline{V}_{lcn} = 0 \qquad\qquad (6.40)$$

If load impedances are Z_{la}, Z_{lb}, Z_{lc}, then the load currents are:

$$
\begin{aligned}
\overline{I}_{la} &= \frac{\overline{V}_{lan}}{Z_{la}} \\
\overline{I}_{lb} &= \frac{\overline{V}_{lbn}}{Z_{lb}} \qquad\qquad (6.41) \\
\overline{I}_{lc} &= \frac{\overline{V}_{lcn}}{Z_{lc}}
\end{aligned}
$$

Also, due to three-phase three-wire configuration,

$$\overline{I}_{la} + \overline{I}_{lb} + \overline{I}_{lc} = 0 \qquad\qquad (6.42)$$

Thus,

$$\frac{\overline{V}_{lan}}{Z_{la}} + \frac{\overline{V}_{lbn}}{Z_{lb}} + \frac{\overline{V}_{lcn}}{Z_{lc}} = 0 \qquad\qquad (6.43)$$

The (6.43) indicates that impedances must be such that the zero sequence component of the current does not exist. In other words, for balanced load voltages, the above equation is true for certain values of load impedances. One of the possibilities is,

$$Z_{la} = Z_{lb} = Z_{lc} = Z_l \tag{6.44}$$

This indicates a three-phase balanced load, and thus the three-phase currents are balanced. Other values for the load impedances can be derived based on the condition (6.43), i.e., by equating its real and imaginary parts to zero.

In this problem, for the given load impedances, the load currents are computed as below.

$$\bar{I}_{la} = \frac{\bar{V}_{lan}}{Z_{la}} = \frac{1\angle 0°}{0.3 + j0.4} = 2\angle -53.13° \text{ p.u.}$$

$$\bar{I}_{lb} = \frac{\bar{V}_{lbn}}{Z_{lb}} = \frac{1\angle -120°}{1.2 + j1.3} = 0.56\angle -167.29° \text{ p.u.}$$

$$\bar{I}_{lc} = \frac{\bar{V}_{lcn}}{Z_{lc}} = \frac{1\angle 120°}{\infty} = 0 \text{ p.u.}$$

Clearly, $\bar{I}_{la} + \bar{I}_{lb} + \bar{I}_{lc} \neq 0$, therefore it is not possible to maintain balanced load voltages with respect to the load neutral. However, it is advisable to maintain balanced load voltages with respect to system neutral so that other balanced loads connected to the load bus do not experience unbalanced voltages with respect to their neutral. Therefore, the load voltages with respect to the system neutral are considered as follows.

$$\begin{aligned}
\bar{V}_{laN} &= 1.0\angle 0° \text{ p.u.} \\
\bar{V}_{lbN} &= 1.0\angle -120° \text{ p.u.} \\
\bar{V}_{lcN} &= 1.0\angle 120° \text{ p.u.}
\end{aligned} \tag{6.45}$$

With these new load voltages with respect to the system neutral, the neutral voltage, $\bar{V}_{nN}$ is computed as follows.

$$\begin{aligned}
\bar{V}_{nN} &= \frac{1}{\frac{1}{Z_{la}} + \frac{1}{Z_{lb}} + \frac{1}{Z_{lc}}} \left[\frac{\bar{V}_{laN}}{Z_{la}} + \frac{\bar{V}_{lbN}}{Z_{lb}} + \frac{\bar{V}_{lcN}}{Z_{lc}} \right] \\
&= \frac{1}{\frac{1}{0.3+j0.4} + \frac{1}{1.2+j1.3} + 0} \left[\frac{1\angle 0°}{0.3 + j0.4} + \frac{1\angle -120°}{1.2 + j1.3} + 0 \right] \\
&= 0.6854 - j0.2167 = 0.7188\angle -17.54° \text{ p.u.}
\end{aligned}$$

The load currents are given as follows.

$$\bar{I}_{la} = \frac{\bar{V}_{laN} - \bar{V}_{nN}}{Z_{la}} = \frac{1\angle 0° - 0.7188\angle - 17.54°}{0.3 + j0.4}$$

$$= 0.764\angle - 18.58° \text{ p.u.}$$

$$\bar{I}_{lb} = \frac{\bar{V}_{lbN} - \bar{V}_{nN}}{Z_{lb}} = \frac{1\angle - 120° - 0.7188\angle - 17.54°}{1.2 + j1.3}$$

$$= 0.764\angle 161.42° \text{ p.u.}$$

$$\bar{I}_{lc} = \frac{\bar{V}_{lcN} - \bar{V}_{nN}}{Z_{lc}} = \frac{1\angle 120° - 0.7188\angle - 17.54°}{\infty} = 0 \text{ p.u.}$$

The angle of phase-a voltage injected by the DVR in this case will be

$$\angle \bar{V}_{fa} = 90° - 18.58° = 71.42°.$$

The values of a_{1a} and b_{1a} is computed as follows:

$$a_{1a} + jb_{1a} = 1.0\angle \bar{V}_{fa} = 0.3186 + j0.9479 \text{ p.u.}$$

Above equation implies that $a_{1a} = 0.3186$ and $b_{1a} = 0.9479$. Further, voltage across the feeder (Z_{sa}) in phase-a is given as

$$Z_{sa}\bar{I}_{la} = (0.01 + j0.1) \times 0.764\angle - 18.58° = 0.0316 + j0.07 \text{ p.u.}$$
$$= a_{2a} + jb_{2a}.$$

From above equation, $a_{2a} = 0.0316$ and $b_{2a} = 0.07$. Now, the voltage V_{fa} is computed by solving the following equation.

$$V_{fa}^2 - 2\left[(1 + a_{2a})a_{1a} + b_{1a}b_{2a}\right]V_{fa} + (1 + a_{2a})^2 + b_{2a}^2 - 1.0 = 0.$$

Substituting variables, the following quadratic equation is obtained.

$$V_{fa}^2 - 0.7899 V_{fa} + 0.069 = 0.$$

Solving the equation, $V_{fa} = 0.6898$ p.u. and 0.1001 p.u. are obtained. However, $V_{fa} = 0.1001$ p.u. is the feasible solution due to the lower value of voltage. Hence, $\bar{V}_{fa} = 0.1001\angle 71.42°$ p.u.

Terminal voltage after compensation:

$$\bar{V}_{taN} = -\bar{V}_{fa} + \bar{V}_{laN}$$
$$= -0.1001\angle 71.42° + 1\angle 0$$
$$= 0.9727\angle - 5.6° \text{ p.u.}$$

Source voltage after compensation:

$$
\begin{aligned}
\overline{V}_{saN} &= \overline{V}_{taN} + \overline{I}_{la}Z_s \\
&= 0.9727\angle - 5.6° + [(0.764\angle - 18.58°) \times (0.01 + j0.1)] \\
&= 1.0\angle - 1.43° \text{ p.u.}
\end{aligned}
$$

Effective source voltage after compensation:

$$
\begin{aligned}
\overline{V}'_{saN} &= \overline{V}_{saN} + \overline{V}_{fa} \\
&= 1.0\angle - 1.43° + 0.1001\angle 71.42° \\
&= 1.034\angle 3.88° \text{ p.u.}
\end{aligned}
$$

Similarly, phase-b DVR voltage can be calculated as following.

$$
\angle \overline{V}_{fb} = 90° + 161.42° = 251.42°.
$$

The values of a_{1b} and b_{1b} is computed as follows:

$$
a_{1b} + jb_{1b} = 1.0\angle \overline{V}_{fb} = -0.3186 - j0.9879 \text{ p.u.}
$$

Above equation implies that a_{1b} = -0.3186 and b_{1b} = -0.9879. Further, voltage across the feeder (Z_{sb}) is given as

$$
\begin{aligned}
Z_{sb}\overline{I}_{lb} &= (0.01 + j0.1) \times 0.764\angle 161.42° = -0.0316 - j0.07 \text{ p.u.} \\
&= a_{2b} + jb_{2b}.
\end{aligned}
$$

The above equation implies that a_{2b} = -0.0316 and b_{2b} = -0.07. Now, the voltage V_{fb} is computed by solving following equation.

$$
V_{fb}^2 - 2(a'_{2b}a_{1b} + b'_{2b}b_{1b})V_{fb} + a'_{2b}{}^2 + b'_{2b}{}^2 - 1 = 0. \qquad (6.46)
$$

where, $a'_{2b} = V_{lbN}\cos\angle \overline{V}_{lbN} + a_{2b} = 1.0 \times \cos(-120°) - 0.0316 = -0.5316, b'_{2b} = V_{lbN}\sin\angle \overline{V}_{lbN} + b_{2b} = 1.0 \times \sin(-120°) - 0.07 = -0.936$.

Substituting variables, the following quadratic equation is used to find out DVR voltage.

$$
V_{fb}^2 - 2.1132V_{fb} + 0.1587 = 0.
$$

Solving the equation, V_{fb} = 2.0352 p.u. and 0.078 p.u. are obtained. However, V_{fb} = 0.078 p.u. is the least voltage and is the feasible solution. Hence, $\overline{V}_{fb} = 0.078\angle - 108.58°$ p.u.

Terminal voltage after compensation:

$$
\begin{aligned}
\overline{V}_{tbN} &= -\overline{V}_{fb} + \overline{V}_{lbN} \\
&= -0.078\angle -108.58° + 1\angle -120° \\
&= 0.9237\angle -120.96°\ \text{p.u.}
\end{aligned}
$$

Source voltage after compensation:

$$
\begin{aligned}
\overline{V}_{sbN} &= \overline{V}_{tbN} + \overline{I}_{lb}Z_s \\
&= 0.9237\angle -120.96° + [(0.764\angle 161.42°) \times (0.01 + j0.1)] \\
&= 1.0\angle -120.45°\ \text{p.u.}
\end{aligned}
$$

Effective source voltage after compensation:

$$
\begin{aligned}
\overline{V}'_{sbN} &= \overline{V}_{sbN} + \overline{V}_{fb} \\
&= 1.0\angle -120.45° + 0.078\angle -108.58° \\
&= 1.0764\angle -119.6°\ \text{p.u.}
\end{aligned}
$$

In phase-c, the load is open-circuited. Therefore, the current in this phase will be zero. Hence, the load voltage will be the same as the source voltage i.e., $V_{lcN} = 1.0\angle 120°$ p.u. Consequently, the voltage drop in the feeder will be zero, and DVR in phase-c does not inject any voltage.

6.9   SUMMARY

Various voltage compensating passive and active devices are used to mitigate voltage disturbances in the power network. A series active power filter denoted as a dynamic voltage restorer (DVR) is commonly used to compensate for voltage sags, swells, and interruptions in the power distribution network. This chapter describes the concept, operation and analysis of single-phase and three-phase DVR compensated systems. A number of examples are given to illustrate the concept, which can be further extended to other methods of DVR control. A mathematical and graphical representation of DVR is interpreted to find the series voltage injected by the DVR. In the case of a non-feasible solution, the maximum compensation capacity of the DVR voltage is computed (without active power support). Transient operation of DVR and switching control are discussed to realize a practical DVR.

6.10   PROBLEMS

P 6.1 Explain the need for voltage compensation. What are the different methods to regulate the voltage in a power distribution system?

P 6.2 Explain the concept of dynamic voltage restorer (DVR). Draw the single-line diagram for DVR compensated system. Discuss its various components.

P 6.3 Discuss the operating principle of DVR. Explain its operation for the following cases.

(a) When line resistance is negligible.

(b) When line resistance is not negligible.

(c) General case.

P 6.4 A single-phase load is supplied from a supply at a given voltage (V_s) through a feeder impedance, $Z_s = R_s + jX_s$. The load voltage is specified as $\overline{V}_t = 1.0\angle 0°$. Draw the circuit diagram for DVR compensated system and explain mathematically how would you compute DVR voltage for a given load.

P 6.5 Explain the transient operation of the DVR for (a) single-phase with supply voltage with harmonics (b) three-phase balanced system without harmonics and (c) three-phase unbalanced system with harmonics.

P 6.6 Explain how would you realize a DVR circuit with a voltage source inverter (VSI) to compensate for the estimated voltage. Draw its equivalent circuit. Discuss the switching operation of the VSI to synthesize the computed voltage.

P 6.7 Is there any maximum compensation capacity for the DVR without real power support? If yes, how do you tackle the situation in case you are not able to meet the required compensation?

P 6.8 Under what conditions it is not possible to maintain the load voltage at the supply voltage (i.e., $V_l = V_s$) with a dynamic voltage regulator (DVR) without real power support? What is the compensation limit under this condition?

P 6.9 If three-phase PCC voltages have both unbalanced and harmonics, how will you generate the reference voltage of the DVR to be compensated? Explain the real-time implementation of dynamic voltage restorer for this system.

P 6.10 Consider a system with rms supply voltage of 230 V (1 p.u.), 50 Hz as shown in Fig. 6.8. The feeder and load impedances of the system are $0.1 + j0.2$ p.u. and $1 + j0.6$ p.u., respectively. For this system,

(a) Compute load current and load voltage when DVR is not in operation.

(b) Compute load current and DVR injected voltage such that $|\overline{V}_l| = |\overline{V}_s|$.

(c) Compute load current and DVR injected voltage such that $|\overline{V}_l| = |\overline{V}_s|$, whereas load is replaced by an inductive load of $j2$ p.u.

P 6.11 For the system given in Fig. 6.8, the feeder and load impedances are $0.1 + j0.3$ p.u. and $0.8 + j0.9$ p.u., respectively. The DVR injects a voltage such that the load voltage magnitude is maintained at 1 p.u. For this compensation system, compute the DVR voltage and rating for the following cases.

(a) Source voltage magnitude is maintained at 1 p.u. i.e., $|\overline{V}_s| = 1$ p.u.

(b) During voltage sag of 20% i.e., $|\overline{V}_s| = 0.8$ p.u.

(c) During voltage swell of 20% i.e., $|\overline{V}_s| = 1.2$ p.u.

P 6.12 As shown in Fig. 6.12, the source is supplying a nonlinear load through a feeder impedance of $0.2 + j0.6$ p.u. The load current, i_l, is represented by the following expression.

$$i_l = \sqrt{2} \sin(\omega t - 30°) + 0.2\sqrt{2}\sin(3\,\omega t - 90°)\,\text{p.u.}$$

(a) Compute load voltage when DVR is not operated, i.e., $v_f = 0$.

(b) If possible, compute DVR injected voltage such that $|\overline{V}_l| = |\overline{V}_s| = 1$ p.u. If not possible, compute the maximum possible voltage that can be maintained at the load terminal by the DVR without taking any real power from the dc bus.

P 6.13 A DVR compensated single-phase system is shown in Fig. 6.17. Determine $\overline{V}_{D2}$ and $\overline{V}_{D1}$ such that the voltage remains 1.0 p.u. at all three buses. Also, determine the value of δ_1 and δ to ensure power flow from source to load. The value of relevant parameters, all in per unit, are given as follows.
$Z_{s1} = 0.01 + j0.2$, $Z_{s2} = 0.02 + j0.5$, $Z_{l1} = 0.35 + j0.6$ and $Z_{l2} = 0.2 + j0.35$.

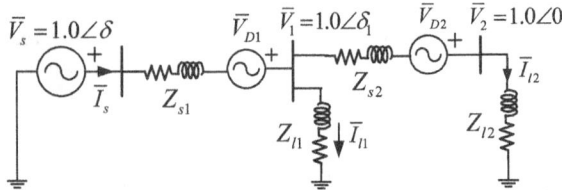

Figure 6.17   A single-phase DVR compensated system

P 6.14 A DVR compensated single-phase system is shown in Fig. 6.17. The values of relevant parameters are shown in per unit. A fault takes place in the power distribution system, which reduces source voltage (V_s) to 0.8 p.u. Due to this and in order to minimize the DVRs rating, the voltages at the two load buses are maintained at 0.9 p.u. Determine $\overline{V}_{D2}$ and $\overline{V}_{D1}$ under above conditions. Assume the following values of the load and feeder impedances.
$Z_{l2} = 0.4 + j0.3$, $Z_{s2} = 0.05 + j0.2$, $Z_{l1} = 0.3 + j0.7$ and $Z_{s1} = 0.02 + j0.4$ p.u.

P 6.15 A single-phase supply, 230 V (rms), 50 Hz has source impedance, $Z_s = R_s + jX_s = 0.2 + j5.0\,\Omega$. It is supplying a load of $Z_{L1} = 25 + j15\,\Omega$. The load $Z_{L2} = 10 + j20\,\Omega$ is connected in parallel with mains for 25 cycles through the switch, as shown in Fig. 6.18.

(a) Calculate the voltage sag/swell at the load terminal under steady state conditions (without DVR).

(b) If the dc bus-based voltage source inverter (without taking any active power) in series with the load to maintain its terminal at rated voltage (230 V, rms), calculate

   (i) The line current

   (ii) DVR voltage, $v_f(t)$

   (iii) DVR rating

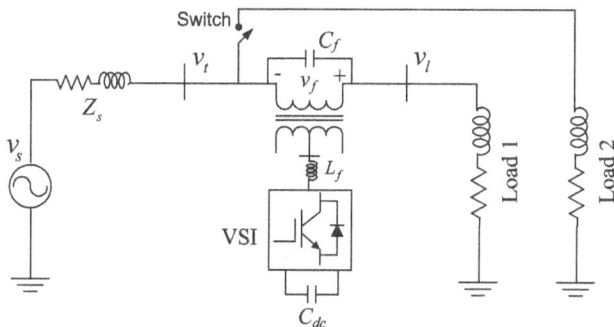

Figure 6.18   A single-phase DVR with load switching

P 6.16 A three-phase balanced system is shown in Fig. 6.14. The source voltages are balanced sinusoids with rms phase value of 230 V (1.0 p.u.). The feeder and load impedances in each phase are $0.01 + j0.1$ and $3 + j1$ p.u., respectively. Compute the currents and DVR injected voltages in all three phases such that $|\overline{V}_l| = |\overline{V}_s| = 1.0$ p.u. in all three phases.

P 6.17 A three-phase four-wire balanced system with a feeder impedance of $0.01 + j0.1$ p.u. in all three phases is supplying to a star connected unbalanced load of $R_a + jX_a = 0.6 + j0.3$ p.u., $R_b + jX_b = 0.4 + j0.4$ p.u. and $R_c + jX_c = 0.2 + j0.5$ p.u. Compute the voltages injected by DVR in terms of magnitude and phase angle in all three phases to maintain $|\overline{V}_l| = |\overline{V}_s| = 1.0$ p.u.

P 6.18 A three-phase three-wire supply with 400 V (line to line, rms), 50 Hz has source impedance of $Z_s = R_s + jX_s = 0.25 + j2.5\,\Omega$. It is supplying a three-phase star connected balanced load of $Z_L = 24 + j18\,\Omega$. A three-phase DVR is connected at the load bus.

   (a) Draw the circuit diagram of the above system.

   (b) Calculate the load voltages without DVR.

   (c) Now a three-phase DVR is employed by using a dc bus-based voltage source inverter (without taking any active power) in series with the load to maintain its terminal at rated voltage (400 V, line to line, rms), calculate the following.

(i) The line currents.

(ii) DVR voltage, $v_{fa}(t), v_{fb}(t), v_{fc}(t)$, effective source voltages.

(iii) DVR rating.

P 6.19 A three-phase four-wire balanced system with feeder impedances of $R_{sa} + jX_{sa} = 0.01 + j0.1$ p.u., $R_{sb} + jX_{sb} = 0.005 + j0.12$ p.u. and $R_{sc} + jX_{sc} = 0.008 + j0.09$ p.u. is supplying to a star connected unbalanced load of $R_a + jX_a = 0.6 + j0.3$ p.u., $R_b + jX_b = 0.4 + j0.4$ p.u. and $R_c + jX_c = 0.2 + j0.5$ p.u. Compute magnitude and phase angle of voltages injected by DVR in all three phases to maintain $|\overline{V}_l| = |\overline{V}_s| = 1.0$ p.u.

P 6.20 A three-phase four-wire balanced source, 1.0 p.u. rms per phase is feeding to three single-phase full-wave rectifiers. Assume the load currents to be constant under varying conditions of the supply voltage, and these are given as follows.

$$i_{la} = 1.0\sqrt{2}\sin(\omega t - 30°) + 0.3\sqrt{2}\sin(3\omega t - 90°)\,\text{p.u.}$$
$$i_{lb} = 1.0\sqrt{2}\sin(\omega t - 150°) + 0.3\sqrt{2}\sin(3\omega t - 90°)\,\text{p.u.}$$
$$i_{lc} = 1.0\sqrt{2}\sin(\omega t - 270°) + 0.3\sqrt{2}\sin(3\omega t - 90°)\,\text{p.u.}$$

The feeder impedances in all three phases are $0.05 + j0.2$ p.u. In the absence of the DVR, the load current is measured with respect to the source voltage, $\overline{V}_s = 1\angle0°$ p.u., assuming the feeder impedance is zero. Although there is a finite feeder impedance given in the problem, this assumption is required to simplify the analytical solution. With the DVR, the given load current is with respect to the load voltage.

(a) Compute load voltage without DVR compensation.

(b) If possible, compute DVR injected voltage such that $|\overline{V}_l| = |\overline{V}_s| = 1.0$ p.u. in three respective phases. If not possible, compute the maximum possible voltage that can be maintained at the load terminal with the DVR without taking any real power from the dc bus.

(c) Compute voltage injected by DVR if there is balanced voltage sag of 20%, i.e., $|\overline{V}_s| = 0.8$ p.u.

(d) Compute voltage injected by DVR if there is a balanced voltage swell of 20%, i.e., $|V_s| = 1.2$ p.u.

[Hint: To simplify the analysis, the current is considered to be constant with respect to the source voltage without DVR and with respect to the load voltage with DVR]

REFERENCES

1. Central Electricity Regulatory Commission, "Grid security – need for tightening of frequency band and others measures," March 2011.

2. Central Electricity Regulatory Commission, "Report of the expert group: Review of indian electricity grid code," 2020.

3. IEEE Standard 1159–2019, "IEEE recommended practice for monitoring electric power quality," 2019.

4. IEEE Std 1547a-2020, "IEEE standard for interconnection and interoperability of distributed energy resources with associated electric power systems interfaces–amendment 1: To provide more flexibility for adoption of abnormal operating performance category iii," pp. 1–16, 2020.

5. IEC 60034-1:2022, "Rotating electrical machines-part 1: Rating and performance," 2022.

6. D. Divan, G. A. Luckjiff, W. E. Brumsickle, J. Freeborg, and A. Bhadkamkar, "A grid information resource for nationwide real-time power monitoring," *IEEE Transactions on Industrial Application*, vol. 40, no. 2, pp. 699–705, March/April 2004.

7. M. H. J. Bollen and E. Styvaktakis, "Characterization of three-phase unbalanced dips (as easy as one-two-three?)," *2000 Power Engineering Society Summer Meeting*, vol. 2, pp. 899–904, 2000.

8. M. H. J. Bollen, "Definitions of voltage unbalance," *IEEE Power Engineering Review*, vol. 22, no. 11, pp. 49–50, 2002.

9. EPRI 2003, "Distribution system power quality assessment: Phase ii—voltage sag and interruption analysis," *Electr. Power Res. Inst. (EPRI)*, 2003.

10. S. Subramanian and Mishra, Mahesh K., "Interphase ac–ac topology for voltage sag supporter," *IEEE Transactions on Power Electronics*, vol. 25, no. 2, pp. 514–518, 2010.

11. G. S. Kumar, B. K. Kumar, and Mahesh K. Mishra, "Mitigation of voltage sags with phase jumps by UPQC with PSO-Based ANFIS," *IEEE Transactions on Power Delivery*, vol. 26, no. 4, pp. 2761–2773, 2011.

12. S. Jothibasu and Mahesh K. Mishra, "A control scheme for storageless DVR based on characterization of voltage sags," *IEEE Transactions on Power Delivery*, vol. 29, no. 5, pp. 2261–2269, 2014.

13. S. Jothibasu and Mahesh K. Mishra, "An improved direct ac–ac converter for voltage sag mitigation," *IEEE Transactions on Industrial Electronics*, vol. 62, no. 1, pp. 21–29, 2015.

14. M. Pradhan and Mahesh K. Mishra, "Dual p-q theory based energy-optimized dynamic voltage restorer for power quality improvement in a distribution system," *IEEE Transactions on Industrial Electronics*, vol. 66, no. 4, pp. 2946–2955, 2019.

15. A. Ghosh and G. Ledwich, "Compensation of distribution system voltage using DVR," *IEEE Transactions on Power Delivery*, vol. 17, no. 4, pp. 1030–1036, Oct. 2002.

16. A. Ghosh and G. Ledwich, "Structures and control of a dynamic voltage regulator (DVR)," in *IEEE Power Engineering Society Winter Meeting*, vol. 3. IEEE, 2001, pp. 1027–1032.

17. A. Ghosh and G. Ledwich, *Power Quality Enhancement Using Custom Power Devices*, ser. Power Electronics and Power Systems. Springer US, 2012.

18. Bhim Singh, Ambrish Chandra, and Kamal Al-Haddad, *Power Quality Problems and Mitigation Techniques*. Wiley, 2015.

19. A. Ghosh, A. K. Jindal, and A. Joshi, "Design of a capacitor-supported dynamic voltage restorer (DVR) for unbalanced and distorted loads," *IEEE Transactions on Power Delivery*, vol. 19, no. 1, pp. 405–413, Jan. 2004.

20. S. Sasitharan and Mahesh K. Mishra, "Constant switching frequency band controller for dynamic voltage restorer," *IET Power Electronics*, vol. 3, no. 5, pp. 657–667, Sept. 2010.

21. S. Sasitharan, Mahesh K. Mishra, B. Kalyan Kumar, and V. Jayashankar, "Rating and design issues of DVR injection transformer," *International Journal of Power Electronics*, vol. 2, no. 2, pp. 143–163, 2010.

7 Unified Power Quality Conditioner

7.1 INTRODUCTION

Modern ac distribution network faces two types of power quality issues. One is about voltage variations in the supply voltage, which may arise due to faults, switching transients, lightning, etc. This may result in voltage sag, swell, interruption, notches, spikes, unbalance, and harmonics in the power system. The second aspect of power quality refers to the quality of load current drawn from the supply under normal and abnormal voltage conditions. The poor quality of load current may be due to saturation of the magnetic core of electric machines, power electronics based loads, faults, switching transients, etc., resulting in poor power factor, unbalance, and harmonics. Under these conditions, the power system utilities are required to provide an uninterrupted power supply to the loads. While loads are sensitive to voltage and current deviations, the utilities are affected by the quality of the current drawn by the load. This puts requirements of voltage compensation from the load's perspective and current compensation from the utility perspective. To meet these aspects of compensation in the power distribution network, a Unified Power Quality Conditioner (UPQC) is used as one of the custom power devices.

Unified power quality conditioner (UPQC) is relatively the latest device in the family of custom power devices [1]–[10]. UPQC is a versatile custom power device that consists of two inverters connected back-to-back with a common DC-link and deals with both load current and supply voltage compensation. One voltage source inverter is connected in series to inject the required voltage, and another voltage source inverter is connected in shunt with the load to compensate for the current. Thus, the main purpose of the UPQC is to compensate for supply voltage flicker/imbalance, reactive power, negative-sequence current, and harmonics. UPQC's capability of correcting the voltage and current at the point of installation on power distribution systems or industrial power systems makes it the most powerful solution to power quality issues. If we consider the functionality of UPQC, it gives the combined features of a shunt active power filter or Distribution Static Compensator (DSTATCOM) and a series active power filter or Dynamic Voltage Restorer (DVR). The combined features of DSTATCOM and DVR fulfill different objectives like maintaining a sinusoidal nominal voltage at the bus at which it is connected, maintaining voltage when there are voltage sags and swells in the system, eliminating harmonics in the source currents and load voltages, load balancing, and compensation of reactive power.

DOI: 10.1201/9781032617305-7

7.2 UPQC STRUCTURE

UPQC can be classified based on supply system (single-phase, three-phase three-wire or three-phase four-wire configuration), converter topology (voltage source inverter or current source inverter), and voltage sag/swell compensation approach such as UPQC-P, UPQC-Q, UPQC-S [9]. A basic topology of UPQC is shown in Fig. 7.1. The UPQC employs voltage source inverters (VSIs), which are connected back-to-back with a common dc energy storage capacitor. One of these two VSIs is connected in series with the feeder, denoted as Dynamic Voltage Restorer (DVR), and the other is connected in parallel to the same feeder, denoted as Distribution Static Compensator (DSTATCOM). The main purpose of the series connected VSI (or series active power filter) is to mitigate the voltage related problems in the distribution systems such as voltage sags/swells, flicker, voltage unbalance and harmonics at the point of common coupling (PCC). It injects voltages so as to maintain the load voltages at a desired level, balanced and distortion free. The ac side of the series VSI is connected in series with the feeder using an injection transformer and a low pass filter. The series low pass filter components, C_{se}, L_{se}, prevent the switching frequency harmonics produced by the series VSI from entering the system. The series injection transformers provide voltage matching and isolation between the network and the VSI.

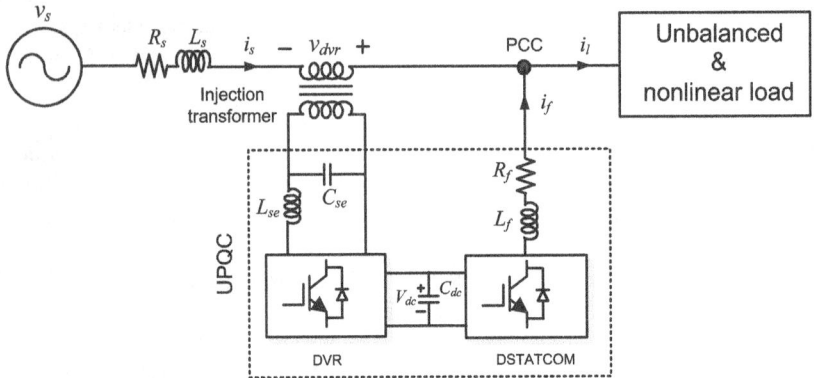

Figure 7.1   Schematic diagram of right shunt UPQC

On the right side of Fig. 7.1, a shunt connected VSI (shunt active power filter) is connected at the Point of Common Coupling (PCC). The shunt active filter is realized by a VSI connected to the common dc storage capacitor on the dc side, and on the ac side it is connected in parallel with the load through the shunt interface inductor (L_f). The purpose of the interfacing inductor is to shape the filter currents injected into the PCC. The shunt VSI injects currents into the PCC, such that the unbalance, reactive, and harmonic components of the load currents are compensated.

The dc link with the dc storage capacitor serves two main purposes: it maintains the dc voltage and it serves as an energy storage element to supply a real power

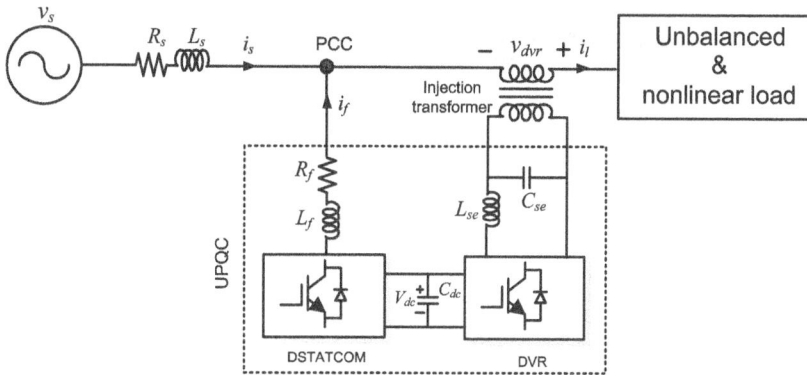

Figure 7.2   Schematic diagram of left shunt UPQC

difference between the load and source during the transient period. The average voltage across the dc capacitor is maintained constant and this is achieved through an appropriate proportional-integral (PI) control by regulating the amount of active current drawn by the shunt active filter from the system. Since the shunt active power filter is connected to the right side, this UPQC topology is called as right shunt UPQC topology. If the shunt VSI is connected to the left side, as shown in Fig. 7.2, the topology is known as left shunt topology. Ideally, a series active power filter is a voltage source, and a shunt active power filter is a current source. Therefore, the series VSI (DVR) and shunt VSI (DSTATCOM) can be shown as voltage and current source, respectively. The right shunt and left shunt UPQC with an ideal voltage source and current source are shown in Fig. 7.3 (a) and (b), respectively.

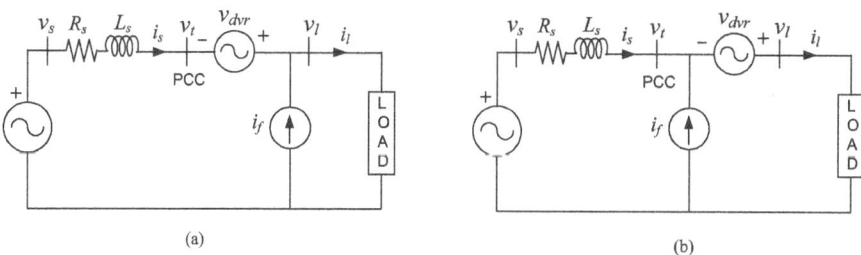

Figure 7.3   An ideal UPQC configuration (a) Right shunt and (b) Left shunt

The right shunt UPQC can operate in zero power injection/absorption mode and can make the power factor unity at the load terminal. For left shunt UPQC, it is not possible to maintain a unity power factor at the load bus, as it depends on load currents. In the right shunt UPQC, the shunt VSI can supply the entire load reactive power requirement, whereas the left shunt configuration can partially supply the load reactive power [8]. Due to the above-mentioned merits, right shunt configuration has

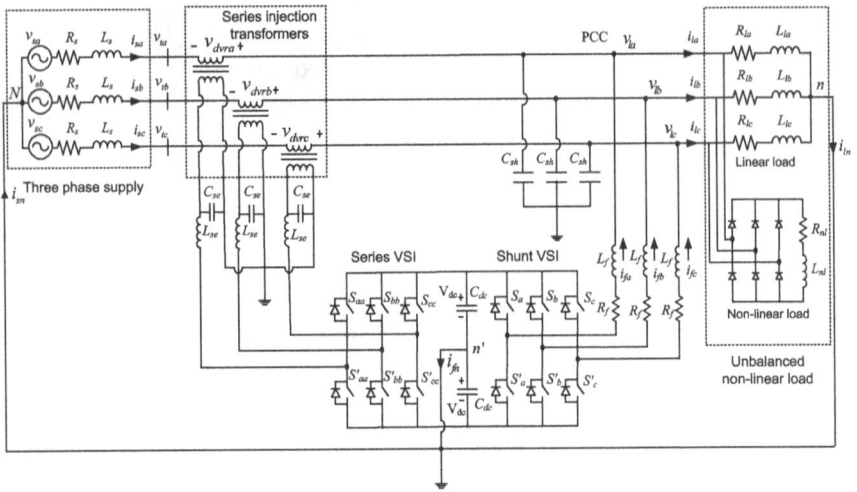

Figure 7.4   Power circuit of three-phase four-wire UPQC

been widely used in applications [7], [8]. In the following discussion, we shall focus on the right shunt UPQC and discuss its design, control, and operational features. Readers can study left shunt UPQC in a similar way [8], [11], [12]. The discussed ideas will be illustrated through solved examples.

Fig. 7.4 shows a complete power circuit of the neutral clamped VSI topology based right shunt UPQC. In this figure, v_{sa}, v_{sb} and v_{sc} are source voltages of phases a, b, and c, respectively. Similarly, v_{ta}, v_{tb} and v_{tc} are the terminal voltages. The voltages v_{dvra}, v_{dvrb}, and v_{dvrc} are the voltage injected by the series active filter. The source currents in three-phase are represented by i_{sa}, i_{sb} and i_{sc}, load currents are represented by i_{la}, i_{lb} and i_{lc}. The shunt active filter currents are denoted by i_{fa}, i_{fb}, i_{fc}, and i_{ln} represents the current in the neutral leg. The parameters, L_s and R_s, represent the feeder inductance and resistance, respectively. The interfacing inductance of the shunt active filter is represented by L_f with its series resistance, R_f, respectively. The shunt filter capacitor is represented by C_{sh}. The interfacing inductance and filter capacitor of the series active filter are represented by L_{se} and C_{se}, respectively. The load is constituted of both linear and nonlinear loads as shown in this figure. The dc link capacitors and voltages across them are represented by $C_{dc1} = C_{dc2} = C_{dc}$ and $V_{dc1} = V_{dc2} = V_{dc}$, respectively. Switches S_a, S_b, S_c are top switches and S'_a, S'_b, S'_c are bottom switches of the shunt VSI. Similarly, S_{aa}, S_{bb}, S_{cc} are top switches and $S'_{aa}, S'_{bb}, S'_{cc}$ are bottom switches of the series VSI.

Some other topological variations of UPQC are shown in Figs. 7.5 and 7.6. Fig. 7.5 illustrates three-phase three-wire UPQC structure and employs only one dc storage capacitor. Since it has one dc capacitor, there is no problem of capacitor imbalance. However, if this UPQC topology is used for a three-phase four-wire system, then the neutral load current due to unbalanced and harmonic components of load

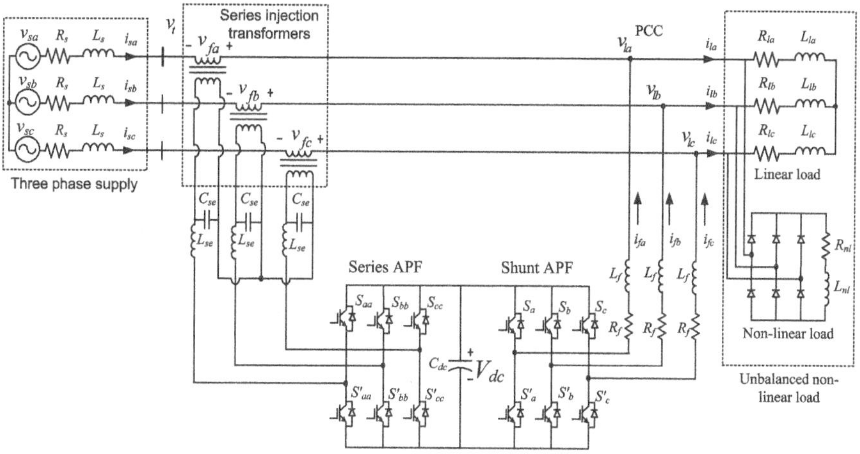

Figure 7.5   Three-phase three-wire UPQC topology

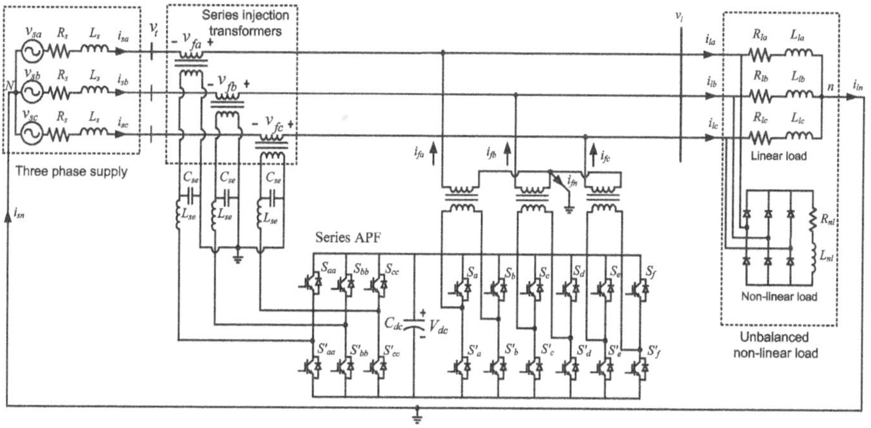

Figure 7.6   Three-phase four-wire H-bridge UPQC topology

can not be compensated by it. UPQC topology shown in Fig. 7.6 represents three-phase four-wire H bridge topology. This allows neutral current compensation, but dc components of load currents can not be compensated due to the presence of a transformer on the shunt VSI circuit. One more interesting UPQC topology with reduced dc link voltage requirement is shown in Fig. 7.7. In this circuit, the common neutral is connected to the negative terminal of the dc bus along with the capacitor, C_f in series with the interfacing inductance of the shunt VSI. The passive capacitor, C_f has the capability to supply a part of the reactive power required by the load, and the active filter will compensate for the remaining reactive power and harmonics present in the load. The addition of a capacitor in series with the interfacing inductor of the shunt active filter will significantly reduce the dc-link voltage requirement and

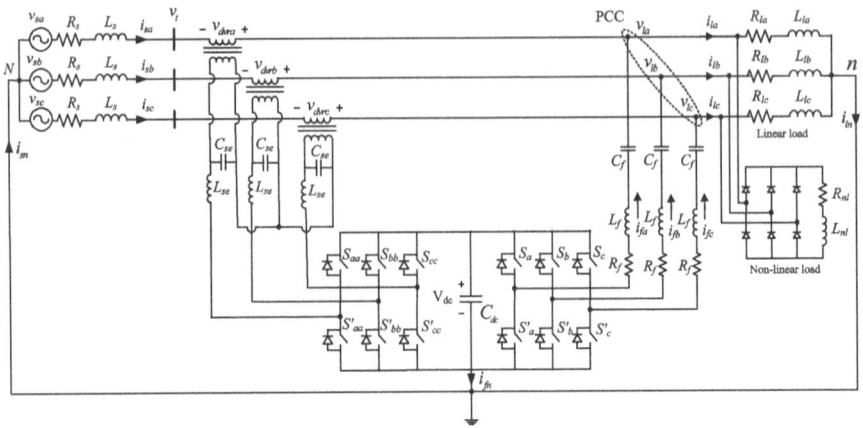

Figure 7.7 Three-phase four-wire reduced dc link voltage UPQC topology

consequently reduces the average switching frequency of the switches [13]. In this chapter, a generic UPQC topology, shown in Fig. 7.4 will be considered for further discussion.

7.3 OPERATION AND CONTROL OF UPQC

The operation of UPQC is achieved by generating the reference voltage and current quantities and tracking them using voltage source inverters. Current reference quantities are tracked using shunt VSI as shown in Fig. 7.4. This will be referred to as shunt VSI or DSTATCOM operation. The voltage reference quantities are synthesized using series VSI and this is termed as series VSI or DVR operation. The shunt and series VSIs are connected back to back through a common dc link supported by dc storage capacitor(s). The capacity and size of dc storage capacitors are chosen to handle the voltage and current disturbances or transients of a given maximum duration. The voltage level of the dc link voltage is chosen based on the requirement of the shunt (DSTATCOM) and series (DVR) operation for which they are designed. Normally the dc link voltage requirement for shunt and series active filters is not the same and it is a challenging task to design common DC-link voltage in order to achieve satisfactory shunt and series compensation [14]. The UPQC structure shown in Fig. 7.4 is for a three-phase four-wire system and represents a generic UPQC [15]. The neutral of load and system is clamped to the midpoint of two dc storage capacitors. This arrangement enables independent control of each leg of both shunt and series VSIs. Since the UPQC topology has two dc capacitors, a capacitor voltage balancing circuit is required to equalize voltage across these dc capacitors [16], [17]. The other variants of UPQC with a three-phase four-wire system are given in [13], [18].

As mentioned above, UPQC is a combination of DSTATCOM and DVR, hence the concepts mentioned in Chapter 5 and Chapter 6, can be used to generate reference current and voltage quantities. These are then tracked using switching schemes of the shunt and series active filters as discussed in these chapters. There are two levels of control in the increasing order of bandwidth. The first is about the control of dc link voltage, which requires relatively less bandwidth. The second level of control concerns to the generation of the reference qualities of the UPQC with higher bandwidth involving the switching control of the converters [19]. These design aspects of the UPQC filter components and parameters are discussed in [10], [13], [20].

The choice of control scheme exhibits the desired operation and behavior of UPQC and determines the reference signals (current for shunt active power filter and voltage for series active power filter). These reference quantities are tracked using switching methods for VSIs, such that the desired performance is achieved. Broadly, there are two types of control schemes in the literature, i.e., frequency domain approach and time domain approach, which have been successfully applied to the UPQC systems. Frequency domain methods such as the Fast Fourier Transform (FFT), are not much popular due to delays in calculating the FFT and relatively large computation time. Time domain based UPQC control methods are based on the instantaneous derivation of reference quantities in the form of current or voltage signals. The three most widely used time domain control techniques for UPQC are the instantaneous active and reactive power theory or pq theory [21], synchronous reference frame method or dq theory [22] and instantaneous symmetrical components theory [23]. Due to the simplicity and ease of real-time implementation, Instantaneous Symmetrical Components Theory (ISCT) is used in the generation of reference voltage and current quantities and will be described in the following section.

7.4 EXTRACTION OF REFERENCE CURRENTS FOR SHUNT VSI USING INSTANTANEOUS SYMMETRICAL COMPONENT THEORY

In Chapter 5, instantaneous reactive power theory and instantaneous symmetrical component theory were discussed to compensate for load reactive power and load current harmonics. Here, instantaneous symmetrical component theory is considered to calculate the reference currents of the shunt active filter, due to its ease of computation and implementation. It uses the instantaneous values of currents and voltages to formulate the reference currents as described below.

7.4.1 GENERATION OF REFERENCE CURRENTS FOR SHUNT VSI

The shunt inverter is connected in parallel with the load at the PCC, as shown in Fig. 7.4. The PCC voltage is the load voltage which is represented as v_{la}, v_{lb} and v_{lc}. Instantaneous symmetrical component theory for load compensation is used to derive the expression for reference filter currents for the shunt VSI. Recalling the three conditions for load compensation from Chapter 5,

1. The source neutral current must be zero. Therefore,

$$i_{sa} + i_{sb} + i_{sc} = 0 \qquad (7.1)$$

2. To have a predefined power factor at the PCC, the relationship between positive sequence load voltage and source current is given below.

$$\angle \bar{v}_{la+} = \angle \bar{i}_{sa+} + \phi_+ \tag{7.2}$$

where, ϕ_+ is the phase angle between $\bar{v}_{la+}$ and $\bar{i}_{sa+}$.

3. The average real power demand by the load should be met by the utility grid. This gives the following equation,

$$v_{la}i_{sa} + v_{lb}i_{sb} + v_{lc}i_{sc} = P_{lavg} = \frac{1}{T}\int_{t_1-T}^{t_1}(v_{la}i_{la} + v_{lb}i_{lb} + v_{lc}i_{lc})dt \tag{7.3}$$

In the above equation, T is the time period of one cycle of voltage or current and t_1 is any arbitrary instant. Using these conditions, the compensation objectives lead to the following reference currents for the shunt VSI.

$$i_{fa}^* = i_{la} - i_{sa}^* = i_{la} - \left(\frac{v_{la}}{\sum_{j=a,b,c} v_{lj}^2}\right)(P_{lavg} + P_{loss})$$

$$i_{fb}^* = i_{lb} - i_{sb}^* = i_{lb} - \left(\frac{v_{lb}}{\sum_{j=a,b,c} v_{lj}^2}\right)(P_{lavg} + P_{loss}) \tag{7.4}$$

$$i_{fc}^* = i_{lc} - i_{sc}^* = i_{lc} - \left(\frac{v_{lc}}{\sum_{j=a,b,c} v_{lj}^2}\right)(P_{lavg} + P_{loss})$$

In the above, i_{sa}^*, i_{sb}^*, and i_{sc}^* are computed source currents, which result in the generation of reference shunt VSI currents, i_{fa}^*, i_{fb}^*, and i_{fc}^*. Specifically, the computed source currents in (7.4) are:

$$i_{sa}^* = \left(\frac{v_{la}}{\sum_{j=a,b,c} v_{lj}^2}\right)(P_{lavg} + P_{loss})$$

$$i_{fb}^* = \left(\frac{v_{lb}}{\sum_{j=a,b,c} v_{lj}^2}\right)(P_{lavg} + P_{loss}) \tag{7.5}$$

$$i_{fc}^* = \left(\frac{v_{lc}}{\sum_{j=a,b,c} v_{lj}^2}\right)(P_{lavg} + P_{loss})$$

The computation of average load power P_{lavg} and P_{loss} were discussed in Chapter 5 and can be referred to for more details. The relations among load, system and filter neutral currents are given as follows.

$$i_{ln} = i_{la} + i_{lb} + i_{lc}$$
$$i_{sn}^* = i_{fn}^* + i_{ln} = 0 \tag{7.6}$$
$$i_{fn}^* = -(i_{la} + i_{lb} + i_{lc})$$

7.4.2   CONTROL OF SHUNT VOLTAGE SOURCE INVERTER

The complete control block for the shunt VSI is shown in Fig. 7.8. The load currents, $i_{l(abc)}$, filter currents, $i_{f(abc)}$, load voltages, $v_{l(abc)}$ and dc link voltage, v_{dc}, are sensed using Hall effect current and voltage transducers and these signal quantities are passed to a digital platform to compute average load power, P_{lavg}, and losses in the shunt VSI, P_{loss}. Then, the reference filter currents are generated using (7.4) and compared against the actual values. Once the reference currents for the shunt filter are known, the hysteresis current controller is used for generating the switching pulses during these conditions so that the current generated by the shunt APF follows the reference current, as discussed in Chapter 5. The shunt interfacing inductor, L_f is used to filter out the switching frequency components produced by the hysteresis controller.

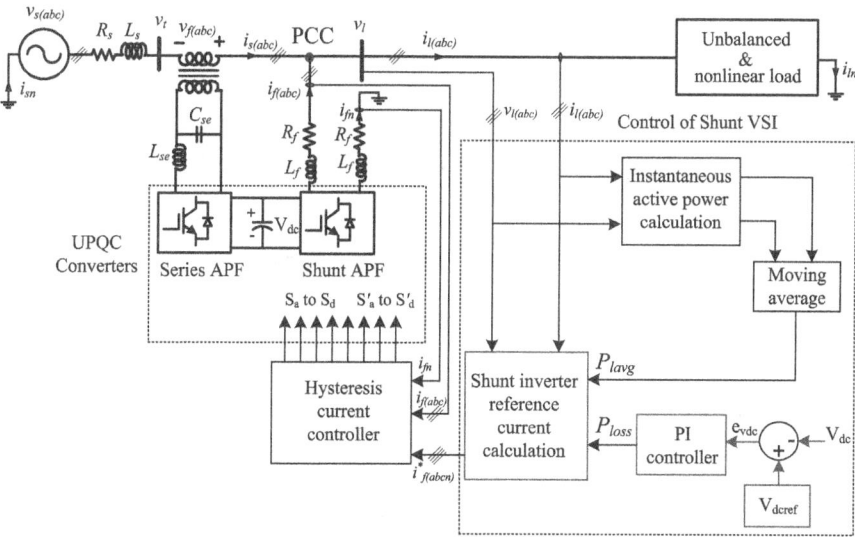

Figure 7.8   Control scheme of shunt VSI

7.5   CONTROL OF SERIES VOLTAGE SERIES INVERTER

The series Voltage Source Inverter (series VSI) or series Active Power Filter (series APF), a part of UPQC, is used for compensating voltage sag and voltage swell due to faults or abnormal conditions in the power system. The control system of series VSI performs sag/swell detection and reference voltage generation. After obtaining reference voltages, these are compared with the actual inverter voltage, and switching gate signals are given to the respective switches of the series VSI. The operation of series VSI is the same as Dynamic Voltage Restorer (DVR) as discussed in Chapter 6. During normal voltage conditions, the injection transformer is bypassed using a switching circuit (not shown in the figure), so that unwanted voltage drop is avoided

[24]. The inclusion of a series VSI injection transformer and the injection of voltage in series with the supply is controlled by the voltage sag/swell detection. Based on the reference voltages, the series VSI injects voltages to maintain the load terminal voltages within the prescribed limits, in all conditions of the source and the load. There are two types of control that can be used to generate the reference voltage signals, i.e., Type-1 and Type-2, as described in Section 6.6. In the Type-1 scheme, the upstream source voltage signals and values of feeder impedance are assumed to be known at the substation or load terminal. In the modern power system, the optical fiber lines based network is used to pass the three-phase voltage and current information to the substations and the load centers. Using the information of supply voltages and feeder impedance of the line, Type-1 control facilitates the generation of reference voltages, as discussed in Section 6.5 of Chapter 6. Type 2 control is based on measurements of the local variables.

7.5.1 TYPE-1 CONTROL FOR REFERENCE VOLTAGE GENERATION OF SERIES VSI

A single-line diagram shown in Fig. 7.1 is referred to for the description. In general, let the load voltage for phase-k ($k = a, b, c$), be maintained at the rated voltage $1.0\angle\delta_{lk}$ p.u. Using KVL in the circuit, we have,

$$\overline{V}_{sk} = V_{lk}\angle\delta_{lk} + (R_s + jX_s)\overline{I}_{sk} - V_{fk}\angle\overline{V}_{fk} \tag{7.7}$$

In the above equation, $(R_s + jX_s)\overline{I}_{sk} = a_{2k} + jb_{2k}$ and the angle of $\overline{V}_{fk}$ is calculated as, $\angle\overline{V}_{fk} = \angle\overline{I}_{sk} + \angle 90° = a_{1k} + jb_{1k}$. With this above equation is simplified to,

$$\begin{aligned} \overline{V}_{sk} &= V_{lk}\angle\delta_{lk} + a_{2k} + jb_{2k} - V_{fk}(a_{1k} + jb_{1k}) \\ &= (V_{lk}\cos\delta_l + a_{2k} - a_{1k}V_{fk}) + j(V_{lk}\sin\delta_{lk} + b_{2k} - b_{1k}V_{fk}) \end{aligned}$$

Denoting $V_{lk}\cos\delta_{lk} + a_{2k} = a'_{2k}$ and $V_{lk}\sin\delta_{lk} + b_{2k} = b'_{2k}$, the above equation can be written as following.

$$V_{sk}\cos\delta_{sk} + jV_{sk}\sin\delta_{sk} = (a'_{2k} - a_{1k}V_{fk}) + j(b'_{2k} - b_{1k}V_{fk}) \tag{7.8}$$

Squaring and adding the real and imaginary terms in (7.8), and arranging the terms in powers of V_{fk}, the following is obtained.

$$V_{fk}^2 - 2[a'_{2k}a_{1k} + b'_{2k}b_{1k}]V_{fk} + a'_{2k}{}^2 + b'_{2k}{}^2 - V_{sk}^2 = 0. \tag{7.9}$$

Once V_{fk} is obtained, the three-phase reference voltages for the series VSI are given as follows.

$$v^*_{fk}(t) = \sqrt{2}V_{fk}\sin(\omega t + \angle\overline{V}_{fk}) \tag{7.10}$$

In (7.10), the superscript, '*' is used to denote the reference quantity. The voltages, v_{fk} are now tracked using switching control schemes. One of the switching control schemes was discussed in Section 6.7. The above method holds true for balanced or unbalanced fundamental source voltages. In the case when source voltages have harmonics due to faults, machine core saturation, oscillatory transients, etc., the fundamental positive sequence voltages, v_{sk1}^+, need to be computed. The fundamental positive sequence voltages of the source are computed as given in the following.

Consider the following three-phase instantaneous source voltages (v_{sa}, v_{sb} and v_{sc}).

$$v_{sa} = \sqrt{2} \sum_{n=1}^{\infty} V_{san} \sin(n\omega t + \theta_{an})$$

$$v_{sb} = \sqrt{2} \sum_{n=1}^{\infty} V_{sbn} \sin\{n(\omega t - 2\pi/3) + \theta_{bn}\} \quad (7.11)$$

$$v_{sc} = \sqrt{2} \sum_{n=1}^{\infty} V_{scn} \sin\{n(\omega t + 2\pi/3) + \theta_{cn}\}$$

In terms of fundamentals and harmonics, the above voltages can be written as in the following.

$$
\begin{aligned}
v_{sa} &= \sqrt{2}V_{sa1} \sin(\omega t + \theta_{a1}) + v_{sah} \\
v_{sb} &= \sqrt{2}V_{sb1} \sin(\omega t - 2\pi/3 + \theta_{b1}) + v_{sbh} \\
v_{sc} &= \sqrt{2}V_{sc1} \sin(\omega t + 2\pi/3 + \theta_{c1}) + v_{sch}
\end{aligned}
\quad (7.12)
$$

where, V_{sa1}, V_{sb1} and V_{sc1} are the fundamental rms value of terminal voltages, θ_{a1}, θ_{b1} and θ_{c1} are the phase angle of fundamental voltages with respect to a reference, v_{sah}, v_{sbh} and v_{sch} are the instantaneous harmonic component of terminal voltages, and ω is the fundamental angular frequency. Using instantaneous symmetrical component theory [25], [26] (refer Appendix), the instantaneous symmetrical components of voltages, v_{sa}, v_{sb} and v_{sc} is expressed as below.

$$
\begin{bmatrix}
\overline{v}_{sa0} \\
\overline{v}_{sa+} \\
\overline{v}_{sa-}
\end{bmatrix}
=
\frac{1}{3}
\begin{bmatrix}
1 & 1 & 1 \\
1 & a & a^2 \\
1 & a^2 & a
\end{bmatrix}
\begin{bmatrix}
v_{sa} \\
v_{sb} \\
v_{sc}
\end{bmatrix}
\quad (7.13)
$$

In the above equation, $\overline{v}_{sa0} = v_{sa0}$ is a real quantity, $\overline{v}_{sa+}$ and $\overline{v}_{sa-}$ are positive and negative sequence components of phase-a voltage and are complex time varying quantities. If the source voltage has harmonic components, then these instantaneous symmetrical components also have harmonics. The extraction of fundamental components using instantaneous symmetrical component theory can be written as,

$$
\begin{bmatrix}
\overline{V}_{sa10} \\
\overline{V}_{sa1+} \\
\overline{V}_{sa1-}
\end{bmatrix}
=
\frac{\sqrt{2}}{T} \int_{t_1-T}^{t_1}
\begin{bmatrix}
\overline{v}_{sa0} \\
\overline{v}_{sa+} \\
\overline{v}_{sa-}
\end{bmatrix}
e^{-j(\omega t - \pi/2)} dt
\quad (7.14)
$$

In the above equation, the time interval, T is one cycle period of supply voltage and t_1 is any arbitrary instant. The averaging can be between any two points and need not be synchronized with the zero crossings of phase voltage. A moving average filter for a window of one cycle can be used for the calculation of fundamental components. The radian frequency, ω is the fundamental angular frequency for extracting the fundamental component of voltage.

Equation (7.14) can be used for extracting fundamental positive sequence phase and magnitude which can be used for sag/swell detection. The fundamental positive sequence component of phase-a terminal voltage can be written as,

$$\overline{V}_{sa1+} = \frac{\sqrt{2}}{T} \int_{t_1-T}^{t_1} \overline{v}_{sa+}\, e^{-j(\omega t - \pi/2)} dt \tag{7.15}$$

$$= \frac{\sqrt{2}}{T} \int_{t_1-T}^{t_1} \frac{v_{sa} + a v_{sb} + a^2 v_{sc}}{3} (\sin \omega t + j\cos \omega t) dt \tag{7.16}$$

substituting (7.12) in (7.16) and simplifying, we get equation (7.17).

$$\overline{V}_{sa1+} = \frac{V_{sa1}\angle\theta_{a1} + a V_{sb1}\angle\theta_{b1} + a^2 V_{sc1}\angle\theta_{c1}}{3} \tag{7.17}$$

Equation (7.17) shows that $\overline{V}_{sa1+}$ is a complex quantity, its magnitude represents the rms value of fundamental positive sequence terminal voltage and the argument represents the phase angle of fundamental positive sequence phase-a terminal voltage. Using (7.17), the fundamental positive sequence voltage phasors, $\overline{V}_{sa1+}, \overline{V}_{sb1+}$, and $\overline{V}_{sc1+}$ can be expressed as following.

$$\begin{aligned}
\overline{V}_{sa1+} &= V_{s1}^+ \angle\theta_{s1}^+ \\
\overline{V}_{sb1+} &= V_{s1}^+ \angle(-120° + \theta_{s1}^+) \\
\overline{V}_{sc1+} &= V_{s1}^+ \angle(120° + \theta_{s1}^+)
\end{aligned} \tag{7.18}$$

The time domain expressions for voltage phasors in (7.18) are as follows.

$$\begin{aligned}
v_{sa1}^+(t) &= \sqrt{2}V_{s1}^+ \sin(\omega t + \theta_{s1}^+) \\
v_{sb1}^+(t) &= \sqrt{2}V_{s1}^+ \sin(\omega t - 120° + \theta_{s1}^+) \\
v_{sc1}^+(t) &= \sqrt{2}V_{s1}^+ \sin(\omega t + 120° + \theta_{s1}^+)
\end{aligned} \tag{7.19}$$

where, V_{s1}^+ is rms value fundamental positive sequence terminal voltage and θ_{s1}^+ is the phase angle of the fundamental positive sequence terminal voltage. Here, for instantaneous, rms and angle values, zero, positive and negative sequence terms are denoted using superscripts 0, +, and −, respectively. For time varying complex quantities such as, $\overline{v}_{sa+}, \overline{v}_{sa-}$ and phasor quantities such as, $\overline{V}_{sa1+}, \overline{V}_{sb1+}, \overline{V}_{sc1+}$, the sequence terms are denoted using subscripts to avoid the closeness between phasor bar notation and "0, +, −" sequence notations. These fundamental positive sequence voltage phasors, $\overline{V}_{sa1+}, \overline{V}_{sb1+}$, and $\overline{V}_{sc1+}$ need to be computed in a closed manner to provide them as input to generate reference voltages to the series VSI. Since the source currents are balanced fundamental after shunt VSI's load compensation, the voltage harmonics do not form any real power. Now the fundamental positive sequence components of DVR voltages, V_{fk1}^+ are computed as follows.

$$V_{fk1}^{+\ 2} - 2[a_{2k}' a_{1k} + b_{2k}' b_{1k}]V_{fk1}^+ + a_{2k}'^{\ 2} + b_{2k}'^{\ 2} - V_{sk1}^{+\ 2} = 0 \qquad (7.20)$$

The angles of fundamental positive sequence DVR voltages are computed as follows.

$$\angle \overline{V}_{fk1+} = \angle \overline{I}_{sk} + 90° \qquad (7.21)$$

The angles of currents, $\angle \overline{I}_{sk}$ are taken from (7.5), which is resulting after load compensation using the shunt VSI. Note that these source currents are balanced fundamental components after load compensation. Once, the rms and angle values of fundamental positive sequence DVR voltages are known, the time domain expression for the DVR voltages are as follows.

$$v_{fk1}^+ = \sqrt{2}V_{fk1}^+ \sin(\omega t + \angle \overline{V}_{fk1}^+) \qquad (7.22)$$

The remaining voltage components of the series VSI are computed as follows.
Let the desired load voltage after compensation has rms value V_l at the fundamental frequency, ω. Then, the reference load voltages can be written as,

$$\begin{aligned} v_{la}^* &= \sqrt{2}V_l \sin \omega t \\ v_{lb}^* &= \sqrt{2}V_l \sin(\omega t - 2\pi/3) \\ v_{lc}^* &= \sqrt{2}V_l \sin(\omega t + 2\pi/3) \end{aligned} \qquad (7.23)$$

These load voltages, v_{la}, v_{lb}, v_{lc} are to be maintained by the series VSI during voltage sag and swell. Based on these desired load voltages and available terminal voltages, the series inverter injects a voltage such that load rms voltage is maintained constant all the time. Since the source voltage harmonics have to be compensated by the series VSI, the injected voltage by the series VSI may have fundamental and harmonics components, depending upon the quality of supply voltages. Accordingly, real power requirement from the dc link may vary [27]–[29].

The rest part of the reference voltage is calculated by equating to the negative sum of the negative and zero sequence fundamental components and harmonic components of terminal voltages. This is given below.

$$v_{fk\,rest} = -[v_{sk1}^- + v_{sk1}^0 + \sqrt{2}\sum_{n=2}^{\infty} V_{skn}\sin(n\omega t + \theta_{skn})] = -(v_{sk} - v_{sk1}^+) \quad (7.24)$$

The final reference voltages for the series VSI are given as follows.

$$\begin{aligned} v_{fk} &= v_{fk1}^+ + v_{fk\,rest} = v_{fk1}^+ - v_{sk1}^- - v_{sk1}^0 - \sqrt{2}\sum_{n=2}^{\infty} V_{skn}\sin(n\omega t + \theta_{skn}) \\ &= v_{fk1} - \sqrt{2}\sum_{n=2}^{\infty} V_{skn}\sin(n\omega t + \theta_{skn}) \end{aligned} \qquad (7.25)$$

The term v_{fk1} is the overall fundamental reference voltage for the series VSI. This voltage may not be in phase quadrature with $i_{sk}(t) = i_{sk1}^+(t)$, hence will result in real power requirement from the series VSI. The real power is supplied from the dc link through P_{loss} in (7.4) using a PI controller as discussed in Chapter 6.

7.5.2 TYPE-2 CONTROL FOR REFERENCE VOLTAGE GENERATION OF SERIES VSI

Type-2 control, as discussed in Section 6.6 of Chapter 6, is based on the local terminal voltages, v_{ta}, v_{tb}, and v_{tc}. The local terminal voltages, v_{ta}, v_{tb}, and v_{tc} are sensed instead of v_{sa}, v_{sb}, and v_{sc}. The control scheme does not require information about the feeder impedance, which is more convenient. The rms value of the reference series injected voltages, v_{fk} for $k = a, b, c$, are found using the following equation.

$$V_{fk1}^{+^2} - 2(a_{1k}\cos\delta_{lk} + b_{1k}\sin\delta_{lk})V_{lk}V_{fk1}^{+} + V_{tk1}^{+^2} = 0 \qquad (7.26)$$

In the above equation, subscript f refers series filter, subscript k refers to phase-k for $k = a, b, c$, and subscript "1" and superscript "+" refer to the fundamental positive sequence component. Here, the load voltages, v_{lk} are balanced fundamental components. The parameter a_{k1} comes from $a_{k1} + jb_{k1} = \angle\overline{V}_{fk1+}$. The angle of reference voltages, $\angle\overline{V}_{fk1+}$ is computed as following.

$$\angle\overline{V}_{fk1+} = \angle\overline{I}_{sk} + 90° \qquad (7.27)$$

The angles of currents, $\angle\overline{I}_{sk}$ are found using (7.5) with the given load compensation scheme. The load voltages at the load terminal can be expressed as,

$$\angle\overline{V}_{lk} = V_{lk}\angle\delta_{lk} \qquad (7.28)$$

These load voltages in phasor form, are expressed in the time domain given by (7.23) for a certain value of δ_{lk}. With rms value of the fundamental positive sequence DVR voltage, V_{fk1}^{+}, and its angle information, $\angle\overline{V}_{fk1}$. The remaining components of DVR voltages are constructed as given in (7.24)–(7.25).

In the following, some examples of UPQC compensation are illustrated. In these examples, the right shunt topology is considered. At the load bus, a power factor of 0.8 to 0.9 is considered rather than a unity power factor. This is done in order to realize the operation of the DVR without the requirement of real power. If a unity power factor is maintained at the PCC, the DVR operation requires some real power to operate. Also, in three-phase UPQC examples, after compensation, source voltages result in some phase shifts for the given rms magnitudes of the source voltages. These phase shifts can be eliminated by allowing the real power flow from the DVR through the dc link.

Example 7.1. Consider a single-phase system with rms voltage of 230 V (1 p.u.) and a 3 kVA base rating. The source supplies a linear inductive load through a feeder impedance, $Z_s = 0.01 + j\,0.05$ p.u. A right shunt UPQC, as shown in Figure 7.1 is placed at the load terminals to make the load voltage to 1 p.u. and the source current to be sinusoidal with a power factor of 0.9 (lag). The load impedance is $Z_l = 0.9 + j\,1.2$ p.u.

(a) Find out the load voltage and current without UPQC compensation.

(b) Compute the shunt compensator reference current i_f^* required to maintain sinusoidal source currents with a power factor of 0.9 lag at the load bus.

(c) Compute the series compensator voltages v_f^* required to maintain load voltage rms magnitude to 1 p.u.

(d) Calculate the rating of the shunt VSI and the series VSI of the UPQC.

Solution: Since here only fundamental frequency components are present, phasors can be used to compute reference quantities. Also, in the following calculations, all quantities are computed in p.u. For reference, the base quantities are given as,
$V_{base} = 230\,\text{V}$, $S_{base} = 3\,\text{kVA}$, $I_{base} = \frac{S_{base}}{V_{base}} = \frac{3000}{230} = 13.04\,\text{A}$

(a) For the case without compensation, the load impedance and feeder impedance are connected in series with the source. Therefore the source voltage is taken as a reference, as given below,

$$\overline{V}_s = 1\angle 0° \,\text{p.u.}$$

The load current in the circuit is given as,

$$\overline{I}_l = \frac{\overline{V}_s}{Z_s + Z_l} = \frac{1}{0.01 + j0.05 + 0.9 + j1.2} = 0.64\angle - 53.94° \,\text{p.u.}$$

The actual value of load current flowing is given as,

$$\overline{I}_{l\,act} = 8.43\angle - 53.94° \,\text{A}$$

The actual load voltages are given as,

$$\overline{V}_l = \overline{I}_{l\,act}Z_l = 223.13\angle - 0.81° \,\text{V}$$

(b) After compensation, the load terminal is required to have a balanced 1 p.u. voltage. Therefore, the desired load voltage after compensation is given as,

$$\overline{V}_l = 1\angle 0° \,\text{p.u.}$$

After compensation, the source voltage is given by,

$$\overline{V}_s - 1\angle\delta° \,\text{p.u.}$$

where δ is the angle of the supply voltage.
With the desired load voltage after compensation, the current drawn by the load is,

$$\overline{I}_l = \frac{\overline{V}_l}{Z_l} = \frac{1\angle 0°}{0.9 + j1.2} = 0.67\angle - 53.13° \,\text{p.u.}$$

To make the source current sinusoidal with the required power factor, a shunt compensator is placed which injects certain current so that the desired currents are drawn from the source. The angle of i_s is such that the source current lags the load voltage with an angle equivalent to the given power factor.

$$\angle \overline{I}_s = \angle \overline{V}_l - pf\,\text{angle} = \angle\left(0° - \cos^{-1}(0.9)\right) = \angle - 25.84°$$

The real power consumed by the load is to be supplied completely by the source, Therefore,

$$Re\{\bar{V}_l\bar{I}s^*\} = Re\{\bar{V}_l\bar{I}_l^*\}$$
$$1 \times Is \times \cos(-25.84°) = 1 \times 0.67 \times \cos(-53.13°)$$
This gives, $I_s = 0.44$ p.u.

The shunt compensator current is given as,

$$\bar{I}_f^* = \bar{I}_l - \bar{I}_s = 0.67\angle -53.13° - 0.44\angle -25.84° = 0.34\angle -90° \text{ p.u.}$$

The actual compensator currents in time domain is given as,

$$i_f^* = 6.27\sin(\omega t - 90°) \text{ A}$$

Since the direction of the shunt compensator current is into the PCC, the angle of current is lagging the voltage by 90°. This is equivalent to a passive compensator, where the current is from the PCC to the compensator. Thus, $\bar{I}_{f\,passive} = 0.34\angle 90°$ p.u. This indicates the passive compensator is a capacitive reactance, $X_f = 51.92\,\Omega$, which gives $C_f = 61.3\,\mu$F. Hence a capacitor, $C_{cf} = 61.3\,\mu$F needs to be connected between the PCC and the ground of the single-phase system. Now Calculating the actual source current,

$$\bar{I}_s = 5.79\angle -25.84° \text{ A}$$

(c) Now computing, DVR series voltage to maintain load voltage at 1 p.u. As we have the information of the feeder impedance, we use Type-1 control to compute DVR voltages as discussed in Chapter 6. The general Type-1 equation is given as,

$$V_f^2 - 2\{(1+a_2)a_1 + b_2b_1\}V_f + (1+a_2)^2 + b_2^2 - V_s^2 = 0$$

where V_f and V_s are the rms magnitudes of DVR voltage and source voltage. The parameters a_1 and b_1 denote the angle of v_f, which is kept 90° leading to the angle of i_s to avoid any real power exchange with the DVR. a_2' and b_2' denote the information of voltage drop across the feeder impedance along with the angle of v_l. The constants a_1 and b_1 are computed as given below.

$$\angle \bar{V}_f = \angle \bar{I}_s + 90° = a_1 + jb_1 = 0.4359 + j0.9 \text{ p.u.}$$

Therefore, $a_1 = 0.4359$ and $b_1 = 0.9$. The feeder voltage drop is computed as,

$$\bar{I}_sZ_s = a_2 + jb_2 = 0.0137 + j0.0221 \text{ p.u.}$$

Now calculating the DVR voltages by solving the quadratic equation given in ((c)) for all the phases as given below,

$$V_f^2 - 2\{(1+a_2)a_1 + b_2b_1\}V_f + (1+a_2)^2 + b_2^2 - V_s^2 = 0$$

Solving for V_f, we get

$$V_f = 0.0315 \quad \text{or} \quad 0.8847$$

$V_{fa} = 0.0315$ is chosen to achieve voltage compensation with a reduced DVR rating. The series compensator reference voltage is given as,

$$\overline{V}_f = 0.0315 \angle 64.16° \, \text{p.u.} = 7.24 \angle 64.16° \, \text{V}$$

The actual reference voltage in the time domain is given as,

$$v_f^* = 10.25 \sin(\omega t + 64.16°) \, \text{V}$$

Computing source voltage to confirm the proper functioning of the series compensator.

$$\overline{V}_s = \overline{V}_l + \overline{I}_s Z_s - \overline{V}_f = 1 \angle -0.59° \, \text{p.u.}$$

The actual source voltage is given as,

$$\overline{V}_{s\,act} = 230 \angle -0.59° \, \text{V}$$

The calculated source voltage is in accordance with ((b)). Therefore, the obtained v_f^* compensates the voltage to maintain $V_s = V_l = 1 \, \text{p.u.}$

(d) Rating of the shunt VSI (DSTATCOM) is calculated as,

$$S_{DSTATCOM} = V_l I_f = 0.3396 \, \text{p.u.}$$

The actual rating of the DSTATCOM is,

$$S_{DSTSTCOM\,act} = 1.02 \, \text{kVA}$$

Rating of the series VSI (DVR) is calculated as,

$$S_{DVR} = V_f I_s = 0.014 \, \text{p.u.}$$

The actual rating of the DVR is,

$$S_{DVR\,act} = 42.03 \, \text{VA}$$

Example 7.2. Consider a three-phase balanced system with rms phase voltage of 230 V (1 p.u.) and a 10 kVA base rating. The source supplies a three-phase unbalanced linear inductive load through a feeder impedance, $Z_s = 0.01 + j0.06$ p.u. A right shunt UPQC, as shown in Fig. 7.4 is placed at the load terminals to make the load voltages to 1 p.u. and the source currents to be sinusoidal balanced with a power factor of 0.9 (lag).

$$Z_{la} = 0.3 + j0.4 \, \text{p.u.}$$
$$Z_{lb} = 0.6 + j0.8 \, \text{p.u.}$$
$$Z_{lc} = 0.5 + j1.2 \, \text{p.u.}$$

(a) Find out the load voltages and currents without UPQC compensation.

(b) Compute the shunt compensator reference currents i_{fa}^*, i_{fb}^*, i_{fc}^* required to maintain balanced sinusoidal source currents.

(c) Compute the series compensator voltages v_{fa}^*, v_{fb}^*, v_{fc}^* required to maintain load voltage rms magnitude at 1 p.u.

(d) Calculate the rating of the shunt VSI and the series VSI of the UPQC.

Solution: Since only fundamental frequency components are present, phasors can be used to compute reference quantities. Also, in the following calculations, all quantities are computed in p.u. and are converted to the actual values whenever required. For reference, the base quantities are given as,

$$V_{base} = 230\,\text{V}, \; S_{base} = 10\,\text{kVA}, \; I_{base} = \frac{S_{base}/3}{V_{base}} = \frac{10000/3}{230} = 14.49\,\text{A}$$

(a) Without UPQC compensation, the load and feeder impedances are connected in series. The source voltages are taken as references and are given below.

$$\overline{V}_{sa} = 1\angle 0° \text{ p.u.}$$
$$\overline{V}_{sb} = 1\angle -120° \text{ p.u.}$$
$$\overline{V}_{sc} = 1\angle 120° \text{ p.u.}$$

The load currents are computed as follows.

$$\overline{I}_{la} = \frac{\overline{V}_{sa}}{Z_s + Z_{la}} = \frac{1}{0.01 + j0.06 + 0.3 + j0.4} = 1.8\angle -56.02° \text{ p.u.}$$

$$\overline{I}_{lb} = \frac{\overline{V}_{sb}}{Z_s + Z_{lb}} = \frac{1\angle -120°}{0.01 + j0.06 + 0.6 + j0.8} = 0.95\angle -174.65° \text{ p.u.}$$

$$\overline{I}_{lc} = \frac{\overline{V}_{sc}}{Z_s + Z_{lc}} = \frac{1\angle 120°}{0.01 + j0.06 + 0.5 + j1.2} = 0.73\angle 52.03° \text{ p.u.}$$

The actual value of load currents are given as,

$$\overline{I}_{la\,act} = 26.12\angle -56.02° \text{ A}$$
$$\overline{I}_{lb\,act} = 13.74\angle -174.65° \text{ A}$$
$$\overline{I}_{lc\,act} = 10.66\angle 52.03° \text{ A}$$

The actual load voltages are given as,

$$\overline{V}_{la} = \overline{I}_{la\,act} Z_{la} = 13.06\angle -2.89° \text{ V}$$
$$\overline{V}_{lb} = \overline{I}_{lb\,act} Z_{lb} = 13.74\angle -121.52° \text{ V}$$
$$\overline{V}_{lc} = \overline{I}_{lc\,act} Z_{lc} = 13.86\angle 119.41° \text{ V}$$

(b) After compensation, the load terminals are required to have a balanced 1 p.u. voltages. Therefore, the desired load voltages after compensation are given as follows.

$$\overline{V}_{la} = 1\angle 0° \text{ p.u.}$$
$$\overline{V}_{lb} = 1\angle -120° \text{ p.u.}$$
$$\overline{V}_{lc} = 1\angle 120° \text{ p.u.}$$

After compensation, the source voltages are represented as,

$$\overline{V}_{sa} = 1\angle \delta \text{ p.u.}$$
$$\overline{V}_{sb} = 1\angle (-120° + \delta) \text{ p.u.}$$
$$\overline{V}_{sc} = 1\angle (120° + \delta) \text{ p.u.}$$

With the required load voltages after compensation, the load currents are computed as follows.

$$\overline{I}_{la} = \frac{\overline{V}_{la}}{Z_{la}} = \frac{1}{0.3 + j0.4} = 2\angle -53.13° \text{ p.u.}$$

$$\overline{I}_{lb} = \frac{\overline{V}_{lb}}{Z_{lb}} = \frac{1\angle -120°}{0.6 + j0.8} = 1\angle -173.13° \text{ p.u.}$$

$$\overline{I}_{lc} = \frac{\overline{V}_{lc}}{Z_{lc}} = \frac{1\angle 120°}{0.5 + j1.2} = 0.77\angle 52.62° \text{ p.u.}$$

From the shunt VSI (DSTATCOM) discussion in Chapter 5, the reference currents for the shunt VSI in time domain are given as follows.

$$i_{fa}^* = i_{la} - i_{sa} = i_{la} - \frac{v_{la} + \beta(v_{lb} - v_{lc})}{\sum_{j=a,b,c} v_{lj}^2} P_{lavg}$$

$$i_{fb}^* = i_{lb} - i_{sb} = i_{lb} - \frac{v_{lb} + \beta(v_{lc} - v_{la})}{\sum_{j=a,b,c} v_{lj}^2} P_{lavg}$$

$$i_{fc}^* - i_{lc} - i_{sc} - i_{lc} - \frac{v_{lc} + \beta(v_{la} - v_{lb})}{\sum_{j=a,b,c} v_{lj}^2} P_{lavg}$$

In the above equation, $\sum_{j=a,b,c} v_{lj}^2 = 3V^2$ (due to balanced fundamental load voltages). The average load power P_{lavg} is given as,

$$P_{lavg} = Re(\overline{V}_{la}\overline{I}_{la}^* + \overline{V}_{lb}\overline{I}_{lb}^* + \overline{V}_{lc}\overline{I}_{lc}^*)$$

Above reference compensator, currents involve only fundamental frequency components. Hence, these expressions can be written in phasor form as given in the following.

$$\bar{I}^*_{fa} = \bar{I}_{la} - \bar{I}_{sa} = \bar{I}_{la} - \frac{\bar{V}_{la} + \beta\left(\bar{V}_{lb} - \bar{V}_{lc}\right)}{3V^2} P_{lavg}$$

$$\bar{I}^*_{fb} = \bar{I}_{lb} - \bar{I}_{sb} = \bar{I}_{lb} - \frac{\bar{V}_{lb} + \beta\left(\bar{V}_{lc} - \bar{V}_{la}\right)}{3V^2} P_{lavg}$$

$$\bar{I}^*_{fc} = \bar{I}_{lc} - \bar{I}_{sc} = \bar{I}_{lc} - \frac{\bar{V}_{lc} + \beta\left(\bar{V}_{la} - \bar{V}_{lb}\right)}{3V^2} P_{lavg}$$

where, $\beta = \frac{\tan\left(\cos^{-1}(pf)\right)}{\sqrt{3}}$, and pf is the desired power factor.

Substituting the value of load voltages and currents phasors, $\beta = \frac{\tan\left(\cos^{-1}(0.9)\right)}{\sqrt{3}} = 0.2796$, $V = 1$ p.u., and $P_{lavg} = 2.0959$ p.u., the reference currents for the DSTATCOM to have balanced currents on the source side with the mentioned power factor are given as follows.

$$\bar{I}^*_{fa} = \bar{I}_{la} - \bar{I}_{sa} = 2\angle -53.13° - 0.77\angle -25.84° = 1.35\angle -68.32° \text{ p.u.}$$

$$\bar{I}^*_{fb} = \bar{I}_{lb} - \bar{I}_{sb} = 1\angle -173.13° - 0.77\angle -145.84° = 0.47\angle 136.94° \text{ p.u.}$$

$$\bar{I}^*_{fc} = \bar{I}_{lc} - \bar{I}_{sc} = 0.77\angle 52.62° - 0.77\angle 94.16° = 0.54\angle -17.29° \text{ p.u.}$$

The actual value of reference currents are as follows.

$$\bar{I}^*_{fa\,act} = 19.67\angle -68.32° \text{ A}$$

$$\bar{I}^*_{fb\,act} = 6.84\angle 137.94° \text{ A}$$

$$\bar{I}^*_{fc\,act} = 7.94\angle -17.29° \text{ A}$$

The source currents after compensation are,

$$\bar{I}_{sa\,act} = 11.25\angle -25.84° \text{ A}$$

$$\bar{I}_{sb\,act} = 11.25\angle -145.84° \text{ A}$$

$$\bar{I}_{sc\,act} = 11.25\angle 94.16° \text{ A}$$

The actual reference currents in the time domain are given as

$$i^*_{fa} = 27.82\sin\left(\omega t - 68.32°\right) \text{ A}$$

$$i^*_{fb} = 9.67\sin\left(\omega t + 137.94°\right) \text{ A}$$

$$i^*_{fc} = 11.23\sin\left(\omega t - 17.29°\right) \text{ A}$$

These currents are realized using switching control of the shunt VSI as discussed in Chapter 5. Now, these currents are considered for the computation of series VSI reference voltages to maintain load voltages at 1 p.u.

(c) As we have the information on the feeder impedances, we use Type-1 control to compute DVR voltages as discussed in Chapter 6. The general Type-1 equation is given as,

$$V_f^2 - 2[a'_2 a_1 + b'_2 b_1]V_f + a'^2_2 + b'^2_2 - V_s^2 = 0$$

where V_f and V_s are the rms magnitudes of DVR voltage and source voltage. The parameters a_1 and b_1 denote the angle information of v_f, which is kept $90°$ leading to angle of i_s to avoid any real power exchange with the DVR. The parameters a_2' and b_2' denote the information of voltage drop across the feeder impedances along with the angle of v_l. The parameters a_1 and b_1, for all three phases, are given below.

$$\angle \overline{V}_{fa} = \angle \overline{I}_{sa} + 90° = a_{1a} + j\,b_{1a} = 0.4359 + j\,0.9 \text{ p.u.}$$
$$\angle \overline{V}_{fb} = \angle \overline{I}_{sb} + 90° = a_{1b} + j\,b_{1b} = 0.5615 - j\,0.8275 \text{ p.u.}$$
$$\angle \overline{V}_{fc} = \angle \overline{I}_{sc} + 90° = a_{1c} + j\,b_{1c} = -0.9974 - j\,0.0725 \text{ p.u.}$$

Therefore, $a_{1a} = 0.4359$, $b_{1a} = 0.9$, $a_{1b} = 0.5615$, $b_{1b} = -0.8275$, $a_{1c} = -0.9974$, and $b_{1c} = -0.0725$.

The feeder voltage drop is computed as,

$$\overline{I}_{sa}Z_s = a_{2a} + j\,b_{2a} = 0.0273 + j\,0.0385 \text{ p.u.}$$
$$\overline{I}_{sb}Z_s = a_{2b} + j\,b_{2b} = 0.0197 - j\,0.0429 \text{ p.u.}$$
$$\overline{I}_{sc}Z_s = a_{2c} + j\,b_{2c} = -0.0470 + j\,0.0044 \text{ p.u.}$$

Parameters a_2' and b_2' for all the phases are calculated by adding the load voltage angle information into a_2 and b_2, as given below,

$$a_{2a}' = a_{2a} + \cos(\angle \overline{V}_{la}) = 1.0273 \text{ p.u.}, \qquad b_{2a}' = b_{2a} + \sin(\angle \overline{V}_{la}) = 0.0385 \text{ p.u.}$$
$$a_{2b}' = a_{2b} + \cos(\angle \overline{V}_{lb}) = -0.4803 \text{ p.u.}, \qquad b_{2a}' = b_{2a} + \sin(\angle \overline{V}_{la})$$
$$= -0.9089 \text{ p.u.}$$
$$a_{2c}' = a_{2c} + \cos(\angle \overline{V}_{lc}) = -0.5470 \text{ p.u.}, \qquad b_{2c}' = b_{2c} + \sin(\angle \overline{V}_{lc}) = 0.8704 \text{ p.u.}$$

Now the DVR voltages are calculated by solving the quadratic equation for all three phases as given below,

$$V_{fa}^2 - 2(a_{2a}'a_{1a} + b_{2a}'b_{1a})V_{fa} + a_{2a}'^2 + b_{2a}'^2 - V_{sa}^2 = 0$$
$$V_{fb}^2 - 2(a_{2b}'a_{1b} + b_{2b}'b_{1b})V_{fb} + a_{2b}'^2 + b_{2b}'^2 - V_{sb}^2 = 0$$
$$V_{fc}^2 - 2(a_{2c}'a_{1c} + b_{2c}'b_{1c})V_{fc} + a_{2c}'^2 + b_{2c}'^2 - V_{sc}^2 = 0$$

Solving for V_{fa}, V_{fb}, V_{fc}, we get

$$V_{fa} = 0.0629 \quad \text{or} \quad 0.9019$$
$$V_{fb} = 0.0629 \quad \text{or} \quad 0.9019$$
$$V_{fc} = 0.0629 \quad \text{or} \quad 0.9019$$

which is evident and three-phase voltages and currents are balanced $V_{fa} = 0.0629$, $V_{fb} = 0.0629$, and $V_{fc} = 0.0629$ are chosen to achieve voltage compensation with reduced DVR rating. The series compensator reference voltages are as follows.

$$\overline{V}_{fa}^* = 0.0629 \angle 64.16° \text{ p.u.}$$
$$\overline{V}_{fb}^* = 0.0629 \angle 55.84° \text{ p.u.}$$
$$\overline{V}_{fc}^* = 0.0629 \angle 175.84° \text{ p.u.}$$

The actual reference voltages in time domain are given as,

$$v_{fa}^* = 20.48 \sin(\omega t + 64.16°) \, V$$
$$v_{fb}^* = 20.48 \sin(\omega t - 55.84°) \, V$$
$$v_{fc}^* = 20.48 \sin(\omega t - 175.84°) \, V$$

The above three-phase DVR voltages are balanced, which is evident as three-phase voltages and currents are balanced, once the load currents are compensated by the shunt VSI. Computing source voltages to confirm the proper functioning of the series compensator.

$$\overline{V}_{sa} = \overline{V}_{la} + \overline{I}_{sa}Z_s - \overline{V}_{fa} = 1\angle - 1.04° \text{ p.u.}$$
$$\overline{V}_{sb} = \overline{V}_{lb} + \overline{I}_{sb}Z_s - \overline{V}_{fb} = 1\angle - 121.04° \text{ p.u.}$$
$$\overline{V}_{sc} = \overline{V}_{lc} + \overline{I}_{sc}Z_s - \overline{V}_{fc} = 1\angle 118.96° \text{ p.u.}$$

The actual source voltages are given as,

$$\overline{V}_{sa \, act} = 230\angle - 1.04° \, V$$
$$\overline{V}_{sb \, act} = 230\angle - 121.04° \, V$$
$$\overline{V}_{sc \, act} = 230\angle 118.96° \, V$$

Therefore, the obtained $v_{fa}^*, v_{fb}^*, v_{fc}^*$ compensates the voltage to maintain $V_s = V_l = 1$ p.u.

(d) Rating of the shunt VSI (DSTATCOM) is calculated as,

$$S_{DSTATCOM} = V_{la}I_{fa} + V_{lb}I_{fb} + V_{lc}I_{fc} = 2.377 \text{ p.u.}$$

The actual rating of the DSTATCOM is,

$$S_{DSTATCOM \, act} = 23.77 \text{ kVA}$$

Rating of the series VSI (DVR) is calculated as,

$$S_{DVR} = V_{fa}I_{sa} + V_{fb}I_{sb} + V_{fc}I_{sc} = 0.1467 \text{ p.u.}$$

The actual rating of the DVR is,

$$S_{DVR \, act} = 1.47 \text{ kVA}$$

Example 7.3. Consider a three-phase four-wire unbalanced system with a base kVA rating of 10 kVA and base rms phase voltage of 230 V. The rms values of the source voltages are given as,

$$V_{sa} = 1 \text{ p.u.}$$
$$V_{sb} = 1.2 \text{ p.u.}$$
$$V_{sc} = 0.85 \text{ p.u.}$$

The source supplies a three-phase unbalanced linear inductive load through a feeder impedance, $Z_s = 0.01 + j0.06$ p.u. A right shunt UPQC, as shown in Figure 7.1 is placed at the load terminals to make the load voltages of all the three phases to 1 p.u. and the source currents to be sinusoidal balanced in nature with a power factor of 0.8 (lag). The load impedances are,

$$Z_{la} = 0.3 + j0.4 \text{ p.u.}$$
$$Z_{lb} = 0.6 + j0.8 \text{ p.u.}$$
$$Z_{lc} = 0.5 + j1.2 \text{ p.u.}$$

(a) Find out the load voltages and currents without any compensation.

(b) Compute the shunt compensator reference currents i_{fa}^*, i_{fb}^*, i_{fc}^* required to maintain balanced sinusoidal source currents.

(c) Compute the series compensator voltages v_{fa}^*, v_{fb}^*, v_{fc}^* required to maintain load voltage rms magnitude at 1 p.u.

(d) Calculate the rating of the shunt VSI and the series VSI of the UPQC.

Solution: Since only fundamental frequency components are present, phasors can be used to compute reference quantities. Also, in the following calculations, all quantities are computed in p.u. For reference, the base quantities are given as,
$V_{base} = 230 \text{ V}$, $S_{base} = 10 \text{ kVA}$, $I_{base} = \frac{S_{base}/3}{V_{base}} = \frac{10000/3}{230} = 14.49 \text{ A}$

(a) Without UPQC compensation, the source voltages are taken as a reference. The source voltages are considered to be only unbalanced in magnitude, as given below,
$$\overline{V}_{sa} = 1\angle 0° \text{ p.u.}$$
$$\overline{V}_{sb} = 1.2\angle -120° \text{ p.u.}$$
$$\overline{V}_{sc} = 0.85\angle 120° \text{ p.u.}$$

The load currents are given as,

$$\overline{I}_{la} = \frac{\overline{V}_{sa}}{Z_s + Z_{la}} = \frac{1}{0.01 + j0.06 + 0.3 + j0.4} = 1.8\angle -56.02° \text{ p.u.}$$

$$\overline{I}_{lb} = \frac{\overline{V}_{sb}}{Z_s + Z_{lb}} = \frac{1.2\angle -120°}{0.01 + j0.06 + 0.6 + j0.8} = 1.14\angle -174.65° \text{ p.u.}$$

$$\overline{I}_{lc} = \frac{\overline{V}_{sc}}{Z_s + Z_{lc}} = \frac{0.85\angle 120°}{0.01 + j0.06 + 0.5 + j1.2} = 0.62\angle 52.03° \text{ p.u.}$$

The actual values of load currents are given as,

$$\overline{I}_{la \, act} = 26.12\angle -56.02° \text{ A}$$
$$\overline{I}_{lb \, act} = 16.49\angle -174.65° \text{ A}$$
$$\overline{I}_{lc \, act} = 9.06\angle 52.03° \text{ A}$$

The actual load voltages are given as,

$$\overline{V}_{la} = \overline{I}_{la\ act} Z_{la} = 13.06\angle -2.89° \text{ V}$$
$$\overline{V}_{lb} = \overline{I}_{lb\ act} Z_{lb} = 16.49\angle -121.52° \text{ V}$$
$$\overline{V}_{lc} = \overline{I}_{lc\ act} Z_{lc} = 11.78\angle 119.41° \text{ V}$$

(b) After compensation, the load terminals are required to have a balanced 1 p.u. voltages. Therefore, the desired load voltages after compensation are given as,

$$\overline{V}_{la} = 1\angle 0° \text{ p.u.}$$
$$\overline{V}_{lb} = 1\angle -120° \text{ p.u.}$$
$$\overline{V}_{lc} = 1\angle 120° \text{ p.u.}$$

After compensation, the source voltages are given by,

$$\overline{V}_{sa} = 1\angle \delta_a \text{ p.u.}$$
$$\overline{V}_{sb} = 1.2\angle \delta_b \text{ p.u.}$$
$$\overline{V}_{sc} = 0.85\angle \delta_c \text{ p.u.}$$

where δ_a, δ_b, δ_c are the grid voltage angles for the three phases.
With the required load voltages after compensation, the currents drawn by the load are,

$$\overline{I}_{la} = \frac{\overline{V}_{la}}{Z_{la}} = \frac{1}{0.3 + j0.4} = 2\angle -53.13° \text{ p.u.}$$

$$\overline{I}_{lb} = \frac{\overline{V}_{lb}}{Z_{lb}} = \frac{1\angle -120°}{0.6 + j0.8} = 1\angle -173.13° \text{ p.u.}$$

$$\overline{I}_{lc} = \frac{\overline{V}_{lc}}{Z_{lc}} = \frac{1\angle 120°}{0.5 + j1.2} = 0.77\angle 52.62° \text{ p.u.}$$

From the shunt VSI (DSTATCOM), as discussed in Chapter 5, the reference currents for the shunt VSI in the time domain are given as follows.

$$i_{fa}^* = i_{la} - i_{sa} = i_{la} - \frac{v_{la} + \beta(v_{lb} - v_{lc})}{\sum_{j=a,b,c} v_{lj}^2} P_{lavg}$$

$$i_{fb}^* = i_{lb} - i_{sb} = i_{lb} - \frac{v_{lb} + \beta(v_{lc} - v_{la})}{\sum_{j=a,b,c} v_{lj}^2} P_{lavg}$$

$$i_{fc}^* = i_{lc} - i_{sc} = i_{lc} - \frac{v_{lc} + \beta(v_{la} - v_{lb})}{\sum_{j=a,b,c} v_{lj}^2} P_{lavg}$$

In the above equation, $\sum_{j=a,b,c} v_{lj}^2 = 3V^2$ (due to balanced fundamental load voltages). The average load power P_{lavg} is given as,

$$P_{lavg} = Re\{\overline{V}_{la}\overline{I}_{la}^* + \overline{V}_{lb}\overline{I}_{lb}^* + \overline{V}_{lc}\overline{I}_{lc}^*\}$$

Compensator reference currents are written in phasor form as,

$$\bar{I}^*_{fa} = \bar{I}_{la} - \bar{I}_{sa} = \bar{I}_{la} - \frac{\bar{V}_{la} + \beta\left(\bar{V}_{lb} - \bar{V}_{lc}\right)}{3V^2}P_{lavg}$$

$$\bar{I}^*_{fb} = \bar{I}_{lb} - \bar{I}_{sb} = \bar{I}_{lb} - \frac{\bar{V}_{lb} + \beta\left(\bar{V}_{lc} - \bar{V}_{la}\right)}{3V^2}P_{lavg}$$

$$\bar{I}^*_{fc} = \bar{I}_{lc} - \bar{I}_{sc} = \bar{I}_{lc} - \frac{\bar{V}_{lc} + \beta\left(\bar{V}_{la} - \bar{V}_{lb}\right)}{3V^2}P_{lavg}$$

where, $\beta = \frac{\tan\left(\cos^{-1}(pf)\right)}{\sqrt{3}}$, and pf is the desired power factor.

Substituting the value of load voltages and currents phasors, $\beta = \frac{\tan\left(\cos^{-1}(0.8)\right)}{\sqrt{3}} = 0.4330$, $V = 1$ p.u., and $P_{lavg} = 2.0959$ p.u., the reference currents for the DSTATCOM with the mentioned power factor are given as follows.

$$\bar{I}^*_{fa} = \bar{I}_{la} - \bar{I}_{sa} = 2\angle - 53.13° - 0.87\angle - 36.87° = 1.18\angle - 65.01° \text{ p.u.}$$

$$\bar{I}^*_{fb} = \bar{I}_{lb} - \bar{I}_{sb} = 1\angle - 173.13° - 0.87\angle - 156.87° = 0.29\angle 130.34° \text{ p.u.}$$

$$\bar{I}^*_{fc} = \bar{I}_{lc} - \bar{I}_{sc} = 0.77\angle 52.62° - 0.87\angle 83.13° = 0.44\angle - 35.2° \text{ p.u.}$$

The actual value of reference currents are as follows.

$$\bar{I}^*_{fa\,act} = 17.2\angle - 65.01° \text{ A}$$
$$\bar{I}^*_{fb\,act} = 4.25\angle 130.34° \text{ A}$$
$$\bar{I}^*_{fc\,act} = 6.43\angle - 35.2° \text{ A}$$

The source currents after compensation are,

$$\bar{I}_{sa\,act} = 12.65\angle - 36.87° \text{ A}$$
$$\bar{I}_{sb\,act} = 12.65\angle - 156.87° \text{ A}$$
$$\bar{I}_{sc\,act} = 12.65\angle 83.13° \text{ A}$$

The actual reference currents in the time domain are given as,

$$i^*_{fa} = 24.33\sin\left(\omega t - 65.01°\right) \text{ A}$$
$$i^*_{fb} = 6\sin\left(\omega t + 130.34°\right) \text{ A}$$
$$i^*_{fc} = 9.09\sin\left(\omega t - 35.2°\right) \text{ A}$$

These currents are realized using switching control of the shunt VSI as discussed in Chapter 5. Now, these currents are considered for the computation of series VSI reference voltages to maintain load voltages at 1 p.u.

(c) Using Type-1 control, DVR voltages are computed as discussed in Chapter 6.

$$V_f^2 - 2(a'_2 a_1 + b'_2 b_1)V_f + a'^2_2 + b'^2_2 - V_s^2 = 0$$

where V_f and V_s are the rms magnitudes of DVR voltage and source voltage. The parameters a_1 and b_1 denote the angle information of v_f, which is kept at $90°$ leading to the angle of i_s to avoid any real power exchange with the DVR. Parameters a_2' and b_2' denote the voltage drop across the feeder impedances along with the angle information of v_l. The constants a_1 and b_1 are calculated as given below.

$$\angle \overline{V}_{fa} = \angle \overline{I}_{sa} + 90° \quad = a_{1a} + jb_{1a} = 0.6 + j0.8$$
$$\angle \overline{V}_{fb} = \angle \overline{I}_{sb} + 90° \quad = a_{1b} + jb_{1b} = 0.3928 - j0.9196$$
$$\angle \overline{V}_{fc} = \angle \overline{I}_{sc} + 90° \quad = a_{1c} + jb_{1c} = -0.9928 + j0.1196$$

Therefore, $a_{1a} = 0.6$, $b_{1a} = 0.8$, $a_{1b} = 0.3928$, $b_{1b} = -0.9196$, $a_{1c} = -0.9928$, and $b_{1c} = 0.1196$.

The feeder voltage drops in each phase are computed as,

$$\overline{I}_{sa}Z_s = a_{2a} + jb_{2a} = 0.0384 + j0.0367 \, \text{p.u.}$$
$$\overline{I}_{sb}Z_s = a_{2b} + jb_{2b} = 0.0126 - j0.0516 \, \text{p.u.}$$
$$\overline{I}_{sc}Z_s = a_{2c} + jb_{2c} = -0.0510 + j0.0149 \, \text{p.u.}$$

Parameters a_2' and b_2' for all the phases are calculated by adding the load voltage angle information into a_2 and b_2, as given below,

$$a_{2a}' = a_{2a} + \cos(\angle \overline{V}_{la}) = 1.0384 \, \text{p.u.}, \quad b_{2a}' = b_{2a} + \sin(\angle \overline{V}_{la}) = 0.0367 \, \text{p.u.}$$
$$a_{2b}' = a_{2b} + \cos(\angle \overline{V}_{lb}) = -0.4874 \, \text{p.u.}, \quad b_{2a}' = b_{2a} + \sin(\angle \overline{V}_{la})$$
$$= -0.9176 \, \text{p.u.}$$
$$a_{2c}' = a_{2c} + \cos(\angle \overline{V}_{lc}) = -0.5510 \, \text{p.u.}, \quad b_{2c}' = b_{2c} + \sin(\angle \overline{V}_{lc}) = 0.8810 \, \text{p.u.}$$

Now, the DVR voltages are calculated using the following quadratic equations for all phases, as given below.

$$V_{fa}^2 - 2(a_{2a}'a_{1a} + b_{2a}'b_{1a})V_{fa} + a_{2a}'^2 + b_{2a}'^2 - V_{sa}^2 = 0$$
$$V_{fb}^2 - 2(a_{2b}'a_{1b} + b_{2b}'b_{1b})V_{fb} + a_{2b}'^2 + b_{2b}'^2 - V_{sb}^2 = 0$$
$$V_{fc}^2 - 2(a_{2c}'a_{1c} + b_{2c}'b_{1c})V_{fc} + a_{2c}'^2 + b_{2c}'^2 - V_{sc}^2 = 0$$

Solving for V_{fa}, V_{fb}, V_{fc}, we get

$$V_{fa} = 0.0642 \quad \text{or} \quad 1.2406$$
$$V_{fb} = -0.2341 \quad \text{or} \quad 1.5389$$
$$V_{fc} = 0.3908 \quad \text{or} \quad 0.9140$$

$V_{fa} = 0.0642$, $V_{fb} = 1.5389$, and $V_{fc} = 0.3908$ are chosen to achieve voltage compensation with reduced DVR rating. Thus, the series compensator reference voltages are as follows.

$$\overline{V}_{fa}^* = 0.06\angle 53.13°$$
$$\overline{V}_{fb}^* = 1.54\angle -66.87°$$
$$\overline{V}_{fc}^* = 0.39\angle 173.13°$$

The actual reference voltages in time domain are given as,

$$v_{fa}^* = 20.88\sin(\omega t + 53.13°)\,\text{V}$$
$$v_{fb}^* = 500.56\sin(\omega t + 66.87°)\,\text{V}$$
$$v_{fc}^* = 127.1\sin(\omega t + 173.13°)\,\text{V}$$

Computing source voltages to confirm the proper functioning of the series compensator.

$$\overline{V}_{sa} = \overline{V}_{la} + \overline{I}_{sa}Z_s - \overline{V}_{fa} = 1\angle-0.84°\,\text{p.u.}$$
$$\overline{V}_{sb} = \overline{V}_{lb} + \overline{I}_{sb}Z_s - \overline{V}_{fb} = 1.2\angle155.50°\,\text{p.u.}$$
$$\overline{V}_{sc} = \overline{V}_{lc} + \overline{I}_{sc}Z_s - \overline{V}_{fc} = 0.85\angle101.06°\,\text{p.u.}$$

The actual source voltages are given as,

$$\overline{V}_{sa\,act} = 230\angle-0.84°\,\text{V}$$
$$\overline{V}_{sb\,act} = 276\angle155.5°\,\text{V}$$
$$\overline{V}_{sc\,act} = 196\angle101.05°\,\text{V}$$

The calculated source voltages have phase shifts, which may be different from those of the grid voltages. These phase shifts can be eliminated by allowing real power flow from the DVR. Thus, the obtained v_{fa}^*, v_{fb}^*, v_{fc}^* DVR voltages compensate to maintain load voltages at 1 p.u. for the given unbalance in the source.

(d) Rating of the shunt VSI (DSTATCOM) is calculated as,

$$S_{DSTATCOM} = V_{la}I_{fa} + V_{lb}I_{fb} + V_{lc}I_{fc} = 1.924\,\text{p.u.}$$

The actual rating of the DSTATCOM is,

$$S_{DSTATCOM\,act} = 19.24\,\text{kVA}$$

Rating of the series VSI (DVR) is calculated as,

$$S_{DVR} = V_{fa}I_{sa} + V_{fb}I_{sb} + V_{fc}I_{sc} = 1.7412\,\text{p.u.}$$

The actual rating of the DVR is,

$$S_{DVR\,act} = 17.41\,\text{kVA}$$

Example 7.4. Consider a three-phase unbalanced system with a base kVA rating of 25 kVA and base rms phase voltage of 230 V, shown in Fig. 7.4. The supply voltages after the feeder impedances are,

$$v_{ta} = \sqrt{2}\sin\omega t + 0.2\sqrt{2}\sin(5\omega t)\,\text{p.u.}$$
$$v_{tb} = \sqrt{2}\sin(\omega t - 120°) + 0.2\sqrt{2}\sin(5(\omega t - 120°))\,\text{p.u.}$$
$$v_{tc} = \sqrt{2}\sin(\omega t + 120°) + 0.2\sqrt{2}\sin(5(\omega t + 120°))\,\text{p.u.}$$

Now a right shunt UPQC, as shown in Fig. 7.4, is placed to make the fundamental balanced load voltages of rms value 1 p.u. The source currents after compensation are of sinusoidal balanced nature with a power factor of 0.866 (lag) at the load bus. The load impedances are:

$$Z_{la} = 0.3 + j0.5 \text{ p.u.}$$
$$Z_{lb} = 0.6 + j1.0 \text{ p.u.}$$
$$Z_{lc} = 0.8 - j1.2 \text{ p.u.}$$

(a) Compute the shunt compensator reference currents i_{fa}^*, i_{fb}^*, i_{fc}^* required to maintain balanced sinusoidal source currents.

(b) Compute the series compensator voltages v_{fa}^*, v_{fb}^*, v_{fc}^* required to maintain the load voltage rms magnitude to 1 p.u., using Type-2 control, as discussed in Chapter 6.

(c) Calculate the rating of the shunt VSI and the series VSI of the UPQC.

Solution:

(a) After compensation, the load terminals are required to have a balanced 1 p.u. voltages. Therefore, the desired load voltages after compensation are given as,

$$\overline{V}_{la} = 1\angle 0° \text{ p.u.}$$
$$\overline{V}_{lb} = 1\angle -120° \text{ p.u.}$$
$$\overline{V}_{lc} = 1\angle 120° \text{ p.u.}$$

With the required load voltages after compensation, the currents drawn by the load are,

$$\overline{I}_{la} = \frac{\overline{V}_{la}}{Z_{la}} = \frac{1}{0.3 + j0.5} = 1.715\angle -59.04° \text{ p.u.}$$

$$\overline{I}_{lb} = \frac{\overline{V}_{lb}}{Z_{lb}} = \frac{1\angle -120°}{0.6 + j1.0} = 0.8575\angle -179.04° \text{ p.u.}$$

$$\overline{I}_{lc} = \frac{\overline{V}_{lc}}{Z_{lc}} = \frac{1\angle 120°}{0.8 - j1.2} = 0.6934\angle 176.31° \text{ p.u.}$$

The reference currents for the shunt VSI in the time domain are given as follows.

$$i_{fa}^* = i_{la} - i_{sa} = i_{la} - \frac{v_{la} + \beta(v_{lb} - v_{lc})}{\sum_{j=a,b,c} v_{lj}^2} P_{lavg}$$

$$i_{fb}^* = i_{lb} - i_{sb} = i_{lb} - \frac{v_{lb} + \beta(v_{lc} - v_{la})}{\sum_{j=a,b,c} v_{lj}^2} P_{lavg}$$

$$i_{fc}^* = i_{lc} - i_{sc} = i_{lc} - \frac{v_{lc} + \beta(v_{la} - v_{lb})}{\sum_{j=a,b,c} v_{lj}^2} P_{lavg}$$

In the above equation, $\sum_{j=a,b,c} v_{lj}^2 = 3V^2 = 3$ p.u. (due to balanced fundamental load voltages). The average load power P_{lavg} is given as,

$$P_{lavg} = Re\{\overline{V}_{la}\overline{I}_{la}^* + \overline{V}_{lb}\overline{I}_{lb}^* + \overline{V}_{lc}\overline{I}_{lc}^*\} = 1.708\,\text{p.u.}$$

The reference compensator currents in phasor form are written as follows.

$$\overline{I}_{fa}^* = \overline{I}_{la} - \overline{I}_{sa} = \overline{I}_{la} - \frac{\overline{V}_{la} + \beta\left(\overline{V}_{lb} - \overline{V}_{lc}\right)}{3V^2}P_{lavg}$$

$$\overline{I}_{fb}^* = \overline{I}_{lb} - \overline{I}_{sb} = \overline{I}_{lb} - \frac{\overline{V}_{lb} + \beta\left(\overline{V}_{lc} - \overline{V}_{la}\right)}{3V^2}P_{lavg}$$

$$\overline{I}_{fc}^* = \overline{I}_{lc} - \overline{I}_{sc} = \overline{I}_{lc} - \frac{\overline{V}_{lc} + \beta\left(\overline{V}_{la} - \overline{V}_{lb}\right)}{3V^2}P_{lavg}$$

where, $\beta = \frac{\tan\left(\cos^{-1}(pf)\right)}{\sqrt{3}}$, and pf is the desired power factor. Substituting the value of load voltages and currents phasors, $\beta = \frac{\tan\left(\cos^{-1}(0.866)\right)}{\sqrt{3}} = 0.3334$, the reference currents for the DSTATCOM for the mentioned power factor are given as follows.

$$\overline{I}_{fa}^* = \overline{I}_{la} - \overline{I}_{sa} = 1.715\angle - 59.04° - 0.65\angle - 30° = 1.18\angle - 65.01°\,\text{p.u.}$$

$$\overline{I}_{fb}^* = \overline{I}_{lb} - \overline{I}_{sb} = 0.8575\angle - 179.04° - 0.65\angle - 150° = 0.29\angle 130.34°\,\text{p.u.}$$

$$\overline{I}_{fc}^* = \overline{I}_{lc} - \overline{I}_{sc} = 0.6934\angle 176.31° - 0.65\angle 90° = 0.44\angle - 35.2°\,\text{p.u.}$$

The actual value of reference currents are as follows.

$$\overline{I}_{fa\,act}^* = 42.89\angle - 74.67°\,\text{A}$$

$$\overline{I}_{fb\,act}^* = 15.44\angle 132.49°\,\text{A}$$

$$\overline{I}_{fc\,act}^* = 33.49\angle - 138.47°\,\text{A}$$

The source currents after compensation are,

$$\overline{I}_{sa\,act} = 23.82\angle - 30°\,\text{A}$$

$$\overline{I}_{sb\,act} = 23.82\angle - 150°\,\text{A}$$

$$\overline{I}_{sc\,act} = 23.82\angle 90°\,\text{A}$$

The actual reference currents in the time domain are given as

$$i_{fa}^* = 60.66\sin\left(\omega t - 74.67°\right)\text{A}$$

$$i_{fb}^* = 21.84\sin\left(\omega t + 132.49°\right)\text{A}$$

$$i_{fc}^* = 47.36\sin\left(\omega t - 138.47°\right)\text{A}$$

These currents are realized using switching control of the shunt VSI as discussed in Chapter 5. Now, these currents are considered for the computation of series VSI reference voltages to maintain load voltages at 1 p.u.

(b) Using Type-2 control, the fundamental DVR voltages are computed as discussed in Chapter 6.

$$V_{fk}^2 - 2(a_{1k}\cos\delta_{lk} + b_{1k}\sin\delta_{lk})V_{lk}V_{fk} + V_{lk}^2 - V_{tk}^2 = 0$$

where V_{fk}, V_{tk} and V_{lk} are the fundamental rms magnitudes of DVR, terminal, and load voltages, respectively. The subscript k denotes the phases a, b, c. The terms a_{1k} and b_{1k} denote the angle information of v_{fk}, which is kept $90°$ leading to the angle of i_{sk} to avoid any real power exchange with the DVR. δ_{lk} is the angle of load voltage v_{lk}. The constants a_1 and b_1 are calculated as given below.

$$\angle \overline{V}_{fa} = \angle \overline{I}_{sa} + 90° \quad = a_{1a} + jb_{1a} = 0.5 + j0.866$$
$$\angle \overline{V}_{fb} = \angle \overline{I}_{sb} + 90° \quad = a_{1b} + jb_{1b} = 0.5 - j0.866$$
$$\angle \overline{V}_{fc} = \angle \overline{I}_{sc} + 90° \quad = a_{1c} + jb_{1c} = -1 + j0$$

Therefore, $a_{1a} = 0.5$, $b_{1a} = 0.866$, $a_{1b} = 0.5$, $b_{1b} = -0.866$, $a_{1c} = -1$, and $b_{1c} = 0$.

Substituting the values of a_{1k}, b_{1k}, V_{lk}, δ_{lk} and V_{tk} in (7.26), we get the following feasible values of DVR voltages V_{fk}.

$$V_{fa} = 1.0$$
$$V_{fb} = 1.0$$
$$V_{fc} = 1.0$$

Thus, the fundamental series compensator reference voltages are as follows.

$$\overline{V}_{fa1}^* = 1.0\angle 60°$$
$$\overline{V}_{fb1}^* = 1.0\angle -60°$$
$$\overline{V}_{fc1}^* = 1.0\angle 180°$$

The actual fundamental reference DVR voltages in time domain are given as,

$$v_{fa1}^* = 230\sqrt{2}\sin(\omega t + 60°)\,V$$
$$v_{fb1}^* = 230\sqrt{2}\sin(\omega t - 60°)\,V$$
$$v_{fc1}^* = 230\sqrt{2}\sin(\omega t + 180°)\,V$$

Based on this the angles of fundamental terminal voltages are calculated as follows.

$$\overline{V}_{ta1} = \overline{V}_{la} - \overline{V}_{fa1}^* = 1\angle 0° - 1\angle 60° = 1\angle -60°$$
$$\overline{V}_{tb1} = \overline{V}_{lb} - \overline{V}_{fb1}^* = 1\angle -120° - 1\angle -60° = 1\angle 180°$$
$$\overline{V}_{tc1} = \overline{V}_{lc} - \overline{V}_{fc1}^* = 1\angle 120° - 1\angle 180° = 1\angle 60°$$

As we can see there is a shift of $-60°$ in the terminal fundamental voltages. This shift should also be considered in the harmonics terms. Therefore, the calculated terminal voltages will be as follows.

$$v_{ta} = \sqrt{2}\sin(\omega t - 60°) + 0.2\sqrt{2}\sin[5(\omega t - 60°)]\,\text{p.u.}$$
$$v_{tb} = \sqrt{2}\sin(\omega t - 180°) + 0.2\sqrt{2}\sin[5(\omega t - 180°)]\,\text{p.u.}$$
$$v_{tc} = \sqrt{2}\sin(\omega t + 60°) + 0.2\sqrt{2}\sin[5(\omega t + 60°)]\,\text{p.u.}$$

In the above equation, the harmonic component of the terminal voltages in the respective phases should be nullified by the DVR to make load voltages free from harmonics. Therefore, the total DVR voltages in time domain are given as following,

$$v_{fa}^* = 230\sqrt{2}\sin(\omega t + 60°) - 46\sqrt{2}\sin[5(\omega t - 60°)]\,\text{V}$$
$$v_{fb}^* = 230\sqrt{2}\sin(\omega t - 60°) - 46\sqrt{2}\sin[5(\omega t - 180°)]\,\text{V}$$
$$v_{fc}^* = 230\sqrt{2}\sin(\omega t + 180°) - 46\sqrt{2}\sin[5(\omega t + 60°)]\,\text{V}$$

(c) Rating of the shunt VSI (DSTATCOM) is calculated as,

$$S_{DSTATCOM} = V_{la}I_{fa} + V_{lb}I_{fb} + V_{lc}I_{fc} = 2.5345\,\text{p.u.}$$

The actual rating of the DSTATCOM is,

$$S_{DSTATCOM\,act} = 63.36\,\text{kVA}$$

Rating of the series VSI (DVR) is calculated as,

$$S_{DVR} = V_{fa}I_{sa} + V_{fb}I_{sb} + V_{fc}I_{sc} = 2.0115\,\text{p.u.}$$

The actual rating of the DVR is,

$$S_{DVR\,act} = 50.28\,\text{kVA}$$

7.6 SUMMARY

The basic configurations and their modes of operation of the UPQC are discussed. It is shown that UPQC is a combination of shunt and series compensators through a common dc link. The shunt compensator, also termed as the shunt voltage source inverter (shunt-VSI), compensates for the unbalance and harmonics in the load currents. The series compensator, termed series-VSI compensates the voltage unbalance and harmonics. Two important configurations of UPQC, i.e., left shunt UPQC and right shunt UPQC configurations are illustrated. Later on, the right shunt UPQC structure is discussed in more detail to compensate for unbalance and harmonics in voltage and currents. The instantaneous symmetrical component theory is used to obtain the reference currents for the shunt compensator to make source currents balanced sinusoidal quantities. For voltage compensation, Type-1 and Type-2 control schemes are discussed to extract reference voltages. These reference currents and voltages are realized using appropriate switching control strategies of shunt and series inverters. The concepts are illustrated through examples with different source voltages and current conditions.

7.7 PROBLEMS

P 7.1 Discuss the applications of UPQC in the power distribution system.

P 7.2 What are the differences between the UPQC and UPFC (Unified Power Flow Controller)?

P 7.3 Give the classification of UPQC, based on various parameters such as topology, number of phases, connections, and arrangement of components.

P 7.4 Draw the schematic diagram of UPQC and explain the operating principle of the UPQC.

P 7.5 Draw the single-line diagram of three-phase right shunt and left shunt topologies. Distinguish the two topologies by mentioning their basic features.

P 7.6 Draw the circuit diagrams of three-phase right shunt UPQC with (i) three-phase four-wire (ii) three-phase three-wire (iii) three-phase four-wire H-bridge (iv) three-phase four-wire reduced dc link voltage.

P 7.7 Discuss the operation and control of the shunt compensator and series compensator of the right shunt UPQC.

P 7.8 Explain how the reference currents are generated for the shunt compensator of right shunt VSI.

P 7.9 Explain how the reference voltages are generated for the series compensator of right shunt VSI under (i) balanced fundamental voltages (ii) Unbalanced fundamental voltage (iii) with unbalance and harmonics in the supply voltages.

P 7.10 Discuss Type-1 and Type-2 control schemes of the series compensator of the UPQC and discuss their merits.

P 7.11 Explain how the losses in the shunt and series compensator of voltage source inverters are replenished to maintain dc link voltage of the required value.

P 7.12 Consider a single-phase system with rms voltage of 230 V (1 p.u.) and a 5 kVA base rating. The source supplies a linear inductive load through a feeder impedance, $Z_s = 0.03 + j0.08$ p.u. A right shunt UPQC, as shown in Fig. 7.1 is placed at the load terminals to make the load voltage to 1 p.u. and the source current to be sinusoidal with a power factor of 0.7 (lag).

$$Z_l = 1.2 + j1.5 \text{ p.u.}$$

(a) Find out the load voltage and current without any compensation.

(b) Compute the shunt compensator reference current i_f^* required to maintain sinusoidal source currents with a power factor of 0.7 lag at the load bus.

(c) Compute the series compensator voltages v_f^* required to maintain load voltage rms magnitude to 1 p.u.

(d) Calculate the rating of the shunt VSI and the series VSI of the UPQC.

P 7.13 Consider a three-phase balanced system with rms voltage of 230 V (1 p.u.) and a 10 kVA base rating. The source supplies a three-phase unbalanced linear inductive load through a feeder impedance, $Z_s = 0.03 + j\,0.09$ p.u.. A right shunt UPQC, as shown in Fig. 7.4 is placed at the load terminals to make the load voltages to 1 p.u. and the source currents to be sinusoidal balanced in nature with a power factor of 0.8 (lag).

$$Z_{la} = 0.3 + j\,0.5 \text{ p.u.}$$
$$Z_{lb} = 0.6 + j\,0.9 \text{ p.u.}$$
$$Z_{lc} = 0.5 - j\,1.3 \text{ p.u.}$$

(a) Find out the load voltages and currents without compensation.

(b) Compute the shunt compensator reference currents i_{fa}^*, i_{fb}^*, i_{fc}^* required to maintain balanced sinusoidal source currents.

(c) Compute the series compensator voltages v_{fa}^*, v_{fb}^*, v_{fc}^* required to maintain load voltage rms magnitude to 1 p.u.

(d) Calculate the rating of the shunt VSI and the series VSI of the UPQC.

P 7.14 Consider a three-phase balanced system with rms voltage of 230 V (1 p.u.) and a 15 kVA base rating. The source supplies a three-phase unbalanced linear inductive load through a feeder impedance, $Z_s = 0.02 + j\,0.09$ p.u.. A right shunt UPQC, as shown in Fig. 7.4 is placed at the load terminals to make the load voltages to 1 p.u. and the source currents to be sinusoidal balanced in nature with a power factor of 0.9 (lag).

$$Z_{la} = 0.4 - j\,0.3 \text{ p.u.}$$
$$Z_{lb} = 0.8 + j\,0.6 \text{ p.u.}$$
$$Z_{lc} = 1.2 - j\,0.5 \text{ p.u.}$$

(a) Find out the load voltages and currents without compensation.

(b) Compute the shunt compensator reference currents i_{fa}^*, i_{fb}^*, i_{fc}^* required to maintain balanced sinusoidal source currents.

(c) Compute the series compensator voltages v_{fa}^*, v_{fb}^*, v_{fc}^* required to maintain load voltage rms magnitude to 1 p.u.

(d) Calculate the rating of the shunt VSI and the series VSI of the UPQC.

P 7.15 Consider a three-phase unbalanced system with a base kVA rating of 20 kVA and base rms voltage of 230 V. The rms values of the source voltages are given as,

$$V_{sa} = 1 \text{ p.u.}$$
$$V_{sb} = 1.05 \text{ p.u.}$$
$$V_{sc} = 0.95 \text{ p.u.}$$

The source supplies a three-phase unbalanced linear inductive load through a feeder impedance, $Z_s = 0.02 + j0.08$ p.u. A right shunt UPQC, as shown in Fig. 7.4 is placed at the load terminals to make the load voltages of all the three phases to 1 p.u. and the source currents to be sinusoidal balanced in nature with a power factor of 0.85 (lag). The load impedances are:

$$Z_{la} = 0.4 + j0.5 \text{ p.u.}$$
$$Z_{lb} = 0.6 - j0.9 \text{ p.u.}$$
$$Z_{lc} = 1.3 - j1.2 \text{ p.u.}$$

(a) Find out the load voltages and currents without any compensation.

(b) Compute the shunt compensator reference currents i_{fa}^*, i_{fb}^*, i_{fc}^* required to maintain balanced sinusoidal source currents.

(c) Compute the series compensator voltages v_{fa}^*, v_{fb}^*, v_{fc}^* required to maintain load voltage rms magnitude to 1 p.u.

(d) Calculate the rating of the shunt VSI and the series VSI of the UPQC.

P 7.16 Consider a three-phase unbalanced system with a base kVA rating of 15 kVA and base rms voltage of 230 V. The rms values of the source voltages are given as,

$$V_{sa} = 1.02 \text{ p.u.}$$
$$V_{sb} = 1.12 \text{ p.u.}$$
$$V_{sc} = 0.88 \text{ p.u.}$$

The source supplies a three-phase unbalanced linear inductive load through a feeder impedance, $Z_s = 0.01 + j0.08$ p.u. A right shunt UPQC, as shown in Fig. 7.4 is placed at the load terminals to make the load voltages of all the three phases to 1 p.u. and the source currents to be sinusoidal balanced in nature with a power factor of 0.85 (lead). The three-phase load has the following impedances.

$$Z_{la} = 0.3 + j0.4 \text{ p.u.}$$
$$Z_{lb} = 0.6 - j0.9 \text{ p.u.}$$
$$Z_{lc} = 0.5 + j1.2 \text{ p.u.}$$

(a) Find out the load voltages and currents without any compensation.

(b) Compute the shunt compensator reference currents i_{fa}^*, i_{fb}^*, i_{fc}^* required to maintain balanced sinusoidal source currents.

(c) Compute the series compensator voltages v_{fa}^*, v_{fb}^*, v_{fc}^* required to maintain load voltage rms magnitude to 1 p.u.

(d) Calculate the rating of the shunt VSI and the series VSI of the UPQC.

P 7.17 Consider a three-phase balanced system with a base kVA rating of 30 kVA and base rms voltage of 230 V. The supply voltages after the feeder impedances are:

$$v_{ta} = 0.98\sqrt{2}\sin\omega t + 0.3\sqrt{2}\sin(7\omega t)\,\text{p.u.}$$
$$v_{tb} = 0.98\sqrt{2}\sin(\omega t - 120°) + 0.3\sqrt{2}\sin(7(\omega t - 120°))\,\text{p.u.}$$
$$v_{tc} = 0.98\sqrt{2}\sin(\omega t + 120°) + 0.3\sqrt{2}\sin(7(\omega t + 120°))\,\text{p.u.}$$

Now a right shunt UPQC, as shown in Fig. 7.4, is placed to make fundamental balanced load voltages of rms value 1 p.u. The source currents after compensation are of sinusoidal balanced nature with a power factor of 0.9 (lag) at the load bus. The loads connected between the phases and the neutral are:

$$Z_{la} = 0.3 - j0.5\,\text{p.u.}$$
$$Z_{lb} = 1.2 + j1.3\,\text{p.u.}$$
$$Z_{lc} = 0.6 + j1.2\,\text{p.u.}$$

(a) Compute the shunt reference currents i_{fa}^*, i_{fb}^*, i_{fc}^* required to maintain balanced sinusoidal source currents.

(b) Compute the series compensator voltages v_{fa}^*, v_{fb}^*, v_{fc}^* required to maintain the load voltage rms magnitude to 1 p.u. using Type-2 control, as discussed in Chapter 6.

(c) Calculate the rating of the shunt VSI and the series VSI of the UPQC.

P 7.18 Consider a three-phase balanced system with a base kVA rating of 24 kVA and base rms voltage of 230 V. The supply voltages after the feeder impedances are:

$$v_{ta} = 0.95\sqrt{2}\sin\omega t + 0.1\sqrt{2}\sin(5\omega t)\,\text{p.u.}$$
$$v_{tb} = 0.95\sqrt{2}\sin(\omega t - 120°) + 0.1\sqrt{2}\sin(5(\omega t - 120°))\,\text{p.u.}$$
$$v_{tc} = 0.95\sqrt{2}\sin(\omega t + 120°) + 0.1\sqrt{2}\sin(5(\omega t + 120°))\,\text{p.u.}$$

Now a right shunt UPQC, as shown in Fig. 7.4, is placed to make fundamental balanced load voltages of rms value 1 p.u. The source currents after compensation are of sinusoidal balanced nature with a power factor of 0.8 (lead) at the load bus. The load impedances are:

$$Z_{la} = 0.4 + j0.6\,\text{p.u.}$$
$$Z_{lb} = 1.3 - j1.2\,\text{p.u.}$$
$$Z_{lc} = 0.5 - j1.2\,\text{p.u.}$$

(a) Compute the shunt reference currents i_{fa}^*, i_{fb}^*, i_{fc}^* required to maintain balanced sinusoidal source currents.

(b) Compute the series compensator voltages v_{fa}^*, v_{fb}^*, v_{fc}^* required to maintain the load voltage rms magnitude to 1 p.u. using Type-2 control, as discussed in Chapter 6.

(c) Calculate the rating of the shunt VSI and the series VSI of the UPQC.

REFERENCES

1. L. Gyugyi, "A unified flow control concept for flexible AC transmission systems," *Proc. IEE*, vol. 139, no. 4, pp. 323–331, July 1992.

2. H. Fujita and H. Akagi, "The unified power quality conditioner: the integration of series and shunt-active filters," *IEEE Transactions on Power Electronics*, vol. 13, no. 2, pp. 315–322, Mar. 1998.

3. M. Aredes, K. Heumann, and E. Watanabe, "An universal active power line conditioner," *IEEE Transactions on Power Delivery*, vol. 13, no. 2, pp. 545–551, 1998.

4. Y. Chen, X. Zha, J. Wang, H. Liu, J. Sun, and H. Tang, "Unified power quality conditioner (UPQC): The theory, modeling and application," in *IEEE Proceedings of International Conference on Power System Technology, 2000 (PowerCon 2000)*, vol. 3, pp. 1329–1333, 2000.

5. L. Tolbert, F. Peng, and T. Habetler, "A multilevel converter-based universal power conditioner," *IEEE Transactions on Industry Applications*, vol. 36, no. 2, pp. 596–603, 2000.

6. B. Mwinyiwiwa, B. Lu, and B. Ooi, "Multiterminal unified power flow controller," *IEEE Transactions on Power Electronics*, vol. 15, no. 6, pp. 1088–1093, 2000.

7. A. Ghosh and G. Ledwich, "A unified power quality conditioner (UPQC) for simultaneous voltage and current compensation," *Electric Power Systems Research*, vol. 59, no. 1, pp. 55–63, 2001.

8. ——, *Power Quality Enhancement Using Custom Power Devices*, ser. Power Electronics and Power Systems. Springer US, 2012.

9. V. Khadkikar, "Enhancing electric power quality using upqc: A comprehensive overview," *IEEE Transactions on Power Electronics*, vol. 27, no. 5, pp. 2284–2297, 2012.

10. Bhim Singh, Ambrish Chandra, and Kamal Al-Haddad, *Power Quality Problems and Mitigation Techniques*. Wiley, 2015.

11. N. Jayanti, M. Basu, M. Conlon, and K. Gaughan, "Performance comparison of a left shunt upqc and a right shunt upqc applied to enhance fault-ride-through capability of a fixed speed wind generator," in *European Conference on Power Electronics and Applications*, 2007, pp. 1–9.

12. B. S. Mohammed, K. S. Rama Rao, R. Ibrahim, and N. Perumal, "Performance evaluation of r-upqc and l-upqc based on a novel voltage detection algorithm," in *2012 IEEE Symposium on Industrial Electronics and Applications*, 2012, pp. 167–172.

13. S. B. Karanki, N. Geddada, Mahesh K. Mishra, and B. K. Kumar, "A modified three-phase four-wire upqc topology with reduced dc-link voltage rating," *IEEE Transactions on Industrial Electronics*, vol. 60, no. 9, pp. 3555–3566, 2013.

14. S. B. Karanki, *Topologies and design of state feedback controller for custom power devices in power distribution system*. Ph.D. Thesis, I.I.T. Madras, India, 2012.

15. M. Brenna, R. Faranda, and E. Tironi, "A new proposal for power quality and custom power improvement: Open UPQC," *IEEE Transactions on Power Delivery*, vol. 24, no. 4, pp. 2107–2116, Oct. 2009.

16. Mahesh K. Mishra, A. Joshi, and A. Ghosh, "Control schemes for equalization of capacitor voltages in neutral clamped shunt compensator," *IEEE Transactions on Power Delivery*, vol. 18, no. 2, pp. 538–544, 2003.

17. A. Shukla, A. Ghosh, and A. Joshi, "Control schemes for DC capacitor voltages equalization in diode-clamped multilevel inverter-based DSTATCOM," *IEEE Transactions on Power Delivery*, vol. 23, no. 2, pp. 1139–1149, 2008.

18. S. B. Karanki, N. Geddada, Mahesh K. Mishra, and B. K. Kumar, "A dstatcom topology with reduced dc-link voltage rating for load compensation with nonstiff source," *IEEE Transactions on Power Electronics*, vol. 27, no. 3, pp. 1201–1211, 2012.

19. S. Kotra and Mahesh K. Mishra, "Design and stability analysis of dc microgrid with hybrid energy storage system," *IEEE Transactions on Sustainable Energy*, vol. 10, no. 3, pp. 1603–1612, 2019.

20. N. M. Ismail and Mahesh K. Mishra, "Study on the design and switching dynamics of hysteresis current controlled four-leg voltage source inverter for load compensation," *IET Power Electronics*, vol. 11, no. 2, pp. 310–319, 2018.

21. H. Akagi, Y. Kanazawa, and A. Nabae, "Instantaneous reactive power compensators comprising switching devices without energy storage components," *Industry Applications, IEEE Transactions on*, vol. IA-20, no. 3, pp. 625–630, May 1984.

22. S. Bhattacharya and D. Divan, "Synchronous frame based controller implementation for a hybrid series active filter system," in *Industry Applications Conference, 1995. Thirtieth IAS Annual Meeting, IAS '95., Conference Record of the 1995 IEEE*, vol. 3, Oct 1995, pp. 2531–2540.

23. A. Ghosh and A. Joshi, "A new approach to load balancing and power factor correction in power distribution system," *Power Delivery, IEEE Transactions on*, vol. 15, no. 1, pp. 417–422, Jan 2000.

24. N. M. Ismail and Mahesh K. Mishra, "Control and operation of unified power quality conditioner with battery-ultracapacitor energy storage system," in *2014 IEEE International Conference on Power Electronics, Drives and Energy Systems (PEDES)*, 2014, pp. 1–6.

25. U. Rao, Mahesh K. Mishra, and A. Ghosh, "Control strategies for load compensation using instantaneous symmetrical component theory under different supply voltages," *Power Delivery, IEEE Transactions on*, vol. 23, no. 4, pp. 2310–2317, 2008.

26. A. Ghosh and G. Ledwich, "Load compensating DSTATCOM in weak AC systems," *Power Delivery, IEEE Transactions on*, vol. 18, no. 4, pp. 1302–1309, Oct 2003.

27. G. S. Kumar, B. K. Kumar, and Mahesh K. Mishra, "Mitigation of voltage sags with phase jumps by UPQC with PSO-Based ANFIS," *IEEE Transactions on Power Delivery*, vol. 26, no. 4, pp. 2761–2773, 2011.

28. M. Basu, S. P. Das, and G. K. Dubey, "Investigation on the performance of UPQC-Q for voltage sag mitigation and power quality improvement at a critical load point," *Inst. Eng. Technol., Gen. Transm. Distrib.*, vol. 2, no. 3, pp. 414–423, May 2008.

29. W. C. Lee, D. M. Lee, and T. K. Lee, "New control scheme for a unified power-quality compensator-q with minimum active power injection," *IEEE Transactions on Power Delivery*, vol. 25, no. 2, pp. 1068–1076, 2010.

Appendix

A Fundamental and Positive Sequence Extraction

In this appendix, extraction of fundamental or fundamental positive sequence components of the electrical quantities such as voltage or current are discussed. This brings more clarity and handy experience in using them while applying the concepts in power system analysis.

A.1 FUNDAMENTAL EXTRACTION

Consider the following waveform of voltage containing dc, fundamental and harmonic components.

$$v(t) = V_{dc} + V_{m1}\sin(\omega t + \phi_1) + V_{m2}\sin(2\omega t + \phi_2) + ... + V_{mn}\sin(n\omega t + \phi_n) \quad (A.1)$$

Now let us define the following term.

$$\overline{C}_1 = \frac{\sqrt{2}}{T}\int_{t_1-T}^{t_1} v(t)\,e^{-j(\omega t - \frac{\pi}{2})}\,dt \quad (A.2)$$

where t_1 is any arbitrary instant and T is the time period of the fundamental voltage. The following is obtained by substituting $v(t)$.

$$\overline{C}_1 = \frac{\sqrt{2}}{T}\int_{t_1-T}^{t_1} \{V_{dc} + V_{m1}\sin(\omega t + \phi_1) + V_{m2}\sin(2\omega t + \phi_2) + ...\}\,e^{-j(\omega t - \frac{\pi}{2})}\,dt \quad (A.3)$$

In the above, the term $e^{-j(\omega t - \frac{\pi}{2})}$ is written as following.

$$e^{-j(\omega t - \frac{\pi}{2})} = \sin \omega t + j\cos \omega t \quad (A.4)$$

Now simplifying (A.3), we get

$$\begin{aligned}
\overline{C}_1(t) = \;& \frac{\sqrt{2}}{T}\int_{t_1-T}^{t_1} \{[V_{dc}\sin \omega t + V_{m1}\sin(\omega t + \phi_1)\sin \omega t \\
& + V_{m2}\sin(2\omega t + \phi_2)\sin \omega t + ...] \\
& + j[V_{dc}\cos \omega t + V_{m1}\sin(\omega t + \phi_1)\cos \omega t \\
& + V_{m2}\sin(2\omega t + \phi_2)\cos \omega t + ...]\}\,dt
\end{aligned} \quad (A.5)$$

DOI: 10.1201/9781032617305-A

In the above equation, all other terms are zero except the multiplication of fundamental sine and cos terms. Therefore,

$$\overline{C}_1(t) = \frac{V_{m1}}{\sqrt{2}}(\cos\phi_1 + j\sin\phi_1) \tag{A.6}$$

$$\overline{C}_1(t) = \frac{V_{m1}}{\sqrt{2}}\angle\phi_1 \tag{A.7}$$

From this, we can construct

$$v_1(t) = \sqrt{2}\,|\overline{C}_1|\sin(\omega t + \angle\overline{C}_1) \tag{A.8}$$

In general, we can extract any harmonic using the following.

$$\overline{C}_n = \frac{1}{T}\int_{t_1-T}^{t_1} v(t)\,e^{-j(n\omega t - \frac{\pi}{2})}dt \tag{A.9}$$

$$\Rightarrow v_n(t) = \sqrt{2}\,|\overline{C}_n|\sin(n\omega t + \angle\overline{C}_n) \tag{A.10}$$

A.2 EXTRACTION OF POSITIVE SEQUENCE FUNDAMENTAL COMPONENTS

Consider the following set of three-phase unbalanced distorted voltages.

$$v_a(t) = V_{dca} + V_{ma1}\sin(\omega t + \phi_{a1}) + V_{ma2}\sin(2\omega t + \phi_{a2}) + \dots$$
$$v_b(t) = V_{dcb} + V_{mb1}\sin(\omega t - 120° + \phi_{b1}) + V_{mb2}\sin\{2(\omega t - 120°) + \phi_{b2}\} + \dots$$
$$v_c(t) = V_{dcc} + V_{mc1}\sin(\omega t + 120° + \phi_{c1}) + V_{mc2}\sin\{2(\omega t + 120°) + \phi_{c2}\} + \dots$$
$$\tag{A.11}$$

We aim to compute positive, negative, and zero sequence fundamental voltages, i.e., $v_{a1+}, v_{b1+}, v_{c1+}, v_{a1-}, v_{b1-}, v_{c1-}, v_{a10}, v_{b10}, v_{c10}$, respectively.

For this, we use the following expression.

$$\overline{C}_{a1+} = \frac{\sqrt{2}}{T}\int_{t_1-T}^{t_1}\overline{v}_{a+}(t)\,e^{-j(\omega t - \frac{\pi}{2})}dt$$

$$\overline{C}_{a1-} = \frac{\sqrt{2}}{T}\int_{t_1-T}^{t_1}\overline{v}_{a-}(t)\,e^{-j(\omega t - \frac{\pi}{2})}dt \tag{A.12}$$

$$\overline{C}_{a10} = \frac{\sqrt{2}}{T}\int_{t_1-T}^{t_1}\overline{v}_{a0}(t)\,e^{-j(\omega t - \frac{\pi}{2})}dt$$

In the above, $\overline{v}_{a+}, \overline{v}_{a-}$, and $\overline{v}_{a0}$ are instantaneous positive, negative, and zero sequence components. These are given by the following equation.

$$\begin{bmatrix} \overline{v}_{a0} \\ \overline{v}_{a+} \\ \overline{v}_{a-} \end{bmatrix} = \frac{1}{3}\begin{bmatrix} 1 & 1 & 1 \\ 1 & a & a^2 \\ 1 & a^2 & a \end{bmatrix}\begin{bmatrix} v_a \\ v_b \\ v_c \end{bmatrix} \tag{A.13}$$

where, $a = e^{j\frac{2\pi}{3}}$. Now let us solve $\overline{C}_{a1+}$ in the above equation.

$$
\begin{aligned}
\overline{C}_{a1+} &= \frac{\sqrt{2}}{3T} \int_{t_1-T}^{t_1} \Big\{ [V_{dca} + V_{ma1}\sin(\omega t + \phi_{a1}) + V_{ma2}\sin(2\omega t + \phi_{a2}) + ...] \\
&+ a[V_{dcb} + V_{mb1}\sin(\omega t - 120° + \phi_{b1}) + V_{mb2}\sin(2(\omega t - 120°) + \phi_{b2}) + ..] \\
&+ a^2[V_{dcc} + V_{mc1}\sin(\omega t + 120° + \phi_{c1}) \\
&+ V_{mc2}\sin(2(\omega t + 120°) + \phi_{c2}) + ..] \Big\}(\sin\omega t + j\cos\omega t)dt \qquad (A.14)
\end{aligned}
$$

The integration of all terms will be zero, except similar frequency terms i.e., ωt terms.

Hence the above equation is written as

$$
\begin{aligned}
\overline{C}_{a1+} &= \frac{\sqrt{2}}{2 \times T \times 3} \int_{t_1-T}^{t_1} \Big\{ [V_{ma1}2\sin(\omega t + \phi_{a1})\sin\omega t] \\
&+ a[V_{mb1}2\sin(\omega t - 120° + \phi_{b1})\sin\omega t] \\
&+ a^2[V_{mc1}2\sin(\omega t + 120° + \phi_{c1})\sin\omega t] \\
&+ j[V_{ma1}2\sin(\omega t + \phi_{a1})\cos\omega t] + a[V_{mb1}2\sin(\omega t - 120° + \phi_{b1})\cos\omega t] \\
&+ a^2[V_{mc1}2\sin(\omega t + 120° + \phi_{c1})\cos\omega t] \Big\}dt \qquad (A.15)
\end{aligned}
$$

This gives,

$$
\begin{aligned}
\overline{C}_{a1+} &= \frac{1}{3}\Big\{ \frac{V_{ma1}}{\sqrt{2}}(\cos\phi_{a1} + j\sin\phi_{a1}) \\
&+ \frac{aV_{mb1}}{\sqrt{2}}[\cos(-120° + \phi_{b1}) + j\sin(-120° + \phi_{b1})] \\
&+ \frac{a^2V_{mc1}}{\sqrt{2}}[\cos(120° + \phi_{c1}) + j\sin(120° + \phi_{c1})] \Big\} \qquad (A.16)
\end{aligned}
$$

$$
\begin{aligned}
\overline{C}_{a1+} &= \frac{\overline{V}_{a1} + a\overline{V}_{b1} + a^2\overline{V}_{c1}}{3} = \overline{V}_{a1+} \\
v_{a1+}(t) &= \sqrt{2}|\overline{C}_{a1+}|\sin(\omega t + \angle\overline{C}_{a1+}) \\
v_{b1+}(t) &= \sqrt{2}|\overline{C}_{a1+}|\sin(\omega t - 120° + \angle\overline{C}_{a1+}) \\
v_{c1+}(t) &= \sqrt{2}|\overline{C}_{a1+}|\sin(\omega t + 120° + \angle\overline{C}_{a1+}) \qquad (A.17)
\end{aligned}
$$

Similarly, we can find out

$$
\overline{C}_{a10} = \frac{\sqrt{2}}{T}\int_{t_1-T}^{t_1} v_{a0}(t)e^{-j(\omega t - \frac{\pi}{2})}dt \qquad (A.18)
$$

with

$$
v_{a10}(t) = v_{b10}(t) = v_{c10}(t) = \sqrt{2}|\overline{C}_{a10}|\sin(\omega t + \angle\overline{C}_{a10}) \qquad (A.19)
$$

The negative sequence quantities can be found as well, which are given below.

$$\overline{C}_{a1-} = \frac{\sqrt{2}}{T} \int_{t_1-T}^{t_1} v_{a-}(t) e^{-j(\omega t - \frac{\pi}{2})} dt \tag{A.20}$$

with

$$v_{a1-}(t) = \sqrt{2} \, |\overline{C}_{a1-}| \, \sin(\omega t + \angle \overline{C}_{a1-})$$
$$v_{b1-}(t) = \sqrt{2} \, |\overline{C}_{a1-}| \, \sin(\omega t + 120° + \angle \overline{C}_{a1-})$$
$$v_{c1-}(t) = \sqrt{2} \, |\overline{C}_{a1-}| \, \sin(\omega t - 120° + \angle \overline{C}_{a1-}) \tag{A.21}$$

B Answers to Numerical Problems

Chapter 2

2.7 $pf = \dfrac{V_{dc}I_{dc} + \sum_{n=1}^{\infty} \frac{V_{mn}I_{mn}}{2} \cos(\phi_{in} - \phi_{vn})}{\left(V_{dc}^2 + \sum_{n=1}^{\infty} V_{mn}^2/2\right)^{\frac{1}{2}} \times \left(I_{dc}^2 + \sum_{n=1}^{\infty} I_{mn}^2/2\right)^{\frac{1}{2}}}$

2.8 $-j6.35\,\Omega$, capacitance, $509.3\,\mu F$, 27.6 A

2.9 (a) $p_{active} = 3983\{1 - \cos 2\omega t\} + 241.48\{1 - \cos 2(3\,\omega t + 30°), p_{reactive} = -2300 \sin 2\omega t + 64.7 \sin 2(3\,\omega t + 30°), p_{rest} = 1150 \cos(2\omega t + 45°) + 1000 \cos(2\omega t + 60°) - 1150 \cos(4\omega t + 45°) - 1000 \cos 4\omega t, P = 4224.48\ W,$ $Q = 2235.3$ VAr, $S = 4852.27$ VA, $pf = 0.87$.

2.10 (a) $i = \frac{v}{R_L}$ (b) $\sum_{n=1}^{\infty} \frac{V_n^2}{R}$ (c) $\sum_{n=1}^{\infty} \frac{V_n^2}{R}, 0, \sum_{n=1}^{\infty} \frac{V_n^2}{R}, 1$ (d) 0, 0

2.11 (a) $i(t) = 44\sqrt{2} \sin(\omega t - 53.13°) + 5.92\sqrt{2} \sin(3\omega t - 75.96°) + 2.17\sqrt{2} \sin(5\omega t - 81.47°)$ A, (b) $p(t) = 5805(1 - \cos 2\omega t) - 7744 \sin 2\omega t + 105.45(1 - \cos 6\omega t) - 421.79 \sin 6\omega t + 14.2(1 - \cos 10\omega t) - 94.67 \sin 10\omega t + \left(\sum_{n=1,3,5} \frac{230\sqrt{2}}{n} \sin n\omega t\right) \left(\sum_{h=1,3,5; h \neq n} \sqrt{2} I_h \sin(h\omega t - \phi_{ih})\right)$

2.12 (a) 1 (b) 0 (lagging) (c) 0 (leading)

(d) $\dfrac{R}{\sqrt{R^2 + (\omega L)^2}} \dfrac{\sqrt{1 + \left(\frac{V_2}{V_1}\right)^2 \frac{R^2 + (\omega L)^2}{R^2 + (2\omega L)^2} + \ldots\ldots + \left(\frac{V_n}{V_1}\right)^2 \frac{R^2 + (\omega L)^2}{R^2 + (n\omega L)^2}}}{\sqrt{1 + \left(\frac{V_2}{V_1}\right)^2 + \ldots\ldots + \left(\frac{V_n}{V_1}\right)^2}}$

(e) $\dfrac{R}{\sqrt{R^2 + (\frac{1}{\omega C})^2}} \dfrac{\sqrt{1 + \left(\frac{V_2}{V_1}\right)^2 \frac{R^2 + (\frac{1}{\omega C})^2}{R^2 + (\frac{1}{2\omega C})^2} + \ldots\ldots + \left(\frac{V_n}{V_1}\right)^2 \frac{R^2 + (\frac{1}{\omega C})^2}{R^2 + (\frac{1}{n\omega C})^2}}}{\sqrt{1 + \left(\frac{V_2}{V_1}\right)^2 + \ldots\ldots + \left(\frac{V_n}{V_1}\right)^2}}$

(f) For an inductive load, the power factor is less than the fundamental power factor ($\cos \phi_1$), and for a capacitive load, the power factor is greater than the fundamental power factor

2.13 (a) $15.89 \sin(\omega t - 12.33°) + 1.63 \sin(3\,\omega t - 53.05°) + 0.65 \sin(5\,\omega, t - 66.55°)$ A (b) 2524.4, 30.74, 2555.1 W (c) 551.85, 44.95, 596.81 VAr (d) 2648.3, 2584, 55.42, 696.44, 358.96 VA (e) 11.03%, 19.44%, 0.9648 (lag)

2.14 Hint: displacement factor of source current for the fully controlled rectifier load, which is fired at an angle α is $\cos \alpha$ and that for semi-controlled rectifier is $(\cos \alpha/2)$, 0.8279 (lag), 0.9235 (lag)

DOI: 10.1201/9781032617305-B

2.15 $389\ \mu F$, $\ 0.976$ (lag)

2.16 1.0, 0.9006, 48.43%

2.17 (a) 8.57 A, 9.08 A (b) 21.15 V, $N_1 = 109$, $N_2 = 10$ (c) 1, 0.9439 (lag)

2.18 $V_s = 249.91$ V, P $= 1125$ W, Q $= 0$, S $= 1249.5$ VA, pf $= 0.9$ (lag)

2.19 $V_s = 288.57$ V, P $= 1125$ W, Q $= 649.5$ VAr, S $= 1442.9$ VA, pf $= 0.78$ (lag)

2.20 (a) $i(t) = 25.23\sqrt{2}\sin(\omega t - 40.33°) + 4.14\sqrt{2}\sin 3\omega t$
$+ 2.48\sqrt{2}\sin 5\omega t$ A (b) P $= 4423.5$ W, Q $= 3756$ VAr (lag), $pf_1 = 0.7623$
(lag), (c) S $= 5908.5$ VA, pf $= 0.7487$ (lag)

2.21 (a) $i(t) = 25.92\sqrt{2}\sin(\omega t + 31.64°) + 6.21\sqrt{2}\sin 3\omega t + 3.73\sqrt{2}\sin 5\omega t$ A
(b) P $= 5073.6$ W, Q $= -3126.3$ VAr (lead), $pf_1 = 0.8514$ (lead), (c) S $=$
6188.2 VA, $pf = 0.8199$ (lead)

2.22 Hint: $i(t) = \sum\limits_{n=1,3,5}^{\infty}(a_n\cos n\omega t + b_n\sin n\omega t)$,
where $a_n = \frac{V_m}{\pi R}\left\{\frac{\cos(n+1)\alpha-1}{n+1} - \frac{\cos(n-1)\alpha-1}{n-1}\right\}$
$b_n = \frac{V_m}{\pi R}\left\{\frac{\sin(n+1)\alpha}{n+1} - \frac{\sin(n-1)\alpha}{n-1}\right\}$ and $a_1 = \frac{V_m}{\pi R}\left(\frac{\cos 2\alpha-1}{2}\right), b_1 = \frac{V_m}{\pi R}\left\{\frac{\sin 2\alpha}{2} + (\pi - \alpha)\right\}$
(a) $i(t) = 4.237\sqrt{2}\sin(\omega t - 4.68°) + 0.346\sqrt{2}\sin(3\omega t + 30°) + 0.305\sqrt{2}\sin$
$(5\omega t - 10.89°)$ (b) $P = P_1 = 971.16$ W, $P_H = 0$ (c) $Q = Q_1 = 79.6$ VAr $Q_H =$
0 (d) $S = 980.18$ VA, $THD_I = 10.89\%$, $pf = 0.9908$ (lag) (e) 0.78 Ω

Chapter 3

3.11 $p(t) = P_{3\phi} = 6084.3$ W, $q(t) = Q_{3\phi} = 6591.4$ VAr

3.12 (a) $\bar{I}_a = 46\angle-53.13°$ A, $\bar{I}_b = 13.0\angle-167.29°$ A, $\bar{I}_c = 23.0\angle173.13°$ A,
$\bar{I}_n = 37.74\angle77.89°$ A (b) $P_a = 6348$ W, $Q_a = 8464$ VAr, $pf_a = 0.6$ (lag),
$P_b = 2028.1$ W, $Q_b = 2197.1$ VAr, $pf_b = 0.6783$ (lag), $P_c = 3174$ W,
$Q_c = -4232$ VAr, $pf_a = 0.6$ (lead), $P_{3\phi} = 11550$ W, $Q_{3\phi} = 6429.1$ VAr
(c) $S_A = 18860$ VA, $S_V = 13219$ VA, $S_e = 21133$, VA, $pf_A = 0.6129$, $pf_V =$
0.8738, $pf_e = 0.5466$

3.13 (a) $\bar{I}_a = 51.99\angle-15.27°$ A, $\bar{I}_b = 10.5\angle151.68°$ A, $\bar{I}_c = 41.82\angle167.98°$ A, $i_a =$
$\sqrt{2} \times 49.06\sin(\omega t + 51.99°)$, $i_b = \sqrt{2} \times 10.5\sin(\omega t + 151.68°)$, $i_c = \sqrt{2} \times$
$41.82\sin(\omega t + 167.98°)$ (b) $P_a = 12743$ W, $Q_a = 3478.9$ VAr, $P_b = 78.38$ W,
$Q_b = 2668$ VAr, $P_c = 7112.9$ W, $Q_c = -7894.3$ VAr, $P_{3\phi} = 19934$ W, $Q_{3\phi} =$
-1747.4 VAr (c) $S_A = 26504$ VA, $S_V = 20011$ VA, $S_e = 29725$, VA, $pf_A =$
0.75, $pf_V = 0.99$, $pf_e = 0.67$

3.14 (a) $\bar{I}_a = 38.5\angle-30.37°$ A, $\bar{I}_b = 12.84\angle139.16°$ A, $\bar{I}_c = 25.98\angle154.78°$ A (b)
$P_a = 7640.5$ W, $Q_a = 4477.1$ VAr, $P_b = 842.71$ W, $Q_b = 2425.3$ VAr, $P_c =$
1992.7 W, $Q_c = -4231.3$ VAr, $P_{3\phi} = 10476$ W, $Q_{3\phi} = 2671.1$ VAr (c) $S_A =$
16100 VA, $S_V = 10811$ VA, $S_e = 17059$, VA, $pf_A = 0.65$, $pf_V = 0.97$, $pf_e =$
0.61

3.15 (a) $\bar{V}_{a0} = 32.7\angle 54.63°$ V, $\bar{V}_{a1} = 196.83\angle -18.77°$ V, $\bar{V}_{a2} = 44.21\angle 56.03°$ V, $\bar{I}_{a0} = 7.59\angle -83.93°$ A, $\bar{I}_{a1} = 20.65\angle -41.15°$ A, $\bar{I}_{a2} = 19.29\angle -54.32°$ A (b) $P_0 = -558.18$ W, $Q_0 = 492.78$ VAr, $pf_0 = 0.75$, $P_1 = 11273$ W, $Q_1 = 4641.7$ VAr, $pf_1 = 0.92$, $P_2 = -889.56$ W, $Q_2 = 2398.8$ VAr, $pf_2 = 0.35$ (c) Total powers are same

3.16 (a) $\bar{I}_a = 46.34\angle -72.35°$ A, $\bar{I}_b = 71.20\angle -144.8°$ A, $\bar{I}_c = 95.94\angle 62.61°$ A (c) $P_{ab} = 5161$ W, $Q_{ab} = 6451.2$ VAr, $\phi_{ab} = 51.34°$ (lag), $P_{bc} = 19354$ W, $Q_{bc} = 15483$ VAr, $\phi_{bc} = 38.66°$ (lag), $P_{ca} = 5472.4$ W, $Q_{ca} = 13681$ VAr, $\phi_{ca} = 68.20°$ (lag) (d) $P_a = 3229.6$ W, $Q_a = 10156$ VAr, $\phi_a = 72.36°$ (lag), $P_b = 14865$ W, $Q_b = 6870$ VAr, $\phi_b = 24.81°$ (lag), $P_c = 11893$ W, $Q_c = 18589$ VAr, $\phi_c = 57.39°$ (lag) (f) $S_A = 49101$ VA, $pf_A = 0.61$, $S_V = 46558$ VA, $pf_V = 0.64$, $S_e = 51051$ VA, $pf_e = 0.59$

3.17 (a) $\bar{I}_a = 81.31\angle 7.31°$ A, $\bar{I}_b = 137.21\angle -111.05°$ A, $\bar{I}_c = 121.81\angle 104.92°$ A (b) $P_a = 18549.4$ W, $Q_a = -2379.9$ VAr, $pf_a = 0.99$ lead, $P_b = 31173.4$ W, $Q_b = -4908.5$ VAr, $pf_b = 0.98$ (lead), $P_c = 27051.1$ W, $Q_c = 7288.4$ VAr, $pf_c = 0.96$ (lag)
(c) $S_A = 78274$ VA, $pf_A = 0.9808$, $S_V = 76774$ VA, $pf_V = 1.0$, $S_e = 79947$ VA, $pf_e = 0.9603$ (d) Net reactive power is zero

3.18 (a) $\bar{I}_a = \frac{\sqrt{3}V}{R}\angle 30°$ A, $\bar{I}_b = \frac{3V}{R}\angle -120°$ A, $\bar{I}_c = \frac{\sqrt{3}V}{R}\angle 90°$ A, $\bar{I}_n = 0$ A (b) $P_a = \frac{3V^2}{2R}$, $Q_a = -\frac{\sqrt{3}V^2}{2R}$, $pf_a = 0.866$ (lag), $P_b = \frac{3V^2}{R}$, $Q_b = 0$, $pf_b = 1.0$, $P_c = \frac{3V^2}{2R}$, $Q_c = \frac{\sqrt{3}V^2}{2R}$, $pf_c = 0.866$ (lead) (c) $P = \frac{6V^2}{R}$ W, $Q = 0$ VAr, $S_v = \frac{6V^2}{R}$ VA, $S_A = \frac{6.464V^2}{R}$ VA, $pf_v = 1.0$, $pf_A = 0.93$ (d) $R_{eq} = \frac{R}{2}$, $\frac{Losses(Unb.)}{Losses(Bal.)} = 0.75$

3.19 (a) $Z_{ab} = 9.17 + j11.46\,\Omega$, $\bar{I}_a = 20.88\angle -24.16°$ A, $\bar{I}_b = -20.88\angle 155.84°$ A, $\bar{I}_c = 0$ A (b) $\bar{V}_a = 196.09\angle -8.67°$ V, $\bar{V}_b = 187.05\angle -114.93°$ V, $\bar{V}_c = 230\angle 120°$ V, $P_{loss} = 872.6$ W, $Q_{loss} = 1745.2$ VAr (c) $S_v = 6403$ VA, $pf_v = 0.6247$, $S_A = 8002.8$ VA, $pf_A = 0.4998$, $S_e = 10499$ VA, $pf_e = 0.381$ (d) $Z_e = 12.59 + j15.74\,\Omega$, $P_{loss} = 317.55$ W, $Q_{loss} = 635.1$ VAr (e) $S_v = S_A = S_e = 6403.1$ VA, $pf_v = pf_A = pf_e = 0.6427$ (f) $Z_{eq} = 4.58 + j10.28\,\Omega$, $S_v = 9818.1$ VA, $pf_v = 0.4074$, $S_A = 9818.1$ VA, $pf_A = 0.4074$, $S_e = 9818.1$ VA, $pf_e = 0.4074$

3.20 (a) $\bar{V}_a = 211.66\angle 0°$ V, $\bar{V}_b = 211.66\angle -120°$ V, $\bar{V}_c = 211.66\angle 120°$ V, $\bar{I}_a = 188.98\angle 0°$ A, $\bar{I}_b = 188.98\angle -120°$ A, $\bar{I}_c = 188.98\angle 120°$ A, $r = 0.0672\,\Omega$ (b) $R_{ab} = 1.0363\,\Omega$, $\bar{S}_a = 60 - j39.13$ kVA, $\bar{S}_b = 60 + j39.13$ kVA, $\bar{S}_c = 0$ kVA, $S_v = 120$ kVA, $pf_v = 1.0$, $S_A = 143.27$ kVA, $pf_A = 0.8376$, $S_e = 181.05$ kVA, $pf_e = 0.6628$ (c) $Z_{eq} = 0.52 + j0.58\,\Omega$, $\bar{I}_a = 277.85\angle -45°$ A (balanced), $\bar{V}_a = 217.2\angle 3.48°$ A (balanced), $S_v = S_A = S_e = 181.05$ kVA, $pf_v = pf_A = pf_e = 0.6628$

3.21 (a) $\bar{I}_a = 29.31\angle -60°$ A, $\bar{I}_b = 29.31\angle 120°$ A, $\bar{I}_c = 18.33\angle 120°$ A, $\bar{I}_n = 18.33\angle -60°$ A, (b) $P_{loss} = 47.81$ W, $Q_{loss} = 239.06$ VAr (c) $P_a = 3224.3$ W, $Q_a = 5584.6$ VAr, $pf_a = 0.5$ (lag), $P_b = -3224.3$ W, $Q_b = 5584.6$ VAr, $pf_b = $

0.5 ($\bar{I}_b$ lags $\bar{V}_b$ by 120°), $P_c = 4033.3\,\text{W}$, $Q_c = 0\,\text{VAr}$, $pf_c = 1.0$, $P = 4033.3\,\text{W}$, $Q = 11169.0\,\text{VAr}$ (d) $S_A = 16930\,\text{VA}$, $pf_A = 0.2382$, $S_v = 11875\,\text{VA}$, $pf_v = 0.3396$, $S_e = 18631\,\text{VA}$, $pf_e = 0.2165$

3.22 (a) $\bar{I}_a = 39.83\angle-23.13°\,\text{A}$, $\bar{I}_b = 42.05\angle-49.87°\,\text{A}$, $\bar{I}_c = 101.23\angle138.01°\,\text{A}$, $\bar{I}_n = 23.0\angle120°\,\text{A}$ (b) $P_a = 8426\,\text{W}$, $Q_a = 3599.2\,\text{VAr}$, $P_b = 3287.9\,\text{W}$, $Q_b = -9096.8\,\text{VAr}$, $P_c = 22142\,\text{W}$, $Q_c = -7198.5\,\text{VAr}$, $P = 33856\,\text{W}$, $Q = -12696$ VAr (c) $S_v = 36158\,\text{VA}$, $S_A = 42118\,\text{VA}$, $S_e = 47358\,\text{VA}$, $pf_v = 0.9363$, $pf_a = 0.8038$, $pf_e = 0.7149$

3.23 (a) $i_a = 24.38\sqrt{2}\sin(\omega t - 32°) + 7.06\sqrt{2}\sin(3\omega t - 91.93°) + 1.52\sqrt{2}\sin(5\omega t - 72.25°)$, $i_b = 24.38\sqrt{2}\sin(\omega t - 152°) + 7.06\sqrt{2}\sin(3\omega t - 91.93°) + 1.52\sqrt{2}\sin(5\omega t + 47.74°)$, $i_c = 24.38\sqrt{2}\sin(\omega t + 87.99°) + 7.06\sqrt{2}\sin(3\omega t - 91.93°) + 1.52\sqrt{2}\sin(5\omega t + 167.74°)$ (b) $P_1 = 14265\,\text{W}$, $Q_1 = 8915.7\,\text{VAr}$, $P_H = 1251.6\,\text{W}$, $Q_H = 2416.4\,\text{VAr}$ (c) $pf_a = 0.775$ (lag), $pf_b = 0.775$ (lag), $pf_c = 0.775$ (lag), $S_v = 20023\,\text{VA}$, $S_A = 20023\,\text{VA}$, $S_e = 22217\,\text{VA}$, $pf_v = 0.775$, $pf_A = 0.775$, $pf_e = 0.6984$ (d) $I_{nN} = 21.18\,\text{A}$

3.24 (a) $i_a = 13\sqrt{2}\sin(\omega t - 47.29°) + 0.98\sqrt{2}\sin(3\omega t - 72.89°) + 0.15\sqrt{2}\sin(5\omega t - 79.54°)$, $i_b = 40\sqrt{2}\sin(\omega t - 156.86°) + 2.03\sqrt{2}\sin(3\omega t - 66.03°) + 0.515\sqrt{2}\sin(5\omega t + 44.93°)$, $i_c = 25\sqrt{2}\sin(\omega t + 173.13°) + 7.61\sqrt{2}\sin(3\omega t + 23.98°) + 1.93\sqrt{2}\sin(5\omega t - 105.06°)$ (b) $P_1 = 12178\,\text{W}$, $Q_1 = 1997.1\,\text{VAr}$, $P_H = 400\,\text{W}$, $Q_H = -80.55\,\text{VAr}$ (c) $pf_a = 0.6676$ (lag), $pf_b = 0.7965$ (lag), $pf_c = 0.6161$ (lead), $S_v = 12872\,\text{VA}$, $S_A = 17792\,\text{VA}$, $S_e = 30514\,\text{VA}$, $pf_v = 0.9771$, $pf_A = 0.7069$, $pf_e = 0.4122$ (d) $I_{nN} = 57.88\,\text{A}$

Chapter 4

4.10 (a) $\bar{V} = 4.62\,\text{kV}$ (phase) $= 8.002\,\text{kV}$ (L-L), $\Delta\bar{V} = 1.8386\angle29.19°\,\text{kV}$, $\bar{E} = 6.35\angle8.11°\,\text{kV}$ (b) $Q_\gamma = -19.84\,\text{MVAr}$ (1-ϕ) $= -59.52\,\text{MVAr}$ (3-ϕ)

4.11 (a) $Q_\gamma = 28.25\,\text{MVAr}$ (phase) $= 84.75\,\text{MVAr}$ (3-ϕ), $C_\gamma = 2230\,\mu\text{F/ph}$ (b) $V = 6.22\,\text{kV/ph} = 10.77\,\text{kV}$ (LL)

4.12 (a) $P_l = 9.8431\,\text{MW/phase}$, $Q_l = 12.168\,\text{MVAr/phase}$ (b) (i) $V_{LL} = 9.1182\,\text{kV}$ (ii) $Q_\gamma = -43.98\,\text{MVAr-3-phase}$, $C_f = 1200\,\mu\text{F/per phase}$ (iii) $Q_\gamma = -36.504\,\text{MVAr-3-phase}$, $V_{LL} = 10.73\,\text{kV}$

4.13 (a) $\bar{I}_{la} = 49.45\angle-9.51°\,\text{A}$, $\bar{I}_{lb} = 92.80\angle-161.66°\,\text{A}$, $\bar{I}_{lc} = 54.24\angle43.54°\,\text{A}$ (b) $B_\gamma^{ab} = 0.0248\,\mho$, $B_\gamma^{bc} = 0.1540\,\mho$, $B_\gamma^{ca} = -0.0011\,\mho$ (c) $\bar{I}_{sa} = 43.60\angle0°\,\text{A}$, $\bar{I}_{sb} = 43.60\angle-120°\,\text{A}$, $\bar{I}_{sc} = 43.60\angle120°\,\text{A}$ (d) $P_s = 30087\,\text{W}$, $Q_s = 0\,\text{VArs}$, $P_f = 0\,\text{W}$, $Q_f = -28197\,\text{VArs}$, $P_l = 30087\,\text{W}$, $Q_l = 28197\,\text{VArs}$

4.14 $B_\gamma^{ab} = 0\,\mho$, $B_\gamma^{bc} = 0.0577\,\mho$, $B_\gamma^{ca} = -0.0577\,\mho$, $P_s = P_l = 16000\,\text{W}$, $Q_s = 0$ VArs, $P_f = 0\,\text{W}$, $Q_f = -Q_l = 0$ VAr, from source side, the circuit is seen as a balanced circuit with each phase resistance of $10\,\Omega$

4.15 (a) $\bar{I}_{la} = 28.88\angle-45.96°$ A, $\bar{I}_{lb} = 20.86\angle154.54°$ A, $\bar{I}_{lc} = 11.86\angle96.02°$ A
(b) $P_l = 7489$ W, $Q_l = 10665$ VArs (c) $B_\gamma^{ab} = 0.0532$ ℧, $B_\gamma^{bc} = 0.007$ ℧, $B_\gamma^{ca} = 0.0069$ ℧,
(d) $\bar{I}_{sa} = 10.85\angle0°$ A, $\bar{I}_{sb} = 10.85\angle-120°$ A, $\bar{I}_{sc} = 10.85\angle120°$ A (e) $P_s = 7489$ W, $Q_s = 0$ VArs, $P_f = 0$ W, $Q_f = -10665$ VArs

4.19 (a) $B_\gamma^{ab} = 0.0189$ ℧, $B_\gamma^{bc} = 0.0078$ ℧, $B_\gamma^{ca} = -0.0078$ ℧, $C_\gamma^{ab} = 60.22\,\mu$F, $C_\gamma^{bc} = 24.83\mu$F, $L_\gamma^{ca} = 0.4$H (b) $\bar{I}_{sa} = 3.108\angle0°$ A, $\bar{I}_{sb} = 3.108\angle-120°$ A, $\bar{I}_{sc} = 3.108\angle120°$ A (c) $P_s = 2144.6$W, $Q_s = 0$ VArs, $P_f = 0$ W, $Q_f = -3002.4$ VAr, $P_l = 2144$ W, $Q_l = 3002.4$ VAr (d) From source side the circuit is seen as balanced circuit with each phase resistance of 74 Ω, from load side, $Z_{la} = j128.1\,\Omega$, $Z_{lb} = -j128.1\Omega$, $Z_{lc} = 222\,\Omega$

4.20 (a) $P_s = P_l = 13278$W, $Q_s = Q_l = 5797.1$ VArs (b) $\bar{I}_{fa} = 7.40\angle-113.17°$ A, $\bar{I}_{fb} = 19.18\angle-13.51$A, $\bar{I}_{fc} = 37.73\angle42.79°$ A (c) $\bar{I}_{la} = 17.69\angle-22.62°$ A, $\bar{I}_{lb} = 23.0\angle-66.87°$ A, $\bar{I}_{lc} = 46.0\angle66.87°$ A, $\bar{I}_{sa} = 19.24\angle0°$ A, $\bar{I}_{sb} = 19.24\angle-120$A, $\bar{I}_{sc} = 19.24\angle120°$ A (d) $P_s = P_l = 13278$W, $Q_s = 0$VArs, $P_f = 0$W, $Q_f = Q_l = 5797.1$ VArs

4.21 (a) $C_{fa} = 94.17\,\mu$F, $L_{fb} = 39.8$mH, $C_{fc} = 509.3\,\mu$F (b) $\bar{I}_{sa} = 16.33\angle0°$ A, $\bar{I}_{sb} = 13.8\angle-120°$ A, $\bar{I}_{sc} = 27.6\angle120°$ A, $\bar{I}_{nN} = 12.72\angle110.08°$ A, currents in the line are reduced in proportion to their load power factor, neutral current is reduced.

Chapter 5

5.9 (a) $i_{fa} = 38.71\sqrt{2}\sin(\omega t - 71.94°)$, $i_{fb} = 18.54\sqrt{2}\sin(\omega t - 1.7°)$, $i_{fc} = 16.24\sqrt{2}\sin(\omega t + 18.60°)$ (b) $Z_{fa} = -1.84 - j5.65\Omega$, $Z_{fb} = 5.88 + j10.92\Omega$, $Z_{fc} = 2.80 - j13.88\Omega$, Impedances Z_{fb} and Z_{fc} are realizable, while Z_{fa} has negative real part hence not realizable.

5.11 (a) $v_{s\alpha} = 281.69\sqrt{2}\sin\omega t$ V, $v_{s\beta} = 281.69\sqrt{2}\sin(\omega t - 90°)$ V, $v_{s0} = 0$ V, $i_{l\alpha} = 6.89\sqrt{2}\sin(\omega t - 45.57°)$A, $i_{l\beta} = 45.7\sqrt{2}\sin(\omega t - 71.23°)$A, $i_{l0} = 30.21\sqrt{2}\sin(\omega t - 55.56°)$A (b) (i) $p_{l\alpha} = 1357.3 - 1939.2\cos(2\omega t - 44.57°)$VA, $p_{l\beta} = 11921 - 12591\cos(2\omega t - 161.22°)$VA, $p_{l\alpha\beta} = 13278 + 11881\cos(2\omega t + 27.23°)$VA, $p_{l0} = 0$ (ii) $q_{l\alpha\beta} = 2666.9 + 11881\cos(2\omega t - 117.23°)$VA (iii) $P_l = 13278$ W, $Q_l = -2666.9$ VAr (c) $i_{f\alpha} = 19.38\sqrt{2}\sin(\omega t - 165.3°)$A, $i_{f\beta} = 23.63\sqrt{2}\sin(\omega t - 52.5°)$A, $i_{f0} = 30.21\sqrt{2}\sin(\omega t - 55.56°)$A, $i_{fa} = 19.18\sqrt{2}\sin(\omega t - 106.48°)$A, $i_{fb} = 37.73\sqrt{2}\sin(\omega t - 42.79°)$A, $i_{fc} = 7.4\sqrt{2}\sin(\omega t + 6.83°)$A (d) $i_{sa} = 19.24\sqrt{2}\sin\omega t$ A, $i_{sb} = 19.24\sqrt{2}\sin(\omega t - 120°)$A, $i_{sc} = 19.24\sqrt{2}\sin(\omega t + 120°)$A

5.13 (b) $i_{fab} = 19.45\sqrt{2}\sin(\omega t - 62.64°)$ A, $i_{fbc} = 12.19\sqrt{2}\sin(\omega t - 165.26°)$ A, $i_{fca} = 15.49\sqrt{2}\sin(\omega t - 111.81°)$ A (c) $i_{sa} = 43.63\sqrt{2}\sin\omega t$ A, $i_{sb} = 43.63\sqrt{2}\sin(\omega t - 120°)$ A, $i_{sc} = 43.63\sqrt{2}\sin(\omega t + 120°)$ A (d) $P_l = 30101$ W, $Q_l = 6332.9$ VAr, $P_s = 30101$ W, $Q_s = 0$, VAr, $P_f = 0$ W, $Q_f = 6332.9$ VAr, (e) $Z_{\gamma ab} = -j24.33\Omega$, $Z_{\gamma bc} = -j31.76\Omega$, $Z_{\gamma ca} = j30.61\Omega$, $i_{\gamma ab} = 16.37\sqrt{2}\sin(\omega t + 120°)$ A, $i_{\gamma bc} = 12.54\sqrt{2}\sin\omega t$ A, $i_{\gamma ca} = 13.05\sqrt{2}\sin(\omega t + 60°)$ A

5.14 (a) Secondary side: $i_a = \frac{\sqrt{6}V}{R}\sin(\omega t + 30°)$ A, $i_b = 0$, $i_c = 0$, Primary side: $i_A = \frac{\sqrt{6}V}{R}\sin(\omega t + 30°)$ A, $i_B = \frac{\sqrt{6}V}{R}\sin(\omega t - 150°)$ A, $i_C = 0$, (b) Net reactive power is zero (c) $P_a = \frac{3V^2}{R}$, $Q_a = 0$, $P_b = 0$, $Q_b = 0$, $P_c = 0$, $Q_c = 0$, $P_A = \frac{3V^2}{2R}$, $Q_A = -\frac{\sqrt{3}V^2}{2R}$, $P_B = \frac{3V^2}{2R}$, $Q_B = \frac{\sqrt{3}V^2}{2R}$, $P_C = 0$, $Q_c = 0$, Overall $P = \frac{3V^2}{R}$, $Q = 0$, (d) The rating of conductors and transformer will reduce.

5.15 (a) Secondary side: $i_a = \frac{\sqrt{2}V}{R}\sin\omega t$ A, $i_b = -i_a i_c = 0$, Primary side: $i_A = \frac{\sqrt{2}V}{R}\sin\omega t$) A, $i_B = 0$, $i_C = 0$ (b) Net reactive power is zero (c) $P_a = \frac{V^2}{2R}$, $Q_a = -\frac{V^2}{2\sqrt{3}R}$, $P_b = \frac{V^2}{2R}$, $Q_b = \frac{V^2}{2\sqrt{3}R}$, $P_c = 0$, $Q_c = 0$, $P_A = \frac{V^2}{R}$, $Q_A = 0$, $P_B = 0$, $Q_B = 0$, $P_C = 0$, $Q_c = 0$, Overall $P = \frac{V^2}{R}$, $Q = 0$ (d) The rating of conductors and transformer will reduce.

5.16 (a) $i_{la} = 96.43\sqrt{2}\sin(\omega t - 7.93°)$ A, $i_{lb} = 106.26\sqrt{2}\sin(\omega t + 171.69°)$ A, $i_{lc} = 9.84\sqrt{2}\sin(\omega t - 12.01°)$ A (b) $i_{fab} = 67.8\sqrt{2}\sin(\omega t - 40.05°)$ A, $i_{fbc} = 33.05\sqrt{2}\sin(\omega t + 164.67°)$ A, $i_{fca} = 30.33\sqrt{2}\sin(\omega t - 91.65°)$ A, $i_{sa} = 42.73\sqrt{2}\sin\omega t$ A, $i_{sb} = 42.73\sqrt{2}\sin(\omega t - 120°)$ A, $i_{sc} = 42.73\sqrt{2}\sin(\omega t + 120°)$ A (b) $P_l = 29482$ W, $Q_l = 27453$ VAr, $P_s = 29482$ W, $Q_s = 0$, VAr, $P_f = 0$ W, $Q_f = 27453$ VAr

5.17 (a) $i_{fa} = 3.12\sqrt{2}\sin(\omega t - 56.48°)$, $i_{fb} = 20.82\sqrt{2}\sin(\omega t - 71.10°)$, $i_{fc} = 3.84\sqrt{2}\sin(\omega t - 111.88°)$ (b) $P_f = 0$ W, $Q_f = -3102.8$ VAr, $P_s = 9475$ W, $Q_s = 0$, VAr, $P_l = 9475$ W, $Q_l = -3102.8$ VAr

5.18 (a) $i_{fa} = 5.12\sqrt{2}\sin(\omega t - 30.74°)$, $i_{fb} = 13.61\sqrt{2}\sin(\omega t - 41.14°)$, $i_{fc} = 18.22\sqrt{2}\sin(\omega t - 75°)$, $i_{sa} = 18.22\sqrt{2}\sin(\omega t - 15°)$, $i_{sb} = 18.22\sqrt{2}\sin(\omega t - 135°)$, $i_{sc} = 18.22\sqrt{2}\sin(\omega t + 105°)$ (b) $P_l = 7744$ W, $Q_l = -2258.7$ VAr, $P_s = 7744$ W, $Q_s = 0$ VAr, $P_f = 0$ W, $Q_f = -2258.7$ VAr, instantaneous power will not be constant because voltages are not balanced.

Chapter 6

6.10 (a) $\overline{I}_l = 0.735\angle-36.027°$ A, $\overline{V}_l = 0.857\angle-5.063°$ V, (b) $\overline{I}_l = 0.857\angle-30.96°$ A, $\overline{V}_{Dv} = 0.3539\angle59.036°$ V, (c) $\overline{I}_l = 0.5\angle-90°$ A, $\overline{V}_D = 0.1013\angle0°$ V

6.11 (a) $\overline{V}_D = 0.3322\angle41.63°$ V, $S_D = 0.2759$ p.u. (b) $\overline{V}_D = 0.7113\angle41.63°$ V, $S_D = 0.5907$ p.u. (c) $\overline{V}_D = 0.0577\angle41.63°$ V, $S_D = 0.0479$ p.u.

6.12 (a) $v_l = 0.6735 \sqrt{2} \sin(\omega t - 38.54°) + 0.3622, \sqrt{2} \sin(3\omega t + 173.66°)$ p.u. (b) $v_D = 0.388\sqrt{2} \sin(\omega t + 60°) + 0.3622\sqrt{2} \sin(3\omega t - 6.34°)$ p.u.

6.13 $\overline{V}_{D2} = 1.2706\angle 29.75°, \overline{V}_{D1} = 0.8065\angle 28.71°, \delta_1 = -3.33°, \delta = -5.916°$

6.14 $\overline{V}_{D2} = 0.5077\angle 53.13°, \overline{V}_{D1} = 1.3856\angle 37.08°$ p.u.

6.15 (a) $\overline{V}_l = 174.34\angle -10.28°$ V (b) $\overline{I}_l = 17.46\angle -49.4°$ A, $v_f = 140.24\sqrt{2}\sin(\omega t + 40.6°)$ V, 2.45 kVA

6.16 $\overline{I}_{la} = 0.316\angle -18.43°$ A, $\overline{V}_{Dva} = 0.0413\angle 71.56°$ V, $\overline{I}_{lb} = 0.316\angle -138.43°$ A, $\overline{V}_{Dvb} = 0.0413\angle -48.44°$ V, $\overline{I}_{lc} = 0.316\angle 101.57°$ A, $\overline{V}_{Dvc} = 0.0413\angle 11.56°$ V

6.17 $\overline{V}_{Dva} = 0.1802\angle 63.43°$ V, $\overline{V}_{Dvb} = 0.1949\angle -75°$ V, $\overline{V}_{Dvc} = 0.1933\angle 141.8°$ V

6.18 (b) $v_{la} = 218.18\sqrt{2}\sin(\omega t - 3.34°)$, $v_{lb} = 218.18\sqrt{2}\sin(\omega t - 123.34°)$, $v_{lc} = 218.18\sqrt{2}\sin(\omega t - 116.66°)$ (c) (i) $\overline{I}_{la} = 7.76\angle -36.87°$ A, $\overline{I}_{lb} = 7.76\angle -156.87°$ A, $\overline{I}_{lc} = 7.76\angle -53.13°$ A (ii) $v_{fa} = 21.75\sqrt{2}\sin(\omega t + 83.13°)$, $v_{fb} = 21.75\sqrt{2}\sin(\omega t - 66.87°)$, $v_{fc} = 21.75\sqrt{2}\sin(\omega t + 173.13°)$, $v'_{sa} = 243.44\sqrt{2}\sin(\omega t + 3.34°)$, $v'_{sb} = 243.44\sqrt{2}\sin(\omega t - 116.66°)$, $v'_{sc} = 243.44\sqrt{2}\sin(\omega t + 123.34°)$ (d) 500.5 VA

6.19 $\overline{V}_{fa} = 0.1802\angle 63.43°$ V, $\overline{V}_{fb} = 0.2211\angle -75°$ V, $\overline{V}_{fc} = 0.1732\angle 141.8°$ V

6.20 (a) $v_{la} = 0.8694\sqrt{2}\sin(\omega t - 9.82°) + 0.1806\sqrt{2}\sin(3\omega t + 175.24°)$ V, balanced (b) $v_{la} = 1.0\sqrt{2}\sin\omega t, v_{fa} = 0.4232\sqrt{2}\sin(\omega t + 60°) + 0.1806\sqrt{2}\sin(3\omega t - 4.76°)$ V, balanced (c) $v_{la} = 0.7774\sqrt{2}\sin\omega t, v_{fa} = 0.3887\sqrt{2}\sin(\omega t + 60°) + 0.1806\sqrt{2}\sin(3\omega t - 4.76°)$ V, balanced (d) $v_{la} = 1.0\sqrt{2}\sin\omega t$, $v_{fa} = -0.1234\sqrt{2}\sin(\omega t + 60°) + 0.1806\sqrt{2}\sin(3\omega t - 4.76°)$ V, balanced

Chapter 7

7.12 (a) $\overline{V}_l = 220.65\angle -0.76°$ V, $\overline{I}_l = 10.85\angle -52.1°$ A
(b) $i_f^* = 1.6246\sqrt{2}\sin(\omega t - 90°)$ A
(c) $v_f^* - 11.75\sqrt{2}\sin(\omega t + 44.43°)$ V
(d) $S_{DSTATCOM} = 373.65$ VA, $S_{DVR} = 118.697$ VA

7.13 (a) $\overline{V}_{la} = 198.38\angle -1.74°$ V, $\overline{V}_{lb} = 212\angle -121.22°$ V, $\overline{V}_{lc} = 215.34\angle 119.83°$ V, $\overline{I}_{la} = 24.44\angle -60.78°$ A, $\overline{I}_{lb} = 12.35\angle -177.53°$ A, $\overline{I}_{lc} = 9.74\angle 50.87°$ A
(b) $i_{fa}^* = 16.06\sqrt{2}\sin(\omega t - 72.59°)$ A, $i_{fb}^* = 5.19\sqrt{2}\sin(\omega t + 143.88°)$ A, $i_{fc}^* = 5.65\sqrt{2}\sin(\omega t - 18.78°)$ A
(c) $v_{fa}^* = 20.83\sqrt{2}\sin(\omega t + 53.13°)$ V, $v_{fb}^* = 20.83\sqrt{2}\sin(\omega t - 66.87°)$ V, $v_{fc}^* = 20.83\sqrt{2}\sin(\omega t + 173.13°)$ V
(d) $S_{DSTATCOM} = 18.56$ kVA, $S_{DVR} = 1.871$ kVA

7.14 (a) $\overline{V}_{la} = 244.90\angle - 10.30°$ V, $\overline{V}_{lb} = 214.61\angle - 123.20°$ V, $\overline{V}_{lc} = 232.31$ $\angle 115.96°$ V, $\overline{I}_{la} = 46.29\angle 26.56°$ A, $\overline{I}_{lb} = 20.28\angle - 160.08°$ A, $\overline{I}_{lc} = 16.89\angle 138.57°$ A

(b) $i_{fa}^* = 38.97\sqrt{2}\sin(\omega t + 71.68°)$ A, $i_{fb}^* = 5.57\sqrt{2}\sin(\omega t + 82.47°)$ A, $i_{fc}^* = 18.74\sqrt{2}\sin(\omega t - 127.74°)$ A

(c) $v_{fa}^* = 35.61\sqrt{2}\sin(\omega t + 64.16°)$ V, $v_{fb}^* = 35.61\sqrt{2}\sin(\omega t - 55.84°)$ V, $v_{fc}^* = 35.61\sqrt{2}\sin(\omega t - 175.84°)$ V

(d) $S_{DSTATCOM} = 43.668$ kVA, $S_{DVR} = 8.026$ kVA

7.15 (a) $\overline{V}_{la} = 205.65\angle - 2.75°$ V, $\overline{V}_{lb} = 254.1\angle - 123.4°$ V, $\overline{V}_{lc} = 223.3\angle 117.6°$ V, $\overline{I}_{la} = 40.47\angle - 54.09°$ A, $\overline{I}_{lb} = 29.60\angle - 67.09°$ A, $\overline{I}_{lc} = 15.90\angle 160.31°$ A

(b) $i_{fa}^* = 25.90\sqrt{2}\sin(\omega t - 67.57°)$ A, $i_{fb}^* = 33.88\sqrt{2}\sin(\omega t - 24.02°)$ A, $i_{fc}^* = 23.39\sqrt{2}\sin(\omega t - 134.24°)$ A

(c) $v_{fa}^* = 19.46\sqrt{2}\sin(\omega t + 58.21°)$ V, $v_{fb}^* = 271.82\sqrt{2}\sin(\omega t - 68.79°)$ V, $v_{fc}^* = 44.52\sqrt{2}\sin(\omega t - 178.21°)$ V

(d) $S_{DSTATCOM} = 57.395$ kVA, $S_{DVR} = 21.8$ kVA

7.16 (a) $\overline{V}_{la} = 205.28\angle - 4.01°$ V, $\overline{V}_{lb} = 272.64\angle - 122.96°$ V, $\overline{V}_{lc} = 190.96\angle 119.1°$ V, $\overline{I}_{la} = 38.81\angle - 57.14°$ A, $\overline{I}_{lb} = 23.82\angle - 66.64°$ A, $\overline{I}_{lc} = 13.88\angle 51.72°$ A

(b) $i_{fa}^* = 45.3\sqrt{2}\sin(\omega t - 75.25°)$ A, $i_{fb}^* = 8.42\sqrt{2}\sin(\omega t - 6.13°)$ A, $i_{fc}^* = 25.77\sqrt{2}\sin(\omega t + 11.62°)$ A

(c) $v_{fa}^* = 20.24\sqrt{2}\sin(\omega t + 121.78°)$ V, $v_{fb}^* = 58.94\sqrt{2}\sin(\omega t + 1.79°)$ V, $v_{fc}^* = 61.57\sqrt{2}\sin(\omega t - 118.22°)$ V

(d) $S_{DSTATCOM} = 54.847$ kVA, $S_{DVR} = 7.23$ kVA

7.17 (a) $i_{fa}^* = 76.68\sqrt{2}\sin(\omega t + 78.57°)$ A, $i_{fb}^* = 9.44\sqrt{2}\sin(\omega t + 106.41°)$ A, $i_{fc}^* = 19.77\sqrt{2}\sin(\omega t + 3.95°)$ A (b) $v_{fa}^* = 11.06\sqrt{2}\sin(\omega t + 64.16°) - 69\sqrt{2}\sin(7\omega t)$ V, $v_{fb}^* = 11.06\sqrt{2}\sin(\omega t - 55.84°) - 69\sqrt{2}\sin(7(\omega t - 120°))$ V, $v_{fc}^* = 11.06\sqrt{2}\sin(\omega t - 175.84°) - 69\sqrt{2}\sin(7(\omega t + 120°))$ V (c) $S_{DSTATCOM} = 73.062$ kVA, $S_{DVR} = 54.629$ kVA

7.18 (a) $i_{fa}^* = 53.87\sqrt{2}\sin(\omega t - 79.74°)$ A, $i_{fb}^* = 2.76\sqrt{2}\sin(\omega t + 50.35°)$ A, $i_{fc}^* = 13.67\sqrt{2}\sin(\omega t - 119.83°)$ A

(b) $v_{fa}^* = 20.16\sqrt{2}\sin(\omega t + 126.87°) - 23\sqrt{2}\sin(5\omega t)$ V, $v_{fb}^* = 20.16\sqrt{2}\sin(\omega t + 6.87°) - 23\sqrt{2}\sin(5(\omega t - 120°))$ V, $v_{fc}^* = 20.16\sqrt{2}\sin(\omega t - 113.13°) - 23\sqrt{2}\sin(5(\omega t + 120°))$ V

(c) $S_{DSTATCOM} = 48.508$ kVA, $S_{DVR} = 42.425$ kVA

Index

For Product Safety Concerns and Information please contact our EU
representative GPSR@taylorandfrancis.com
Taylor & Francis Verlag GmbH, Kaufingerstraße 24, 80331 München, Germany

www.ingramcontent.com/pod-product-compliance
Lightning Source LLC
Chambersburg PA
CBHW060755220326
41598CB00022B/2438